AF780909

Modeling and Computation in Vibration Problems, Volume 1

Numerical and semi-analytical methods

Modeling and Computation in Vibration Problems, Volume 1

Numerical and semi-analytical methods

Edited by

S Chakraverty

Department of Mathematics, National Institute of Technology Rourkela, Odisha, India

F Tornabene

Department of Innovation Engineering, University of Salento, Lecce, Italy

J N Reddy

J. Mike Walker'66 Department of Mechanical Engineering, Texas A&M University, College Station, USA

IOP Publishing, Bristol, UK

© IOP Publishing Ltd 2021

All rights reserved. No part of this publication may be reproduced, stored in a retrieval system or transmitted in any form or by any means, electronic, mechanical, photocopying, recording or otherwise, without the prior permission of the publisher, or as expressly permitted by law or under terms agreed with the appropriate rights organization. Multiple copying is permitted in accordance with the terms of licences issued by the Copyright Licensing Agency, the Copyright Clearance Centre and other reproduction rights organizations.

Permission to make use of IOP Publishing content other than as set out above may be sought at permissions@ioppublishing.org.

S Chakraverty, F Tornabene and J N Reddy have asserted their right to be identified as the editor of this work in accordance with sections 77 and 78 of the Copyright, Designs and Patents Act 1988.

ISBN 978-0-7503-3483-9 (ebook)
ISBN 978-0-7503-3481-5 (print)
ISBN 978-0-7503-3484-6 (myPrint)
ISBN 978-0-7503-3482-2 (mobi)

DOI 10.1088/978-0-7503-3483-9

Version: 20211201

IOP ebooks

British Library Cataloguing-in-Publication Data: A catalogue record for this book is available from the British Library.

Published by IOP Publishing, wholly owned by The Institute of Physics, London

IOP Publishing, Temple Circus, Temple Way, Bristol, BS1 6HG, UK

US Office: IOP Publishing, Inc., 190 North Independence Mall West, Suite 601, Philadelphia, PA 19106, USA

Contents

Preface

The primary objective of this two-volume book is to bring together leading researchers of vibrations who are involved in the development of new computational and analytical/semi-analytical techniques. Vibration problems are commonly encountered in various systems of applied mathematics, physics, aeronautical, civil, architectural, marine, mechanical, nuclear, biological and other areas of science and technology. These systems need to be analyzed by easy, fast and efficient computational approaches. Different mathematical theories of vibration, numerical simulation, machine intelligence techniques, physical experiments with computational investigations and their various engineering and science applications are included in these two volumes. Accordingly, these volumes will provide an outstanding opportunity to learn from the contributions of well-known researchers of their ideas, experiences, and advances.

Volume 1 of this book contains numerical and semi-analytical methods including analytical methods in solving various vibration problems. A total of 14 chapters are included in this volume. Chapters 1–3 contain theoretical and modelling investigations of different vibration problems. Studies on nano-structural members are incorporated into chapters 4–9. Finally, chapters 10–14 address various applications of structural vibrations. More details of each chapter are outlined next.

Chapter 1 is contributed by Tornabene, Viscoti, and Dimitri. They have investigated an interesting problem of 'Higher-order theory for the modal analysis of doubly-curved shells with lattice layers and honeycomb cores.' They proposed an innovative method based on Higher Order Shear Deformation Theories (HSDTs) to study the free vibration response of composite lattice structures, with a single or double curvature. A general method for the homogenization of lattice layers made of honeycomb cells and grid patterns is presented, which predicts accurately the anisotropic equivalent elastic constants for a wide range of cell configurations with different heights, wall thicknesses, and geometric layups. The reliability of the strategy is successfully verified against solutions from finite element analyses, both in terms of frequencies and mode shapes, with a very good agreement.

The second chapter is contributed by Ahmad, Gupta, Sujata, and Poomani, titled, 'Particle impact damping technology: modelling and applications.' In this chapter a modelling technique is introduced to capture the interactions between the dynamics of the damping particles and the dynamics of the host structure with a particular focus on the discrete element method (DEM). Hertz's dissipative contact model for normal contact forces and Coulomb's friction model for tangential force are used to derive the governing equation. The vibration attenuation trend has been studied for the location of particle dampers, size of particles, fill fraction, and excitation energy.

Transverse vibration of thick functionally graded (FG) skew plates with various skew angles has been investigated by Pradhan and Chakraverty in chapter 3. Material properties of FG constituents may vary spatially along thickness direction in power-law form. A shear deformation plate theory is considered here to define the constitutive relation, and the generalized eigenvalue problem is obtained by means of the Ritz method. The obtained results for natural frequencies are validated with those from the existing literature.

Chapter 4 is authored by Tornabene, Dimitri, and Brischetto, where mechanical modelling of functionally graded carbon nanotubes-reinforced composite materials and structures is studied. Variation of the material properties, such as thermal resistance, thermal conductivity and coefficient of thermal expansion, is included. Several higher-order shear deformation theories are used to investigate the problems of the vibration response, the dynamic stability, and the critical speed evaluation of thin and moderately thick structures with high values for the in-plane and transverse anisotropy; the governing equations are solved numerically by means of the Generalized Differential Quadrature (GDQ) method.

In chapter 5, contributed by Saxena and Sarkar, vibration of micro/nano structural members is investigated by discrete energy-based formulation. The authors focused on deriving a molecular-dynamics like discrete framework for different local and non-local continuum models such as beams and plates so that the models can simulate discontinuities. A major aspect of the formulation is that it unifies different local and non-local theories.

Chapter 6 includes the problem of 'Effect of thermal environment on nonlinear flutter of laminated composite plates reinforced with graphene nanoplates,' authored by Guo, Yang, Żur, Reddy, and Ferreira. The authors employed the element-free IMLS-Ritz method to obtain the flutter behavior of matrix cracked functionally graded multilayer graphene nanoplatelets (GPLs) reinforced composites (GPLRCs).

Malikan and Eremeyev have studied forced vibrations of piezo-flexomagnetic nano-actuator beams in chapter 7. The effect of excitation frequency on the piezomagnetic Euler–Bernoulli nanobeam taking the flexomagnetic material phenomenon is addressed here. The attained linear differential equation is transferred into an algebraic equation by using the Galerkin method. Then, the resulting linear algebraic equation is solved to determine the numerical values of dynamic deflections.

Nano-electromechanical systems have made significant advances in various sciences such as mechanics, medicine, and chemistry, due to their unique features and properties, especially in the fields of sensors and actuators. As such, vibration of size-dependent carbon nanotube-based biosensors in liquid is investigated by Sheikhmamoo, Mohammad-Sedighi, and Shishesaz in chapter 8. The equations and boundary conditions are derived using the Hamiltonian principle. The electrostatic and Casimir forces are the source of the nonlinearity in this problem. In this work, an analytical modified Adomian decomposition method is used to investigate the static response of the system. In the static section, the instability of the system and the pull-in voltage under the influence of surface tension effect and size effect are investigated.

Chapter 9, authored by Brischetto, Tornabene, and Dimitri, incorporates continuum 3D and 2D shell models for free vibration analysis of single-walled and double-walled carbon nanotubes. 3D shell solutions based on the exponential matrix methodology for the study of the free vibration response of simply supported

Single- and Double-Walled Carbon NanoTubes (SWCNTs and DWCNTs) are presented. A continuum approach based on the three-dimensional theory of elasticity is employed to represent discrete elements such as SWCNTs and DWCNTs. Proposed analytical 3D shell models are compared with some classical and refined 2D models based on the Generalized Differential Quadrature Method (GDQM) in the case of free vibration study of SWCNTs and DWCNTs with different lengths, diameters and equivalent elastic properties. Moreover, comparisons with well-known continuum beam models from the literature are proposed to analyze the differences between 3D, 2D and 1D approaches, depending on the carbon nanotube length and on the thickness of its walls.

Crack and interface interaction under quasi-static and dynamic loading is addressed by Pranavi, Reddy, Rajagopal, and Reddy in chapter 10. A thermodynamically consistent phase-field formulation for modelling the interactions between interfacial damage and bulk fracture in heterogeneous materials having matrix and inclusion phases with a matrix-inclusion interface is presented. A regularization scheme is considered for both the interface and the crack phase field. A coupled exponential cohesive zone law is adopted to model the interface which has the contributions of both normal and tangential displacement jump components. A novel nonlocal approach is devised to evaluate the smoothed values of jump at the regularized interface using element specific geometric information. The effects of stiff and soft interface on the mechanical response and the crack propagation is studied.

Industrial robotic systems have two essential functional sub-systems namely, gripper and wrist, that are prone to vibration-induced characteristics. This vibration is randomized in real-time and can be *in situ* and/or external impulse-based. Compliant Robotic Gripper (CRG) belongs to a selected niche of the first subsystem and these grippers are modular, semi-flexible and often small-enveloped with multi-task enabled ability. Accordingly, chapter 11 authored by Roy addresses vibration of compliant robotic grippers and wrists. The chapter includes modelling of vibration of CRG & CRW *ab initio*, for both *in situ* as well as external excitation. The author also discusses the local effects of vibration in this section, which are in the form of vibration accumulation at CRW and the consequences of external forcing on the robotic manipulator. Case-studies supported by test results are reported for a table-top small-sized semi-flexible robotic system, augmented with a tailor-made CRG.

Reinforced concrete (RC) beams are one of the critical structural elements in buildings. These elements suffer from distress and primarily cracks due to many reasons resulting into degradation of stiffness and strength. The mode shape-based approaches are very efficient techniques in damage identification in structural elements. As such, in chapter 12, authored by Panigrahi, Chourasia, and Bisht, a study of mode shape based approaches for health monitoring of reinforced concrete beams under transverse loading is presented. An analytical approach has been developed for identification of damage on a beam model considering damage in the beam. The mode shapes and modal curvatures have been computed both for an undamaged and damaged beam structure for damage identification. For validation of the approach, simply-supported RC beams were subjected to an incrementally increasing static two-point loading in steps till ultimate failure. After each load step, vibration measurements were performed using sensors mounted at critical locations. The mode shapes and modal curvature graphs have been drawn to find out the location of developed crack pattern on the beam.

Chapter 13, authored by Gupta, Ahmad, Aditya, and Poomani, addresses the analysis and testing of honeycomb sandwich structures for spacecraft. Sandwich panels generally consist of three significant components, two thin face sheets and a thick core. The adhesive films are placed between face sheets and core to bond them. The honeycomb sandwich structure is widely used as a primary cylinder, payload mounting panels, shear webs and other support structures in a spacecraft. The experimental setup used in this work is described and then responses obtained through experiments and finite element computations are compared.

Finally, in chapter 14, numerical analysis of Qutb Minar (New Delhi, India) using non-linear plastic-damage macro model for constituent masonry is investigated by Chourasia and Panigrahi. A plastic–damage macro model, originally proposed to model plasticity in concrete, has been adapted to simulate the behavior of masonry present in Qutb Minar. The natural frequencies obtained from an idealized finite element model show better correlation with experimental values from previous studies. The material model implemented is characterized by a bi-dissipative and isotropic degradation of material during cyclic loading. A seismic simulation has been carried out for comparison between response in two cases—elastic and non-elastic material models for constituent masonry. The need to take account of the dynamic parameters and material non-linearity for a realistic seismic response prevision has also been established.

The Editors are certain that the contents of this book will be useful for academic researchers as well as engineers in industry. In academics, this book will be useful for graduate students and researchers of vibration problems of different fields. The Editors believe the integrated and holistic analytical, new theories, computationally efficient approaches presented in various chapters will certainly benefit readers for their future studies and research. The editors thank all the chapter contributors for their effort and support in preparing and submitting on time. Finally, the editors also thank the IOP team for their help and support throughout this project.

May 2021
S Chakraverty
F Tornabene
J N Reddy
Editors

Editor biographies

S Chakraverty

Professor S Chakraverty has 30 years of experience as a researcher and teacher. Presently he is working in the Department of Mathematics (Applied Mathematics Group), National Institute of Technology Rourkela, Odisha as a Senior (Higher Administrative Grade) Professor. Prior to this, he was with CSIR-Central Building Research Institute, Roorkee, India. After completing Graduation from St. Columba's College (Ranchi University), his career started from the University of Roorkee (now, Indian Institute of Technology Roorkee) and did MSc (Mathematics) and MPhil (Computer Applications) from there securing the first positions in the university. Dr Chakraverty received his PhD from IIT Roorkee in 1993. Thereafter he did his post-doctoral research at the Institute of Sound and Vibration Research (ISVR), University of Southampton, UK and at the Faculty of Engineering and Computer Science, Concordia University, Canada. He was also a visiting professor at Concordia and McGill universities, Canada, during 1997–1999 and visiting professor of University of Johannesburg, South Africa during 2011–2014. He has authored/edited 24 books, published 392 research papers (till date) in journals and conferences and two books are ongoing. He is in the Editorial Boards of various International Journals, Book Series and Conferences. Professor Chakraverty is the Chief Editor of *International Journal of Fuzzy Computation and Modelling* (IJFCM), Inderscience Publisher, Switzerland (http://www.inderscience.com/ijfcm), Associate Editor of *Computational Methods in Structural Engineering, Frontiers in Built Environment* and happens to be the Editorial Board member of Springer Nature Applied Sciences, IGI Research Insights Books, Springer Book Series of Modeling and Optimization in Science and Technologies, *Coupled Systems Mechanics* (Techno Press), *Curved and Layered Structures* (De Gruyter), *Journal of Composites Science* (MDPI), *Engineering Research Express* (IOP) and *Applications and Applied Mathematics: An International Journal.* He is also the reviewer of around 50 national and international Journals of repute and he was the President of the Section of Mathematical Sciences (including Statistics) of Indian Science Congress (2015–2016) and was the Vice President Orissa Mathematical Society (2011–2013). Professor Chakraverty is a recipient of prestigious awards viz. Indian National Science Academy (INSA) nomination under International Collaboration/Bilateral Exchange Program (with the Czech Republic), Platinum Jubilee ISCA Lecture Award (2014), CSIR Young Scientist Award (1997), BOYSCAST Fellow. (DST), UCOST Young Scientist Award (2007, 2008), Golden Jubilee Director's (CBRI) Award (2001), INSA International Bilateral Exchange Award ([2010–11 (selected but could not undertake), 2015 (selected)], Roorkee University Gold Medals (1987, 1988) for first positions in MSc and MPhil (Comp. Appl.) etc.

He is in the list of 2% world scientists recently (2020) in Artificial Intelligence & Image Processing category based on an independent study done by Stanford University scientists. His world rank is 1862 out of 215114 researchers throughout the globe.

IOP Publishing Top Cited Paper Award for one of the most cited articles from India, published across the entire IOP Publishing journal portfolio within the past three years (2018 to 2020). It also features in the top 1% of most cited papers in the Materials subject category. This data is from the citations recorded in Web of Science.

He has already guided nineteen (19) PhD students and twelve are ongoing. Professor Chakraverty has undertaken around 16 research projects as Principle Investigator funded by international and national agencies totaling about Rs.1.5 crores. He has hoisted around 8 international students with different international/national fellowships to work in his group as PDF, PhD, visiting researchers for different periods. A good number of international and national Conferences, Workshops, and Training programs have also been organised by him. His present research areas include differential equations (ordinary, partial and fractional), numerical analysis and computational methods, structural dynamics (FGM, Nano) and fluid dynamics, mathematical and uncertainty modeling, soft computing and machine intelligence (artificial neural network, fuzzy, interval and affine computations).

Following are his Google, Scopus, other related and Vidwan (Institute) links:

Google Scholar : https://scholar.google.co.in/citations?user=4EeFYt4AAAAJ&hl=en

Scopus ID: http://www.scopus.com/authid/detail.url?origin=resultslist&authorId=7005011457&zone=

ORCID ID: https://orcid.org/0000-0003-4857-644X

Publons: https://publons.com/researcher/685949/snehashish-chakraverty/

Researchgate: https://www.researchgate.net/profile/S_Chakraverty

Vidwan (Institute) Link: http://nitrkl.irins.org/profile/62073

My VIDEO Lectures: https://www.youtube.com/channel/UCEsAtWBMIvDTnrPH7Yn82wA/videos?view_as=subscriber

F Tornabene

Francesco Tornabene is an Assistant Professor and Lecturer at the School of Engineering, Department of Innovation Engineering, University of Salento. He was born on 13 January 1978 in Bologna, where he received the high school degree at Liceo Classico *San Luigi*, in 1997. In 2001 he achieved a National Patent Bologna (Italy) for the Industrial Invention: *Friction Clutch for High Performance Vehicles* Question BO2001A00442. He received from the University of Bologna - Alma Mater Studiorum, an MSc degree in *Mechanical Engineering* (Curriculum in *Structural Mechanics*), on 23 July 2003, discussing a thesis entitled: *Dynamic Behavior of Cylindrical Shells: Formulation and Solution.* In December 2003, he was admitted to the PhD course in *Structural Mechanics*, at the University of Bologna, reaching the first position in the competitive admission. In 2004, he received from the University of Bologna a Thesis prize in memory of *Carlo Felice Jodi*; in 2007 he received the PhD degree in *Structural Mechanics* at the University of Bologna, discussing the Thesis entitled *Modeling and Solution of Shell Structures Made of Anisotropic Materials*. From 2007 to 2009 he received a research fellowship by the University of Bologna, working on the *Unified Formulation of Shell Structures Made of Anisotropic Materials. Numerical Analysis Using the Generalized Differential Quadrature Method and the Finite Element Method*. From 2011 to 2012 he became a junior researcher within the research program entitled *Advanced Numerical Schemes for Anisotropic Materials*; from 2012 to 2018 he was an Assistant Professor and Lecturer at the Alma Mater Studiorum - University of Bologna; from 2018 up to date he has been an Assistant Professor in Structural Mechanics and Lecturer at the University of Salento, Department of Innovation Engineering (Lecce). For a long time his scientific interests have included structural mechanics, solid mechanics, innovative and smart materials, computational mechanics and numerical techniques, damage and fracture mechanics. He is author of more than 260 scientific publications, and collaborates with many national or international researchers and professors all over the world, as visible from his scientific production. He is author of 11 books, see e.g. *Meccanica delle Strutture a Guscio in Materiale Composito. Il metodo Generalizzato di Quadratura Differenziale* (2012); *Mechanics of Laminated Composite Doubly-Curved Shell Structures. The Generalized Differential Quadrature Method and the Strong Formulation Finite Element Method* (2014); *Laminated Composite Doubly-Curved Shell Structures I. Differential Geometry. Higher-Order Structural Theories* (2016); *Laminated Composite Doubly-Curved Shell Structures II. Differential and Integral Quadrature. Strong Formulation Finite Element Method* (2016), *Anisotropic Doubly-Curved Shells. Higher-Order Strong and Weak Formulations for Arbitrarily Shaped Shell Structures* (2018), among many. He is member of the Editorial Board for 42 international journals (see, e.g. *Journal of Engineering*, *International Journal of Engineering & Applied Sciences*, *Composite Structures*, *Technologies*, *Journal of Applied and Computational Mechanics*, *Journal of Composites Science*, *Advanced Materials and Technologies*, *Heliyon*, *International Scholarly Research Notices*, *Mathematical Problems in Engineering*, *ISRN Mechanical Engineering*, *Journal of Computational Engineering*, *Advances in Aircraft and Spacecraft Science*, among others). He is also Editor-in-Chief for two international journals: *Curved and Layered Structures and Journal of Composites Science*; from 2019 he has been Associate Editor for the international journal *Mechanics Based Design of Structures and Machines*. In recent years he received different important awards, see e.g. *Highly Cited Researcher by Clarivate Analytics* (years 2018, 2019, 2020), *Ambassador of Bologna Award* for the organization of 21st International Conference on Composite Structures ICCS21, 4–7 September 2018, Bologna, Italy (2019), Member of the European Academy of Sciences (since 2018). He collaborates as a reviewer with more than 240 prestigious international journals in the structural mechanics field. From 2012, his teaching activities have included: dynamics of structures; computational mechanics; plates and shells; theory of structures; structural mechanics or mechanics of solids and structures. He is habilitated as Associate Professor and Full Professor in the area 08/B2 (Mechanics of Solids and Structures) and as Associate

Professor in Area 09/A1 (Aeronautical and Aerospace Engineering and Naval Architecture).

Assistant Professor at the Department of Innovation Engineering - University of Salento, Via per Monteroni, 73100, Lecce, Italy
E-mail Address: francesco.tornabene@unisalento.it

J N Reddy

Dr Reddy, the *O'Donnell Foundation Chair IV Professor* in J. Mike Walker '66 Department of Mechanical Engineering at Texas A &M University, is a highly-cited researcher, author of a large number of journal papers and 24 books (several with second, third, and fourth editions) on energy principles, variational methods, plates and shells, composite materials, mechanics of solids, and the finite element method (linear and nonlinear) and its applications. Dr Reddy has delivered over 180 plenary, keynote, and invited lectures at international conferences; and taught 116 short courses; he advised 50 postdoctoral fellows and research visitors, and guided and co-guided 120 graduate students (74 PhD and 46 MS students).

In research, Dr Reddy is known worldwide for his significant contributions to the field of applied mechanics through the authorship of widely used textbooks on the linear and nonlinear finite element analysis, variational methods, and composite materials and structures. His pioneering works on the development of shear deformation theories (that bear his name in the literature as the *Reddy third-order plate theory and the Reddy layerwise theory*) have had a major impact and have led to new research developments and applications. Some of the ideas on shear deformation theories and penalty finite element models of fluid flows have been implemented into commercial finite element computer programs like ABAQUS, NISA, and HyperXtrude.

Dr Reddy is the recipient of numerous professional awards from various professional organizations and societies. Recent significant national and international awards include: member of the US National Academy of Engineering, and foreign member of the Indian National Academy of Engineering, Canadian Academy of Engineering, Brazilian National Academy of Engineering, the Royal Academy of Engineering of Spain, the Chinese Academy of Engineering, a member of the European Academy of Sciences and Arts, Honorary Member of the European Association of Sciences, and the Academia Scientiarum et Artium Europaea (the European Academy of Sciences and Arts). He also received the *John von Neumann Medal* from the US Association of Computational Mechanics, the *Theodore von Karman Medal* from the American Society of Civil Engineers, and the *Stephan P. Timoshenko Medal* from American Society of Mechanical Engineers.

List of contributors

Akash Aditya
Structures Group, U. R. Rao Satellite Centre, Bangalore, India

Nazeer Ahmad
Structures Group, U. R. Rao Satellite Centre, Bangalore, India

R S Bisht
CSIR-Central Building Research Institute, Roorkee, India

Salvatore Brischetto
Department of Mechanical and Aerospace Engineering, Politecnico di Torino, Turin, Italy

S Chakraverty
Department of Mathematics, National Institute of Technology Rourkela, Rourkela, Odisha, India

Ajay Chourasia
CSIR-Central Building Research Institute, Roorkee, India

Rossana Dimitri
Department of Innovation Engineering, School of Engineering, University of Salento, Lecce, Italy

Victor A Eremeyev
Department of Mechanics of Materials and Structures, Gdansk University of Technology, 80-233 Gdansk, Poland
and
DICAAR, Università degli Studi di Cagliari, Via Marengo, 2, 09123 Cagliari, Italy

A J M Ferreira
Departamento de Engenharia Mecanica, Universidade do Porto, Porto, Portugal

Ankur Kumar Gupta
Structures Group, U. R. Rao Satellite Centre Bangalore, India

Hulun Guo
Department of Mechanics, Tianjin University, Tianjin, China

Mohammad Malikan
Department of Mechanics of Materials and Structures, Gdansk University of Technology, 80-233 Gdansk, Poland
and
DICAAR, Università degli Studi di Cagliari, Via Marengo, 2, 09123 Cagliari, Italy

S K Panigrahi
CSIR-Central Building Research Institute, Roorkee, India

D Poomani
Structures Group, U. R. Rao Satellite Centre, Bangalore, India

K K Pradhan
Department of Basic Science, Parala Maharaja Engineering College, Berhampur, Sitalapalli Ganjam, Odisha, India

Dhaladhuli Pranavi
IIT Hyderabad, India

Amirtham Rajagopal
IIT Hyderabad, India

J N Reddy
J. Mike Walker'66 Department of Mechanical Engineering, Texas A&M University, College Station, USA

K S S Reddy
IIT Hyderabad, India

Debanik Roy
Division of Remote Handling & Robotics, Bhabha Atomic Research Centre & Homi Bhabha National Institute, Department of Atomic Energy, Mumbai, India

Mukul Saxena
Discipline of Civil Engineering, IIT Indore, India

Saikat Sarkar
Discipline of Civil Engineering, IIT Indore, India

Hamid Mohammad Sedighi
Mechanical Engineering Department, Faculty of Engineering, Shahid Chamran University of Ahvaz, Ahvaz, Iran

Fatemeh Sheikhmamoo
Mechanical Engineering Department, Faculty of Engineering, Shahid Chamran University of Ahvaz, Ahvaz, Iran

Mohammad Shishesaz
Mechanical Engineering Department, Faculty of Engineering, Shahid Chamran University of Ahvaz, Ahvaz, Iran

Sujata
Structures Group, U. R. Rao Satellite Centre Bangalore, India

Francesco Tornabene
Department of Innovation Engineering, University of Salento, Lecce, Italy

Matteo Viscoti
Department of Innovation Engineering, School of Engineering, University of Salento, Lecce, Italy

Tianzhi Yang
School of Mechanical Engineering and Automation, Northeastern University, Shenyang, China

Krzysztof Kamil Żur
Faculty of Mechanical Engineering, Bialystok University of Technology, Bialystok, Poland

Chapter 1

Higher order theory for the modal analysis of doubly-curved shells with lattice layers and honeycomb cores

Francesco Tornabene, Matteo Viscoti and Rossana Dimitri

This chapter proposes an innovative method based on Higher Order Shear Deformation Theories (HSDTs) to study the free vibration response of composite lattice structures, with a single or double curvature. We develop a general method for the homogenization of lattice layers made of honeycomb cells and grid patterns, which predicts accurately the anisotropic equivalent elastic constants for a wide range of cell configurations with different heights, wall thicknesses and geometric layups. A general lamination scheme is modelled following an Equivalent Single Layer (ESL) methodology, whereas the Generalized Differential Quadrature (GDQ) method is proposed to solve accurately the eigenvalue problems with a reduced computational cost. A systematic investigation is performed to check for the sensitivity of the dynamic response for shells with a varying honeycomb and grid layups, along with different geometries and boundary conditions. The reliability of the present strategy is successfully verified against solutions from classical finite elements, both in terms of frequencies and mode shapes, with a very good agreement among results.

1.1 Introduction

In recent decades, new frontiers in many engineering applications have seen a lot of endeavour in searching for optimum parameters of structures with respect to their mechanical response [1–3]. In order to seek the best performance, the design can be oriented looking at structural shapes, as well as a proper selection of the material constituents, keeping in mind the manufacturing issues [4, 5]. As suggested by Vasiliev [6, 7], one of the most challenging structures among doubly-curved shells can be found in latticed composite structures. They are conceived so that a particular shape can be assessed starting from a correct location of some stiffeners along the

doi:10.1088/978-0-7503-3483-9ch1

© IOP Publishing Ltd 2021

parametric lines of a surface. These kinds of structures are characterized by large versatility. Looking up the applications of these structures, several grid architectures can be found, with open and closed cell lattices, hierarchical cells, disordered or randomized. Some regular patterns can be additionally found in some applications, with a different location of hoops and helical ribs, as well as a varying distance between stiffeners and inclination angle. In this way, each beam element of a three-dimensional structure is essentially subjected to a compressive load.

Another interesting way for infilling doubly curved structures is represented by honeycomb cells. According to this technology, a periodic infill is defined within a layer, characterized by some geometric parameters like the internal cell angle, and geometry of the cell panels [8]. Different mechanical properties can be obtained for the same structure, with a proper selection of its geometrical parameters [9]. Nevertheless, the honeycomb cell can be easily applied to layered structures so that a general orthotropic softcore can be obtained starting from an isotropic raw material. The application of optimization design algorithms usually requires simple but accurate formulations based on iterative processes that check for the best periodic unit cell in the whole structure, and its equivalent elastic properties [10]. Among continuum models, it is possible to apply a homogenization process on the so-called Representative Volume Element (RVE), to determine its equivalent elasticity constants, independently of the unit cell scale. In other words, the so-called separation of scales principle must be guaranteed [11–13], as guaranteed by a Multiscale Aggregating Discontinuities (MAD) method [14] or a Continuum Strong Discontinuity Approach (MSDA) [15–17]. One of the most adopted tools for the homogenization of the unit cell is the well-known Finite Element Method (FEM) [18]. A comprehensive literature overview about a finite element modeling of lattice structures can be found in [19–21]. Since the strategy at issue is based on a structural decomposition, the homogenized properties must be properly assessed independently from the selected mesh [22]. A finite element-based homogenization of lattice structures must account for local deformation effects in line with the predefined shape functions per each element [23].

For honeycomb unit patterns, we can employ a large variety of formulations, based on different initial hypotheses. Starting from the pioneer works by Gibson and Ashby (G&A) [24], various homogenization theories have been developed in literature [25–27] that account for the actual microstructural nature. Most of them, however, cannot provide a unique value of the equivalent elasticity constants, since they follow the energy procedure based on the minimum potential energy principle, as suggested by Kelsey *et al* [28]. In order to determine a unique value of homogenized stiffness constants, Grediac [29] proposed a general procedure based on a linear interpolation between the extreme values of stiffness based on the out-of-plane cell slenderness. Scarpa and Tomlin [30] measured the equivalent constants for re-entrant cell units starting from finite element simulations. Malek and Gibson (M&G) [31] embedded a rigid node assumption in the definition of the homogenized cell model. Each element in the unit cell was modelled by means of a Timoshenko Beam Theory (TBT) [32], or an Euler–Bernoulli Beam Theory [33]. An improvement of the G&A formulae was also proposed by Sorohan *et al* [34], who introduced some correction factors in order to account for the shear effects in the cell walls during a deformation process, as observed from FEM simulations, as well as experimental predictions.

Further shear correction factors were introduced by Fu and Yin [35] to improve the original formulation by G&A. In the work by Torabi *et al* [36] the above-mentioned formulation was applied for the flutter analysis of a cantilever trapezoidal honeycomb plate, employing the Differential Quadrature Method. Regarding the out-of-plane elastic properties, their mean value was computed by the authors between the extreme ones. Sorohan *et al* [37, 38] also provided an improved version of the M&G homogenization theory, based on an accurate geometrical description of the unit cell, with special attention to the geometry of nodes. In the original M&G model a rigid node assumption was embedded in the formulation. In this case, instead, its contribution to the total strain energy was computed by introducing an equivalent nodal length, based on the actual value of the node area. Following the same approach, a revision of G&A homogenization has been provided. Based on the recent work by Tornabene *et al* [39], a reliable formulation was proposed by the authors to determine the out-of-plane elastic constants for a large variety of honeycomb-based cells.

On the other hand, when a smearing technique is applied to anisogrid structures, after the proper definition of a generally-curved geometry, an important aspect is related to the periodicity of a unit pattern within the entire shell, or of a lattice layer. Differently from honeycomb cells, the cell geometry within a latticed structure comes out from the superimposition of various families of ribs which are assumed to be independent from each other. The overall mechanical properties for similar cell units are computed by considering the axial behaviour of each beam, accounting for its orientation with respect to the geometric principal axes of the shell structure. According to [4], each single rib contribution is computed independently from the presence of other beams within the unit cell.

In this way, cells with different numbers of rib families are obtained without considering the nodal stiffness, greater than the free area of the single rib. Based on this smearing technique it is also possible to assess a homogenized model by employing two different cell configurations with a different number of internal intersections. If only one node is embedded in the unit cell, a flake (F) unit pattern is identified, otherwise a star (S) cell layup is obtained. Actually, the rib family geometric parameters embedded in the model refer to the angle orientation and interspace between two adjacent beams [40].

The homogenized model of a shell structure can be described geometrically by means of the differential geometry basics, and a proper definition of the orthogonal reference system, together with the well-known Lamè Parameters and principal curvatures. The pioneering works by Calladine [41], Kraus [42] and Gould [43] applied this strategy to thin shells. The governing equations for the dynamic problem of curved shell structures are built on a bi-dimensional model that accounts for the through-the-thickness structural behaviour. In this perspective, two different strategies are proposed in this chapter: the Equivalent Single Layer (ESL) theory and the Layer-Wise (LW) approach [44]. In the former, we define the equivalent mechanical properties, in terms of elasticity constants, stress and strain field, for the entire shell thickness. A key aspect of this method is related to an accurate computation of such quantities, especially in the presence of very complex lamination schemes and possible coupling effects. On the other hand, the LW approach can be considered as a generalization of the ESL theory, since the fundamental equations of the problem are written separately at each layer of the lamination scheme by introducing some general thickness functions [45], in respect of the compatibility conditions and of a consistent distribution of the stress/strain field. A recovery procedure based on the equilibrium equations provides the three-dimensional stress distribution within the shell structure [46]. Starting with the classical shell theories of Kirchhoff–Love [47, 48], Reissner and Mindlin [49–51], or Reddy [52–54], more advanced models in the recent literature embed possible warping and stretching functions. In such a context, a first attempt can be found in the works by Washizu [55] and Reddy [56], where the authors provide a general description of the displacement field in shell members with a single constant curvature, and introduce for the first time a unified notation for a compact computation of thickness functions. In the theory developed by Tornabene *et al* [57], a generalized ESL approach based on HSDTs has been recently employed for a static and dynamic analysis of doubly-curved structures with anisotropic materials. It is interesting to say that the introduction of anisotropy within a structure enhances its shear strain field, that can be well-captured only under a proper assumption of thickness functions. In the further works by Tornabene *et al* [58–60] the same set of fundamental equations has been also written for each layer of anisotropic structures with very complex geometries.

The dynamic behavior of lattice-based structures is a key aspect that must be accounted for optimum design purposes of shell members with high stiffness/weight ratios. In this perspective, some optimization algorithms can be found in [61–63], which compute numerically the modal frequencies of arbitrarily-shaped structures. Among many possible high-performance computational methods, in this chapter we propose the Generalized Differential Quadrature (GDQ) [64–69] as efficient tool to solve the dynamics of complicated shell geometries made of innovative materials [70–77], in a strong form. The proposed method stems from classical quadrature methods [78] and expresses the derivative of a smooth function as a weighted sum of the values assumed by the same function in a set of points. As stated in the pioneering work by Shu *et al* [79], the main performances of this method, in terms of accuracy, convergence, and stability, are related to the domain discrete point distribution and weighting coefficients selection. In a recent work [80], the GDQ method was successfully compared to a classical Ritz formulation for the modal analysis of isotropic and non-homogeneous thin-walled structures, while revealing a great accuracy even with a reduced number of degrees of freedom (DOFs). The high level of accuracy was also demonstrated in many convergence studies [81–84] involving the statics/dynamics and fracture mechanics nonlinear problems [85–94].

In the present chapter, we propose a homogenized model based on Higher Order Theories for the dynamic study of doubly-curved sandwich structures with a softcore layer made of latticed/honeycomb configurations. The governing equations of the problem together with the boundary conditions are here determined by applying the Hamiltonian Principle in a variational form, combined with different homogenization methods for both honeycomb layers and the latticed cores, to derive the equivalent anisotropic material properties. The GDQ method is, thus, applied to discretize the governing equations and boundary conditions of the problems, whose solutions are validated systematically with respect to refined 3D finite element-based solutions.

The rate of convergence analyses also demonstrate the stability and efficiency of the response, even for complex coupled mode shapes and a reduced computational effort, as implemented in the MATLAB code DiQuMASPAB, written by Tornabene *et al* [95].

1.2 Equivalent single layer shell theory

In this section we introduce all the theoretical basics of an ESL approach for lattice curved shells according to higher order theories. Particular attention is given to the structural geometry, the field variable assumptions, the RVE homogenization methods, and the derivation of the fundamental equations. According to an ESL strategy, all the geometric, kinematic and equilibrium features of the structure are referred to an equivalent middle surface, starting from a general three-dimensional formulation of the problem.

1.2.1 Geometrical description of the shell

We start considering the theoretical issues related to the geometry of a lattice structure with different curvatures. Generally speaking, pantographic structures and honeycomb shells present a three-dimensional extension, and their mathematical description usually turns out to be a challenging issue, especially for complicated shapes. In such cases, an orthogonal curvilinear reference system must be properly introduced to simplify the geometrical description of a generally-shaped structure, based on the definition of a principal coordinate system along the parametric directions α_1 and α_2, in the reference mid-surface, and the outward coordinate ζ in the thickness direction. Based on an indicial notation, it is $\zeta = \alpha_3$. As a consequence, the position vector $\mathbf{R}(\alpha_1, \alpha_2, \zeta)$ of a generic point of the structure can be written as follows [57]

$$\mathbf{R}(\alpha_1, \alpha_2, \zeta) = \mathbf{r}(\alpha_1, \alpha_2) + \frac{h(\alpha_1, \alpha_2)}{2} z\, \mathbf{n}(\alpha_1, \alpha_2) \tag{1.1}$$

where $\mathbf{r}(\alpha_1, \alpha_2)$ is the projection of the arbitrary point on the reference surface, $\mathbf{n}(\alpha_1, \alpha_2)$ the normal unit vector, and $h(\alpha_1, \alpha_2)$ the shell thickness (see figure 1.1). In equation (1.1), the position along the thickness coordinate ζ with respect to the middle surface is defined with a dimensionless coordinate $z = 2\zeta/h(\alpha_1, \alpha_2) \in [-1, 1]$. By computing the partial derivatives $\mathbf{r}_{,1}$ and $\mathbf{r}_{,2}$ of the reference surface position vector with respect to α_1 and α_2, respectively, it is possible to define at each point the Lamè Parameters of the middle layer $A_1(\alpha_1, \alpha_2)$ and $A_2(\alpha_1, \alpha_2)$

$$A_1(\alpha_1, \alpha_2) = \sqrt{\mathbf{r}_{,1} \cdot \mathbf{r}_{,1}}, \quad A_2(\alpha_1, \alpha_2) = \sqrt{\mathbf{r}_{,2} \cdot \mathbf{r}_{,2}} \tag{1.2}$$

The normal unit vector $\mathbf{n}(\alpha_1, \alpha_2)$ is thus defined as

$$\mathbf{n}(\alpha_1, \alpha_2) = \frac{\mathbf{r}_{,1} \times \mathbf{r}_{,2}}{A_1 A_2} \tag{1.3}$$

whereas, the principal radii of curvature $R_1(\alpha_1, \alpha_2)$ and $R_2(\alpha_1, \alpha_2)$ are computed as

$$R_1(\alpha_1, \alpha_2) = -\frac{\mathbf{r}_{,1} \cdot \mathbf{r}_{,1}}{\mathbf{r}_{,11} \cdot \mathbf{n}}, \quad R_2(\alpha_1, \alpha_2) = -\frac{\mathbf{r}_{,2} \cdot \mathbf{r}_{,2}}{\mathbf{r}_{,22} \cdot \mathbf{n}} \tag{1.4}$$

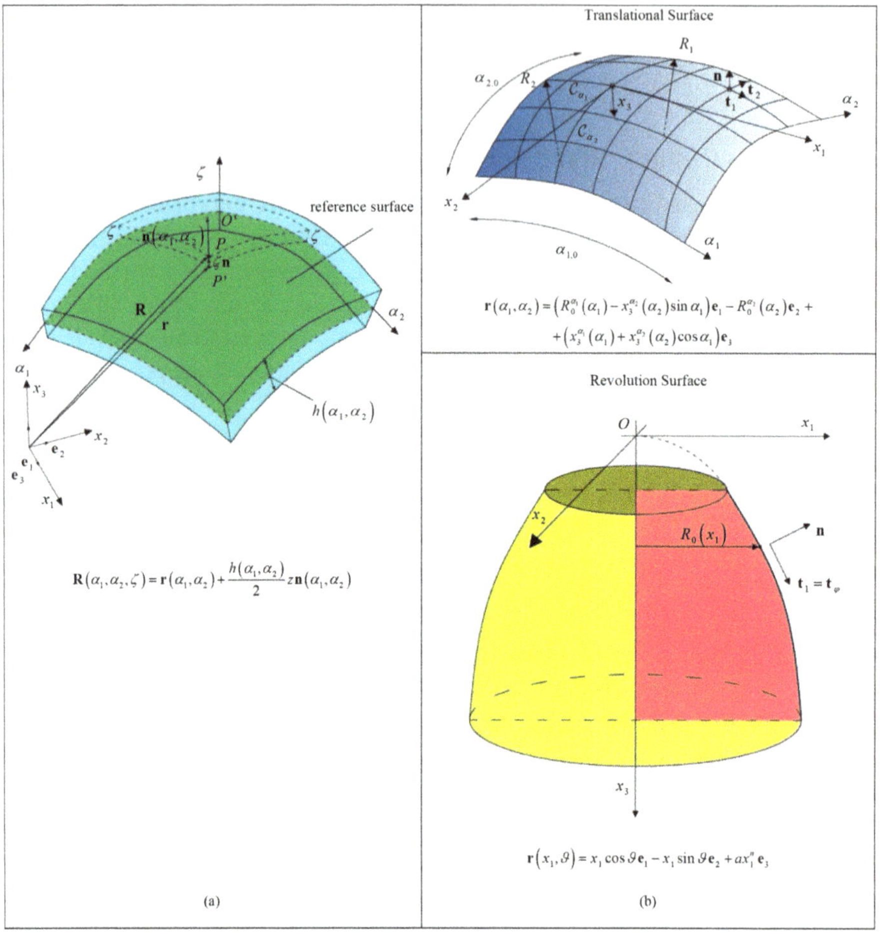

Figure 1.1. Geometric representation of a generic doubly-curved shell.

Note that the in-plane coordinates $\alpha_i \in [\alpha_i^0, \alpha_i^1]$, for $i = 1, 2$, whereas ζ is limited in the thickness direction as $\zeta(\alpha_1, \alpha_2) \in [-h(\alpha_1, \alpha_2)/2, h(\alpha_1, \alpha_2)/2]$. For a laminated structure, the overall thickness $h(\alpha_1, \alpha_2)$ is determined as [57]

$$h(\alpha_1, \alpha_2) = \sum_{k=1}^{l} h_k(\alpha_1, \alpha_2) \tag{1.5}$$

where $h_k(\alpha_1, \alpha_2) = \zeta_{k+1}(\alpha_1, \alpha_2) - \zeta_k(\alpha_1, \alpha_2)$ is the thickness for the kth layer of a l-laminate, with $k = 1, \ldots, l$.

1.2.2 Kinematic formulation

In what follows the displacement field represents the primary unknown. Therefore, we introduce a procedure for defining equivalent quantities in the mid-structure, and we determine the 3D displacement field vector $\mathbf{U}(\alpha_1, \alpha_2, \zeta, t) = [U_1 \ U_2 \ U_3]^T$ in the curvilinear reference system O', α_1, α_2, ζ in the following form [57]

$$\begin{aligned} U_1(\alpha_1, \alpha_2, \zeta, t) &= \sum_{\tau=0}^{N+1} F_\tau^{\alpha_1}(\zeta) u_1^{(\tau)}(\alpha_1, \alpha_2, t) \\ U_2(\alpha_1, \alpha_2, \zeta, t) &= \sum_{\tau=0}^{N+1} F_\tau^{\alpha_2}(\zeta) u_2^{(\tau)}(\alpha_1, \alpha_2, t) \\ U_3(\alpha_1, \alpha_2, \zeta, t) &= \sum_{\tau=0}^{N+1} F_\tau^{\alpha_3}(\zeta) u_3^{(\tau)}(\alpha_1, \alpha_2, t) \end{aligned} \tag{1.6}$$

or in the compact notation

$$\mathbf{U}(\alpha_1, \alpha_2, \zeta, t) = \sum_{\tau=0}^{N+1} \mathbf{F}_\tau \mathbf{u}^{(\tau)} \tag{1.7}$$

where $\mathbf{F}_\tau = \mathbf{F}_\tau(\zeta)$ is a diagonal matrix collecting the thickness functions $F_\tau^{\alpha_i}(\zeta)$, with $i = 1, 2, 3$

$$\mathbf{F}_\tau(\zeta) = \begin{bmatrix} F_\tau^{\alpha_1} & 0 & 0 \\ 0 & F_\tau^{\alpha_2} & 0 \\ 0 & 0 & F_\tau^{\alpha_3} \end{bmatrix} \tag{1.8}$$

In equation (1.6) a generalized displacement field component vector $\mathbf{u}^{(\tau)}(\alpha_1, \alpha_2, t) = [u_1^{(\tau)} \ u_2^{(\tau)} \ u_3^{(\tau)}]^T$ with $\tau = 0, \ldots, N+1$ is defined for each τth order of the kinematic expansion, defined on the reference surface. In order to take into account the interaction between adjacent layers, for $\tau = N+1$ we introduce the well-known Murakami's function, together with a power series expansion for $\tau = 0, \ldots, N$

$$F_\tau = \begin{cases} \zeta^\tau & \text{for} \quad \tau = 0, 1, \ldots, N \\ (-1)^k z_k & \text{for} \quad \tau = N+1 \end{cases} \tag{1.9}$$

where $z_k = z_k(\zeta) \in [-1, \ 1]$ is defined for the kth layer as follows

$$z_k = \frac{2}{\zeta_{k+1} - \zeta_k}\zeta - \frac{\zeta_{k+1} + \zeta_k}{\zeta_{k+1} - \zeta_k} \tag{1.10}$$

In what follows we introduce a compact notation, as already proposed in [57] in order to identify easily higher order theories, namely

$$\begin{aligned} &[F_0^{\alpha_1}][F_1^{\alpha_1}]\ldots[F_N^{\alpha_1}][Z] \\ \text{ED} - &[F_0^{\alpha_2}][F_1^{\alpha_2}]\ldots[F_N^{\alpha_2}][Z] \\ &[F_0^{\alpha_3}][F_1^{\alpha_3}]\ldots[F_N^{\alpha_3}][Z] \end{aligned} \tag{1.11}$$

where E stands for the ESL approach, D refers to the displacement field, Z is the Murakami's function (1.9) and N is the variable expansion order.

The kinematic relations for the ESL higher order theory are, thus, defined as

$$\boldsymbol{\varepsilon} = \mathbf{D}\,U = \mathbf{D}_\zeta\left(\sum_{i=1}^{3}\mathbf{D}_\Omega^{\alpha_i}\right)\mathbf{U} = \mathbf{D}_\zeta(\mathbf{D}_\Omega^{\alpha_1} + \mathbf{D}_\Omega^{\alpha_2} + \mathbf{D}_\Omega^{\alpha_3})\mathbf{U} \tag{1.12}$$

where $\boldsymbol{\varepsilon}(\alpha_1, \alpha_2, \alpha_3, t) = [\varepsilon_1\ \ \varepsilon_2\ \ \gamma_{12}\ \ \gamma_{13}\ \ \gamma_{23}\ \ \varepsilon_3]^T$ is the three-dimensional strain vector [57] and $\mathbf{D}$ is the differential operator split in two different parts $\mathbf{D}_\zeta$ and $\mathbf{D}_\Omega^{\alpha_i}$ $i = 1, 2, 3$, such that

$$\mathbf{D}_\zeta = \begin{bmatrix} \frac{1}{H_1} & 0 & 0 & 0 & 0 & 0 & 0 & 0 & 0 \\ 0 & \frac{1}{H_2} & 0 & 0 & 0 & 0 & 0 & 0 & 0 \\ 0 & 0 & \frac{1}{H_1} & \frac{1}{H_2} & 0 & 0 & 0 & 0 & 0 \\ 0 & 0 & 0 & 0 & \frac{1}{H_1} & 0 & \frac{\partial}{\partial\zeta} & 0 & 0 \\ 0 & 0 & 0 & 0 & 0 & \frac{1}{H_2} & 0 & \frac{\partial}{\partial\zeta} & 0 \\ 0 & 0 & 0 & 0 & 0 & 0 & 0 & 0 & \frac{\partial}{\partial\zeta} \end{bmatrix} \tag{1.13}$$

$$\mathbf{D}_\Omega^{\alpha_1} = \begin{bmatrix} \frac{1}{A_1}\frac{\partial}{\partial\alpha_1} & 0 & 0 \\ \frac{1}{A_1A_2}\frac{\partial A_2}{\partial\alpha_1} & 0 & 0 \\ -\frac{1}{A_1A_2}\frac{\partial A_1}{\partial\alpha_2} & 0 & 0 \\ \frac{1}{A_2}\frac{\partial}{\partial\alpha_2} & 0 & 0 \\ -\frac{1}{R_1} & 0 & 0 \\ 0 & 0 & 0 \\ 1 & 0 & 0 \\ 0 & 0 & 0 \\ 0 & 0 & 0 \end{bmatrix} \quad \mathbf{D}_\Omega^{\alpha_2} = \begin{bmatrix} 0 & \frac{1}{A_1A_2}\frac{\partial A_1}{\partial\alpha_2} & 0 \\ 0 & \frac{1}{A_2}\frac{\partial}{\partial\alpha_2} & 0 \\ 0 & \frac{1}{A_1}\frac{\partial}{\partial\alpha_1} & 0 \\ 0 & -\frac{1}{A_1A_2}\frac{\partial A_2}{\partial\alpha_1} & 0 \\ 0 & 0 & 0 \\ 0 & -\frac{1}{R_2} & 0 \\ 0 & 0 & 0 \\ 0 & 1 & 0 \\ 0 & 0 & 0 \end{bmatrix} \quad \mathbf{D}_\Omega^{\alpha_3} = \begin{bmatrix} 0 & 0 & \frac{1}{R_1} \\ 0 & 0 & \frac{1}{R_2} \\ 0 & 0 & 0 \\ 0 & 0 & 0 \\ 0 & 0 & \frac{1}{A_1}\frac{\partial}{\partial\alpha_1} \\ 0 & 0 & \frac{1}{A_2}\frac{\partial}{\partial\alpha_2} \\ 0 & 0 & 0 \\ 0 & 0 & 0 \\ 0 & 0 & 1 \end{bmatrix} \tag{1.14}$$

By introducing the through-the-thickness assumptions (1.7) for the displacement field within three-dimensional congruence relations (1.12), the ESL formulation of kinematic equation comes out

$$\boldsymbol{\varepsilon} = \sum_{\tau=0}^{N+1}\sum_{i=1}^{3}\mathbf{D}_\zeta\mathbf{D}_\Omega^{\alpha_i}\mathbf{F}_\tau\mathbf{u}^{(\tau)} = \sum_{\tau=0}^{N+1}\sum_{i=1}^{3}\mathbf{Z}^{(\tau)\alpha_i}\mathbf{D}_\Omega^{\alpha_i}\mathbf{u}^{(\tau)} = \sum_{\tau=0}^{N+1}\sum_{i=1}^{3}\mathbf{Z}^{(\tau)\alpha_i}\boldsymbol{\varepsilon}^{(\tau)\alpha_i} \tag{1.15}$$

which is redefined for the τth order as

$$\boldsymbol{\varepsilon}^{(\tau)\,\alpha_i} = \mathbf{D}_\Omega^{\alpha_i}\mathbf{u}^{(\tau)} \quad \text{for} \quad \tau = 0, 1, 2, \ldots, N, N+1 \quad i = 1, 2, 3 \tag{1.16}$$

where

$$\mathbf{Z}^{(\tau)\alpha_i} = \begin{bmatrix} \frac{F_\tau^{\alpha_i}}{H_1} & 0 & 0 & 0 & 0 & 0 & 0 & 0 & 0 \\ 0 & \frac{F_\tau^{\alpha_i}}{H_2} & 0 & 0 & 0 & 0 & 0 & 0 & 0 \\ 0 & 0 & \frac{F_\tau^{\alpha_i}}{H_1} & \frac{F_\tau^{\alpha_i}}{H_2} & 0 & 0 & 0 & 0 & 0 \\ 0 & 0 & 0 & 0 & \frac{F_\tau^{\alpha_i}}{H_1} & 0 & \frac{\partial F_\tau^{\alpha_i}}{\partial \zeta} & 0 & 0 \\ 0 & 0 & 0 & 0 & 0 & \frac{F_\tau^{\alpha_i}}{H_2} & 0 & \frac{\partial F_\tau^{\alpha_i}}{\partial \zeta} & 0 \\ 0 & 0 & 0 & 0 & 0 & 0 & 0 & 0 & \frac{\partial F_\tau^{\alpha_i}}{\partial \zeta} \end{bmatrix} \tag{1.17}$$

1.2.3 Homogenization of the lattice core and equivalent elastic behaviour

Let us now introduce the constitutive elastic relations defining the equivalent behavior of the shells based on an ESL approach and a homogenization procedure. Thus, the elastic relationships both in a static and kinematic sense are consistently supported by relation (1.15), defined in the curvilinear geometric reference system $O'\alpha_1\alpha_2\zeta$. The assessment of the elasticity constants is essentially based on possible material symmetries, which define the so-called material reference system denoted with $O'\hat{\alpha}_1^{(k)}\hat{\alpha}_2^{(k)}\hat{\zeta}^{(k)}$, for the kth lamina. In composite laminates, the outward geometric unit vector $\hat{\zeta}^{(k)}$ for the kth lamina corresponds to the one for the whole structure ζ, in the global geometric reference system [57]. For this reason, a simplified notation can be adopted, such that $O'\hat{\alpha}_1^{(k)}\hat{\alpha}_2^{(k)}\hat{\zeta}^{(k)} = O'\hat{\alpha}_1^{(k)}\hat{\alpha}_2^{(k)}\zeta$. As mentioned before, for a completely anisotropic material in the kth lamina, the constitutive behavior in the material reference system can be expressed as follows

$$\begin{bmatrix} \hat{\sigma}_1^{(k)} \\ \hat{\sigma}_2^{(k)} \\ \hat{\tau}_{12}^{(k)} \\ \hat{\tau}_{13}^{(k)} \\ \hat{\tau}_{23}^{(k)} \\ \hat{\sigma}_3^{(k)} \end{bmatrix} = \begin{bmatrix} C_{11}^{(k)} & C_{12}^{(k)} & C_{16}^{(k)} & C_{14}^{(k)} & C_{15}^{(k)} & C_{13}^{(k)} \\ C_{12}^{(k)} & C_{22}^{(k)} & C_{26}^{(k)} & C_{24}^{(k)} & C_{25}^{(k)} & C_{23}^{(k)} \\ C_{16}^{(k)} & C_{26}^{(k)} & C_{66}^{(k)} & C_{64}^{(k)} & C_{65}^{(k)} & C_{63}^{(k)} \\ C_{14}^{(k)} & C_{24}^{(k)} & C_{64}^{(k)} & C_{44}^{(k)} & C_{45}^{(k)} & C_{43}^{(k)} \\ C_{15}^{(k)} & C_{25}^{(k)} & C_{65}^{(k)} & C_{45}^{(k)} & C_{55}^{(k)} & C_{53}^{(k)} \\ C_{13}^{(k)} & C_{23}^{(k)} & C_{63}^{(k)} & C_{43}^{(k)} & C_{53}^{(k)} & C_{33}^{(k)} \end{bmatrix} \begin{bmatrix} \hat{\varepsilon}_1^{(k)} \\ \hat{\varepsilon}_2^{(k)} \\ \hat{\gamma}_{12}^{(k)} \\ \hat{\gamma}_{13}^{(k)} \\ \hat{\gamma}_{23}^{(k)} \\ \hat{\varepsilon}_3^{(k)} \end{bmatrix} \tag{1.18}$$

where $\hat{\boldsymbol{\sigma}}^{(k)}(\hat{\alpha}_1^{(k)}, \hat{\alpha}_2^{(k)}, \zeta, t)$ and $\hat{\boldsymbol{\varepsilon}}^{(k)}(\hat{\alpha}_1^{(k)}, \hat{\alpha}_2^{(k)}, \zeta, t)$ refer to the 3D stress and strain state for the kth lamina, respectively. The constitutive relation (1.18) can be written in a compact matrix form as [57]

$$\hat{\boldsymbol{\sigma}}^{(k)} = \mathbf{C}^{(k)}\,\hat{\boldsymbol{\varepsilon}}^{(k)} \tag{1.19}$$

where $\mathbf{C}^{(k)}$ stands for the elastic stiffness matrix. When the constitutive equations are written in the geometric reference system, the possible discrepancy in (1.18) between the material symmetry axes and the principal reference system of the shell requires a proper transformation of the material properties with respect to the geometric reference system $O'\alpha_1\alpha_2\zeta$ by means of matrix $\mathbf{T}^{(k)}$. Defining with θ_k the angle between α_1 and $\hat{\alpha}_1^{(k)}$, $\mathbf{T}^{(k)}$ can be written as [57]

$$\mathbf{T}^{(k)} = \begin{bmatrix} \cos^2\theta_k & \sin^2\theta_k & -2\sin\theta_k\cos\theta_k & 0 & 0 & 0 \\ \sin\theta_k & \cos^2\theta_k & 2\sin\theta_k\cos\theta_k & 0 & 0 & 0 \\ \sin\theta_k\cos\theta_k & -\sin\theta_k\cos\theta_k & \cos^2\theta_k - \sin^2\theta_k & 0 & 0 & 0 \\ 0 & 0 & 0 & \cos\theta_k & -\sin\theta_k & 0 \\ 0 & 0 & 0 & \sin\theta_k & \cos\theta_k & 0 \\ 0 & 0 & 0 & 0 & 0 & 1 \end{bmatrix} \tag{1.20}$$

The transformed stiffness matrix $\bar{\mathbf{C}}^{(k)}$ in the reference system $O'\alpha_1\alpha_2\zeta$ is defined as

$$\bar{\mathbf{C}}^{(k)} = \mathbf{T}^{(k)}\mathbf{C}^{(k)}\mathbf{T}^{(k)T} \tag{1.21}$$

or in extended matrix form as [57]

$$\bar{\mathbf{C}}^{(k)} = \begin{bmatrix} \bar{C}_{11}^{(k)} & \bar{C}_{12}^{(k)} & \bar{C}_{16}^{(k)} & \bar{C}_{14}^{(k)} & \bar{C}_{15}^{(k)} & \bar{C}_{13}^{(k)} \\ \bar{C}_{12}^{(k)} & \bar{C}_{22}^{(k)} & \bar{C}_{26}^{(k)} & \bar{C}_{24}^{(k)} & \bar{C}_{25}^{(k)} & \bar{C}_{23}^{(k)} \\ \bar{C}_{16}^{(k)} & \bar{C}_{26}^{(k)} & \bar{C}_{66}^{(k)} & \bar{C}_{64}^{(k)} & \bar{C}_{65}^{(k)} & \bar{C}_{63}^{(k)} \\ \bar{C}_{14}^{(k)} & \bar{C}_{24}^{(k)} & \bar{C}_{64}^{(k)} & \bar{C}_{44}^{(k)} & \bar{C}_{45}^{(k)} & \bar{C}_{43}^{(k)} \\ \bar{C}_{15}^{(k)} & \bar{C}_{25}^{(k)} & \bar{C}_{65}^{(k)} & \bar{C}_{45}^{(k)} & \bar{C}_{55}^{(k)} & \bar{C}_{53}^{(k)} \\ \bar{C}_{13}^{(k)} & \bar{C}_{23}^{(k)} & \bar{C}_{63}^{(k)} & \bar{C}_{43}^{(k)} & \bar{C}_{53}^{(k)} & \bar{C}_{33}^{(k)} \end{bmatrix} \tag{1.22}$$

Thus, the constitutive relation between the 3D stress field $\boldsymbol{\sigma}^{(k)}(\alpha_1, \alpha_2, \zeta, t) = [\sigma_1^{(k)}\ \sigma_2^{(k)}\ \tau_{12}^{(k)}\ \tau_{13}^{(k)}\ \tau_{23}^{(k)}\ \sigma_3^{(k)}]^T$ and the strain field $\boldsymbol{\varepsilon}^{(k)}(\alpha_1, \alpha_2, \zeta, t) = [\varepsilon_1^{(k)}\ \varepsilon_2^{(k)}\ \gamma_{12}^{(k)}\ \gamma_{13}^{(k)}\ \gamma_{23}^{(k)}\ \varepsilon_3^{(k)}]^T$ in the global reference system $O'\alpha_1\alpha_2\zeta$ reads as

$$\boldsymbol{\sigma}^{(k)} = \bar{\mathbf{C}}^{(k)}\boldsymbol{\varepsilon}^{(k)} \quad \text{for} \quad k = 1, 2, \ldots, l \tag{1.23}$$

To define the equivalent elastic constants for a lattice material, we require some homogenization procedures capable of studying a wide range of cell and pantographic grid configurations in a general and efficient manner. The homogenization of a lattice layer made of various orders of ribs can be assessed taking into account the axial contribution of each single rib in the RVE. Namely, some simple load cases must be defined starting from the elastic constants definition (1.18) for the kth lamina, in the material reference system. In addition, each beam contribution to the overall stiffness can be computed for the whole pattern independently from the coupling effects of further stiffeners. As stated by Vasiliev *et al* [6, 7], each rib in a unit cell is very slender in all the configurations of manufacturing interest, such that the influence of the rigid node zone is not charged with the role of physical constraint for the adjacent area. Therefore, the equivalent elastic properties of each unit cell can be written considering the superimposition of each single beam stiffness in the same reference system. As a consequence, it is convenient to define the single axial contribution assuming that each beam is made of an isotropic medium. This means that we have to define the Young's modulus E_i, density ρ_i and shear modulus G_i, for each ith element, while assuming a global orthotropic behaviour due to the slenderness of the single frame within the cell. If a local reference system $\hat{x}_1^{(i)}\hat{x}_2^{(i)}\hat{x}_3^{(i)}$ is defined for each stiffener along its principal axes, an overall constitutive elastic relationship can be written as

$$\begin{bmatrix} \sigma_1^{(i)} \\ \sigma_2^{(i)} \\ \tau_{12}^{(i)} \\ \tau_{13}^{(i)} \\ \tau_{23}^{(i)} \\ \sigma_3^{(i)} \end{bmatrix} = \begin{bmatrix} C_{11}^{(i)} & C_{12}^{(i)} & C_{16}^{(i)} & 0 & 0 & C_{13}^{(i)} \\ C_{12}^{(i)} & C_{22}^{(i)} & C_{26}^{(i)} & 0 & 0 & C_{23}^{(i)} \\ C_{16}^{(i)} & C_{26}^{(i)} & C_{66}^{(i)} & 0 & 0 & C_{36}^{(i)} \\ 0 & 0 & 0 & C_{44}^{(i)} & C_{45}^{(i)} & 0 \\ 0 & 0 & 0 & C_{45}^{(i)} & C_{55}^{(i)} & 0 \\ C_{13}^{(i)} & C_{23}^{(i)} & C_{36}^{(i)} & 0 & 0 & C_{33}^{(i)} \end{bmatrix} \begin{bmatrix} \varepsilon_1^{(i)} \\ \varepsilon_2^{(i)} \\ \gamma_{12}^{(i)} \\ \gamma_{13}^{(i)} \\ \gamma_{23}^{(i)} \\ \varepsilon_3^{(i)} \end{bmatrix} \tag{1.24}$$

where $\boldsymbol{\sigma}^{(i)}(\hat{x}_1^{(i)}, \hat{x}_2^{(i)}, \hat{x}_3^{(i)}, t) = [\sigma_1^{(i)}\ \sigma_2^{(i)}\ \tau_{12}^{(i)}\ \tau_{13}^{(i)}\ \tau_{23}^{(i)}\ \sigma_3^{(i)}]^T$ and $\boldsymbol{\varepsilon}^{(i)}(\hat{x}_1^{(i)}, \hat{x}_2^{(i)}, \hat{x}_3^{(i)}, t) = [\varepsilon_1^{(i)}\ \varepsilon_2^{(i)}\ \gamma_{12}^{(i)}\ \gamma_{13}^{(i)}\ \gamma_{23}^{(i)}\ \varepsilon_3^{(i)}]^T$ are the stress and strain vector in the local reference system. Nevertheless, the equivalent density ρ of the unit cell can be computed starting from the width of each rib, δ_i, together with the interspace a_i between two adjacent frames characterized by same inclination angle with respect to the α_1-direction. Eventually, one gets

$$\rho = \sum_{i=1}^{n} \rho_i \frac{\delta_i}{a_i} \tag{1.25}$$

with n the number of rib families within the unit pattern. Accounting for the stretching effect acting on each single rib, we define the non-reduced elastic constants of the constitutive relation (1.24) as

$$\begin{aligned} C_{11}^{(i)} &= C_{33}^{(i)} = E_i \\ C_{44}^{(i)} &= G_i = \frac{E_i}{2(1+\nu_i)} \\ C_{22}^{(i)} &= C_{12}^{(i)} = C_{13}^{(i)} = C_{23}^{(i)} = C_{55}^{(i)} = C_{66}^{(i)} = 0 \end{aligned} \tag{1.26}$$

Starting from the orthotropic relation (1.24), the actual contribution of the single frame to the equivalent stiffness of the unit cell can be derived from the application of a rotation transformation by an angle ϕ_i, which corresponds to the rib inclination with respect to the geometric principal reference direction α_1. Thus, the equivalent elastic properties of a unit cell made of n rib families can be written as

$$\begin{aligned}
C_{11}^{(k)} &= \sum_{i=1}^{n} E_i \frac{\delta_i}{a_i} \cos^4 \phi_i, & C_{22}^{(k)} &= \sum_{i=1}^{n} E_i \frac{\delta_i}{a_i} \sin^4 \phi_i,\\
C_{12}^{(k)} = C_{66}^{(k)} &= \sum_{i=1}^{n} E_i \frac{\delta_i}{a_i} \sin^2 \phi_i \cos^2 \phi_i, & C_{16}^{(k)} &= \sum_{i=1}^{n} E_i \frac{\delta_i}{a_i} \cos^3 \phi_i \sin \phi_i,\\
C_{26}^{(k)} &= \sum_{i=1}^{n} E_i \frac{\delta_i}{a_i} \cos \phi_i \sin^3 \phi_i, & C_{44}^{(k)} &= \sum_{i=1}^{n} G_i \frac{\delta_i}{a_i} \cos^2 \phi_i,\\
C_{45}^{(k)} &= \sum_{i=1}^{n} G_i \frac{\delta_i}{a_i} \cos \phi_i \sin \phi_i, & C_{55}^{(k)} &= \sum_{i=1}^{n} G_i \frac{\delta_i}{a_i} \sin^2 \phi_i,\\
C_{33}^{(k)} &= \sum_{i=1}^{n} E_i \frac{\delta_i}{a_i}, & C_{13}^{(k)} &= C_{23}^{(k)} = C_{36}^{(k)} = 0
\end{aligned} \tag{1.27}$$

Hereafter we introduce the following effective notation to identify the main properties of a unit cell

$$\mathrm{S/F}\left[(\phi_i)_a^\delta\right] \quad i = 1, \ldots, n \tag{1.28}$$

where (S) and (F) refer to the star and flake pattern configuration, respectively, depending on the number of rigid nodes in the RVE; n is the number of rib families, and angle ϕ_i refers to the ith frame family.

For a laminated structure with a latticed softcore, different elastic properties can be determined, depending on the cellular geometry of the unit pattern. This means that it is possible to obtain a large variety of elastic performances by simply tailoring the geometric parameters even for an isotropic medium implementation. More specifically, the overall properties of a lattice cell unit are mainly related to the wall slenderness determined by the following dimensionless parameters

$$\alpha = \frac{s_2}{l}, \quad \beta = \frac{s}{l}, \quad \gamma = \frac{b}{l} \tag{1.29}$$

A schematic representation of the geometric quantities that defines a honeycomb cell can be found in figure 1.2. Each unit cell is intended to be made of an isotropic medium with bulk density ρ_s, Young modulus E_s, Poisson's ratio ν_s, and shear stiffness $G_s = E_s/(2(1 + \nu_s))$.

To account for the geometric characteristics of a lattice honeycomb cell, the following nomenclature can be introduced

$$\mathrm{H}\ c/C\ \vartheta\ (l, l_2, s) \tag{1.30}$$

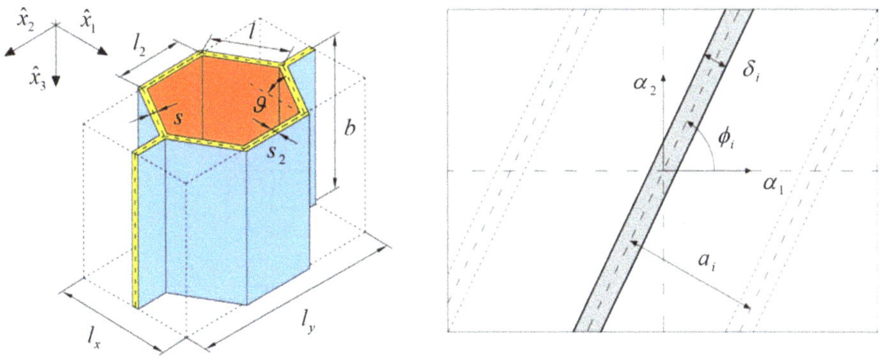

(a) (b)

Figure 1.2. Lattice core grid patterns. (a) Honeycomb cell, (b) Frame infill.

where c/C stands for a classic ($s_2 = s$) or commercial ($s_2 = 2s$) geometric layup. These two cell configurations refer to lattices obtained by two different manufacturing processes. In detail, a commercial cell configuration features a double thickness vertical wall, since it comes out from the juxtaposition of two corrugated sheets. The geometric parameters of the honeycomb cell are reported in figure 1.2 for the sake of clarity. Generally speaking, according to [24], a general cellular material can be modelled as an equivalent orthotropic continuum, where a homogenization procedure should express the unit cell elasticity by means of the engineering constants. In such a context, a local material reference system is defined, as visible in figure 1.2, where the local axis $\hat{x}_1$ is taken perpendicularly to the vertical cell wall, $\hat{x}_2$ is parallel to the same ribbon, and $\hat{x}_3$ coincides with the opposite of the outward unit vector $\mathbf{n}(\alpha_1, \alpha_2)$ of the shell defined in (1.3), so that a right-handed system is obtained. Equivalent elastic properties of the honeycomb cell consider the in-plane moduli E_1 and E_2 in the $\hat{x}_1$ and $\hat{x}_2$ directions, respectively, the shear modulus G_{12}, the stretching stiffness E_3, the out-of-plane shear moduli G_{13} and G_{23}, together with the Poisson's ratios ν_{12}, ν_{21}, ν_{13}, ν_{31}, ν_{23} and ν_{32}. In what follows, the equivalent density of the cell is determined as proposed by Gibson and Ashby [24], namely

$$\rho = \frac{A_s}{A} = \rho_s \frac{\beta(\alpha + 2)}{2\cos\vartheta(\alpha + \sin\vartheta)} \tag{1.31}$$

where A_s is the effective area of the cell and A its bulk value. At the same time, the elastic moduli are determined as proposed by Sorohan *et al* [34]

$$E_1 = k_1 E_s \beta^3 \frac{\cos\vartheta}{(\alpha + \sin\vartheta)\sin^2\vartheta}, \quad E_2 = k_2 E_s \beta^3 \frac{\alpha + \sin\vartheta}{\cos^3\vartheta}, \quad E_3 = E_s \frac{\rho}{\rho_s} \tag{1.32}$$

and the equivalent shear stiffnesses are defined as

$$G_{12} = k_{12} E_s \beta^3 \frac{\alpha + \sin\vartheta}{\alpha^2(1 + 2\alpha)\cos\vartheta}, \quad G_{13} = G_s \beta \frac{\cos\vartheta}{\alpha + \sin\vartheta} \tag{1.33}$$

with

$$
\begin{aligned}
k_1 &= \frac{1}{1 + \beta^2(2.4 + 1.5\nu_s + \cot^2\vartheta)} \\
k_2 &= \frac{1}{1 + \beta^2\left(2.4 + 1.5\nu_s + \tan^2\vartheta + \dfrac{2\alpha}{\cos^2\vartheta}\right)} \\
k_{12} &= \frac{1 + 2\alpha}{1 + 2\alpha + \beta^2\left(\dfrac{2.4 + 1.5\nu_s}{\alpha}(2 + \alpha + \sin\vartheta) + \dfrac{\alpha + \sin\vartheta}{\alpha^2}((\alpha + \sin\vartheta)\tan^2\vartheta + \sin\vartheta)\right)}
\end{aligned} \tag{1.34}
$$

Moreover, the in-plane Poisson's ratios are computed as

$$
v_{12} = c_{12}\frac{\cos^2\vartheta}{(\alpha + \sin\vartheta)\sin\vartheta}, \quad v_{21} = c_{21}\frac{(\alpha + \sin\vartheta)\sin\vartheta}{\cos^2\vartheta} \tag{1.35}
$$

The out-of-plane transverse deformation coefficients are calculated taking into account the symmetry of the stiffness matrix (1.18) of the homogenized material

$$
\begin{aligned}
\nu_{13} &= \frac{E_1}{E_s}\nu_s, \quad \nu_{31} = \nu_s \\
\nu_{23} &= \frac{E_2}{E_s}\nu_s, \quad \nu_{32} = \nu_s
\end{aligned} \tag{1.36}
$$

In addition, the shear modulus G_{23} follows the approach by Kelsey *et al* [28] based on the minimum potential energy's principle, for which an upper-bound value G_{23}^U and a lower-bound value G_{23}^L can be computed as

$$
G_{23}^U = G_s\beta\frac{\alpha + \sin^2\vartheta}{(\alpha + \sin\vartheta)\cos\vartheta} \tag{1.37}
$$

$$
G_{23}^L = G_s\beta\frac{\alpha + \sin\vartheta}{(1 + \alpha)\cos\vartheta} \tag{1.38}
$$

where G_s is the shear modulus of the constituent material. On the other hand, the linear interpolation procedure proposed by Grediac [29] between extreme values G_{23}^U and G_{23}^L has been demonstrated to be inadequate for re-entrant cells, and it is reliable only for a limited type of cell configurations [30, 39]. Due to this limitation, as highlighted in the work by Fu and Yin [35], we apply the following interpolation formula

$$
G_{23} = G_{23}^L + \frac{A\gamma}{B\gamma^2 + C}(G_{23}^U - G_{23}^L) \tag{1.39}
$$

where $A = 2/5$, $B = 3/4$ and $C = 1/4$ are some calibrated constants. It is worth noticing that the interpolation (1.39) fits Torabi's formula [36], when $\gamma \to 0.5$, whereas for $\gamma \to 10$ a lower bound value of shear modulus is steadily reached.

An ESL model is developed hereafter, such that the three-dimensional constitutive relation (1.23) is computed for each single layer of the lamination scheme. Based on the kinematic assumptions (1.7), we must consider the strain field coming from the adopted thickness functions matrix $\mathbf{F}_\tau = \mathbf{F}_\tau(\zeta)$ for each τth order of expansion. Nevertheless, a generalized stress component vector $\mathbf{S}^{(\tau)\alpha_i}(\alpha_1, \alpha_2, t) = [N_1^{(\tau)\alpha_i}\ N_2^{(\tau)\alpha_i}\ N_{12}^{(\tau)\alpha_i}\ N_{21}^{(\tau)\alpha_i}\ T_1^{(\tau)\alpha_i}\ T_2^{(\tau)\alpha_i}\ P_1^{(\tau)\alpha_i}\ P_2^{(\tau)\alpha_i}\ S_3^{(\tau)\alpha_i}]^T$ must be defined for each unknown variable order of the displacement field, by computing all the stress components along the thickness coordinate ζ, while keeping in mind the presence of possible coupling effects, due to the anisotropic behavior. In addition, the equivalent constitutive relation must account for the interlaminar phenomena. In this way, the ESL generalized strain component vector $\boldsymbol{\varepsilon}^{(\tau)\alpha_i}(\alpha_1, \alpha_2, t)$ is directly related to $\mathbf{S}^{(\tau)\alpha_i}(\alpha_1, \alpha_2, t)$ thanks to a generalized constitutive matrix. A formulation independent from the thickness coordinate is obtained, since ζ is computed in the definition of equivalent elastic and kinematic quantities. Based on the equivalence between the actual elastic strain energy and the homogenized one, the generalized anisotropic ESL Hooke's relation comes out [57]

$$\mathbf{S}^{(\tau)\alpha_i} = \sum_{\eta=0}^{N+1} \sum_{j=1}^{3} \mathbf{A}^{(\tau\eta)\alpha_i\alpha_j} \boldsymbol{\varepsilon}^{(\eta)\alpha_j} \quad \text{for} \quad \tau = 0, 1, 2, \ldots, N, N+1, \ \alpha_i = \alpha_1, \alpha_2, \alpha_3 \tag{1.40}$$

where the generalized elastic matrix $\mathbf{A}^{(\tau\eta)\alpha_i\alpha_j}$, for $\tau = 0, 1, 2, \ldots, N, N+1$ and $\alpha_i = \alpha_1, \alpha_2, \alpha_3$ reads as follows

$$\mathbf{A}^{(\tau\eta)\alpha_i\alpha_j} = \left[\begin{array}{cccccc|ccc}
A_{11(20)}^{(\tau\eta)[00]\alpha_i\alpha_j} & A_{12(11)}^{(\tau\eta)[00]\alpha_i\alpha_j} & A_{16(20)}^{(\tau\eta)[00]\alpha_i\alpha_j} & A_{16(11)}^{(\tau\eta)[00]\alpha_i\alpha_j} & A_{14(20)}^{(\tau\eta)[00]\alpha_i\alpha_j} & A_{15(11)}^{(\tau\eta)[00]\alpha_i\alpha_j} & A_{14(10)}^{(\tau\eta)[01]\alpha_i\alpha_j} & A_{15(10)}^{(\tau\eta)[01]\alpha_i\alpha_j} & A_{13(10)}^{(\tau\eta)[01]\alpha_i\alpha_j} \\
A_{12(11)}^{(\tau\eta)[00]\alpha_i\alpha_j} & A_{22(02)}^{(\tau\eta)[00]\alpha_i\alpha_j} & A_{26(11)}^{(\tau\eta)[00]\alpha_i\alpha_j} & A_{26(02)}^{(\tau\eta)[00]\alpha_i\alpha_j} & A_{24(11)}^{(\tau\eta)[00]\alpha_i\alpha_j} & A_{25(02)}^{(\tau\eta)[00]\alpha_i\alpha_j} & A_{24(01)}^{(\tau\eta)[01]\alpha_i\alpha_j} & A_{25(01)}^{(\tau\eta)[01]\alpha_i\alpha_j} & A_{23(01)}^{(\tau\eta)[01]\alpha_i\alpha_j} \\
A_{16(20)}^{(\tau\eta)[00]\alpha_i\alpha_j} & A_{26(11)}^{(\tau\eta)[00]\alpha_i\alpha_j} & A_{66(20)}^{(\tau\eta)[00]\alpha_i\alpha_j} & A_{66(11)}^{(\tau\eta)[00]\alpha_i\alpha_j} & A_{46(20)}^{(\tau\eta)[00]\alpha_i\alpha_j} & A_{56(11)}^{(\tau\eta)[00]\alpha_i\alpha_j} & A_{46(10)}^{(\tau\eta)[01]\alpha_i\alpha_j} & A_{56(10)}^{(\tau\eta)[01]\alpha_i\alpha_j} & A_{36(10)}^{(\tau\eta)[01]\alpha_i\alpha_j} \\
A_{16(11)}^{(\tau\eta)[00]\alpha_i\alpha_j} & A_{26(02)}^{(\tau\eta)[00]\alpha_i\alpha_j} & A_{66(11)}^{(\tau\eta)[00]\alpha_i\alpha_j} & A_{66(02)}^{(\tau\eta)[00]\alpha_i\alpha_j} & A_{46(11)}^{(\tau\eta)[00]\alpha_i\alpha_j} & A_{56(02)}^{(\tau\eta)[00]\alpha_i\alpha_j} & A_{46(01)}^{(\tau\eta)[01]\alpha_i\alpha_j} & A_{56(01)}^{(\tau\eta)[01]\alpha_i\alpha_j} & A_{36(01)}^{(\tau\eta)[01]\alpha_i\alpha_j} \\
A_{14(20)}^{(\tau\eta)[00]\alpha_i\alpha_j} & A_{24(11)}^{(\tau\eta)[00]\alpha_i\alpha_j} & A_{46(20)}^{(\tau\eta)[00]\alpha_i\alpha_j} & A_{46(11)}^{(\tau\eta)[00]\alpha_i\alpha_j} & A_{44(20)}^{(\tau\eta)[00]\alpha_i\alpha_j} & A_{45(11)}^{(\tau\eta)[00]\alpha_i\alpha_j} & A_{44(10)}^{(\tau\eta)[01]\alpha_i\alpha_j} & A_{45(10)}^{(\tau\eta)[01]\alpha_i\alpha_j} & A_{34(10)}^{(\tau\eta)[01]\alpha_i\alpha_j} \\
A_{15(11)}^{(\tau\eta)[00]\alpha_i\alpha_j} & A_{25(02)}^{(\tau\eta)[00]\alpha_i\alpha_j} & A_{56(11)}^{(\tau\eta)[00]\alpha_i\alpha_j} & A_{56(02)}^{(\tau\eta)[00]\alpha_i\alpha_j} & A_{45(11)}^{(\tau\eta)[00]\alpha_i\alpha_j} & A_{55(02)}^{(\tau\eta)[00]\alpha_i\alpha_j} & A_{45(01)}^{(\tau\eta)[01]\alpha_i\alpha_j} & A_{55(01)}^{(\tau\eta)[01]\alpha_i\alpha_j} & A_{35(01)}^{(\tau\eta)[01]\alpha_i\alpha_j} \\
\hline
A_{14(10)}^{(\tau\eta)[10]\alpha_i\alpha_j} & A_{24(01)}^{(\tau\eta)[10]\alpha_i\alpha_j} & A_{46(10)}^{(\tau\eta)[10]\alpha_i\alpha_j} & A_{46(01)}^{(\tau\eta)[10]\alpha_i\alpha_j} & A_{44(10)}^{(\tau\eta)[10]\alpha_i\alpha_j} & A_{45(01)}^{(\tau\eta)[10]\alpha_i\alpha_j} & A_{44(00)}^{(\tau\eta)[11]\alpha_i\alpha_j} & A_{45(00)}^{(\tau\eta)[11]\alpha_i\alpha_j} & A_{34(00)}^{(\tau\eta)[11]\alpha_i\alpha_j} \\
A_{15(10)}^{(\tau\eta)[10]\alpha_i\alpha_j} & A_{25(01)}^{(\tau\eta)[10]\alpha_i\alpha_j} & A_{56(10)}^{(\tau\eta)[10]\alpha_i\alpha_j} & A_{56(01)}^{(\tau\eta)[10]\alpha_i\alpha_j} & A_{45(10)}^{(\tau\eta)[10]\alpha_i\alpha_j} & A_{55(01)}^{(\tau\eta)[10]\alpha_i\alpha_j} & A_{45(00)}^{(\tau\eta)[11]\alpha_i\alpha_j} & A_{55(00)}^{(\tau\eta)[11]\alpha_i\alpha_j} & A_{35(00)}^{(\tau\eta)[11]\alpha_i\alpha_j} \\
A_{13(10)}^{(\tau\eta)[10]\alpha_i\alpha_j} & A_{23(01)}^{(\tau\eta)[10]\alpha_i\alpha_j} & A_{36(10)}^{(\tau\eta)[10]\alpha_i\alpha_j} & A_{36(01)}^{(\tau\eta)[10]\alpha_i\alpha_j} & A_{34(10)}^{(\tau\eta)[10]\alpha_i\alpha_j} & A_{35(01)}^{(\tau\eta)[10]\alpha_i\alpha_j} & A_{34(00)}^{(\tau\eta)[11]\alpha_i\alpha_j} & A_{35(00)}^{(\tau\eta)[11]\alpha_i\alpha_j} & A_{33(00)}^{(\tau\eta)[11]\alpha_i\alpha_j}
\end{array}\right] \tag{1.41}$$

As visible in (1.41), the stiffness matrix is conveniently split into blocks in order to collect the zeroth and first order derivatives of $\mathbf{F}_\tau$ components with respect to ζ. The definition of all the terms of $\mathbf{A}^{(\tau\eta)\alpha_i\alpha_j}$ is strictly related to the basic material assumptions. Generally speaking, we employ the three-dimensional constitutive coefficients $\bar{C}_{nm}^{(k)}$ with $n, m = 1, \ldots 6$, which are redefined as reduced elastic constants

$\bar{Q}_{nm}^{(k)}$, under in-plane stress assumptions. Moreover, by assuming a linear shape function throughout the thickness, for the in-plane displacement field, we introduce the well-known shear correction factor $\kappa(\zeta) = 5/6$ [49, 50], such that the equivalent elastic constants for the homogenized material $\bar{B}_{nm}^{(k)}$ take the following form [57]

$$\bar{B}_{nm}^{(k)} = \begin{cases} \bar{E}_{nm}^{(k)} & \text{for} \quad n, m = 1, 2, 3, 6 \\ \kappa(\zeta)\bar{E}_{nm}^{(k)} & \text{for} \quad n, m = 4, 5 \end{cases} \tag{1.42}$$

where $\bar{E}_{nm}^{(k)} = \bar{C}_{nm}^{(k)}$ or $\bar{E}_{nm}^{(k)} = \bar{Q}_{nm}^{(k)}$, in agreement with the stress state hypotheses.

Thus, an effective expression for each generalized elastic constant of (1.41) is summarized as follows [57]

$$\begin{aligned} &A_{nm(pq)}^{(\tau\eta)[fg]\alpha_i\alpha_j} = \sum_{k=1}^{l} \int_{\zeta_k}^{\zeta_{k+1}} \bar{B}_{nm}^{(k)} \frac{\partial^f F_\eta^{\alpha_j}}{\partial \zeta^f} \frac{\partial^g F_\tau^{\alpha_i}}{\partial \zeta^g} \frac{H_1 H_2}{H_1^p H_2^q} d\zeta && \text{for } \tau, \eta = 0, 1, 2, \ldots, N, N+1 \\ & && \text{for } n, m = 1, 2, 3, 4, 5, 6 \\ & && \text{for } p, q = 0, 1, 2 \\ &\frac{\partial^0 F_\tau^{\alpha_i}}{\partial \zeta^0} = F_\tau^{\alpha_i}, \quad \frac{\partial^0 F_\eta^{\alpha_j}}{\partial \zeta^0} = F_\eta^{\alpha_j} && \text{for } \alpha_i, \alpha_j = \alpha_1, \alpha_2, \alpha_3 \\ & && \text{for } f, g = 0, 1 \end{aligned} \tag{1.43}$$

1.2.4 Governing equations

We determine now the fundamental equations of motion for shells with a lattice core, in an ESL setting. These equations are written taking into account the elastic strain energy and inertial contribution of the structure, to solve a dynamic eigenvalue problem. The dynamic equilibrium can be computed from the Hamiltonian Principle in its variational form [57] within a time interval $[t_1, t_2]$

$$\delta \int_{t_1}^{t_2} (T - \Phi) dt = 0 \quad \rightarrow \quad \int_{t_1}^{t_2} (\delta T - \delta\Phi) dt = 0 \tag{1.44}$$

with T the kinetic energy and Φ the shell elastic strain energy. Based on the ESL approach, we account for the thickness functions set (1.7), while defining a differential operator $\mathbf{D}_\Omega^{*\alpha_i}$ for each direction in the reference system $\alpha_1, \alpha_2, \alpha_3$

$$\mathbf{D}_\Omega^{*\alpha_1} = \begin{bmatrix} (\mathbf{D}_\Omega^{*\alpha_1})_{11} & (\mathbf{D}_\Omega^{*\alpha_1})_{12} & (\mathbf{D}_\Omega^{*\alpha_1})_{13} & (\mathbf{D}_\Omega^{*\alpha_1})_{14} & (\mathbf{D}_\Omega^{*\alpha_1})_{15} & (\mathbf{D}_\Omega^{*\alpha_1})_{16} & (\mathbf{D}_\Omega^{*\alpha_1})_{17} & (\mathbf{D}_\Omega^{*\alpha_1})_{18} & (\mathbf{D}_\Omega^{*\alpha_1})_{19} \\ 0 & 0 & 0 & 0 & 0 & 0 & 0 & 0 & 0 \\ 0 & 0 & 0 & 0 & 0 & 0 & 0 & 0 & 0 \end{bmatrix} \tag{1.45}$$

$$\mathbf{D}_\Omega^{*\alpha_2} = \begin{bmatrix} 0 & 0 & 0 & 0 & 0 & 0 & 0 & 0 & 0 \\ (\mathbf{D}_\Omega^{*\alpha_2})_{21} & (\mathbf{D}_\Omega^{*\alpha_2})_{22} & (\mathbf{D}_\Omega^{*\alpha_2})_{23} & (\mathbf{D}_\Omega^{*\alpha_2})_{24} & (\mathbf{D}_\Omega^{*\alpha_2})_{25} & (\mathbf{D}_\Omega^{*\alpha_2})_{26} & (\mathbf{D}_\Omega^{*\alpha_2})_{27} & (\mathbf{D}_\Omega^{*\alpha_2})_{28} & (\mathbf{D}_\Omega^{*\alpha_2})_{29} \\ 0 & 0 & 0 & 0 & 0 & 0 & 0 & 0 & 0 \end{bmatrix} \tag{1.46}$$

$$\mathbf{D}_{\Omega}^{*\alpha_3} = \begin{bmatrix} 0 & 0 & 0 & 0 & 0 & 0 & 0 & 0 & 0 \\ 0 & 0 & 0 & 0 & 0 & 0 & 0 & 0 & 0 \\ (\mathbf{D}_{\Omega}^{*\alpha_3})_{31} & (\mathbf{D}_{\Omega}^{*\alpha_3})_{32} & (\mathbf{D}_{\Omega}^{*\alpha_3})_{33} & (\mathbf{D}_{\Omega}^{*\alpha_3})_{34} & (\mathbf{D}_{\Omega}^{*\alpha_3})_{35} & (\mathbf{D}_{\Omega}^{*\alpha_3})_{36} & (\mathbf{D}_{\Omega}^{*\alpha_3})_{37} & (\mathbf{D}_{\Omega}^{*\alpha_3})_{38} & (\mathbf{D}_{\Omega}^{*\alpha_3})_{39} \end{bmatrix} \quad (1.47)$$

The equilibrium equations can be written in terms of generalized stress resultants $\mathbf{S}^{(\tau)\alpha_i}(\alpha_1, \alpha_2, t)$ and generalized displacement field $\mathbf{u}^{(\tau)}(\alpha_1, \alpha_2, t)$ located on the reference surface, for each τth order of expansion, namely

$$\sum_{i=1}^{3} \mathbf{D}_{\Omega}^{*\alpha_i} \mathbf{S}^{(\tau)\alpha_i} = \sum_{\eta=0}^{N+1} \mathbf{M}^{(\tau\eta)} \ddot{\mathbf{u}}^{(\eta)} \quad \text{for} \quad \tau = 0, 1, 2, \ldots, N, N+1 \tag{1.48}$$

The inertial shell properties are collected in the τth order generalized mass matrix $\mathbf{M}^{(\tau\eta)}$

$$\mathbf{M}^{(\tau\eta)} = \begin{bmatrix} I^{(\tau\eta)\alpha_1\alpha_1} & 0 & 0 \\ 0 & I^{(\tau\eta)\alpha_2\alpha_2} & 0 \\ 0 & 0 & I^{(\tau\eta)\alpha_3\alpha_3} \end{bmatrix} \quad \text{for} \quad \tau, \eta = 0, 1, 2, \ldots, N, N+1 \tag{1.49}$$

where the mass matrix terms $I^{(\tau\eta)\alpha_i\alpha_j}$ in each principal direction are defined as [57]

$$I^{(\tau\eta)\alpha_i\alpha_j} = \sum_{k=1}^{l} \int_{\zeta_k}^{\zeta_{k+1}} \rho^{(k)} F_{\tau}^{\alpha_i} F_{\eta}^{\alpha_j} H_1 H_2 d\zeta \qquad \text{for} \quad \alpha_i, \alpha_j = \alpha_1, \alpha_2, \alpha_3 \tag{1.50}$$

accounting for the weight distribution along the shell thickness. On the other hand, the differential operators in (1.45)–(1.47) are defined as

$$\begin{gathered}
(\mathbf{D}_{\Omega}^{*\alpha_1})_{11} = (\mathbf{D}_{\Omega}^{*\alpha_2})_{23} = (\mathbf{D}_{\Omega}^{*\alpha_3})_{35} = \frac{1}{A_1}\frac{\partial}{\partial\alpha_1} + \frac{1}{A_1A_2}\frac{\partial A_2}{\partial\alpha_1}, \\
(\mathbf{D}_{\Omega}^{*\alpha_1})_{12} = -(\mathbf{D}_{\Omega}^{*\alpha_2})_{24} = -\frac{1}{A_1A_2}\frac{\partial A_2}{\partial\alpha_1}, \\
(\mathbf{D}_{\Omega}^{*\alpha_1})_{14} = (\mathbf{D}_{\Omega}^{*\alpha_2})_{22} = (\mathbf{D}_{\Omega}^{*\alpha_3})_{36} = \frac{1}{A_2}\frac{\partial}{\partial\alpha_2} + \frac{1}{A_1A_2}\frac{\partial A_1}{\partial\alpha_2}, \\
(\mathbf{D}_{\Omega}^{*\alpha_1})_{13} = -(\mathbf{D}_{\Omega}^{*\alpha_2})_{21} = \frac{1}{A_1A_2}\frac{\partial A_1}{\partial\alpha_2}, \\
(\mathbf{D}_{\Omega}^{*\alpha_1})_{17} = (\mathbf{D}_{\Omega}^{*\alpha_2})_{28} = (\mathbf{D}_{\Omega}^{*\alpha_3})_{39} = -1, \\
(\mathbf{D}_{\Omega}^{*\alpha_1})_{15} = -(\mathbf{D}_{\Omega}^{*\alpha_3})_{31} = \frac{1}{R_1}, \\
(\mathbf{D}_{\Omega}^{*\alpha_1})_{16} = (\mathbf{D}_{\Omega}^{*\alpha_1})_{18} = (\mathbf{D}_{\Omega}^{*\alpha_1})_{19} = 0, \\
(\mathbf{D}_{\Omega}^{*\alpha_2})_{26} = -(\mathbf{D}_{\Omega}^{*\alpha_3})_{32} = \frac{1}{R_2}, \\
(\mathbf{D}_{\Omega}^{*\alpha_3})_{33} = (\mathbf{D}_{\Omega}^{*\alpha_3})_{34} = (\mathbf{D}_{\Omega}^{*\alpha_3})_{37} = (\mathbf{D}_{\Omega}^{*\alpha_3})_{38} = 0 \\
(\mathbf{D}_{\Omega}^{*\alpha_2})_{25} = (\mathbf{D}_{\Omega}^{*\alpha_2})_{27} = (\mathbf{D}_{\Omega}^{*\alpha_2})_{29} = 0,
\end{gathered} \tag{1.51}$$

If the generalized stress resultants $\mathbf{S}^{(\tau)\alpha_i}$ in (1.48) are expressed in terms of generalized strain resultants $\boldsymbol{\varepsilon}^{(\tau)\alpha_i}$ via the anisotropic Hooke's law (1.40) and ESL approach, it is possible to define the equilibrium problem directly by means of the kinematic primary unknowns $\mathbf{u}^{(\eta)}(\alpha_1, \alpha_2, t) = [u_1^{(\eta)}\ u_2^{(\eta)}\ u_3^{(\eta)}]^T$ with $\eta = 0, \dots, N+1$. Taking into account the definition (1.16) of the strain component vector $\boldsymbol{\varepsilon}^{(\tau)\alpha_i}$ for each principal direction $\alpha_i = \alpha_1, \alpha_2, \alpha_3$ coming from relations (1.15), the fundamental equations are derived for a doubly-curved shell with a generally-oriented lattice core [57]

$$\sum_{\eta=0}^{N+1} \mathbf{L}^{(\tau\eta)}\mathbf{u}^{(\eta)} = \sum_{\eta=0}^{N+1} \mathbf{M}^{(\tau\eta)}\ddot{\mathbf{u}}^{(\eta)} \qquad \text{for} \quad \tau = 0, 1, 2, \dots, N, N+1 \tag{1.52}$$

where

$$\mathbf{L}^{(\tau\eta)} = \sum_{i=1}^{3}\sum_{j=1}^{3} \mathbf{D}_\Omega^{*\alpha_i}\mathbf{A}^{(\tau\eta)\alpha_i\alpha_j}\mathbf{D}_\Omega^{\alpha_j} = \begin{bmatrix} L_{11}^{(\tau\eta)\alpha_1\alpha_1} & L_{12}^{(\tau\eta)\alpha_1\alpha_2} & L_{13}^{(\tau\eta)\alpha_1\alpha_3} \\ L_{21}^{(\tau\eta)\alpha_2\alpha_1} & L_{22}^{(\tau\eta)\alpha_2\alpha_2} & L_{23}^{(\tau\eta)\alpha_2\alpha_3} \\ L_{31}^{(\tau\eta)\alpha_3\alpha_1} & L_{32}^{(\tau\eta)\alpha_3\alpha_2} & L_{33}^{(\tau\eta)\alpha_3\alpha_3} \end{bmatrix} \tag{1.53}$$

and $\tau, \eta = 0, \dots, N+1$.

The fundamental equation (1.52) can be written in extended form as

$$\begin{bmatrix} \mathbf{L}^{(00)} & \mathbf{L}^{(01)} & \cdots & \cdots & \mathbf{L}^{(0(N))} & \mathbf{L}^{(0(N+1))} \\ \mathbf{L}^{(10)} & \mathbf{L}^{(11)} & \cdots & \cdots & \mathbf{L}^{(1(N))} & \mathbf{L}^{(1(N+1))} \\ \vdots & \vdots & \ddots & & \vdots & \vdots \\ \vdots & \vdots & & \ddots & \vdots & \vdots \\ \mathbf{L}^{((N)0)} & \mathbf{L}^{((N)1)} & \cdots & \cdots & \mathbf{L}^{((N)(N))} & \mathbf{L}^{((N)(N+1))} \\ \mathbf{L}^{((N+1)0)} & \mathbf{L}^{((N+1)1)} & \cdots & \cdots & \mathbf{L}^{((N+1)(N))} & \mathbf{L}^{((N+1)(N+1))} \end{bmatrix} \begin{bmatrix} \mathbf{u}^{(0)} \\ \mathbf{u}^{(1)} \\ \vdots \\ \vdots \\ \mathbf{u}^{(N)} \\ \mathbf{u}^{(N+1)} \end{bmatrix} = \\ = \begin{bmatrix} \mathbf{M}^{(00)} & \mathbf{M}^{(01)} & \cdots & \cdots & \mathbf{M}^{(0(N))} & \mathbf{M}^{(0(N+1))} \\ \mathbf{M}^{(10)} & \mathbf{M}^{(11)} & \cdots & \cdots & \mathbf{M}^{(1(N))} & \mathbf{M}^{(1(N+1))} \\ \vdots & \vdots & \ddots & & \vdots & \vdots \\ \vdots & \vdots & & \ddots & \vdots & \vdots \\ \mathbf{M}^{((N)0)} & \mathbf{M}^{((N)1)} & \cdots & \cdots & \mathbf{M}^{((N)(N))} & \mathbf{M}^{((N)(N+1))} \\ \mathbf{M}^{((N+1)0)} & ^{((N+1)1)} & \cdots & \cdots & \mathbf{M}^{((N+1)(N))} & \mathbf{M}^{((N+1)(N+1))} \end{bmatrix} \begin{bmatrix} \ddot{\mathbf{u}}^{(0)} \\ \ddot{\mathbf{u}}^{(1)} \\ \vdots \\ \vdots \\ \ddot{\mathbf{u}}^{(N)} \\ \ddot{\mathbf{u}}^{(N+1)} \end{bmatrix} \tag{1.54}$$

The complete expressions of the fundamental coefficients can be found in [57].

For a free vibration study of doubly-curved shell, we select an harmonic solution for the fundamental equation (1.52), such that the unknown vector $\mathbf{u}^{(\tau)} = [u_1^{(\tau)}\ u_2^{(\tau)}\ u_3^{(\tau)}]^T$ is described in terms of the mode shape $\mathbf{U}^{(\tau)} = [U_1^{(\tau)}(\alpha_1, \alpha_2)\ \ U_2^{(\tau)}(\alpha_1, \alpha_2)\ \ U_3^{(\tau)}(\alpha_1, \alpha_2)]^T$ and the corresponding circular frequency $\omega = 2\pi f$ as

$$\mathbf{u}^{(\tau)}(\alpha_1, \alpha_2, t) = \mathbf{U}^{(\tau)}(\alpha_1, \alpha_2)e^{i\omega t} \tag{1.55}$$

It is worth remembering that the $(N+1)$th expansion order is embedded in the model if we adopt the Murakami's strategy (1.9) for describing possible interlaminar effects, otherwise the maximum order is equal to N. Thus, the free vibration equations can be set as [57]

$$\sum_{\eta=0}^{N+1} \mathbf{L}^{(\tau\eta)}\mathbf{U}^{(\eta)} + \omega^2 \sum_{\eta=0}^{N+1} \mathbf{M}^{(\tau\eta)}\mathbf{U}^{(\eta)} = \mathbf{0} \qquad \text{for} \quad \tau = 0,\, 1,\, 2,\, \dots,\, N,\, N+1 \tag{1.56}$$

For the definition of the external boundary constraints, it is useful to express the generalized stress resultants set $\mathbf{S}^{(\tau)\alpha_i}$ in terms of the DOFs of (1.52), taking into account the constitutive elastic relation (1.40), as well as the compatibility equation (1.16), i.e.

$$\begin{bmatrix} N_1^{(\tau)\alpha_i} \\ N_2^{(\tau)\alpha_i} \\ N_{12}^{(\tau)\alpha_i} \\ N_{21}^{(\tau)\alpha_i} \\ T_1^{(\tau)\alpha_i} \\ T_2^{(\tau)\alpha_i} \\ P_1^{(\tau)\alpha_i} \\ P_2^{(\tau)\alpha_i} \\ S_3^{(\tau)\alpha_i} \end{bmatrix} = \sum_{\eta=0}^{N+1} \begin{bmatrix} O_{11}^{(\tau\eta)\alpha_i\alpha_1} & O_{21}^{(\tau\eta)\alpha_i\alpha_2} & O_{31}^{(\tau\eta)\alpha_i\alpha_3} \\ O_{12}^{(\tau\eta)\alpha_i\alpha_1} & O_{22}^{(\tau\eta)\alpha_i\alpha_2} & O_{32}^{(\tau\eta)\alpha_i\alpha_3} \\ O_{13}^{(\tau\eta)\alpha_i\alpha_1} & O_{23}^{(\tau\eta)\alpha_i\alpha_2} & O_{33}^{(\tau\eta)\alpha_i\alpha_3} \\ O_{14}^{(\tau\eta)\alpha_i\alpha_1} & O_{24}^{(\tau\eta)\alpha_i\alpha_2} & O_{34}^{(\tau\eta)\alpha_i\alpha_3} \\ O_{15}^{(\tau\eta)\alpha_i\alpha_1} & O_{25}^{(\tau\eta)\alpha_i\alpha_2} & O_{35}^{(\tau\eta)\alpha_i\alpha_3} \\ O_{16}^{(\tau\eta)\alpha_i\alpha_1} & O_{26}^{(\tau\eta)\alpha_i\alpha_2} & O_{36}^{(\tau\eta)\alpha_i\alpha_3} \\ O_{17}^{(\tau\eta)\alpha_i\alpha_1} & O_{27}^{(\tau\eta)\alpha_i\alpha_2} & O_{37}^{(\tau\eta)\alpha_i\alpha_3} \\ O_{18}^{(\tau\eta)\alpha_i\alpha_1} & O_{28}^{(\tau\eta)\alpha_i\alpha_2} & O_{38}^{(\tau\eta)\alpha_i\alpha_3} \\ O_{19}^{(\tau\eta)\alpha_i\alpha_1} & O_{29}^{(\tau\eta)\alpha_i\alpha_2} & O_{39}^{(\tau\eta)\alpha_i\alpha_3} \end{bmatrix} \begin{bmatrix} u_1^{(\eta)} \\ u_2^{(\eta)} \\ u_3^{(\eta)} \end{bmatrix} \tag{1.57}$$

with $\tau = 0, \dots, N+1$. Based on the Hamiltonian Principle (1.44), and the application of the Gauss' Theorem for integration purposes, one obtains the equilibrium relations (1.48). Actually, since two different integrals are derived at this stage in any time interval $[t_1, t_2]$, it is possible to derive both the kinematic and static external constraints. Hereafter, we consider two different sets of BCs, namely a fully clamped (C) and a free (F) restraint, defined as [57]

Clamped (C)

$$\begin{aligned} &u_1^{(\tau)} - u_2^{(\tau)} = u_3^{(\tau)} = 0 \qquad \text{for} \quad \tau = 0,\, 1,\, 2,\, \dots,\, N,\, N+1, \\ &\qquad \text{at} \quad \alpha_1 = \alpha_1^0 \quad or \quad \alpha_1 = \alpha_1^1, \qquad \alpha_2^0 \leqslant \alpha_2 \leqslant \alpha_2^1 \\ &u_1^{(\tau)} = u_2^{(\tau)} = u_3^{(\tau)} = 0 \qquad \text{for} \quad \tau = 0,\, 1,\, 2,\, \dots,\, N,\, N+1, \\ &\qquad \text{at} \quad \alpha_2 = \alpha_2^0 \quad or \quad \alpha_2 = \alpha_2^1, \qquad \alpha_1^0 \leqslant \alpha_1 \leqslant \alpha_1^1 \end{aligned} \tag{1.58}$$

Free (C)

$$\begin{aligned} &N_1^{(\tau)} = 0, \quad N_{12}^{(\tau)} = 0, \quad T_1^{(\tau)} = 0 \qquad \text{for} \quad \tau = 0,\, 1,\, 2,\, \dots,\, N,\, N+1, \\ &\qquad \text{at} \quad \alpha_1 = \alpha_1^0 \quad \text{or} \quad \alpha_1 = \alpha_1^1, \qquad \alpha_2^0 \leqslant \alpha_2 \leqslant \alpha_2^1 \\ &N_{21}^{(\tau)} = 0, \quad N_2^{(\tau)} = 0, \quad T_2^{(\tau)} = 0 \qquad \text{for} \quad \tau = 0,\, 1,\, 2,\, \dots,\, N,\, N+1, \\ &\qquad \text{at} \quad \alpha_2 = \alpha_2^0 \quad \text{or} \quad \alpha_2 = \alpha_2^1, \qquad \alpha_1^0 \leqslant \alpha_1 \leqslant \alpha_1^1 \end{aligned} \tag{1.59}$$

For a smart definition of shell edges, we adopt the following nomenclature

$$\begin{array}{lll} \text{Westedge (W)} \rightarrow & \alpha_1^0 \leqslant \alpha_1 \leqslant \alpha_1^1, & \alpha_2 = \alpha_2^0 \\ \text{Southedge (S)} \rightarrow & \alpha_1 = \alpha_1^1, & \alpha_2^0 \leqslant \alpha_2 \leqslant \alpha_2^1 \\ \text{Eastedge (E)} \rightarrow & \alpha_1^0 \leqslant \alpha_1 \leqslant \alpha_1^1, & \alpha_2 = \alpha_2^1 \\ \text{North edge(N)} \rightarrow & \alpha_1 = \alpha_1^0, & \alpha_2^0 \leqslant \alpha_2 \leqslant \alpha_2^1 \end{array} \tag{1.60}$$

1.2.5 Assembly procedure of the discrete governing equations

Once the analytical expression of the fundamental equation (1.56) has been derived for the eigenvalue problem, we provide a discrete version to solve the problem numerically. In the present formulation, we employ the GDQ method to solve the problem in a strong form. First of all, a set of $I_N \times I_M$ discrete points is identified within the computational domain. For each point, a coordinate location $(\alpha_{1i}, \alpha_{2j})$ with $i = 1, \dots, I_N$ and $j = 1, \dots, I_M$ must be provided with respect to the curvilinear reference system (1.1). In the present work, a non-uniform bidimensional Chebyshev–Gauss–Lobatto's (CGL) distribution has been selected [57], due to its capability to yield accurate and stable results, as verified in [57, 59, 60, 65, 70], namely

$$\alpha_{rp} = \left(1 - \cos\left(\frac{p-1}{I_P - 1}\pi\right)\right)\frac{(\alpha_r^1 - \alpha_r^0)}{2} + \alpha_r^0 \qquad \begin{array}{ll} \text{for} & I_P = I_N, I_M, \\ \text{for} & r = 1, 2, \\ \text{for} & p = 1, 2, \dots, I_P, \\ \text{for} & \alpha_{rp} \in \left[\alpha_{rp}^0, \alpha_{rp}^1\right] \end{array} \tag{1.61}$$

The GDQ method allows one to express directly the nth order derivative of a function as a linear combination of its values assumed in a fixed point distribution. If $f(x)$ is a one-variable function with $x \in [x_0, x_1]$ and x_i is a discrete point belonging to a discrete distribution of I_N points within its domain, it gives

$$\left.\frac{d^n f(x)}{dx^n}\right|_{x=x_i} \cong \sum_{j=1}^{I_N} \varsigma_{ij}^{(n)} f(x_j) \qquad i = 1, 2, \dots, I_N \tag{1.62}$$

The weighting coefficients are computed from the recursive relation provided by Shu [64]

$$\begin{array}{ll} \varsigma_{ij}^{(1)} = \dfrac{\mathcal{L}^{(1)}(x_i)}{(x_i - x_j)\mathcal{L}^{(1)}(x_j)}, \quad \varsigma_{ij}^{(n)} = n\left(\varsigma_{ij}^{(1)}\varsigma_{ii}^{(n-1)} - \dfrac{\varsigma_{ij}^{(n-1)}}{x_i - x_j}\right) & i \neq j \\ \varsigma_{ii}^{(n)} = -\displaystyle\sum_{j=1 j\neq i}^{N} \varsigma_{ij}^{(n)} & i = j \end{array} \tag{1.63}$$

depending on the properties of the interpolating polynomials $\mathcal{L}$. On the other hand, based on the integral fundamental theorems, from equation (1.62) we can similarly derive the well-known Generalized Integral Quadrature (GIQ) procedure [57], which approximates the integral of a function as follows

$$\int_{x_i}^{x_j} f(x)dx = \sum_{k=1}^{I_T}(w_{jk} - w_{ik})f(x_k) \tag{1.64}$$

in the closed interval $[x_i, x_j]$. In equation (1.64) the weighting coefficients w_{ik} and w_{jk} can be derived from (1.63).

Accordingly, the partial derivation of a bi-dimensional function $f(\alpha_1, \alpha_2)$ with respect to α_1 and α_2 up to the nth and mth order within the bi-dimensional discretization (α_i, α_j) is defined numerically as

$$\begin{aligned} \left.\frac{\partial^n f(\alpha_1, \alpha_2)}{\partial \alpha_1^n}\right|_{\alpha_1=\alpha_{1i},\, \alpha_2=\alpha_{2j}} &\cong \sum_{k=1}^{I_N} \varsigma_{\alpha_1(ik)}^{(n)} f(\alpha_{1k}, \alpha_{2j}) \\ & i = 1, 2, \dots, I_N, \quad j = 1, 2, \dots, I_M \\ \left.\frac{\partial^m f(\alpha_1, \alpha_2)}{\partial \alpha_2^m}\right|_{\alpha_1=\alpha_{1i},\, \alpha_2=\alpha_{2j}} &\cong \sum_{l=1}^{I_M} \varsigma_{\alpha_2(jl)}^{(m)} f(\alpha_{1i}, \alpha_{2j}) \\ & i = 1, 2, \dots, I_N, \quad j = 1, 2, \dots, I_M \end{aligned} \tag{1.65}$$

with $i = 1, \dots, I_N$ and $j = 1, \dots, I_M$. At the same time, the mixed derivative of $(n+m)$th order of $f(\alpha_1, \alpha_2)$ is expressed as [57]

$$\begin{aligned} \left.\frac{\partial^{n+m} f(\alpha_1, \alpha_2)}{\partial \alpha_1^n \partial \alpha_2^m}\right|_{\alpha_1=\alpha_{1i},\, \alpha_2=\alpha_{2j}} &\cong \sum_{k=1}^{I_N} \varsigma_{\alpha_1(ik)}^{(n)} \left(\sum_{l=1}^{I_M} \varsigma_{\alpha_2(jl)}^{(m)} f(\alpha_{1k}, \alpha_{2l}) \right) \\ & i = 1, 2, \dots, I_N, \quad j = 1, 2, \dots, I_M \end{aligned} \tag{1.66}$$

Now, the discrete form of the fundamental equation (1.56) for the eigenvalue problem is obtained in a strong formulation, taking into account each expansion order of the kinematic ESL assumption (1.7)

$$\mathbf{K}\boldsymbol{\delta} = \omega^2 \mathbf{M}\boldsymbol{\delta} \tag{1.67}$$

$\mathbf{K}$ and $\mathbf{M}$ being the stiffness and inertia matrix written for the whole domain. Nevertheless, $\boldsymbol{\delta}$ is the global DOFs array, which is conveniently separated as $\boldsymbol{\delta} = [\boldsymbol{\delta}_b \ \boldsymbol{\delta}_d]^T$ to account for the external points of the structure denoted with 'b', and those of the internal ones of the computational domain labelled with 'd'. As a consequence, also $\mathbf{K}$ and $\mathbf{M}$ are properly rearranged, such that equation (1.67) becomes [57]

$$\begin{bmatrix} \mathbf{K}_{bb} & \mathbf{K}_{bd} \\ \mathbf{K}_{db} & \mathbf{K}_{dd} \end{bmatrix} \begin{bmatrix} \boldsymbol{\delta}_b \\ \boldsymbol{\delta}_d \end{bmatrix} = \omega^2 \begin{bmatrix} \mathbf{0} & \mathbf{0} \\ \mathbf{0} & \mathbf{M}_{dd} \end{bmatrix} \begin{bmatrix} \boldsymbol{\delta}_b \\ \boldsymbol{\delta}_d \end{bmatrix} \tag{1.68}$$

Employing a kinematic condensation, it is possible to reduce the size of (1.68) that yields the following expression

$$(\mathbf{M}_{dd}^{-1}(\mathbf{K}_{dd} - \mathbf{K}_{db}\mathbf{K}_{bb}^{-1}\mathbf{K}_{bd}) - \omega^2\mathbf{I})\boldsymbol{\delta}_d = \mathbf{0} \tag{1.69}$$

Equation (1.68) points out also a problem related to the definition of BCs near the corners of the shell. According to the identification of edges in (1.60), the discrete corner point belongs to two different sides of the structure. Therefore, the definition of external constraints is not immediate, as the discrete form of BCs is univocally defined at each point. To this end, a possible solution is to consider in these points the sum of the external constraints defined in (1.59). In this way, the free corner condition can be easily set, since it comes from a combination of some basic constraint cases.

1.3 Numerical applications

In the present section we perform some numerical examples to validate the proposed approach against some reference 3D finite element-based predictions. The focus has been on various parameters occurring in the generalized problem at issue, namely the presence of variable curvatures within a structure, the introduction of latticed materials, and different stacking sequences along the shell thickness.

As far as the geometry is concerned, the structure is described with respect to the reference surface (1.1), in line with the ESL strategy. As summarized in figure 1.1, two categories of shells are distinguished among the proposed examples. First, a helicoidal panel is described as a translational surface if a principal coordinate system is set. It leads to the following position vector [57]

$$\begin{aligned}\mathbf{r}(\alpha_1, \alpha_2) = {} & -a\cos(\alpha_1 + \alpha_2)\sinh(\alpha_1 - \alpha_2)\mathbf{e}_1 - a\sin(\alpha_1 + \alpha_2)\sinh(\alpha_1 - \alpha_2)\,\mathbf{e}_2 \\ & + a(\alpha_1 - \alpha_2)\,\mathbf{e}_3\end{aligned} \tag{1.70}$$

This kind of structure consists of a curve described with the principal coordinate α_1 sliding over another one described from the other principal line α_2. On the other hand, a revolution surface is obtained from a curve along α_1 direction getting around a reference rotation axis. The parallel direction denotes a principal parametric set of lines along α_2. In the following, the principal coordinate reference surface equation is reported for a revolution paraboloid of nth order [57]

$$\mathbf{r}(x_1, \vartheta) = x_1\cos\vartheta\,\mathbf{e}_1 - x_1\sin\vartheta\,\mathbf{e}_2 + a x_1^n\,\mathbf{e}_3 \tag{1.71}$$

In equations (1.70)–(1.71), the parameter a is used to fix the geometry scaling. This parameter assumes a different meaning depending on the selected shell structure. As a matter of fact, a rectangular plate can be described as a translational surface (1.70) in which both generatrix and directrix curves consist of a straight line.

As extensively discussed in the previous sections, the main advantage of employing a lattice medium in the structure is to come across a general orthotropic behaviour of the structure, even though an isotropic material with spherical elastic symmetry is assumed. In this way, it is possible to perform a topological optimization of the cell unit to assess *a priori* the dynamic properties of the shell, by simply tuning the unit cell geometry rather than the entire structure or its BCs. Moreover, a lattice material turns into a very lightweight performing structure. In the present chapter two different approaches have been applied to define the central layer of shells. In figure 1.2 one can find all the geometric quantities employed in the model for both a honeycomb cell (a) and single grid family (a), based on thin-walled hypotheses.

The accuracy of the approach is affected by the complexity of the selected stacking sequence. Two main aspects must be considered, namely, the complexity of the single layer stiffness matrix (1.22) and the relative thickness of adjacent laminae. According to [24], a lattice honeycomb structure can be obtained in a lot of engineering applications, embedding both nanomaterials, aerospace, and civil applications. For this reason, the validation of the dynamic behaviour is compulsory for both thin and thick layers within a laminate.

As also stated by Vasiliev *et al* [6, 7], the manufacturing process of a lattice grid structure consists of a central thick core made by various stiffeners, together with two outer thinner isotropic sheets, which cannot be considered as structural contributions. However, since the dynamic behaviour must be assessed within the whole structure, in the present work we employ two isotropic thin skins for a lattice grid layer.

The first set of examples studies the accuracy of the proposed homogenization method to model the honeycomb core. A rectangular plate is considered with in-plane dimensions $L_x \simeq L_y \simeq 0.4$ m, which is assumed to be completely clamped (C) at each edge. The plate is made of an isotropic aluminum material, also for outer external skins, whereas a three-layer configuration is considered for each case.

As far as the honeycomb unit cell layup is concerned, examples with three different RVEs have been developed employing various internal cell angles ϑ, as depicted in figure 1.3, together with a conceptual three-dimensional representation of the sandwich panel. In particular, the hexagonal honeycomb cell is characterized by $\vartheta = 30°$, the rectangular one is denoted with $\vartheta = 0°$. Note that we revert to an auxetic behaviour in the case of re-entrant honeycomb, with $\vartheta = -30°$. In order to assess the smearing technique with a right set of cell configurations, different plates have been developed with various cell wall thicknesses, accounting for both the case of thin and thick layer. In all the analyses, the central honeycomb core lies within two external layers with $h_1 = h_3 = 0.001$ m. For each case, a refined 3D FEM model is employed for a large number of DOFs, and the first ten mode frequencies are calculated, together with their corresponding eigenvectors. The FEM-based outcomes are treated as reference solutions for the validation check of the proposed homogenized GDQ model.

The influence of the through-the-thickness axiomatic assumptions are, then, investigated by adopting different higher order theories. The results are summarized in tables 1.1–1.3, while a geometric representation of the FEM model can be found in figure 1.4, along with its homogenized scheme. The finite element model discretizes the outer layers with parabolic 20-nodes bricks, while the internal grid assumes higher order shell elements, to yield a conforming mesh, and an optimized interpolation of the problem.

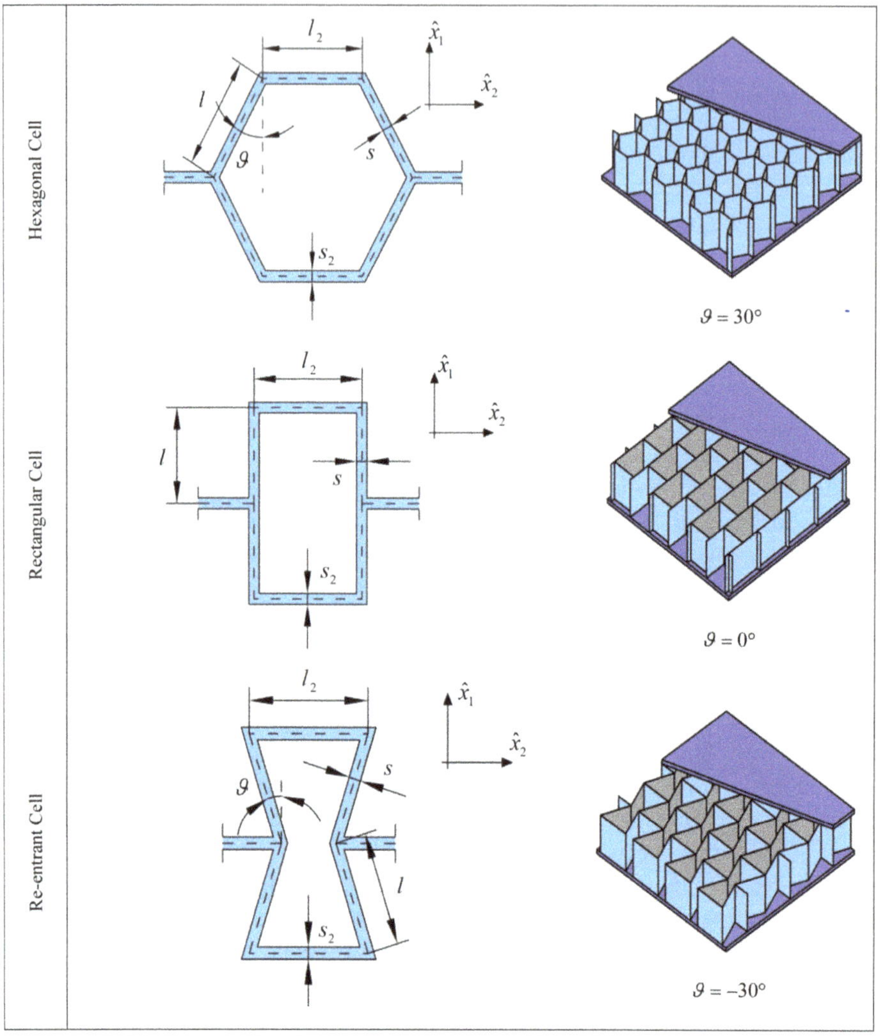

Figure 1.3. Honeycomb cell layup.

In table 1.1 we summarize the first ten natural frequencies of the rectangular plate with a central infill of hexagonal honeycomb ($\vartheta = 30°$). In this case, the equivalent elastic properties, generally orthotropic, degenerate into a transverse isotropic material, due to the perfect correspondence between the in-plane elastic moduli along α_1 and α_2. Accordingly, the symmetry of the domain and BCs is reflected on the vibration modes, that share four pairs of values, with few discrepancies related to the geometrical dimensions of L_x and L_y.The same analyses are then repeated in table 1.2 for a rectangular honeycomb unit cell ($\vartheta = 0°$), by employing different HSDTs. The 3D FEM model is developed taking into account the exact geometry of the core infill, as well as perfectly-bonded interfaces. All the internal cell walls also present the same thickness at any arbitrary point. A twofold numerical investigation

Table 1.1. Free vibration analysis of a CCCC rectangular sandwich plate with a lattice core made of hexagonal honeycomb cells, different lattice layers and layer thickness.

CCCC rectangular plate—hexagonal honeycomb								
Mode f [Hz]	3D FEM	FSDT	FSDTZ	TSDTZ	EDZ1$_{(\kappa = 5/6)}$	EDZ2$_{(\kappa = 5/6)}$	EDZ3$_{(\kappa = 1)}$	EDZ4$_{(\kappa = 1)}$
Thin layer ($b = h_2 = 5$ mm)								
DOFs	1700 490	5046	7569	12 615	7569	10 092	12 615	15 138
Thin cell wall ($s_2 = s = 0.1$ mm)								
1	497.420	562.023	499.049	499.029	540.705	489.755	499.995	499.665
2	937.430	1138.712	942.166	942.110	993.995	915.807	944.363	943.308
3	937.520	1139.725	942.875	942.819	994.694	916.480	945.074	944.017
4	1306.000	1670.009	1314.580	1314.487	1366.662	1271.120	1318.049	1316.132
5	1530.300	2024.507	1541.770	1541.649	1587.401	1485.261	1546.278	1543.554
6	1544.200	2035.337	1555.703	1555.581	1602.999	1498.889	1560.178	1557.527
7	1844.400	2524.945	1859.401	1859.243	1899.250	1786.186	1865.188	1861.512
8	1845.000	2526.375	1860.222	1860.065	1900.031	1786.955	1866.014	1862.334
9	2230.400	3203.529	2251.152	2250.939	2274.575	2153.314	2259.064	2253.690
10	2232.600	3207.605	2253.342	2253.129	2276.648	2155.362	2261.269	2255.882
Thick cell wall ($s_2 = s = 0.4$ mm)								
1	507.570	524.448	508.950	508.925	572.870	507.287	510.782	510.281
2	1005.400	1063.046	1011.036	1010.954	1124.097	1003.509	1014.852	1013.707
3	1005.400	1063.993	1011.897	1011.815	1125.027	1004.355	1015.716	1014.570
4	1446.400	1559.666	1458.681	1458.527	1606.803	1443.244	1464.401	1462.553
5	1733.900	1891.147	1751.459	1751.247	1918.189	1729.388	1758.520	1756.148
6	1745.900	1901.205	1763.338	1763.128	1932.833	1741.578	1770.423	1768.068
7	2131.300	2359.446	2156.675	2156.377	2346.730	2124.760	2165.609	2162.455
8	2132.100	2360.785	2157.807	2157.508	2347.908	2125.858	2166.746	2163.591
9	2655.400	2994.680	2694.166	2693.729	2906.987	2646.190	2705.767	2701.494
10	2658.900	2998.499	2697.293	2696.855	2910.197	2649.209	2708.911	2704.630

(*Continued*)

Table 1.1. (*Continued*)

CCCC rectangular plate—hexagonal honeycomb								
Mode f [Hz]	3D FEM	FSDT	FSDTZ	TSDTZ	EDZ1$_{(\kappa=5/6)}$	EDZ2$_{(\kappa=5/6)}$	EDZ3$_{(\kappa=1)}$	EDZ4$_{(\kappa=1)}$
Thick layer ($b = h_2 = 10$ mm)								
DOFs	2606 202	5046	7569	12 615	7569	10 092	12 615	15 138
Thin cell wall ($s_2 = s = 0.1$ mm)								
1	803.600	985.369	805.519	805.493	843.671	781.480	808.063	806.722
2	1448.800	1971.116	1453.922	1453.859	1476.916	1394.118	1459.179	1455.889
3	1449.000	1972.823	1454.895	1454.832	1477.829	1395.023	1460.157	1456.863
4	1973.000	2858.961	1981.914	1981.817	1985.140	1890.367	1989.567	1984.422
5	2273.900	3444.450	2285.447	2285.328	2270.406	2172.296	2294.853	2288.098
6	2296.100	3466.138	2307.803	2307.682	2292.862	2193.156	2317.238	2310.528
7	2708.200	4257.295	2722.957	2722.808	2684.757	2580.924	2734.491	2725.981
8	2708.900	4259.593	2724.018	2723.869	2685.738	2581.904	2735.559	2727.042
9	3214.900	5346.405	3234.577	3234.388	3161.738	3054.516	3249.293	3237.800
10	3217.100	5352.810	3237.389	3237.200	3164.333	3057.114	3252.125	3240.611
Thick cell wall ($s_2 = s = 0.4$ mm)								
1	830.190	874.702	831.356	831.314	919.592	826.643	837.355	833.971
2	1608.700	1753.613	1614.633	1614.505	1750.915	1593.991	1626.554	1619.743
3	1608.800	1755.139	1615.937	1615.809	1752.262	1595.257	1627.868	1621.051
4	2278.600	2548.175	2291.807	2291.581	2453.451	2251.944	2308.985	2299.068
5	2702.900	3073.353	2721.993	2721.690	2888.812	2665.760	2742.705	2730.662
6	2725.200	3092.216	2744.324	2744.024	2915.804	2688.612	2765.219	2753.114
7	3288.700	3804.057	3315.411	3315.005	3491.702	3237.807	3340.899	3325.977
8	3289.700	3806.127	3317.016	3316.609	3493.285	3239.336	3342.518	3327.586
9	4037.900	4785.462	4078.135	4077.568	4248.043	3965.327	4110.172	4091.213
10	4041.800	4791.252	4082.468	4081.900	4252.269	3969.438	4114.544	4095.558

Geometric inputs $L_x = 0.39999$ mm, $L_y = 0.40029$ mm

Honeycomb cell configuration $l_2 = 5.1320$ mm, $l = 5.1320$ mm, $\vartheta = 30\,°$, $E_s = 70$ GPa, $\nu_s = 0.33$, $\rho_s = 2700$ kg/m^3

Table 1.2. Free vibration analysis of a CCCC rectangular sandwich plate with a lattice core made of rectangular honeycomb cells, different lattice layers and layer thickness.

CCCC rectangular plate—rectangular honeycomb								
Mode f [Hz]	3D FEM	FSDT	FSDTZ	TSDTZ	EDZ1$_{(\kappa = 5/6)}$	EDZ2$_{(\kappa = 5/6)}$	EDZ3$_{(\kappa = 1)}$	EDZ4$_{(\kappa = 1)}$
Thin layer ($b = h_2 = 5$ mm)								
DOFs	1938 258	5046	7569	12 615	7569	10 092	12 615	15 138
Thin cell wall ($s_2 = s = 0.1$ mm)								
1	489.070	563.844	490.427	490.408	527.369	480.055	491.460	491.072
2	907.740	1142.038	910.100	910.051	949.921	881.306	912.576	911.288
3	917.610	1143.791	923.906	923.853	967.788	895.704	926.137	924.976
4	1264.300	1675.085	1271.842	1271.760	1310.527	1225.830	1275.503	1273.322
5	1468.000	2030.916	1472.951	1472.848	1500.407	1413.434	1477.901	1474.691
6	1497.900	2042.286	1512.476	1512.364	1547.605	1453.457	1516.999	1514.112
7	1765.400	2532.091	1775.230	1775.096	1795.881	1699.354	1781.394	1777.181
8	1782.100	2534.576	1798.432	1798.291	1823.093	1722.734	1804.363	1800.338
9	2111.900	3212.577	2119.889	2119.716	2119.663	2020.124	2128.436	2122.178
10	2160.900	3219.668	2188.170	2187.976	2197.918	2088.356	2196.133	2190.412
Thick cell wall ($s_2 = s = 0.4$ mm)								
1	508.240	529.315	510.338	510.314	572.585	508.471	512.570	511.983
2	997.660	1070.243	1004.884	1004.809	1111.149	996.637	1010.104	1008.551
3	1004.400	1076.640	1015.091	1015.007	1123.566	1006.080	1018.935	1017.819
4	1432.300	1572.959	1451.033	1450.887	1589.229	1433.496	1457.642	1455.519
5	1708.800	1905.299	1731.007	1730.817	1881.658	1706.488	1740.433	1737.236
6	1733.000	1923.285	1761.302	1761.086	1919.247	1735.920	1768.177	1765.905
7	2092.100	2375.762	2127.882	2127.612	2297.935	2092.174	2138.748	2134.845
8	2106.100	2384.841	2145.765	2145.475	2320.176	2109.555	2155.042	2151.747
9	2587.400	3012.452	2636.728	2636.347	2817.787	2583.340	2651.770	2646.014
10	2631.300	3038.338	2691.050	2690.601	2885.119	2636.624	2701.948	2697.904

(Continued)

Table 1.2. *(Continued)*

CCCC rectangular plate—rectangular honeycomb

Mode f [Hz]	3D FEM	FSDT	FSDTZ	TSDTZ	EDZ1$_{(\kappa=5/6)}$	EDZ2$_{(\kappa=5/6)}$	EDZ3$_{(\kappa=1)}$	EDZ4$_{(\kappa=1)}$
Thick layer ($b = h_2 = 10$ mm)								
DOFs	2791 698	5046	7569	12 615	7569	10 092	12 615	15 138
Thin cell wall ($s_2 = s = 0.1$ mm)								
1	780.090	990.465	780.884	780.860	809.609	754.781	783.532	782.090
2	1372.400	1978.452	1372.119	1372.072	1377.581	1309.840	1377.578	1373.965
3	1404.400	1985.579	1410.881	1410.814	1421.952	1348.584	1415.929	1412.610
4	1879.200	2872.050	1884.797	1884.713	1870.984	1791.332	1892.271	1886.946
5	2128.800	3459.404	2128.235	2128.153	2092.395	2014.528	2137.641	2130.485
6	2207.800	3486.323	2222.846	2222.714	2194.780	2107.097	2231.762	2225.060
7	2539.300	4272.727	2544.365	2544.251	2486.528	2403.124	2555.539	2546.757
8	2584.600	4282.427	2599.232	2599.088	2544.961	2456.548	2610.211	2601.633
9	2965.300	5362.606	2965.050	2964.926	2871.387	2789.709	2979.294	2967.410
10	3094.000	5389.763	3120.984	3120.773	3035.469	2940.949	3134.883	3123.467
Thick cell wall ($s_2 = s = 0.4$ mm)								
1	831.070	886.808	832.312	832.272	915.420	827.104	839.603	835.493
2	1582.100	1766.097	1587.074	1586.999	1706.210	1565.126	1603.054	1593.658
3	1609.600	1790.702	1622.815	1622.641	1746.890	1598.066	1634.648	1627.922
4	2242.600	2578.107	2263.274	2263.076	2402.732	2218.896	2282.641	2271.194
5	2628.300	3095.904	2647.735	2647.584	2780.954	2587.668	2674.604	2658.151
6	2706.400	3155.220	2740.575	2740.127	2886.638	2675.864	2760.124	2748.738
7	3189.500	3832.376	3226.338	3226.080	3364.440	3142.414	3256.136	3237.797
8	3237.100	3866.003	3282.220	3281.761	3427.349	3195.598	3307.949	3292.521
9	3864.000	4806.716	3909.938	3909.684	4025.016	3790.163	3949.227	3924.150
10	4004.800	4900.745	4072.182	4071.299	4206.283	3945.444	4101.454	4083.899

Geometric inputs $L_x = 0.39999$ mm, $L_y = 0.40029$ mm

Honeycomb cell configuration $l_2 = 5.1320$ mm, $l = 5.1320$ mm, $\vartheta = 0^\circ$, $E_s = 70$ GPa, $\nu_s = 0.33$, $\rho_s = 2700$ kg/m^3

Table 1.3. Free vibration analysis of a CCCC rectangular sandwich plate with a lattice core made of re-entrant honeycomb cells, different lattice layers and layer thickness.

CCCC rectangular plate—re-entrant honeycomb								
Mode f [Hz]	3D FEM	FSDT	FSDTZ	TSDTZ	EDZ1$_{(\kappa=5/6)}$	EDZ2$_{(\kappa=1)}$	EDZ3$_{(\kappa=1)}$	EDZ4$_{(\kappa=1)}$
Thin layer ($b = h_2 = 5$ mm)								
DOFs	1224 690	5046	7569	12 615	7569	10 092	12 615	15 138
Thin cell wall ($s_2 = s = 0.1$ mm)								
1	490.960	578.921	490.972	490.954	523.359	492.172	492.155	491.673
2	901.600	1172.339	906.894	906.847	939.042	909.561	909.522	908.038
3	911.940	1174.200	907.855	907.808	939.910	910.541	910.503	909.009
4	1249.600	1719.257	1250.808	1250.733	1276.431	1254.922	1254.864	1252.265
5	1444.900	2084.327	1454.799	1454.703	1471.169	1460.100	1460.028	1456.391
6	1474.800	2096.070	1468.428	1468.331	1485.540	1473.699	1473.625	1470.064
7	1733.100	2598.238	1743.312	1743.190	1749.508	1750.008	1749.917	1745.117
8	1749.600	2600.864	1744.290	1744.168	1750.360	1751.008	1750.917	1746.101
9	2064.000	3296.048	2090.335	2090.174	2077.613	2099.392	2099.274	2092.384
10	2112.000	3303.525	2092.846	2092.686	2079.793	2101.961	2101.844	2094.910
Thick cell wall ($s_2 = s = 0.4$ mm)								
1	513.320	542.005	518.332	518.308	579.408	521.202	521.180	520.414
2	997.700	1098.256	1019.797	1019.719	1121.776	1025.419	1025.355	1023.661
3	1005.100	1099.689	1020.997	1020.919	1123.097	1026.729	1026.665	1024.938
4	1424.500	1609.960	1460.305	1460.165	1588.688	1468.487	1468.376	1465.677
5	1687.700	1954.013	1746.590	1746.395	1885.705	1756.422	1756.269	1752.840
6	1717.500	1964.699	1759.521	1759.327	1901.521	1769.460	1769.307	1765.883
7	2059.900	2434.693	2138.509	2138.243	2292.610	2150.587	2150.386	2145.919
8	2078.500	2436.723	2140.001	2139.736	2294.152	2152.206	2152.004	2147.493
9	2523.800	3092.747	2654.785	2654.394	2816.916	2669.877	2669.582	2663.627
10	2583.500	3098.528	2658.812	2658.422	2820.995	2674.234	2673.940	2667.858

(Continued)

Table 1.3. (*Continued*)

CCCC rectangular plate—re-entrant honeycomb

Mode f [Hz]	3D FEM	FSDT	FSDTZ	TSDTZ	EDZ1$_{(\kappa=5/6)}$	EDZ2$_{(\kappa=1)}$	EDZ3$_{(\kappa=1)}$	EDZ4$_{(\kappa=1)}$
Thick layer ($b = h_2 = 10$ mm)								
DOFs	1764 306	5046	7569	12 615	7569	10 092	12 615	15 138
Thin cell wall ($s_2 = s = 0.1$ mm)								
1	761.020	1015.222	762.585	762.567	780.903	765.468	765.454	763.814
2	1322.100	2029.063	1324.113	1324.073	1315.261	1329.545	1329.518	1325.723
3	1343.800	2031.749	1350.842	1350.798	1345.735	1356.361	1356.332	1352.543
4	1792.000	2940.656	1799.278	1799.218	1767.993	1806.926	1806.887	1801.167
5	2030.900	3543.703	2034.329	2034.258	1983.218	2043.488	2043.440	2036.149
6	2085.300	3566.557	2100.316	2100.237	2053.782	2109.697	2109.645	2102.306
7	2408.300	4375.306	2416.547	2416.460	2341.646	2427.588	2427.532	2418.457
8	2438.400	4378.700	2453.955	2453.863	2381.229	2465.140	2465.081	2455.947
9	2809.300	5494.487	2814.552	2814.445	2706.470	2828.333	2828.262	2816.332
10	2894.500	5503.676	2920.896	2920.774	2817.532	2935.070	2934.990	2922.884
Thick cell wall ($s_2 = s = 0.4$ mm)								
1	822.070	903.605	833.190	833.162	909.920	842.329	842.313	837.082
2	1527.500	1810.490	1582.541	1582.461	1683.080	1599.107	1599.061	1589.186
3	1575.300	1811.307	1597.730	1597.644	1705.041	1614.416	1614.368	1604.484
4	2160.400	2623.223	2228.163	2228.038	2340.948	2250.824	2250.757	2236.928
5	2477.100	3170.463	2614.965	2614.780	2713.583	2640.959	2640.853	2624.587
6	2618.900	3189.702	2666.129	2665.926	2780.013	2692.817	2692.702	2676.209
7	3010.000	3913.371	3162.324	3162.110	3259.794	3193.289	3193.173	3173.485
8	3100.100	3913.836	3192.164	3191.936	3298.109	3223.563	3223.440	3203.602
9	3567.700	4933.278	3828.070	3827.738	3893.056	3864.508	3864.315	3840.410
10	3834.600	4933.800	3918.820	3918.444	4006.026	3956.664	3956.450	3931.999

Geometric inputs $L_x = 0.39999$ mm, $L_y = 0.40029$ mm, $h_1 = h_3 = 0.001$ m

Honeycomb cell configuration $l_2 = 5.1320$ mm, $l = 5.1320$ mm, $\vartheta = -30\,°$, $E_s = 70$ GPa, $\nu_s = 0.33$, $\rho_s = 2700$ kg/m^3

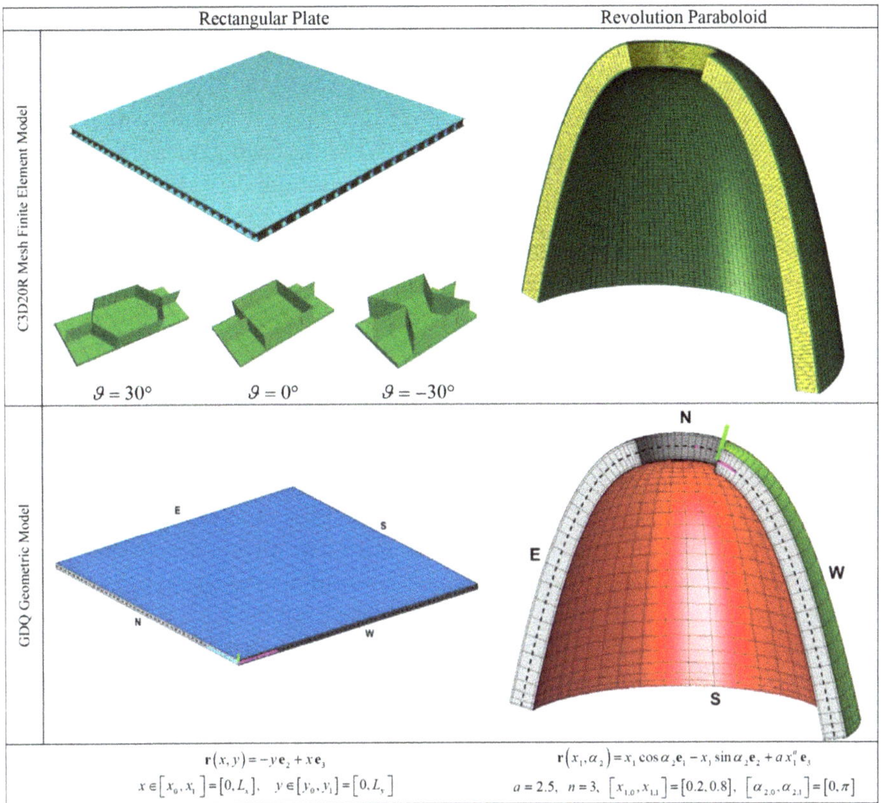

Figure 1.4. Finite element models and shell representation of the panels reinforced with a lattice infill.

is now performed for sandwich panels with thin and/or thick layers. For each case, two different thickness of the internal cell walls are selected. Based on the comparative results in table 1.2, it is worth observing an increased accuracy of GDQ-based predictions with respect to the finite element reference solution, when the zig-zag function (1.9) is embedded in the ESL displacement field (1.7). This is particularly evident when the Murakami's strategy is combined with a classical FSDT assumption. Any further increase in accuracy is observed, when implementing a third order theory. A similar behaviour can be also observed in the case of re-entrant honeycomb lattice layer ($\vartheta = -30°$). In table 1.3 we report the first ten mode frequencies for a completely clamped (CCCC) rectangular plate with a central layer made of a honeycomb re-entrant pattern. The good accordance between our GDQ-based results and finite element predictions reveal the great capability of the proposed homogenization approach and higher order theories to handle the problem. Unlike other classical homogenization methods, the present formulation is able to predict accurately the actual value of the shear modulus, G_{23}, which is usually computed by interpolating the values provided by a static and a kinematic admissible problem. The non-linear interpolation proposed in equation (1.39) is calibrated on a series of validation examples conducted on a rectangular plate [39] and represents a valid alternative to the original Grediac formulation [29] assessed by Scarpa *et al* [30]. In this way, also for the re-entrant case, the eigenfrequencies are properly predicted for a large variety of cell cases, taking into account different cell configurations in terms of wall thickness and layer height. From a direct comparison between tables 1.1–1.3, it can be stated that the honeycomb grid does not affect the dynamic behaviour, for lower frequencies, whereas higher modes depend significantly on the cell geometric configuration. In particular, the internal cell angle ϑ is a key design parameter. For re-entrant honeycomb angles, indeed, a meaningful increase in vibration frequencies can be noticed for any cell configuration. On the other hand, since a re-entrant honeycombs turns into a negative equivalent in plane Poisson's ratio, ν_{12}, the auxetic behaviour of the central core features complicated interfacial phenomena. As a consequence, only the EDZ4 theory seems to agree with the FEM-based solution in terms of vibration modes, especially for thick layer ($h_2 = 0.001$ m) layups.

As also visible in figure 1.5, the proposed approach is capable of providing very complex in-plane and out-of-plane deformation phenomena. In addition, for a honeycomb cell, the variation of the cell geometry can affect significantly the mode shapes, especially for higher modes. Despite the symmetry of the geometry and BCs, different eigenvectors are associated with the problem in figure 1.5, by simply tuning the internal cell angle ϑ. For instance, if we consider three different unit cells with $\vartheta = 30°, 0°, -30°$, the vibration modes 5 and 6 move from a central-symmetric deformation with four waves for the hexagonal honeycomb, to a three-waves mode in the case of rectangular cell, and eventually to a four-waves mode in the case of auxetic configuration. For lower modes, the orthogonality of the deformation field is induced from different constitutive lattices and in-plane properties.

Another validation analysis is performed on a paraboloidal panel, whose mode frequencies are summarized in tables 1.4, 1.5. In figure 1.4 we report the representative geometric parameters for each selected shell structure. In this case we model the central core with a continuum model based on the equivalent elastic properties of the grid pattern computed as in equation (1.27). A fine structured mesh is used for the discretization of the domain, here modelled with C3D20R parabolic elements. The laminate consists of two external thin layers made of isotropic aluminum and a thick central core made of a lattice layer rather than a cellular material. The unit cell is made by four rib families: two of them ($\phi_i = 0°$ and $\phi_i = 90°$) are stretched along the α_1 and α_2 principal lines, whereas the other two classes of frames are symmetrically sorted along directions $\phi_i = \pm 45°$ with respect to α_1. Employing the effective compact notation (1.28), the unit cell can be identified univocally with the notation $F[-45^{\delta_1}_{a_1=0.1414\,\mathrm{m}}/0^{\delta_2}_{a_2=0.1\,\mathrm{m}}/45^{\delta_3}_{a_3=0.1414\,\mathrm{m}}/90^{\delta_4}_{a_4=0.1\,\mathrm{m}}]$, where a_i is the interspace between two adjacent ribs and δ_i the width of each frame. A modal analysis is adopted for a paraboloidal panel with two different BCs and different values of δ_i. A cantilevered configuration (FCFF) of the structure is first analysed, whose mode frequencies are listed in table 1.4. The same study is, thus, repeated for the same structure constrained on the two lateral edges along two different meridian lines (CFCF), as summarized in table 1.5. The 3D FEM solution has been

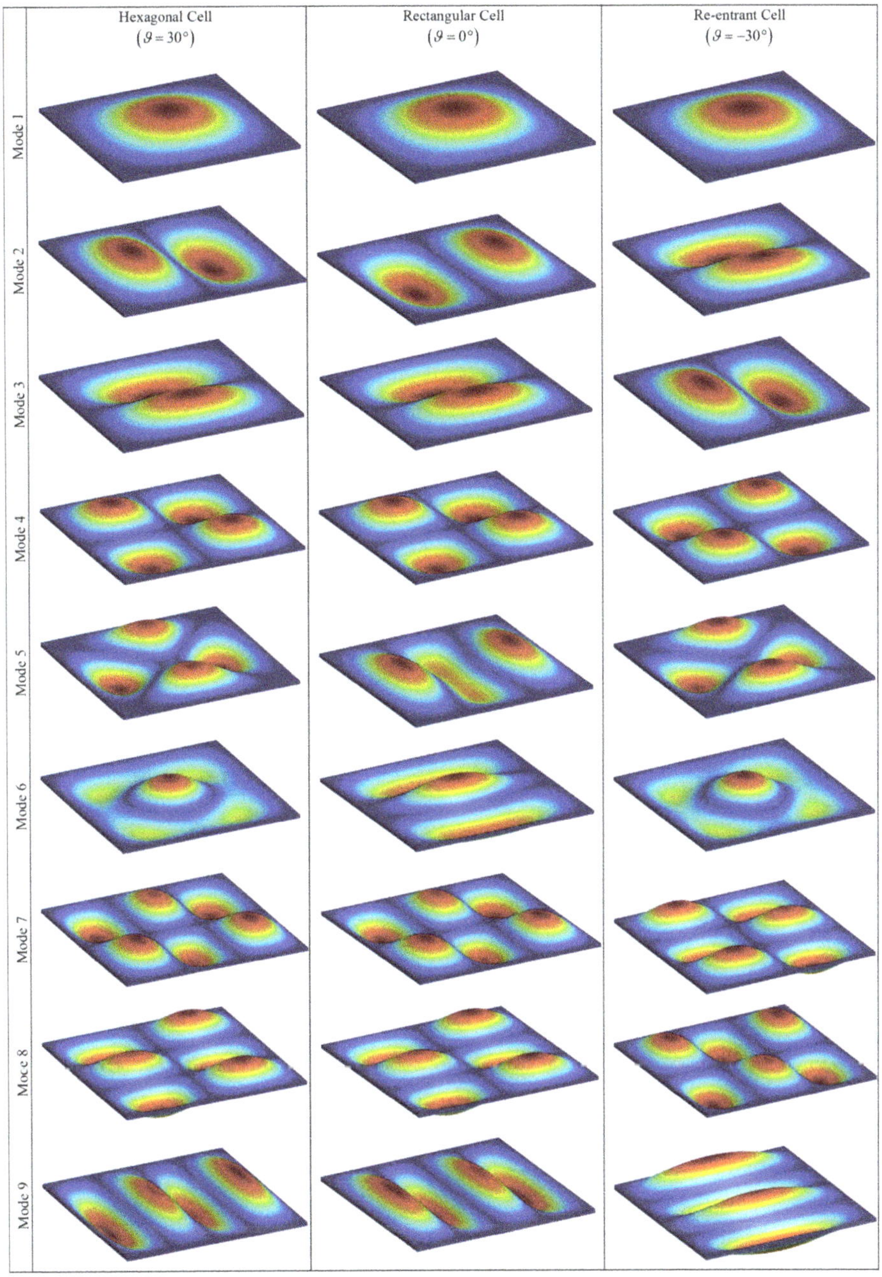

Figure 1.5. Mode shapes of a rectangular plate with a central lattice honeycomb layer.

compared with predictions from the homogenized model based on different orders of expansion. Differently from the previous cases, the influence of the Murakami's function (1.9) is trivial, since the central core is predominant within the thickness profile with respect to the external skins, leading to reduced

Table 1.4. Natural frequencies of a FCFF revolution paraboloid with a central core made of a grid layer with four families, and two external skins of aluminum. Both helical and meridian loops have been considered, together with stiffeners in two opposite directions. The shell reference surface equation and homogenized geometric representation, can be found in figure 1.4.

FCFF revolution paraboloid—pantographic central layer							
f [Hz]	3D FEM	FSDTZ	TSDTZ	EDZ1$_{(\kappa = 5/6)}$	EDZ2$_{(\kappa = 5/6)}$	EDZ3$_{(\kappa = 1)}$	EDZ4$_{(\kappa = 1)}$
DOFs	1310 499	7569	12 615	7569	10 092	12 615	15 138
Thin stiffeners ($\delta_1 = \delta_2 = \delta_3 = \delta_4 = 0.005$ m)							
1	197.410	197.545	197.523	198.567	196.350	198.363	197.651
2	267.397	267.396	267.384	268.052	267.185	267.955	267.664
3	385.044	385.684	385.495	389.842	381.401	387.663	385.697
4	539.214	539.390	539.194	535.283	531.698	540.389	539.511
5	588.531	588.636	588.501	587.027	584.747	590.020	589.122
6	670.093	670.288	670.080	672.600	666.002	671.916	670.612
7	727.782	729.362	728.983	741.227	719.380	734.569	729.612
8	788.498	789.735	789.328	783.922	773.627	793.757	789.912
9	918.067	918.195	918.126	919.194	918.061	918.982	918.554
10	981.492	982.141	982.006	976.770	971.487	982.675	981.844
Moderately thick stiffeners ($\delta_1 = \delta_2 = \delta_3 = \delta_4 = 0.01$ m)							
1	177.717	177.774	177.753	179.442	177.624	178.837	178.001
2	244.693	244.677	244.660	245.543	244.793	245.284	244.930
3	349.296	349.507	349.371	356.036	349.063	352.767	350.147
4	496.037	496.012	495.857	495.626	492.764	497.359	496.324
5	540.241	540.228	540.073	540.893	537.864	542.315	540.896
6	615.749	615.709	615.542	620.390	614.837	618.163	616.347
7	673.654	674.456	674.147	692.963	673.221	683.205	676.118
8	738.421	738.963	738.558	742.612	731.677	745.374	740.238
9	844.757	844.817	844.739	846.000	844.973	845.612	845.142
10	898.704	898.953	898.848	898.927	898.372	899.212	898.877

Thick stiffeners ($\delta_1 = \delta_2 = \delta_3 = \delta_4 = 0.02$ m)							
1	155.691	155.731	155.687	157.298	156.151	156.846	156.000
2	222.326	222.312	222.284	223.181	222.627	222.959	222.552
3	304.759	304.880	304.716	311.167	306.699	308.713	305.823
4	442.910	442.984	442.708	444.390	441.770	445.228	443.543
5	484.702	484.688	484.521	486.227	484.458	486.506	485.247
6	556.241	556.238	556.044	561.056	557.242	559.375	557.097
7	598.588	599.171	598.740	617.705	604.198	610.047	601.713
8	659.484	659.961	659.351	667.906	659.631	667.754	661.789
9	776.035	776.038	775.961	777.040	776.386	776.739	776.327
10	819.391	819.495	819.379	819.645	819.184	819.798	819.506
Geometric inputs $a = 2.5$, $n = 3$, $[\alpha_{1,0}, \alpha_{1,1}] = [0.2, 0.8]$, $[\alpha_{2,0}, \alpha_{2,1}] = [0, \pi]$, $h_1 = h_3 = 0.01m$, $h_2 = 0.1$m							
Grid unit pattern layup $\mathrm{F}\left[-45^{\delta_1}_{a_1=0.1414\mathrm{m}}/0^{\delta_2}_{a_2=0.1\mathrm{m}}/45^{\delta_3}_{a_3=0.1414\mathrm{m}}/90^{\delta_4}_{a_4=0.1\mathrm{m}}\right]$, $E_s = 70\mathrm{GPa}$, $\nu_s = 0.3$, $\rho_s = 2707\mathrm{kg/m^3}$							

Table 1.5. Natural frequencies of a CFCF revolution paraboloid with a central core made of a grid layer with four families, and two external skins of aluminum. Both helical and meridian loops have been considered, together with stiffeners in two opposite directions. The shell reference surface equation and homogenized geometric representation, can be found in figure 1.4.

CFCF revolution paraboloid—pantographic central layer

f [Hz]	3D FEM	FSDTZ	TSDTZ	$EDZ1_{(\kappa=5/6)}$	$EDZ2_{(\kappa=1)}$	$EDZ3_{(\kappa=1)}$	$EDZ4_{(\kappa=1)}$
DOFs	1310 499	7569	12 615	7569	10 092	12 615	15 138
Thin stiffeners ($\delta_1 = \delta_2 = \delta_3 = \delta_4 = 0.005$ m)							
1	295.670	296.285	295.985	299.202	291.347	298.450	296.300
2	409.346	409.822	409.505	407.900	402.856	411.461	409.722
3	628.255	629.058	628.661	627.579	621.357	631.094	628.805
4	708.349	709.402	708.859	705.684	698.032	711.923	709.090
5	859.502	859.725	859.525	856.245	841.490	861.486	859.944
6	861.205	862.800	861.794	860.760	855.388	867.819	862.506
7	893.055	894.837	894.185	896.128	881.368	898.637	894.444
8	1115.651	1116.622	1116.284	1119.101	1109.399	1118.722	1116.311
9	1137.401	1138.870	1138.308	1145.233	1126.499	1143.560	1138.928
10	1202.283	1204.644	1203.908	1196.650	1181.066	1209.728	1204.120
Moderately thick stiffeners ($\delta_1 = \delta_2 = \delta_3 = \delta_4 = 0.01$ m)							
1	272.760	272.909	272.748	279.271	272.213	276.565	273.680
2	381.659	381.742	381.550	383.314	378.910	384.328	382.239
3	580.204	580.441	580.192	583.573	577.412	583.950	581.016
4	660.616	660.986	660.676	663.923	656.660	665.070	661.634
5	791.469	791.474	791.354	795.846	790.848	794.140	792.094
6	817.706	818.170	817.482	822.272	807.693	826.796	819.710
7	835.960	836.730	836.343	843.632	831.064	842.713	837.623
8	1019.993	1020.486	1020.240	1026.127	1019.000	1023.307	1020.685
9	1059.821	1060.638	1060.241	1074.950	1056.831	1068.969	1062.067
10	1123.590	1124.650	1124.117	1132.625	1117.776	1132.854	1125.842

Thick stiffeners ($\delta_1 = \delta_2 = \delta_3 = \delta_4 = 0.02$ m)							
1	240.014	240.073	239.882	246.970	242.083	244.572	241.247
2	343.452	343.503	343.239	346.847	343.432	346.921	344.361
3	520.225	520.334	520.027	525.276	520.772	524.621	521.332
4	596.731	596.985	596.586	602.645	597.044	602.235	598.170
5	720.441	720.458	720.306	725.039	721.646	723.532	721.250
6	736.676	737.099	736.161	747.189	735.965	748.075	739.631
7	758.181	758.752	758.205	768.046	758.879	766.526	760.426
8	918.893	919.153	918.879	924.957	920.201	922.586	919.765
9	960.832	961.519	960.976	978.244	964.884	972.360	963.913
10	1017.534	1018.357	1017.567	1031.658	1019.558	1029.504	1020.747

Geometric inputs $a = 2.5$, $n = 3$, $[\alpha_{1,0}, \alpha_{1,1}] = [0.2, 0.8]$, $[\alpha_{2,0}, \alpha_{2,1}] = [0, \pi]$, $h_1 = h_3 = 0.01m$, $h_2 = 0.1\text{m}$

Grid unit pattern layup $\text{F}\left[-45^{\delta_1}_{a_1=0.1414\text{m}}/0^{\varepsilon_2}_{c_2=0.1\text{m}}/45^{\delta_3}_{a_3=0.1414\text{m}}/90^{\delta_4}_{a_4=0.1\text{m}}\right]$, $E_s = 70\text{GPa}$, $\nu_s = 0.3$, $\rho_s = 2707\text{kg/m}^3$

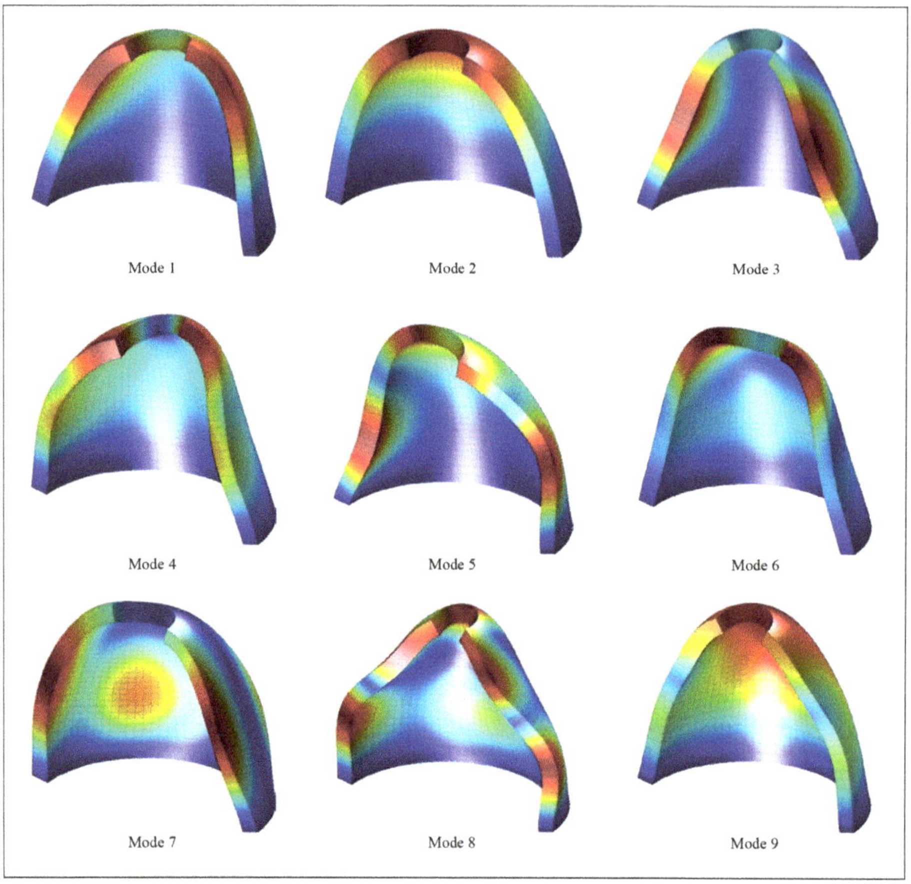

Figure 1.6. Mode shapes of a cantilever (FCFF) revolution paraboloidal panel with a lattice core made of a grid with four symmetric moderately thick rib families.

interlaminar phenomena. For this reason, even a classical third order theory by Reddy [53] behaves well together with the zig-zag strategy (1.9). The accuracy of the proposed model for grid structures seems to be unaffected by the width δ_i for each frame, as visible in both tables 1.4 and 1.5, except for CFCF shell structures with thick stiffeners $\delta_i = 0.02$ m, for which the EDZ4 formulation seems to provide the highest level of accuracy among results. Figure 1.6 depicts the first nine mode shapes of a moderately thick cantilevered paraboloidal panel. In this case, the influence of geometry becomes crucial for the overall response, because of the transversely-isotropic behaviour of the unit cell made of four ribs families. Actually, the proposed higher order model is capable of providing both symmetric and antisymmetric modal deformations, as well as all possible complex stretching effects that occur along the shell thickness, due to the interlaminar interactions.

The last set of examples considers a doubly-curved shell structure reinforced by a honeycomb central layer with $h = 5$ mm. More specifically, a helicoidal panel is modelled according to an ESL method, whose geometry definition in the principal

Helicoidal Panel

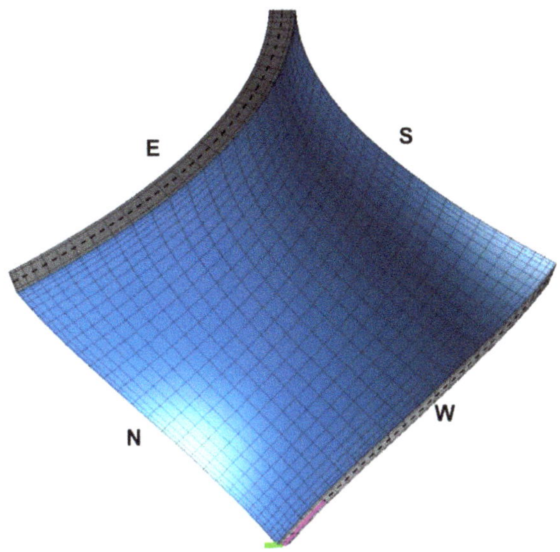

$$\mathbf{r}(\alpha_1,\alpha_2) = -a\cos(\alpha_1+\alpha_2)\sinh(\alpha_1-\alpha_2)\mathbf{e}_1 - a\sin(\alpha_1+\alpha_2)\sinh(\alpha_1-\alpha_2)\mathbf{e}_2 + a(\alpha_1-\alpha_2)\mathbf{e}_3$$

$$\alpha_1 \in [\alpha_1^0, \alpha_1^1], \alpha_2 \in [\alpha_2^0, \alpha_2^1], \alpha_1^0 = \alpha_2^0 = 0[\text{rad}], \alpha_1^1 = \alpha_2^1 = 1[\text{rad}], a = 0.07[\text{m}]$$

Figure 1.7. Geometric representation of a helicoidal panel in the principal reference system.

reference system can be found in figure 1.7. Also, in this case we compute the first ten mode frequencies of the structure, while keeping a CFFC boundary restraint. Three different unit cells are here implemented, with geometrical properties as defined in figure 1.3. For each selected case, the helicoidal panel assumes both thin ($s_2 = s = 0.1$ mm) and thick ($s_2 = s = 0.4$ mm) cell walls, whose results are summarized in table 1.6 under the assumption of hexagonal ($\vartheta = 30\,°$) honeycomb softcore. We recall that a transversely isotropic behaviour is provided by the cell unit at issue. For both thin cells or thick walls, the numerical results for each mode, always converge to the frequency predictions by the EDZ4 theory. Similarly, in table 1.7 we summarize the natural frequencies for the helicoidal panel (figure 1.7) with a lattice layer consisting of a series of rectangular thin-walled cells, as predicted by different HSDTs. Assuming the EDZ4-based solution as the reference one, it is worth noticing that lower order theories embedding a reduced number of DOFs provide results with a lower precision, and a higher dispersion. The same behaviour can be observed in table 1.8 for a helicoidal structure embedding a lattice layer with re-entrant cells. Based on results in this table, the accuracy of the proposed model depends on the shear correction factor assumption, $k(\zeta)$, when lower EDZ-N theories are taken on. In this case, results based on the EDZ1 theory are very accurate (if compared to EDZ4 predictions), when keeping $k(\zeta) = 5/6$. On the other hand, an EDZ2 theory does not require any shear correction factor, which corresponds to the assumption $k(\zeta) = 1$. Figure 1.8 plots the first nine mode shapes for the helicoidal shell with a central layer infilled by re-entrant cells, as computed

Table 1.6. Natural frequencies of a CFFC helicoid with a central layer made of a hexagonal honeycomb pattern. Both lattice layer and laminated structure are obtained using an isotropic aluminum material. The shell reference surface equation and homogenized geometric representation, can be found in figure 1.7.

CFFC helicoid—hexagonal honeycomb ($\vartheta = 30\,°$)

f [Hz]	FSDT	FSDTZ	TSDTZ	EDZ1$_{(\kappa = 5/6)}$	EDZ2$_{(\kappa = 1)}$	EDZ3$_{(\kappa = 1)}$	EDZ4$_{(\kappa = 1)}$
DOFs	5046	7569	12 615	7569	10 092	12 615	15 138
Thin cell wall ($s_2 = s = 0.1$ mm)							
1	1901.469	1384.682	1369.790	1397.930	1372.781	1372.574	1371.561
2	5440.802	3075.498	3024.351	2978.919	3040.084	3039.625	3033.426
3	7200.263	5736.547	5619.536	5510.106	5678.348	5677.392	5616.769
4	9808.352	6071.326	6028.289	5984.357	6048.407	6048.172	6025.510
5	12 458.479	6650.659	6597.278	6503.599	6636.379	6635.752	6598.467
6	12 500.523	7120.994	7039.499	6992.425	7079.268	7078.641	7038.838
7	15 075.318	8079.437	7913.276	7776.675	8043.259	8041.805	7918.052
8	15 129.797	8276.879	8171.826	8057.320	8245.237	8244.043	8173.664
9	17 601.664	8796.553	8664.116	8458.876	8777.101	8775.519	8665.228
10	17 672.505	9601.631	9515.206	9415.605	9588.348	9586.988	9515.458
Thick cell wall ($s_2 = s = 0.4$ mm)							
1	1831.307	1619.522	1617.284	1685.304	1621.252	1621.129	1615.236
2	5245.850	4069.307	4057.935	4036.202	4069.460	4069.004	4066.557
3	6951.404	6427.256	6421.781	6407.768	6429.068	6428.884	6426.882
4	9459.418	7943.127	7921.035	7801.382	7941.569	7940.807	7924.483
5	12 027.494	8216.985	8190.301	8109.838	8226.726	8225.500	8192.869
6	12 080.949	8826.772	8803.462	8664.301	8841.335	8839.626	8806.206
7	14 555.770	10 927.142	10 893.238	10 758.833	10 939.184	10 937.711	10 898.382
8	14 619.663	11 387.890	11 343.713	11 088.905	11 410.372	11 408.197	11 348.666
9	17 022.254	11 947.596	11 890.926	11 566.973	11 969.830	11 967.607	11 897.558
10	17 079.829	12 867.759	12 822.489	12 544.019	12 894.360	12 891.985	12 832.541

Geometric inputs $a = 0.07$, $[\alpha_{1,0}, \alpha_{1,1}] = [0, 1]$, $[\alpha_{2,0}, \alpha_{2,1}] = [0, 1]$, $h_1 = h_3 = 0.001\ m$, $h_2 = 0.005$ m

Honeycomb cell configuration $l_2 = 5.1320$ mm, $l = 5.1320$ mm, $E_s = 70$ GPa, $\nu_s = 0.33$, $\rho_s = 2700$ kg/m^3

Table 1.7. Natural frequencies of a CFFC helicoid with a central layer made of rectangular honeycomb patterns. Both lattice layer and laminated structure are obtained using an isotropic aluminum material. The shell reference surface equation and homogenized geometric representation, can be found in figure 1.7.

CFFC helicoid—rectangular honeycomb ($\vartheta = 0\,°$)

f [Hz]	FSDT	FSDTZ	TSDTZ	EDZ1$_{(\kappa = 5/6)}$	EDZ2$_{(\kappa = 1)}$	EDZ3$_{(\kappa = 1)}$	EDZ4$_{(\kappa = 1)}$
DOFs	5046	7569	12 615	7569	10 092	12 615	15 138
Thin cell wall ($s_2 = s = 0.1$ mm)							
1	1899.474	1398.420	1384.941	1413.976	1387.832	1387.623	1387.372
2	5435.346	3115.892	3068.028	3018.652	3083.371	3082.947	3076.455
3	7194.844	5859.476	5747.672	5622.272	5804.487	5803.542	5743.433
4	9798.266	6082.516	6042.589	5997.343	6061.101	6060.878	6039.218
5	12 446.595	6680.345	6628.924	6529.338	6668.063	6667.481	6628.443
6	12 489.314	7168.996	7093.391	7045.314	7129.120	7128.514	7092.721
7	15 058.076	8319.927	8170.666	8013.312	8272.609	8271.356	8170.866
8	15 147.805	8417.657	8305.618	8162.537	8403.019	8401.691	8311.268
9	17 586.229	8723.656	8593.832	8395.159	8705.914	8704.431	8594.762
10	17 655.802	9686.337	9608.275	9501.697	9672.919	9671.624	9606.560
Thick cell wall ($s_2 = s = 0.4$ mm)							
1	1824.397	1623.346	1621.351	1691.256	1625.614	1625.479	1621.495
2	5226.970	4102.912	4092.635	4074.371	4104.457	4104.008	4101.463
3	6932.722	6427.125	6422.306	6407.380	6428.664	6428.481	6425.164
4	9424.258	7996.494	7979.099	7901.775	7995.142	7994.445	7984.782
5	11 983.870	8430.859	8404.748	8301.309	8446.400	8444.920	8406.706
6	12 044.475	8756.603	8734.862	8579.431	8770.994	8769.354	8734.729
7	14 493.179	11 078.160	11 051.632	10 949.204	11 095.344	11 093.916	11 057.247
8	14 688.463	11 474.772	11 432.019	11 184.910	11 495.588	11 493.472	11 436.203
9	16 959.363	12 341.060	12 290.581	11 945.627	12 371.351	12 368.905	12 299.895
10	17 038.859	12 887.904	12 845.014	12 473.943	12 918.498	12 915.807	12 849.926

Geometric inputs $a = 0.07$, $[\alpha_{1,0}, \alpha_{1,1}] = [0, 1]$, $[\alpha_{2,0}, \alpha_{2,1}] = [0, 1]$, $h_1 = h_3 = 0.001\ m$, $h_2 = 0.005$ m

Honeycomb cell configuration $l_2 = 5.1320$ mm, $l = 5.1320$ mm, $E_s = 70$ GPa, $\nu_s = 0.33$, $\rho_s = 2700$ kg/m^3

Table 1.8. Natural frequencies of an helicoid with general constraints (CFFC) with a central layer made of re-entrant honeycomb patterns. Cell configurations and geometric layups are reported in the table. Both the lattice layer and the laminated structure are obtained from isotropic aluminum. Shell reference surface equation, as well as an homogenized geometric representation, can be found in figure 1.7.

CFFC helicoid—re-entrant honeycomb ($\vartheta = -30°$)

f [Hz]	FSDT	FSDTZ	TSDTZ	EDZ1$_{(\kappa = 5/6)}$	EDZ2$_{(\kappa = 1)}$	EDZ3$_{(\kappa = 1)}$	EDZ4$_{(\kappa = 1)}$
DOFs	5046	7569	12 615	7569	10 092	12 615	15 138
Thin cell wall ($s_2 = s = 0.1$ mm)							
1	1882.924	1453.037	1443.648	1479.484	1446.649	1446.475	1446.692
2	5389.094	3310.940	3274.628	3216.800	3288.546	3288.170	3281.727
3	7132.567	6166.235	6143.234	6067.371	6153.501	6153.330	6140.107
4	9714.606	6318.167	6228.764	6103.897	6279.411	6278.451	6223.805
5	12 345.413	6972.306	6921.356	6777.417	6963.239	6962.501	6917.637
6	12 385.037	7387.383	7335.057	7288.318	7358.028	7357.434	7334.676
7	14 937.200	8828.222	8727.181	8545.861	8801.618	8800.386	8727.314
8	15 032.210	9175.225	9070.433	8827.154	9172.266	9170.825	9073.699
9	17 430.734	9200.075	9076.879	8831.276	9182.041	9180.536	9076.947
10	17 510.058	10 028.989	9962.868	9828.811	10 017.484	10 016.177	9957.399
Thick cell wall ($s_2 = s = 0.4$ mm)							
1	1768.134	1611.534	1610.122	1686.200	1615.997	1615.806	1613.245
2	5069.486	4171.982	4164.742	4173.226	4178.608	4178.079	4174.547
3	6719.991	6315.639	6311.870	6307.958	6317.130	6316.699	6309.437
4	9138.789	7997.866	7987.234	7982.568	8001.452	8000.628	7995.157
5	11 643.630	8761.084	8740.231	8692.208	8788.224	8786.395	8743.833
6	11 683.228	8909.241	8893.312	8708.274	8931.379	8929.270	8891.586
7	14 098.480	11 227.478	11 208.324	11 170.876	11 253.354	11 251.477	11 210.687
8	14 282.876	11 851.278	11 819.548	11 610.178	11 882.096	11 879.703	11 828.328
9	16 438.322	12 954.898	12 920.986	12 691.910	13 000.381	12 997.732	12 939.300
10	16 519.338	13 336.736	13 303.328	12 874.136	13 376.037	13 372.435	13 304.305

Geometric inputs $a = 0.07$, $[\alpha_{1,0}, \alpha_{1,1}] = [0, 1]$, $[\alpha_{2,0}, \alpha_{2,1}] = [0, 1]$, $h_1 = h_3 = 0.001\ m$, $h_2 = 0.005$ m

Honeycomb cell configuration $l_2 = 5.1320$ mm, $l = 5.1320$ mm, $E_s = 70$ GPa, $\nu_s = 0.33$, $\rho_s = 2700$ kg/m^3

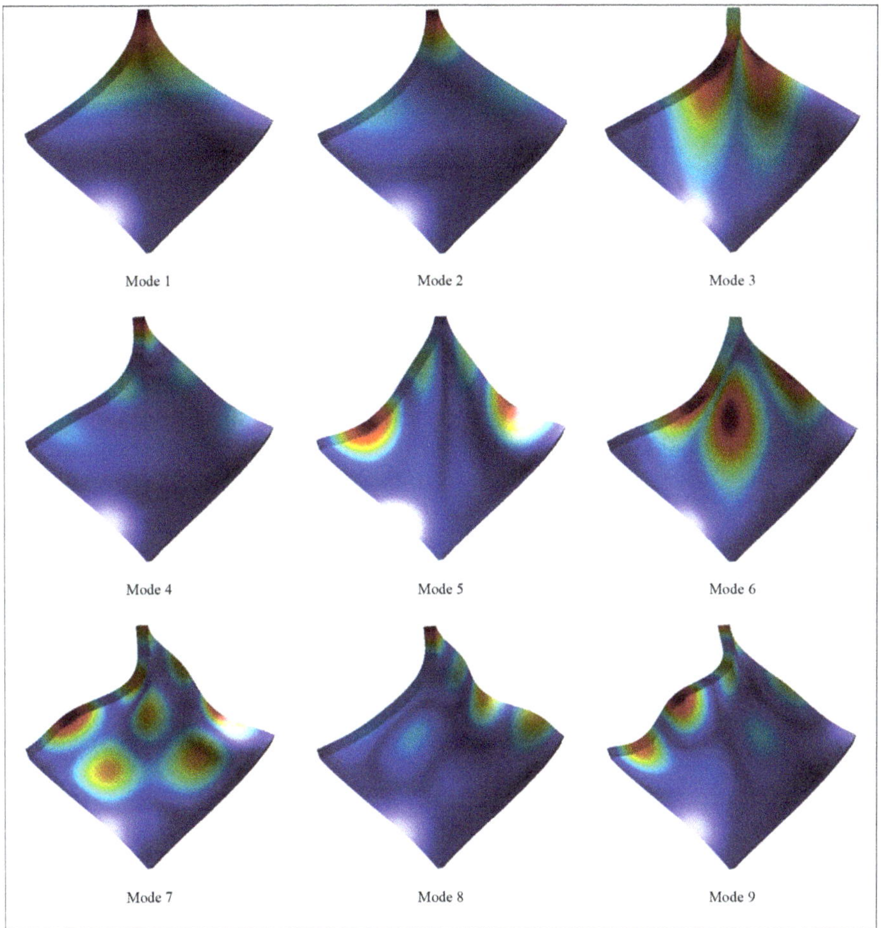

Figure 1.8. Mode shapes of a CFFC helicoidal panel with a re-entrant ($\vartheta = -30$ °) lattice core.

with an EDZ4 theory. Based on these plots, the present approach is confirmed to be capable of getting the bending deformations in each principal direction. For higher vibration modes, the out-of-plane waves are visible in both principal directions, together with the warping effects.

1.4 Conclusions

In the present chapter an innovative strategy based on the GDQ method has been proposed for the free vibration analysis of sandwich and laminated structures with variable curvatures with a central layer made of a honeycomb pattern or stiffeners system. The fundamental equations have been derived employing the ESL approach, where a generalized through-the-thickness assumption has been considered along with different higher order theories. Within each layer, a completely anisotropic elastic behaviour has been modelled, once a general rotation of the material reference system has been provided. Based on a homogenization method, a smeared orthotropic continuum has been provided, depending on the material properties and geometric features. A different in-plane constitutive behaviour has been provided, based on the variation of the internal cell angle. After the definition of a repetitive unit, the rib unit cell has been homogenized by assuming a thin-walled hypothesis, taking into account the axial contribution of each frame and the independence from the number of grid nodes. The fundamental equation of the problem together with the BCs are derived from the Hamiltonian Principle in its variational formulation. After the assembly procedure of the governing equations by means of the GDQ method, a free vibration analysis has been provided for different lattice and honeycomb shells. The results have been systematically compared to those provided by highly computational demanding finite elements, with a great accordance in terms of mode frequencies and shapes. The proposed higher order formulation based on a homogenized continuum modelling has been revealed to be a reliable design tool to treat complicated structures with latticed and honeycomb cores, with respect to high demanding 3D finite element simulations.

References

[1] Bauer J, Meza L R, Schaedler T A, Schwaiger R and Valdevit L 2017 Nanolattices: an emerging class of mechanical metamaterials *Adv. Mater.* **29** 1701850

[2] Dell'Isola F, Seppecher P, Alibert J J and Lekszycki T *et al* 2019 Pantographic metamaterials: an example of mathematically driven design and of its technological challenges *Continuum Mech. Thermodyn.* **31** 854–84

[3] Liddell I 2015 Frei Otto and the development of gridshells *Case Stud. Struct. Eng.* **4** 39–49

[4] Haftka R T and Grandhi R V 1986 Structural shape optimization—a survey *Comput. Meth. Appl. Mech. Eng.* **57** 91–106

[5] Ashby M F and Cebon D 1993 Materials selection in mechanical design *J. Phys. IV* **3** C7–1

[6] Vasiliev V V, Barynin V A and Rasin A F 2001 Anisogrid lattice structures—survey of development and application *Compos. Struct.* **54** 361–70

[7] Vasiliev V V, Barynin V A and Rasin A F 2009 Anisogrid composite lattice structures—development and space applications *Compos. Nanostruct.* **3** 38–50

[8] Van Quyen N, Van Thanh N, Quan T Q and Duc N D 2021 Nonlinear forced vibration of sandwich cylindrical panel with negative Poisson's ratio auxetic honeycombs core and CNTRC face sheets *Thin-Walled Struct.* **162** 107571

[9] Kalamkarov A L, Andrianov I V and Danishevsâ V V 2009 Asymptotic homogenization of composite materials and structures *Appl. Mech. Rev.* **62** 030802

[10] Gitman I, Askes H and Sluys L 2007 Representative volume: existence and size determination *Eng. Fract. Mech.* **74** 2518–34

[11] Tornabene F, Dimitri R and Brischetto S 2019 Higher-order formulation for the mechanical analysis of laminated and latticed shells with complex geometries and materials *Advances in Boundary Element and Meshless Techniques XX* ed I Benedetti, A Milazzo and F M H Aliabadi (Eastleigh (UK): EC Ltd))

[12] Tornabene F and Dimitri R 2020 Higher-order modelling of anisogrid lattice shell structures with complex geometries *18th Int. Conf. of Numerical Analysis and Applied Mathematics (ICNAAM2020) Proc. (Rhodes, Greece, 17–23 September)*

[13] Khisaeva Z F and Ostoja-Starzewski M 2006 On the size of RVE in finite elasticity of random composites *J. Elast.* **85** 153
[14] Belytschko T, Loehnert S and Song J H 2008 Multiscale aggregating discontinuities: a method for circumventing loss of material stability *Int. J. Numer. Methods Eng.* **73** 869–94
[15] Oliver J 1996 Modelling strong discontinuities in solid mechanics via strain softening constitutive equations. Part 1: Fundamentals *Int. J. Numer. Methods Eng.* **39** 3575–600
[16] Oliver J and Huespe A E 2004 Continuum approach to material failure in strong discontinuity settings *Comput. Meth. Appl. Mech. Eng.* **193** 3195–220
[17] Oliver J and Huespe A E 2004 Theoretical and computational issues in modelling material failure in strong discontinuity scenarios *Comput. Meth. Appl. Mech. Eng.* **193** 2987–3014
[18] Burton W S and Noor A K 1997 Assessment of continuum models for sandwich panel honeycomb cores *Comput. Meth. Appl. Mech. Eng.* **145** 341–60
[19] Lee J, Choi J B and Choi K 1996 Application of homogenization FEM analysis to regular and re-entrant honeycomb structures *J. Mater. Sci.* **31** 4105–10
[20] Potluri R and Rao U K 2017 Determination of elastic properties of reverted hexagonal honeycomb core: FEM approach *Mater. Today Proc.* **4** 8645–53
[21] Burlayenko V N and Sadowski T 2010 Effective elastic properties of foam-filled honeycomb cores of sandwich panels *Compos. Struct.* **92** 2890–900
[22] Réthoré J, Dang T B T and Kaltenbrunner C 2017 Anisotropic failure and size effects in periodic honeycomb materials: a gradient-elasticity approach *J. Mech. Phys. Solids* **99** 35–49
[23] Catapano A and Montemurro M 2014 A multi-scale approach for the optimum design of sandwich plates with honeycomb core. Part I: homogenization of core properties *Compos. Struct.* **118** 664–76
[24] Gibson L J and Ashby M F 1999 *Cellular Solids: Structure and Properties* (Cambridge: Cambridge University Press)
[25] Chen D H, Horii H and Ozaki S 2009 Analysis of in-plane elastic modulus for a hexagonal honeycomb core: analysis of Young's modulus and shear modulus *J. Comput. Sci. Technol.* **3** 1–12
[26] Sorohan S, Sandu M, Constantinescu D M and Sandu A G 2016 Estimation of out of plane shear moduli for honeycomb cores with modal finite element analyses *Romanian J. Tech. Sci. Appl. Mech.* **61** 71–88
[27] Xu X F and Qiao P 2002 Homogenized elastic properties of honeycomb sandwich with skin effect *Int. J. Solids Struct.* **39** 2153–88
[28] Kelsey S, Gellatly R and Clark B 1958 The shear modulus of foil honeycomb cores *Aircr. Eng. Aerosp. Technol.* **30** 294–302
[29] Grediac M 1993 A finite element study of the transverse shear in honeycomb cores *Int. J. Solids Struct.* **30** 1777–88
[30] Scarpa F and Tomlin P 2000 On the transverse shear modulus of negative Poisson ratio honeycomb structures *Fatigue Fract. Eng. Mater. Struct.* **23** 717–20
[31] Malek S and Gibson L 2015 Effective elastic properties of periodic hexagonal honeycombs *Mech. Mater.* **91** 226–40
[32] Timoshenko S and Goodier J 2011 *Theory of Elasticity* (New York: Mc-Graw Hill)
[33] Roark R J, Young W C and Plunkett R 1976 Formulas for stress and strain *J. Appl. Mech.* **43** 522
[34] Sorohan S, Sandu M, Constantinescu D M and Sandu A G 2015 On the evaluation of mechanical properties of honeycombs by using finite element analyses *INCAS Bull.* **7** 135–50

[35] Fu M-H and Yin J R 1999 Equivalent elastic parameters of the honeycomb core *Acta Mech. Sin.* **15** 113–18

[36] Torabi K, Afshari H and Aboutalebi F H 2019 Vibration and flutter analyses of cantilever trapezoidal honeycomb sandwich plates *J. Sandw. Struct. Mater.* **21** 2887–920

[37] Sorohan S, Constantinescu D M, Sandu M and Sandu A G 2018 On the homogenization of hexagonal honeycombs under axial and shear loading. Part I: analytical formulation for free skin effect *Mech. Mater.* **119** 74–91

[38] Sorohan S, Constantinescu D M, Sandu M and Sandu A G 2018 On the homogenization of hexagonal honeycombs under axial and shear loading. Part II: comparison of free skin and rigid skin effects on effective core properties *Mech. Mater.* **119** 92–108

[39] Tornabene F, Viscoti M, Dimitri R and Aiello M A 2021 Higher Order Formulations for doubly-curved shell structures with a honeycomb core *Thin-Walled Struct.* **164** 107789

[40] Tornabene F, Viscoti M, Dimitri R and Aiello M A 2021 Higher-order modelling of anisogrid composite lattice structures with complex geometries *Eng. Struct.* **244** 112686

[41] Calladine C R 1989 *Theory of Shell Structures* (Cambridge: Cambridge University Press)

[42] Kraus H 1967 *Thin Elastic Shells* (New York: Wiley)

[43] Gould P L 1999 *Analysis of Plates and Shells* (Upper Saddle River, NJ: Prentice-Hall)

[44] Li D 2020 Layer-wise theories of laminated composite structures and their applications: A review *Arch. Comput. Meth. Eng.* **28** 577–600

[45] Reddy J N 1993 An evaluation of equivalent-single-layer and layerwise theories of composite laminates *Compos. Struct.* **25** 21–35

[46] Tornabene F, Fantuzzi N, Viola E and Batra R C 2015 Stress and strain recovery for functionally graded free-form and doubly-curved sandwich shells using higher-order equivalent single layer theory *Compos. Struct.* **119** 67–89

[47] Love A E H 1944 *A Treatise on the Mathematical Theory of Elasticity* (New York: Dover Publication)

[48] Germain S 2013 *Recherches sur la théorie des surfaces élastiques* (Cambridge: Cambridge University Press)

[49] Reissner E 1945 The effect of transverse shear deformation on the bending of elastic plates *J. Appl. Mech. ASME* **12** 66–77

[50] Mindlin R D 1951 Thickness-shear and flexural vibrations of crystal plates *J. Appl. Phys.* **22** 316–23

[51] Arefi M, Mohammadi M, Tabatabaeian A, Dimitri R and Tornabene F 2018 Two-dimensional thermo-elastic analysis of FG-CNTRC cylindrical pressure vessels *Steel Compos. Struct.* **27** 525–36

[52] Reddy J N 2003 *Mechanics of Laminated Composite Plates and Shells* (Boca Raton, FL: CRC Press)

[53] Reddy J N 1986 A refined shear deformation theory for the analysis of laminated plates *NASA Contractor Report* 3955

[54] Mohammadi M, Arefi M, Dimitri R and Tornabene F 2019 Higher-order thermo-elastic analysis of FG-CNTRC cylindrical vessels surrounded by a Pasternak foundation *Nanomaterials* **9** 79

[55] Washizu K 1975 *Variational Methods in Elasticity and Plasticity* (Oxford: Pergamon)

[56] Reddy J N 1987 A generalization of two-dimensional theories of laminated composite plates *Commun. Appl. Numer. Methods* **3** 173–80

[57] Tornabene F and Bacciocchi M 2018 *Anisotropic Doubly-Curved Shells. Higher-Order Strong and Weak Formulations for Arbitrarily Shaped Shell Structures* (Bologna: Società Editrice Esculapio)
[58] Tornabene F, Bacciocchi M, Fantuzzi N and Viola E 2016 *Laminated Composite Doubly-Curved Shell Structures: Differential Geometry Higher-Order Structural Theories* (Bologna: Società Editrice Esculapio)
[59] Tornabene F 2016 General higher-order layer-wise theory for free vibrations of doubly-curved laminated composite shells and panels *Mech. Adv. Mater. Struct.* **23** 1046–67
[60] Tornabene F, Viscoti M and Dimitri R 2022 Generalized higher order layerwise theory for the dynamic study of anisotropic doubly-curved shells with a mapped geometry *Eng. Anal. Bound. Elem.* **134** 147–83
[61] Lopatin A V, Morozov E V and Shatov A V 2016 An analytical expression for fundamental frequency of the composite lattice cylindrical shell with clamped edges *Compos. Struct.* **141** 232–39
[62] Lopatin A V, Morozov E V and Shatov A V 2018 Fundamental frequency of a composite anisogrid lattice cylindrical panel with clamped edges *Compos. Struct.* **201** 200–7
[63] Morozov E V, Lopatin A V and Nesterov V A 2011 Finite-element modelling and buckling analysis of anisogrid composite lattice cylindrical shells *Compos. Struct.* **93** 308–23
[64] Shu C 2000 *Differential Quadrature and Its Application in Engineering* (London: Springer)
[65] Shu C, Ding H and Yeo K S 2003 Local radial basis function-based differential quadrature method and its application to solve two-dimensional incompressible Navier–Stokes equations *Comput. Meth. Appl. Mech. Eng.* **192** 941–54
[66] Bellman R and Casti J 1971 Differential quadrature and long-term integration *J. Math. Anal. Appl.* **34** 235–38
[67] Zong Z and Zhang Y 2009 *Advanced Differential Quadrature Methods* (Boca Raton, FL: CRC Press)
[68] Tornabene F, Fantuzzi N, Ubertini F and Viola E 2015 Strong formulation finite element method based on differential quadrature: a survey *Appl. Mech. Rev.* **67** 20801
[69] Quan J R and Chang C T 1989 New insights in solving distributed system equations by the quadrature method—I. Analysis *Comput. Chem. Eng.* **13** 779–88
[70] Tornabene F, Liverani A and Caligiana G 2011 FGM and laminated doubly curved shells and panels of revolution with a free-form meridian: a 2-D GDQ solution for free vibrations *Int. J. Mech. Sci.* **53** 446–70
[71] Tornabene F and Reddy J N 2013 FGM and laminated doubly-curved and degenerate shells resting on nonlinear elastic foundations: a GDQ solution for static analysis with *a posteriori* stress and strain recovery *J. Indian Inst. Sci.* **93** 635–88
[72] Tornabene F and Bacciocchi M 2018 Effect of curvilinear reinforcing fibers on the linear static behavior of soft-core sandwich structures *J. Compos. Sci.* **2** 14
[73] Tornabene F, Bacciocchi M, Fantuzzi N and Reddy J N 2019 Multiscale approach for three-phase CNT/polymer/fiber laminated nanocomposite structures *Polym. Compos.* **40** E102–26
[74] Zare Jouneghani F, Dimitri R, Bacciocchi M and Tornabene F 2017 Free vibration analysis of functionally graded porous doubly-curved shells based on the first-order shear deformation theory *Appl. Sci.* **7** 1252
[75] Shen J, Wang H and Zheng S 2018 Size-dependent pull-in analysis of a composite laminated micro-beam actuated by electrostatic and piezoelectric forces: generalized differential quadrature method *Int. J. Mech. Sci.* **135** 353–61
[76] Hua L I and Lam K Y 1998 Frequency characteristics of a thin rotating cylindrical shell using the generalized differential quadrature method *Int. J. Mech. Sci.* **40** 443–59

[77] Dastjerdi S, Malikan M, Dimitri R and Tornabene F 2021 Nonlocal elasticity analysis of moderately thick, porous functionally graded plates in a hygro-thermal environment *Compos. Struct.* **255** 112925
[78] Bert C W and Malik M 1996 Differential quadrature method in computational mechanics: a review *Appl. Mech. Rev.* **49** 1–28
[79] Shu C, Chen W, Xue H and Du H 2001 Numerical study of grid distribution effect on accuracy of DQ analysis of beams and plates by error estimation of derivative approximation *Int. J. Numer. Methods Eng.* **51** 159–79
[80] Fazzolari F A, Viscoti M, Dimitri R and Tornabene F 2020 1D-hierarchical Ritz and 2D-GDQ formulations for the free vibration analysis of circular/elliptical cylindrical shells and beam structures *Compos. Struct.* **258** 113338
[81] Tornabene F, Fantuzzi N and Bacciocchi M 2018 Strong and weak formulations based on differential and integral quadrature methods for the free vibration analysis of composite plates and shells: convergence and accuracy *Eng. Anal. Boundary Elem.* **92** 3–37
[82] Fantuzzi N, Tornabene F, Bacciocchi M and Ferreira A J M 2018 On the convergence of laminated composite plates of arbitrary shape through finite element models *J. Compos. Sci.* **2** 16
[83] Shu C and Wee K H A 2002 Numerical simulation of natural convection in a square cavity by simple generalized differential quadrature method *Comput. Fluids* **31** 209–26
[84] Chen C N 2000 A generalized differential quadrature element method *Comput. Meth. Appl. Mech. Eng.* **188** 553–66
[85] Dimitri R, Fantuzzi N, Tornabene F and Zavarise G 2016 Innovative numerical methods based on SFEM and IGA for computing stress concentrations in isotropic plates with discontinuities *Int. J. Mech. Sci.* **118** 166–87
[86] Dimitri R, Tornabene F and Reddy J N 2020 Numerical study of the mixed-mode behaviour of generally-shaped composite interfaces *Compos. Struct.* **237** 111935
[87] Dimitri R, Fantuzzi N, Li Y and Tornabene F 2017 Numerical computation of the crack development and SIF in composite materials with XFEM and SFEM *Compos. Struct.* **160** 468–90
[88] Hong C C 2017 GDQ analysis of a beam-plate with delamination *Compos. Struct.* **182** 237–41
[89] Viola E, Dilena M and Tornabene F 2007 Analytical and numerical results for vibration analysis of multi-stepped and multi-damaged circular arches *J. Sound Vib.* **299** 143–63
[90] Shu C and Du H 1997 A generalized approach for implementing general boundary conditions in the GDQ free vibration analysis of plates *Int. J. Solids Struct.* **34** 837–46
[91] Shu C and Wang C M 1999 Treatment of mixed and nonuniform boundary conditions in GDQ vibration analysis of rectangular plates *Eng. Struct.* **21** 125–34
[92] Brischetto S, Tornabene F, Fantuzzi N and Bacciocchi M 2017 Interpretation of boundary conditions in the analytical and numerical shell solutions for mode analysis of multi-layered structures *Int. J. Mech. Sci.* **122** 18–28
[93] Tornabene F and Viola E 2013 Static analysis of functionally graded doubly-curved shells and panels of revolution *Meccanica* **48** 901–30
[94] Tornabene F, Viscoti M, Dimitri R and Reddy J N 2021 Higher order theories for the vibration study of doubly-curved anisotropic shells with a variable thickness and isogeometric mapped geometry *Compos. Struct.* **267** 113829
[95] Tornabene F, Fantuzzi N and Bacciocchi M 2018 *DiQuMASPAB: Differential Quadrature for Mechanics of Anisotropic Shells, Plates, Arches and Beams* (Bologna: Società Editrice Esculapio)

IOP Publishing

Modeling and Computation in Vibration Problems, Volume 1
Numerical and semi-analytical methods
S Chakraverty, F Tornabene and J N Reddy

Chapter 2

Particle impact damping technology: modelling and applications

Nazeer Ahmad, Ankur Kumar Gupta, Sujata and Poomani D

2.1 Introduction

Particle impact damping (PID) technology is essentially a passive damping technique where the collision of damping particles dissipates the vibrational energy of a host structure. The damping particles are metallic or non-metallic granules. The damping particles are kept in specially designed cavities at the strategically chosen location in the vibrating structure. In some cases, a container filled with damping particles is attached. When the host structure vibrates, it transfers kinetic energy to the damping particles through impact, friction and momentum exchange. The mode of transfer could be single or more at the same time. The damping particles dissipate the energy gained from the vibrating structure through particle–particle and particle–wall impact and friction. The energy dissipation happens in the form of heat, sound and elastic waves, and is manifested as enhanced damping and reduction in the response of the structure. The energy dissipation mechanism in PID is nonlinear and depends mainly on the size, shape, and material of the damping particles; the level of response, amplitude and frequency of input excitation. The dominant mode of dissipation depends on the relative proportion of different parameters. The particle impact damper design involves either drilling a hole or attaching a container to the vibrating structure and filling the container or the holes with damping particles. The damper is so simple and is known to perform well in wide frequency bands and a harsh environment where other damping technologies are challenging to use and are not effective. The harsh environments may include very high temperatures, radiation fields, etc. This technology has been used in a skyscraper to restrict the large displacement of the upper floors due to strong wind loads and earthquakes [1]. This is also effective in limiting the vibration induced in the machine tools. One of the early applications of particle impact damping technology is in turbomachinery blades. A hole is drilled in the blade and filled

doi:10.1088/978-0-7503-3483-9ch2

© IOP Publishing Ltd 2021

with single/multiple damping particles that are effective in reducing the vibrations. It has found its application in the aerospace industry as well. Space and aerospace industries where the primary structural member is a honeycomb sandwich are particularly suitable for this technology. The empty cells of the honeycomb core are readily available for filling the damping particles [2, 3]. The damping characteristic can be tailored according to the target mode shapes and frequency bandwidth. Another novel application of this technology that has attracted research attention is its use on printed circuit boards and solar panels [4] for space applications to reduce very high vibration generated by a launch vehicle.

In this chapter, we introduce a modeling technique to capture the interactions between the dynamics of the damping particles and the dynamics of the host structure with a particular focus on the discrete element method (DEM) [5]. The DEM has been successfully used by many researchers to analyze the dynamics of the assemblage of particles in vibration mitigation applications. Tracking the motion and contact history of each particle allows estimation of energy loss and damping introduced by the collision of the particle against the wall of the container. The DEM formulation is based on Newton's laws of motion derived considering the contact forces and moments by immediate neighboring particles. The accuracy of the model depends on the model used for normal and tangential contact forces. Time step size is another very crucial parameter in the success of the DEM computation as the new contacts are established and old ones are broken in a very short duration. Subsequently, the governing equations of motion of the particles are updated. Hertz's dissipative contact model for normal contact forces and Coulomb's friction model for tangential force are used to derive the equation used in DEM. The vibration attenuation trend has been studied for the location of particle dampers, size of particles, fill fraction and excitation energy. The attenuation has been quantified using the model described.

2.2 Mathematical formulations

In the particle impact dampers, the energy of the vibrating structure is transferred to damping particles via momentum exchange. The damping particles dissipate the kinetic energy by hysteresis and friction due to collisions and rubbing actions among particles and particle-container walls. A mathematical model to capture the complex interactions between the dynamics of damping particles and the host structure is presented in this section. Figure 2.1 shows the particles i and j colliding each other with mass centers having position vectors $\mathbf{P}_i$ and $\mathbf{P}_j$ and linear velocity $\mathbf{V}_i$ and $\mathbf{V}_j$, respectively. The particles have angular velocity $\boldsymbol{\omega}_i$ and $\boldsymbol{\omega}_j$, mass m_i and m_j, respectively.

The contact force evolving in the low-velocity impacts that take place among the damping particles can be resolved into normal and tangential components, as shown in figure 2.1

$$\mathbf{f}_{ij} = \mathbf{f}_{ij}^{n} + \mathbf{f}_{ij}^{t} \tag{2.1}$$

The inter-relation between the normal contact force $\mathbf{f}_{ij}^{n}$ and the local deformation δ_{ij}^{n} depends on the geometry, material, and the relative velocity of the colliding bodies. The $\mathbf{f}_{ij}^{n}$ is modeled using a nonlinear elastic-dissipative model as most of the

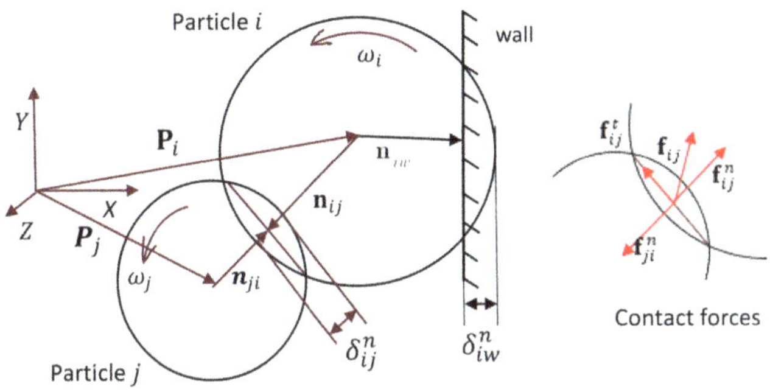

Figure 2.1. Impact of damping particles.

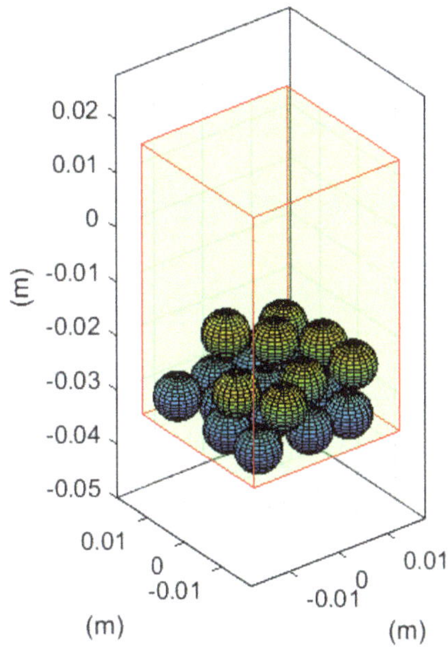

Figure 2.2. Damping particles in a rectangular container.

collisions are low-velocity impact resulting in predominantly elastic deformations. Kelvin–Voigt [6], mathematically expressed it as

$$\mathbf{f}^n_{ij} = k\delta^n_{ij} + D\dot{\delta}^n_{ij} \tag{2.2}$$

This simple model, basically a parallel combination of linear spring and the viscous damper, is widely used. However, this model does not capture correctly many phenomena associated with impact. The Kelvin–Voigt model gives a uniform dissipation of energy during the approach and restitution phase against the fact that most of the energy dissipation happens during the approach. It is established that most of the impact energy dissipation occurs during the approach. An improved model governing the $\mathbf{f}^n_{ij}$ was obtained by replacing the elastic term of the Kelvin–Voigt model with a more accurate Hertz nonlinear model and used in many particle damping problems. However, the improved model retained all the weaknesses of the Kelvin–Voigt model. The model proposed by Tsuji *et al* [7] that is a nonlinear Hertzian-dissipative contact theory as given by equation (2.3) is reported to give reasonably good results [2, 3, 8, 9] in predicting the dynamics of damping particles and their effect on vibrating structures.

$$\mathbf{f}_{ij}^{n} = -\left(k_n\left(\delta_{ij}^{n}\right)^{\frac{3}{2}} + \alpha\sqrt{m_{ij}^{*}k_n}\left(\delta_{ij}^{n}\right)^{\frac{1}{4}}\dot{\delta}_{ij}\right)\mathbf{n}_{ij} \tag{2.3}$$

where α_n is a nonlinear damping constant and a function of normal restitution coefficient e_n given as

$$\alpha_n = -ln(e_n)\sqrt{\frac{5}{ln(e_n)^2 + \pi^2}} \tag{2.4}$$

the Hertz constant k_n for the pair particle–particle and particle–wall is given by equations (2.5) and (2.6) [10], respectively.

$$k_n = \frac{4}{3}\sqrt{\frac{r_i r_j}{r_i + r_j}}\frac{E_i E_j}{\left(1 - \nu_j^2\right)E_i + \left(1 - \nu_i^2\right)E_j} \tag{2.5}$$

$$k_n = \frac{4}{3}\sqrt{r_i}\frac{E_i E_w}{(1 - \nu_w^2)E_i + (1 - \nu_i^2)E_w} \tag{2.6}$$

Where Young's moduli of particle and wall are, respectively, E_i and E_w and Poisson's ratios ν_i and ν_w. The effective mass m^* of the damping particles given in equation (2.3) can be written as

$$m_{ij}^{*} = \frac{m_i + m_j}{m_i m_j} \tag{2.7}$$

The modeling of the tangential contact force $\mathbf{f}_{ij}^{t}$ is a recent research area. Mindlin–Deresiewicz [11] proposed a model which could relate the tangential force $\mathbf{f}_{ij}^{t}$ to tangential displacement using an incremental tangential stiffness based on a series of loading and unloading assumptions. This model adequately addresses the problem of sticking and gross-sliding phenomenon in the contact process. Maw *et al* [12] proposed a theory which not only relaxed many assumptions of the Mindlin–Deresiewicz theory but also accurately predicted the post-contact dynamics of colliding bodies, including the phenomenon of velocity reversal. It also identifies the phases of micro-slip and gross sliding. There are many other models that are the modification of Mindlin–Deresiewicz and Maw's model. Coulomb's law of sliding friction [10] to govern the tangential force in the particle damping problem is the most straightforward and most efficient among all the models discussed. However, it fails to predict the negative bounce, which happens when the incidence angle is less than a critical value that depends on the material properties of impacting bodies; for ordinary material, it is usually around 30 degrees. The choice of the model is based on efficiency as a large number of particles and contacts are involved in impact damping applications. The Coulombs model is the fastest, but it is only moderately accurate. The model has been used by many researchers [1–3, 13] in DEM applications reporting encouraging results. In this work, the tangential force is modeled using Coulomb's law, given as

$$\mathbf{f}_{ij}^{t} = -\mu \left| \mathbf{f}_{ij}^{n} \right| \frac{\mathbf{V}_{ij}^{t}}{\left| \mathbf{V}_{ij}^{t} \right|} \tag{2.8}$$

where μ is the coefficient of sliding friction. The relative velocity $\mathbf{V}_{ij}$ and relative tangential velocity $\mathbf{V}_{ij}^{t}$ at the point of contact can be written as

$$\mathbf{V}_{ij} = \mathbf{V}_i - \mathbf{V}_j + (\mathbf{r}_i\boldsymbol{\omega}_i + \mathbf{r}_j\boldsymbol{\omega}_j) \times \mathbf{n}_{ij} \tag{2.9}$$

$$\mathbf{V}_{ij}^{t} = \mathbf{V}_{ij} - \left(\mathbf{V}_{ij}\mathbf{n}_{ij}\right)\mathbf{n}_{ij} \tag{2.10}$$

The equation of motion of the particle i that is in contact with n_l number of surrounding particles and n_2 points with cell walls can be written as:

$$m_i\ddot{\mathbf{P}}_i = -m_i\mathbf{g} + \sum_{j=1}^{n_1}\mathbf{f}_{ij} + \sum_{w=1}^{n_2}\mathbf{f}_{iw} \tag{2.11}$$

$$\mathbf{I}_i\ddot{\boldsymbol{\Theta}}_i = \sum_{j=1}^{n_1}\left(r_i - \frac{\delta_{ij}}{2}\right)\mathbf{n}_{ij} \times \mathbf{f}_{ij} + \sum_{w=1}^{n_2}(r_i - \delta_{iw}))\mathbf{n}_{iw} \times \mathbf{f}_{iw} \tag{2.12}$$

where $\mathrm{m_i}$, r_i, and $\mathbf{I}_i$ are the mass, radius and mass moment of inertia of the particle i, respectively; $\mathbf{P}_\mathrm{i}$ is the position vector of the center of mass and $\boldsymbol{\Theta}_\mathrm{i}$ the angular displacement of the particle i. The unit vectors $\mathbf{n}_{ij}$ and $\mathbf{n}_\mathrm{iw}$ point from the center of particle i towards the center of particle j and towards the point of contact with the cell wall. The acceleration due to gravity is represented by $\mathbf{g}$, whereas $\mathbf{f}_{ij}$ and $\mathbf{f}_{iw}$ are the forces on particle i due to interaction of particle j and by the wall of the cell, respectively. The normal relative displacements of the center of mass of particle i with respect to particle j when they are in contact with each other or with the wall are represented by, δ_{ij} and δ_{iw}, respectively. The local indentation or the normal component of the displacement vector is governed by equation (2.3).

Let us assume that the damping particles are inside a container, and the container is connected to the shaker through a spring-damper system. The equation of motion of the container can be written as

$$m_c\ddot{\mathbf{w}} + c\dot{\mathbf{w}} = \mathbf{F}^d + \mathbf{F}^e \tag{2.13}$$

where m_c is the mass of the container and c is the damping. w represents the displacement of the container, and a dot over a variable corresponds to the time derivative. $\mathbf{F}^d$ and $\mathbf{F}^e$ are particle damping force and external excitation force. The particle damping force $\mathbf{F}^d$ is acting on the container due to the interaction of damping particles with the container. It is this force that links the dynamics of the container with damping particles. This force is assumed to be acting at the center of the container, and it is the vector sum of the damping force due to each damping particle. In the next subsection, computation of impact damping force is explored

2.2.1 Impact damping force and its computation

When the container is excited with the shaker, the damping particle in the container exchanges momentum with the container and gains kinematic energy. Subsequently, damping particles dissipate the energy through collision and friction.

Let there be n particles in the container and n_c particles are in contact with the walls of the container. And also, let the contact force on the container due to the interaction of particle i be $-\mathbf{f}_{iw}$, then the impact damping force $\mathbf{F}^d$ assumed to be acting on the geometric center of the container due to impact and friction of all the damping particles that are in direct contact with the container can be written as

$$\mathbf{F}^d = \sum_{i=1}^{n_c} -\mathbf{f}_{iw} \tag{2.14}$$

The $\mathbf{F}^d$ is expected to be predominantly in the excitation direction as the major contribution is from the damping particle hitting the top and bottom walls of the container and a part of it from the friction due to sidewalls. The impact force depends on the relative velocity of damping particles and container walls as well as the materials. The contact forces due to the impact of damping particle being a function of local deformation along with other material and geometric properties. Its computation requires knowledge of the time history of each particle and walls of the container. The normal local deformation between colliding damping particles and particle-wall, δ_{ij}^n and δ_{iw}^n, respectively can be calculated using equations (2.15) and (2.16).

$$\delta_{ij}^n = r_i + r_j - \left| \mathbf{P}_i - \mathbf{P}_j \right| \tag{2.15}$$

$$\delta_{iw}^n = r_i - | \mathbf{P}_i - \mathbf{P}_w | \tag{2.16}$$

where the $\mathbf{P}_w$ is the position vector of the center of the contact at the wall of the container and r_i and r_j are the radii of the colliding particles i and j, respectively. The evaluation of δ_{iw}^n requires positions of the walls of the container. Assuming a coordinate system (xyz) attached to the geometric center of the container, the equations of the wall can be written in the local coordinate system as

$$\bar{z} \pm \frac{h}{2} = 0$$

$$y \pm b = 0$$

$$\bar{x} \pm a = 0 \tag{2.17}$$

where h is the height of the container, a is the length and b is the width of the container. The damping particles are constrained to move inside the container when it is vibrated on the shaker. The equations of motion of the damping particles given by equations (2.11) and (2.12) along with the equation of motion of the container given by equation

(2.13) can be integrated to get the complete time history of damping particles and container using numerical techniques. In this work, the Runga–Kutta method of order 4 has been used to generate all the results discussed in subsequent sections. The entire numerical implementation of the algorithm has been carried out in the MATLAB environment. The key to the successful computation is the time step chosen to integrate the equations of motion. The desired time step is chosen one order less than the contact duration. A very small time step is desirable; however, it imposes enormous computation cost as the contact detection is performed at each time step that requires calculating the distance from the center of each damping particle to all damping particles and walls of the container. A larger time step risks missing the new contacts formed and old contacts, which are broken and subsequent relevant equation of motions even though the DEM formulation assumes that motion of a damping particle is affected by only its immediate neighborhood particles: in contact with it.

2.3 Numerical simulations

2.3.1 Impact of a single particle with the container wall

A single damping particle with an initial position in a local coordinate system (xyz) $\left[P_x, P_y, P_z\right] = [0.0, 0.0, -0.02]$ is imparted an initial velocity of $\left[V_x, V_y, V_z\right] = [0.0, 0.5, -0.5]$ m s^{-1}. It hits obliquely the bottom plane of the container defined by $z = -h/2 = 12.5$ mm, which is at rest. The changes in linear and angular velocities and the contact forces and moments in the impact of a single damping particle with the wall of a container are presented in figures 2.3–2.6.

The effects of the nonlinear dissipative term in equation (2.3) on contact forces and moments can be seen in figure 2.3. The relative velocities reach zero before the end of the contact process resulting in the small loop in force and moment plots, as shown at the end of the restitution phase. The velocity and force evolving in another collision of the same damping particle with a smaller incidence angle that is less than 30 degrees with the container wall are presented in figure 2.5(a)–(b) and figure 2.6(a)–(b), respectively. For smaller incidence angles when the friction force is predominantly sticking rather than sliding, the prediction of friction force using Coulomb's model is not accurate. The oscillation of tangential force at the end of the restitution phase shown in figure 2.6(c) is due to Coulomb's friction model, which is not a real phenomenon. Despite the shortcomings of Coulomb's friction model, particularly in low incidence angle impacts, it is extensively used in impact damping particle related vibration problems due to the faster speed and more straightforward implementation in DEM computations.

2.3.2 Dissipation in impact damping device

The energy dissipation by damping particles that are in the container due to normal dissipative and Coulomb's tangential forces in N impacts can be written as

$$E_{\mathrm{d}} = \sum_{\text{contact}=1}^{N} \int_0^{t_c} \left(\underbrace{\alpha\sqrt{\mathrm{m}_{ij}^{*} k_n}\,\delta^{\frac{1}{4}} \dot{\delta}_{ij} \dot{\delta}_{ij}}_{\text{Term-1}} + \underbrace{\mathbf{f}_{ij}^{t} \cdot \mathbf{V}_{ij}^{t}}_{\text{Term-2}} \right) dt \quad i, j \ldots 1, 2 \tag{2.18}$$

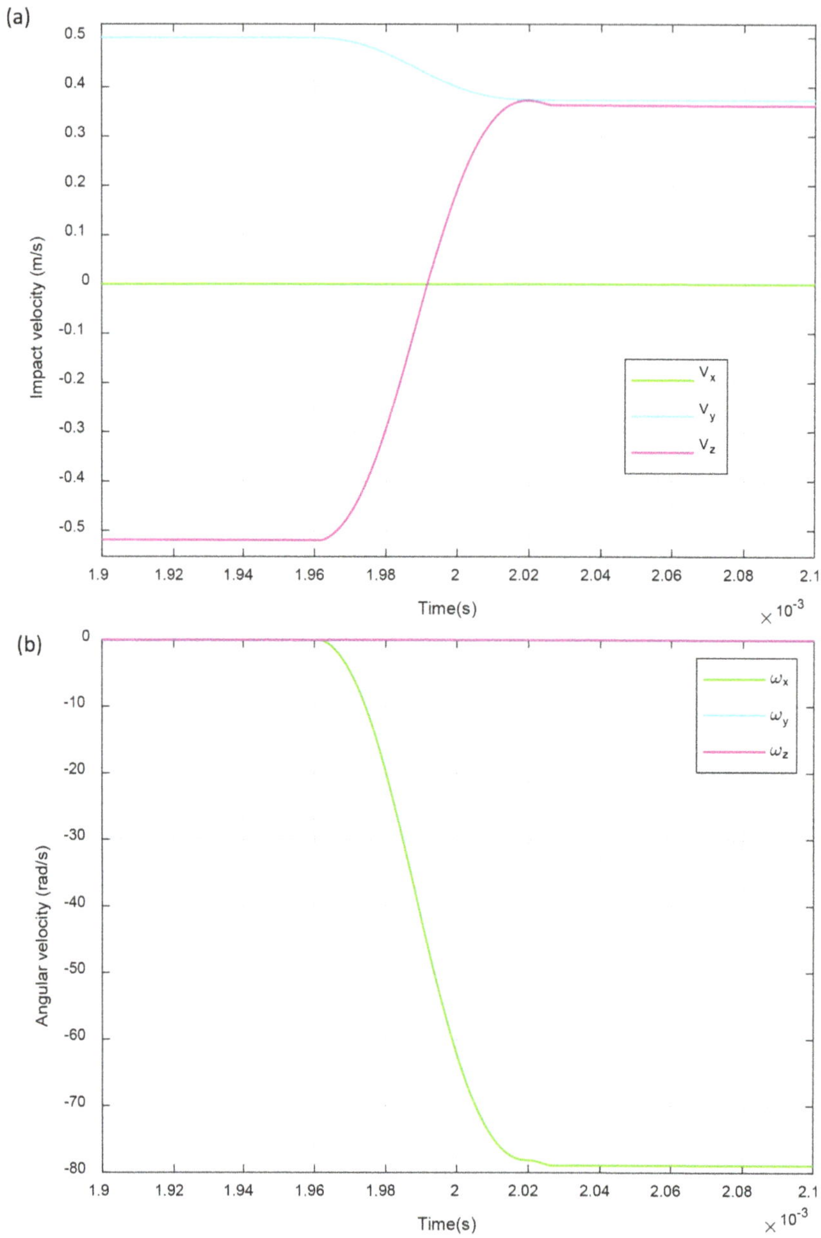

Figure 2.3. Change in velocity in the contact process.

where E_{d} is the total energy dissipated and t_c is the contact duration. The first term in equation (2.18) represents the dissipation in hysteresis due to normal contact force and the second term corresponds to the energy dissipated in Coulomb's friction. The energy change and dissipation happening in a typical impact that is described in the previous section is shown in figure 2.7. Having computed the energy dissipated by

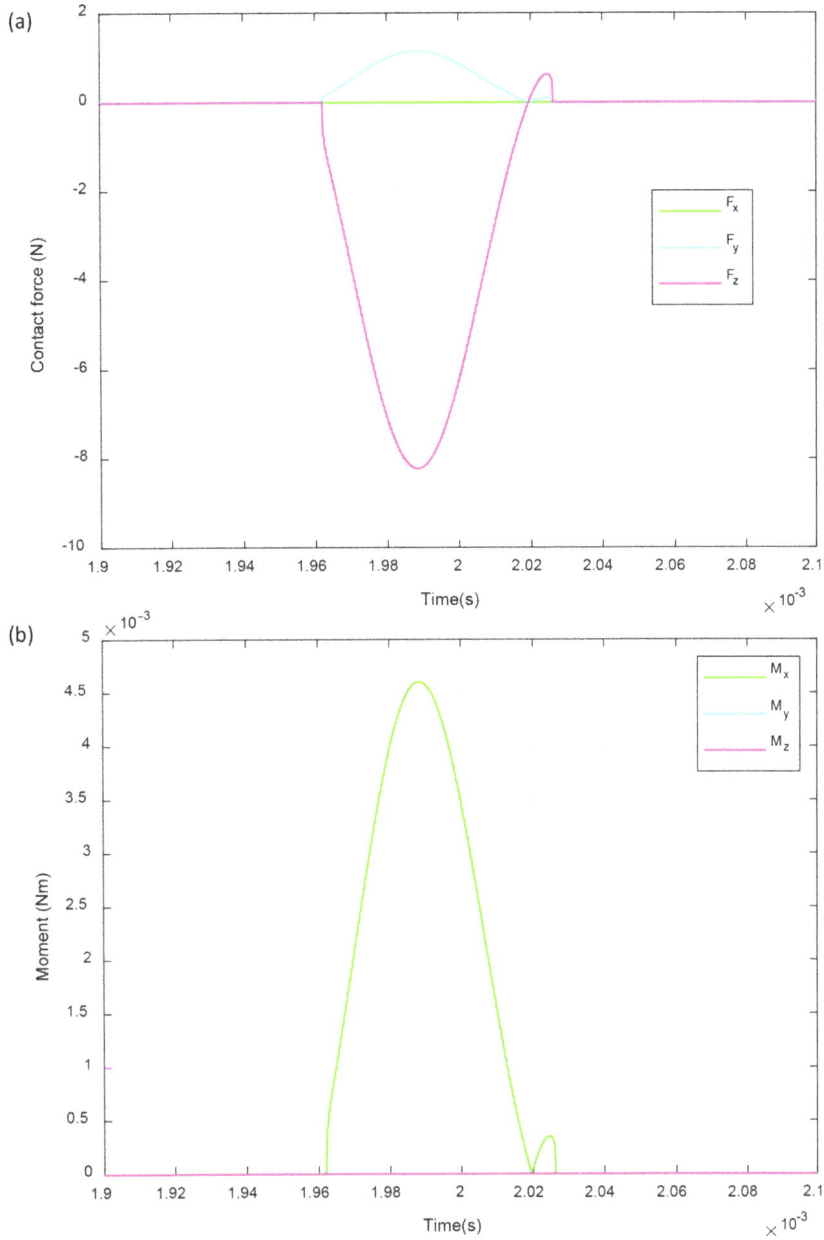

Figure 2.4. Contact forces and moments.

the impact damper, the equivalent viscous damping coefficient can be calculated using the relation

$$C_{\text{eq}} = \frac{E_d}{\pi \omega W^2} \tag{2.19}$$

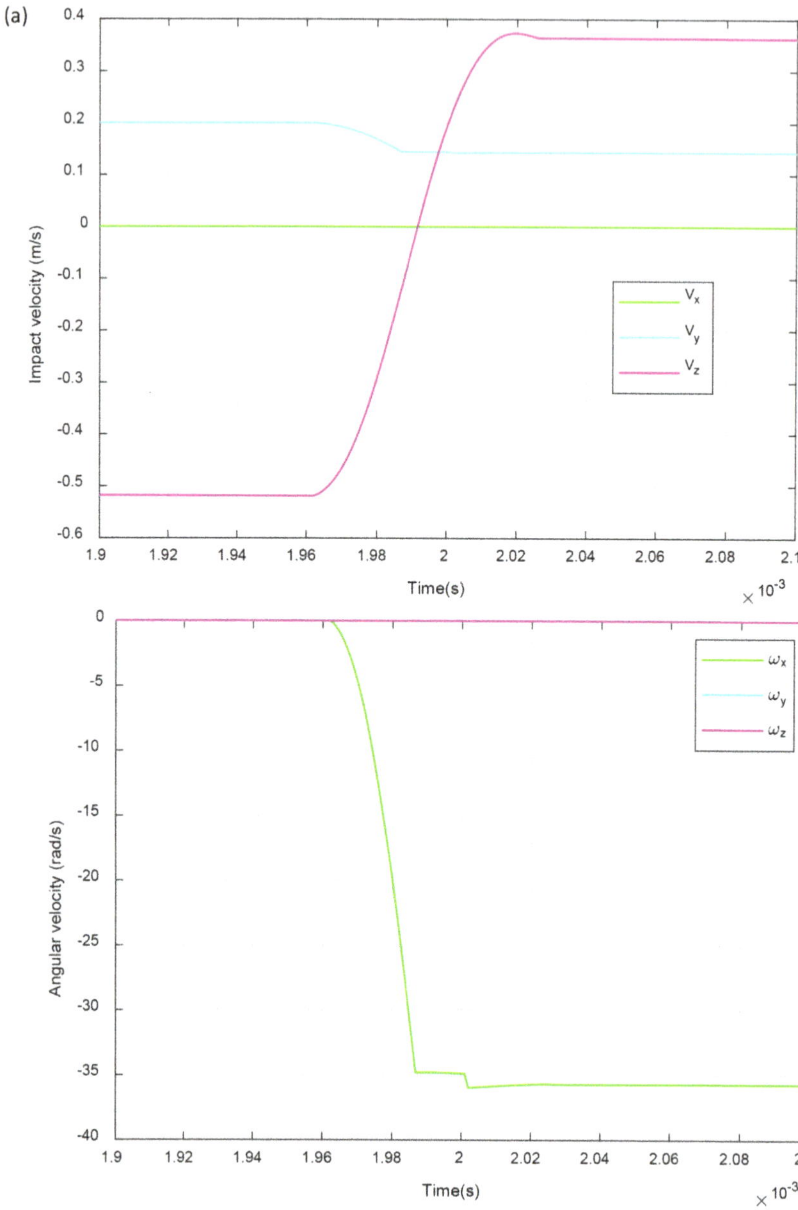

Figure 2.5. Change in velocity in low incidence angle contact process.

Where C_{eq}, ω and W are the equivalent viscous damping, excitation frequency and amplitude of excitation, respectively. A container made up of aluminum of dimensions 30 mm × 30 mm × 110 mm was filled with spherical acrylic particles as damping particles. When the container is filled with damping particles, it contains approximately 15 000 particles calculated from the mass of the damping particles.

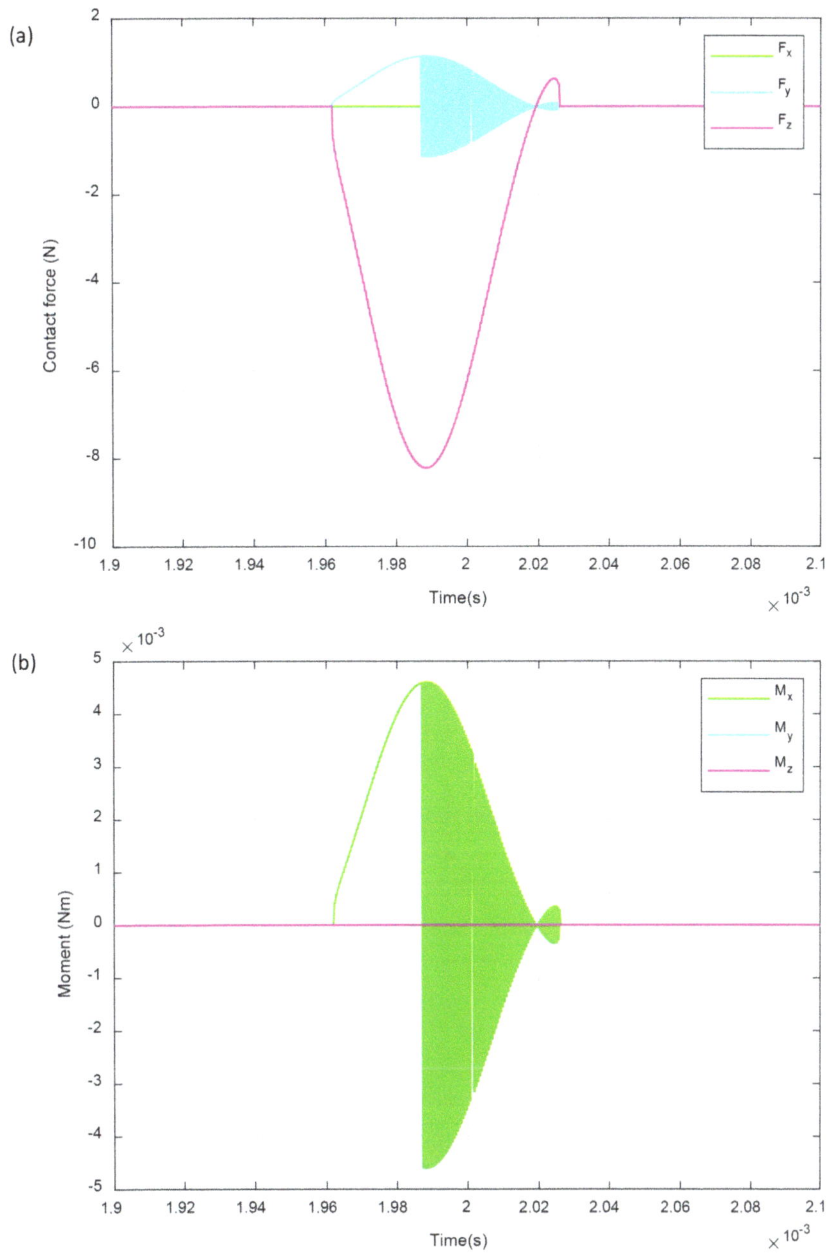

Figure 2.6. Contact force and moments in low incidence angle contact process.

The material and geometric properties of acrylic damping particles and containers are given in tables 2.1 and 2.2.

The energy dissipation by the damping particle and subsequently the equivalent viscous damping are calculated using equation (2.19) while the container is excited

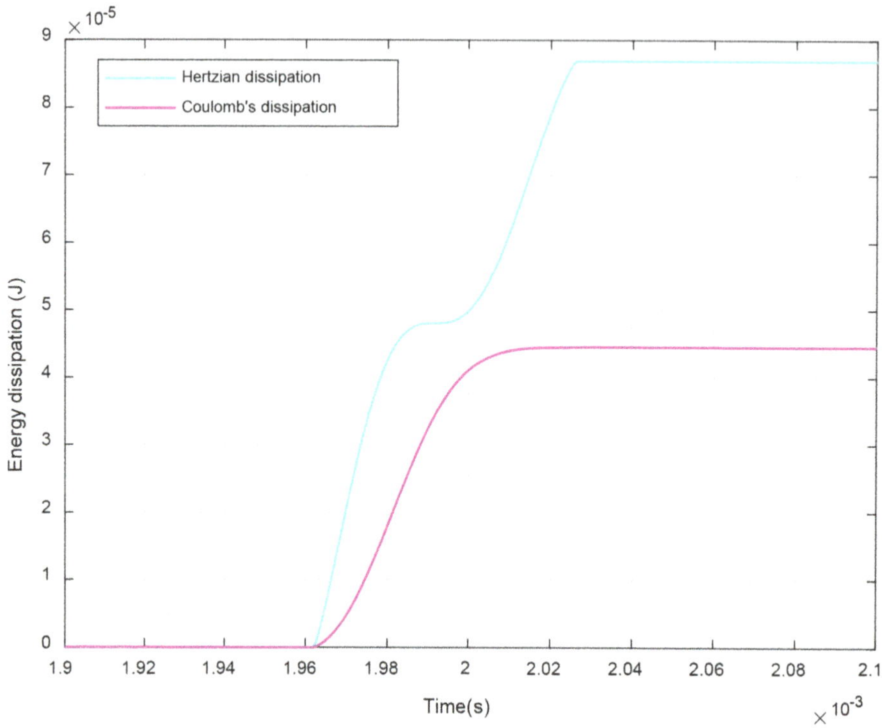

Figure 2.7. The change in energy and dissipation in typical impact process.

Table 2.1. Geometry and material properties.

Properties	Aluminum	Acrylic
Radius		1.25 mm
Density	2850 kg m^{-3}	1180 mm
Young's modulus	70 $\times 10^9$ Pa	2.84 $\times 10^9$ Pa
Poisson's ratio	0.33	0.402

Table 2.2. Friction and restitution coefficient.

Material pairs	Coulomb's coefficient of friction	Restitution coefficient (normal)
Aluminum–aluminium	0.50	0.85
Acrylic–acrylic	0.096	0.70
Aluminum–acrylic	0.14	0.70

by harmonic acceleration for different packing ratios, which is defined as the ratio of the volume of the container filled with damping particle to the total volume of the container. The numerical simulations for the equivalent viscous damping have been studied with respect to the many parameters; they were prominently dependent on

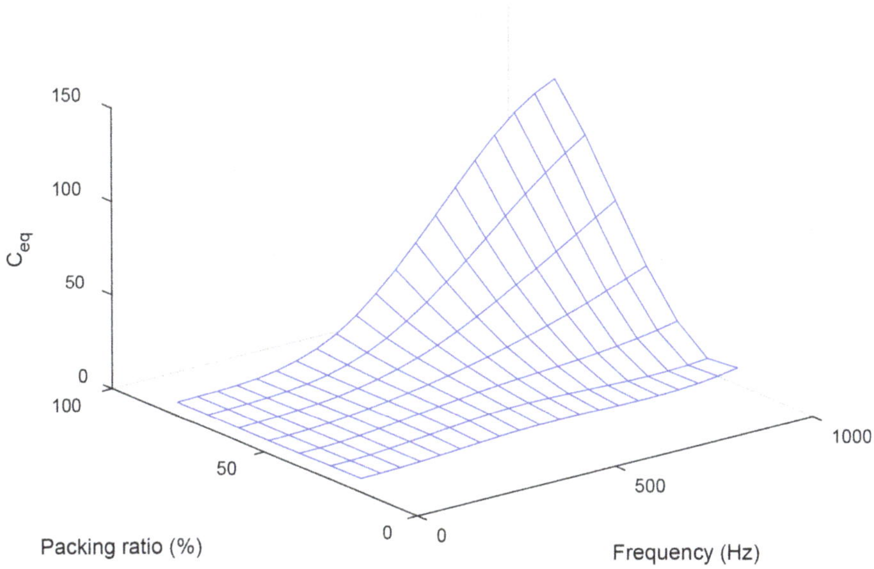

Figure 2.8. Amplitude of sine input acceleration of 2g.

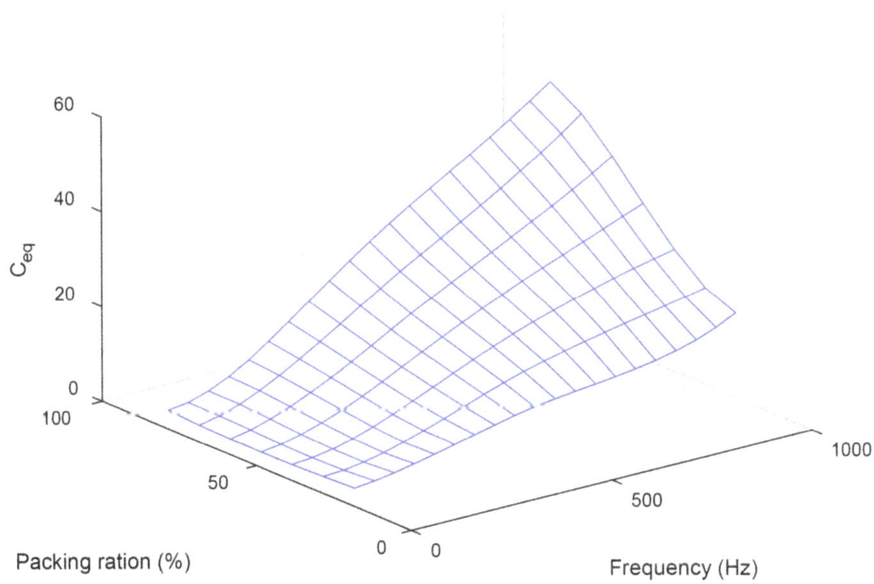

Figure 2.9. Amplitude of sine input acceleration of 5g.

packing fraction, excitation level, and excitation frequency. Figures 2.8–2.10 presents the variation of C_{eq} with respect to excitation frequency and packing ratio for three levels of amplitude of input excitation: 2g, 5g and 8g. The equivalent viscous damping was more effective at a higher frequency and higher packing ratio

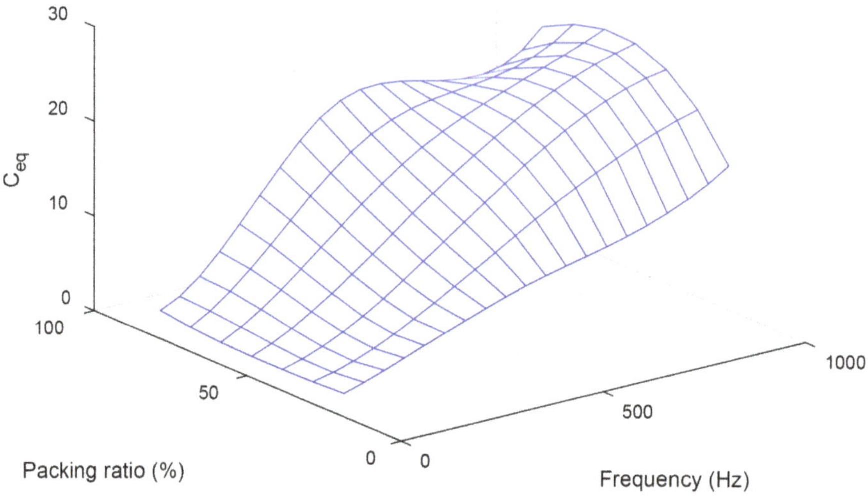

Figure 2.10. Amplitude of sine input acceleration of 8g.

when the excitation amplitude is low. There is a monotonous increase in equivalent damping with packing ratio and excitation frequency. However, this trend seems to change when excitation input is increased to 8g. Beyond 500 Hz, the C_{eq} is reaching a maximum and decresing afterwad with increasing packing ratio. It is observed that at a higher packing ratio, more than 80, the equivalent damping decreases. This behaviour could be due to the fact that at a higher packing ratio, the space for the movement of particles is less and hence less collision and dissipation take place, which is reflected as a decrease in the damping constant.

Figures 2.11–2.13 present the variation of C_{eq} with input excitation acceleration and frequency for a fixed packing ratio. This study was carried for three sets of packing ratios: 25%, 50% and 80%. The C_{eq} is less for lower ratios, 25%, at all frequencies and acceleration due to the fact that the number of collisions that result in the dissipation of energy is significantly lower. In a higher frequency regime, it is found to be decreasing with increasing excitation level after reaching maxima. This decrease could be due to more particle–particle collision compared to the particle and container collision. However, at a higher fill fraction of 80%, as shown in figure 2.13, it is more effective at a higher frequency and lower input acceleration.

2.4 Conclusions

Particle impact damping technology is a passive vibration mitigation strategy where the energy of the system is dissipated by the collision and rubbing action of damping particles. The damping particles embedded in the structure get kinetic energy through momentum transfer from the vibration and dissipate as heat. To quantify the effect of damping particles, a mathematical model governing the dynamics of the

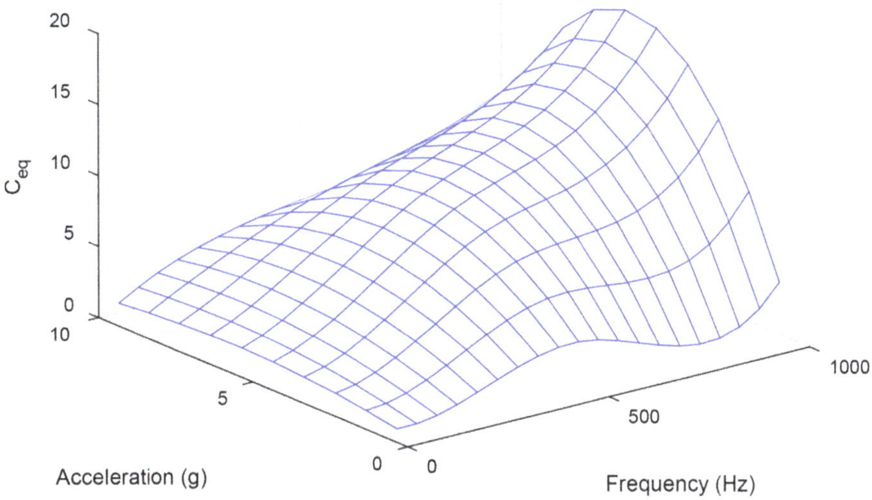

Figure 2.11. Packing ratio of 25%.

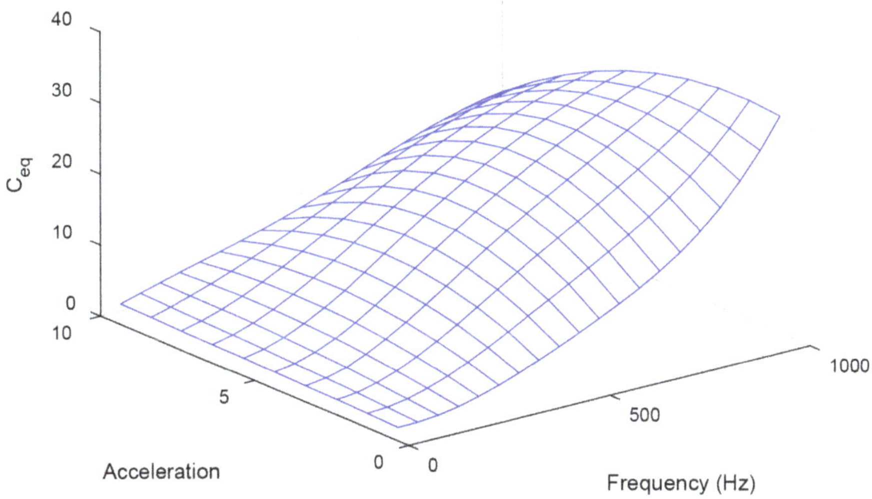

Figure 2.12. Packing ratio of 50%.

particles is developed using the Newton second law in conjunction with the nonlinear Hertzian dissipative contact force model for normal impact forces along with the Coulombs model friction force resulting in the oblique impacts and rubbing actions of the particles. The model described herein is used to predict the dynamic behaviour of a single particle and the contact forces and velocities evolving during

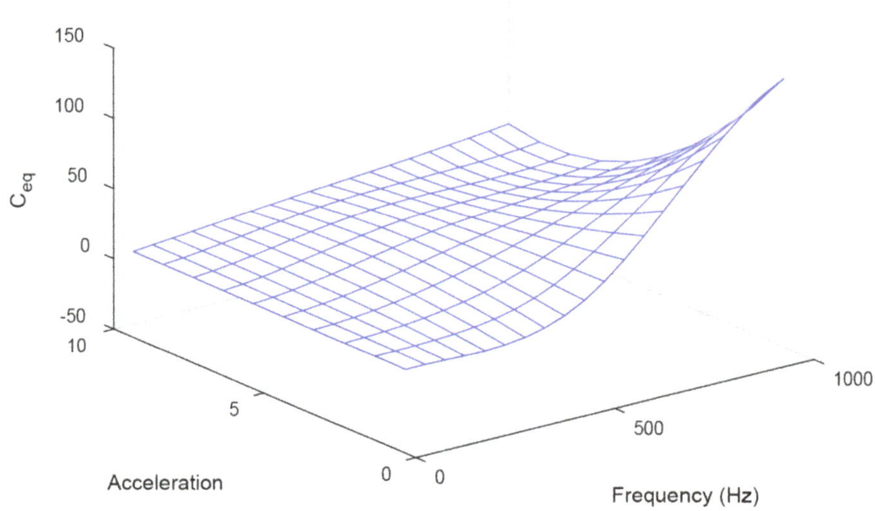

Figure 2.13. Packing ratio of 80%.

the impact process. Finally, using the model, energy dissipated by the damping particle is estimated. The major parameters on which the dissipative efficiency of the damping particles depends were studied. A fill fraction of around 80% was found to be more effective. The effectiveness was seen to be also dependent on the frequency and magnitude of the excitation.

References

[1] Lu Z, Lu X, Lu W and Masri S F 2012 Experimental studies of the effects of buffered particle dampers attached to a multi-degree-of-freedom system under dynamic loads *J. Sound Vib.* **331** 2007–22

[2] Ahmad N, Ranganath R and Ghosal A 2019 Modeling of the coupled dynamics of damping particles filled in the cells of a honeycomb sandwich plate and experimental validation *J. Vib. Control* **25** 1706–19

[3] Ahmad N, Ranganath R and Ghosal A 2017 Modeling and experimental study of a honeycomb beam filled with damping particles *J. Sound Vib.* **391** 20–34

[4] Ahmad N, Ranganath R and Ghosal A 2016 Assessment of particle damping device for large laminated structures under acoustic excitations Presented at the Proc. of 14th ISAMPE National Conf. on Composites (INCCOM-14) *(Hyderabad)*

[5] Cundall P A and Strack O D L 1979 A discrete numerical model for granular assemblies *Géotechnique* **29** 47–65

[6] Goldsmith W 2001 *Impact(Dover Civil and Mechanical Engineering)* (Mineola, NY: Dover Publications Inc.)

[7] Tsuji Y, Tanaka T and Ishida T 1992 Lagrangian numerical simulation of plug flow of cohesionless particles in a horizontal pipe *Powder Technol.* **71** 239–50

[8] Fang X, Tang J and Luo H 2007 Granular damping analysis using an improved discrete element approach *J. Sound Vib.* **308** 112–31

[9] Saeki M 2005 Analytical study of multi-particle damping *J. Sound Vib.* **281** 1133–44
[10] Johnson K L 1985 *Contact Mechanics.* (Cambridge: Cambridge University Press)
[11] Mindlin R D and Deresiewicz H 1953 Elastic spheres in contact under varying oblique forces *ASME J. Appl. Mech* **20** 327–44
[12] Maw N, Barber J R and Fawcett J N 1976 The oblique impact of elastic spheres *Wear* **38** 101–14
[13] Olson S E 2003 An analytical particle damping model *J. Sound Vib.* **264** 1155–66

IOP Publishing

Modeling and Computation in Vibration Problems, Volume 1
Numerical and semi-analytical methods
S Chakraverty, F Tornabene and J N Reddy

Chapter 3

Vibration of thick functionally graded materials skew plates based on a new shear deformation plate theory

K K Pradhan and S Chakraverty

Parts of this chapter have been adapted from [39] and [33], with permission from Springer and Elsevier.

In this chapter, transverse vibration of thick functionally graded (FG) skew plates with various skew angles has been investigated. Material properties of FG constituents may vary spatially along thickness direction in power-law form. A newly proposed shear deformation plate theory is considered here to define the constitutive relation and the generalized eigenvalue problem is obtained by means of Rayleigh–Ritz method. Trial functions denoting the transverse displacement may be expressed as a linear combination of simple algebraic polynomials (generated from Pascal's triangle). The present results are validated with the existing literature after checking the convergence to report the natural frequencies. Consequently, the objective is to find the effect of aspect ratio, skew angles and volume fractions on the natural frequencies.

3.1 Introduction

Functionally graded (FG) composites are generally a combination of ceramic and metal constituents. In the recent era, these composites have played a significant role in structural dynamics and architectural design because of their thermal resistant property (especially for aircraft, space vehicles and other extreme engineering structures). The material properties in these advanced FG composites vary gradually from one interface to another in a specific mathematical pattern. Investigating static and dynamic characteristics on FG rectangular (regular geometry) plates have been carried out rigorously in recent decades, whereas very few research efforts can be found on FG skew plates. In this context, existing literature based on dynamic

doi:10.1088/978-0-7503-3483-9ch3 © IOP Publishing Ltd 2021

behaviors of isotropic and FG skew plates using different computational techniques are surveyed here.

Nair and Durvasula [1] have considered the variational Ritz method to solve vibration problems of skew plates with different edge conditions. Sathyamoorthy [2] has studied the large amplitude free vibration of elastic and isotropic skew plates. Mizusawa *et al* [3] have discussed free vibration of skew plates by Rayleigh–Ritz method using B-spline trial functions. The application of Rayleigh–Ritz method with Lagrange multipliers to analyze skew plate problems with various constraints has been presented by Mizusawa *et al* [4]. The spline element method is implemented by Mizusawa and Katija [5] to study vibration of skew plates resting on point supports. The flexural vibration of skew plates is investigated by Liew and Lam [6] using Rayleigh–Ritz method in terms of two-dimensional orthogonal plate functions. Liew *et al* [7] have assumed Mindlin's shear deformation theory along with *pb*-2 Rayleigh–Ritz method to investigate free flexural vibration of thick skew plates. Meanwhile, Singh and Chakraverty [8] have used boundary characteristic orthogonal polynomials in two variables to determine first five frequencies of transverse vibration of rectangular and skew plate under different boundary conditions. Free vibration of skew fiber-reinforced composite laminates is analyzed by Wang [9] based on first order shear deformation plate theory. Later, B-spline based Rayleigh–Ritz method is applied in [10] to study transverse vibration of rectangular and skew plates. A general high precision triangular plate bending finite element has been developed by Krishna Reddy and Palaninathan [11] to study the free vibration analysis of laminated skew plates. The *p*-Ritz method is used by Wang *et al* [12] to investigate free vibration of skew sandwich plates composed of an orthotropic core and laminated facings. Vibration of thick skew plates has been studied by Mizusawa and Kondo [13] with the application of spline element method based on Mindlin plate theory. Upadhyay and Shukla [14] have investigated geometrically nonlinear static and dynamic behavior of FG skew plates using higher order shear deformation theory in conjunction with von-Karman's nonlinear kinetics. Chakraverty and Pradhan [15] have studied flexural vibration of FG skew plates based on classical plate theory. Bending and vibration behaviors of carbon nanotube (CNT) with reinforced FG composite skew plates are analyzed in [16] using the isogeometric analysis. Kiani *et al* [17] have studied free vibration characteristics of a skew cylindrical panel made of FG carbon nanotube reinforced composites (FG-CNTRCs). Three-dimensional free vibration behavior of skew magneto-electro-elastic plates is addressed by Vinyas *et al* [18] under the framework of a higher order shear deformation theory. Zhang [19] has investigated the effect of in-plane forces on the vibration behavior of CNT reinforced composite skew plates. Free vibration behaviors of FG plates are discussed by Xue *et al* [20] considering in-plane material inhomogeneity.

Developing new shear deformation beam (or plate) theories to handle aforesaid studies has also been a trend in recent literature. In this respect, Reissner [21] has given a two-dimensional plate theory (2-DPT) to the problem of transverse bending of homogeneous plates. A higher-order nonlinear shear deformation theory has been proposed by Reddy [22] to obtain exact solutions of simply supported plates. Again in another article, Reddy [23] has used this theory for the elastic plates to determine the natural frequencies and buckling loads. Reddy [24] has found exact closed-form solutions of symmetric cross-ply laminates by using the first-order shear deformation theory. Nayak *et al* [25] have used Reddy's higher-order theory to obtain the natural frequencies of composite and sandwich plates using two C^0 assumed finite element formulations. The exact closed form characteristic equations for the Mindlin plates is instigated by Hosseini-Hashemi and Arsanjani [26] with two opposite edges simply supported. Two-variable refined plate theory is implemented by Shimpi and Patel [27] for free vibration analysis of plates. Shimpi *et al* [28] have introduced two new displacement based first-order shear deformation plate theories to study the dynamic problems. Aydogdu [29] has proposed a new higher order shear deformation laminated composite plate theory from 3-D elasticity solutions by using an inverse method. Various shear deformation theories have been implemented by Xiang *et al* [30] for the modeling of isotropic, sandwich and laminated plates. Flexural vibration of Lévy-type rectangular plates has been studied by Hosseini-Hashemi *et al* [31] within the framework of third-order shear deformation theory. A new inverse tangent shear deformation theory (ITSDT) was presented by Thai *et al* [32] for the static, free vibration and buckling analysis of laminated composite and sandwich plates. Inverse trigonometric shear deformation plate theories were developed by Pradhan and Chakraverty [33] to handle transverse vibration of isotropic thick rectangular plates. In another article, Pradhan and Chakraverty [34] have proposed a generalized power-law exponent based shear deformation beam theory to investigate vibration of FG beams.

In view of the above, studies related to vibration of FG skew plates based on new shear deformation theory are still open and these problems are essential in engineering design and architecture. As such, this chapter assumes the shear deformation theory mentioned in [34] to study transverse vibration of thick FG skew plates with various skew angles. The numerical formulation is carried out by means of Rayleigh–Ritz method to generate the generalized eigenvalue problem. Trial functions denoting the displacement components may be expressed as a linear combination of simple algebraic polynomials. On this note, the objective is to find the effect of various physical parameters and volume fractions of FG constituents on the natural frequencies. Accordingly, the present results are validated with the existing literature after checking the convergence to report the new results.

3.2 Shear deformation plate theory

Let us consider an FG skew plate defined by three parameters such as length (a), adjacent side (b) and skew angle (α) as shown in figure 3.1(a). The rectangular plate is its special case by taking $\alpha = 90°$. A transformation (or mapping) is considered here, which maps this skew plate into a unit square, as given in figure 3.1 [8].

$$x = a\xi + b(\cos\alpha)\eta, \quad y = b(\sin\alpha)\eta \tag{3.1}$$

where ξ and η are new coordinates as demonstrated in figure 3.1(b).

Assuming the deformation of the plate to be in the middle surface (x–y) plane, one can represent the displacement fields of isotropic plate based on generalized

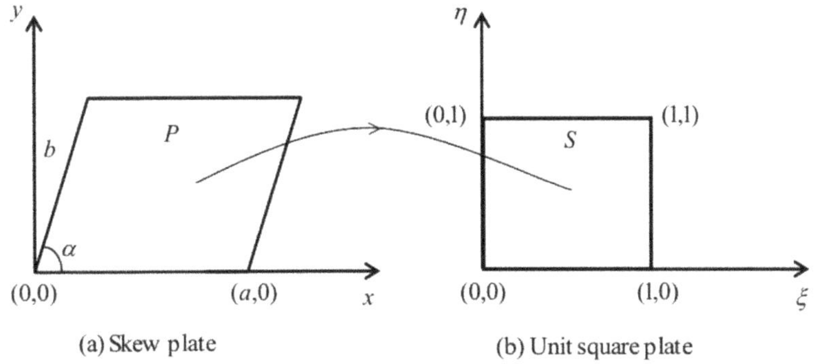

Figure 3.1. Transformation of coordinates from skew plate to a unit square (see [15], copyright (2017) with permission of Springer).

higher-order shear deformation plate theory with similar assumptions to [29, 30] as follows:

$$\begin{aligned} u_x(x, y, z; t) &= u(x, y; t) - z\frac{\partial w}{\partial x} + f(z)\phi_x(x, y; t) \\ u_y(x, y, z; t) &= v(x, y; t) - z\frac{\partial w}{\partial y} + f(z)\phi_y(x, y; t) \\ u_z(x, y, z; t) &= w(x, y; t) \end{aligned} \tag{3.2}$$

where u, v, w, ϕ_x and ϕ_y are five unknown displacement components of the middle plane of the plate and $f(z)$ denotes the transverse shear function to determine the parabolic distribution of the transverse shear strains and stresses across the thickness. Different researchers have proposed various shear deformation plate theories (SDPTs) by suitably selecting the shape function, $f(z)$. In this study, the shear function is assumed as

$$f(z) = h\left(\frac{z}{h}\right)^{2n+1} - (2n+1)z\left(\frac{1}{2}\right)^{2n}.$$

Major criteria to choose the shape functions ($f(z)$) for these models are given as follows [22, 29]:

1. Suitable $f(z)$ function should approximately satisfy parabolic shear deformation distribution, which can be easily observed in figure 3.2(b).
2. Boundary conditions should be satisfied on the bottom and top surfaces of the plate.

In the developed model, n is a non-negative integer which will play a key role for the present investigation. It can be viewed that classical or Euler–Bernoulli beam theory is a special case of the present model. If we assume $n = 0$, then PESDPT will follow the assumptions of classical beam theory. The effects of power-law exponents (n) of the present model along the thickness coordinate can be observed in figure 3.2. It is also evident that the shape function gradually coincides with the horizontal (x-) axis with

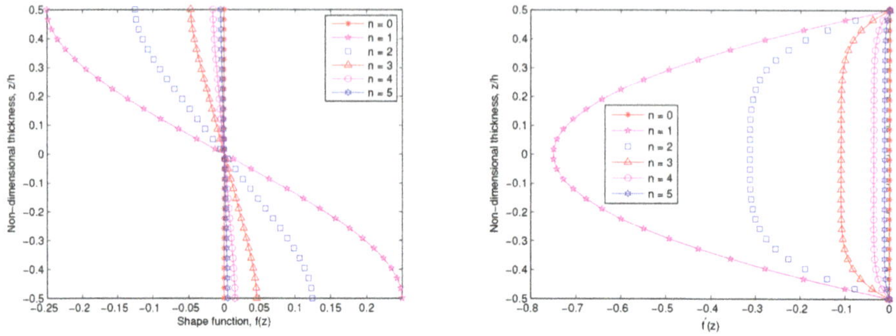

(*a*) Shape functions (*b*) Derivatives of shape functions

Figure 3.2. Behavior of shape functions and their derivatives along thickness of thick skew plate.

increase in the value of n. In this figure, one can also see that $f(z)$ in the case of PESDPT is continuous in the interval $\left[-\frac{h}{2}, +\frac{h}{2}\right]$. Derivative of $f(z)$ in the given theory yields

$$\frac{\partial f}{\partial z} = (2n + 1)\left(\frac{z}{h}\right)^{2n} - (2n + 1)\left(\frac{1}{2}\right)^{2n}$$

and equation (3.3) satisfies the boundary conditions on the bottom and top surfaces of the beam.

$$\tau_{xz}\left(x, \mp\frac{h}{2}\right) = 0 \tag{3.3}$$

In the above displacement field, the shape functions concerned with these SDPTs are demonstrated in figure 3.2. In this regard, figure 3.2(a) depicts the behavior of shape functions, while their respective derivatives are plotted in figure 3.2(b) with reference to the thickness coordinate.

3.3 Constitutive relations

Using the displacement field stated in equation (3.2), the non-zero strain-displacement relations can be expressed as (see [33])

$$\begin{aligned}
\epsilon_{xx} &= \frac{\partial u_x}{\partial x} = \frac{\partial u}{\partial x} - z\frac{\partial^2 w}{\partial x^2} + f(z)\frac{\partial \phi_x}{\partial x} \\
\epsilon_{yy} &= \frac{\partial u_y}{\partial y} = \frac{\partial v}{\partial y} - z\frac{\partial^2 w}{\partial y^2} + f(z)\frac{\partial \phi_y}{\partial y} \\
\gamma_{xy} &= \frac{\partial u_x}{\partial y} + \frac{\partial u_y}{\partial x} = \left(\frac{\partial u}{\partial y} + \frac{\partial v}{\partial x}\right) - 2z\frac{\partial^2 w}{\partial x \partial y} + f(z)\left(\frac{\partial \phi_x}{\partial y} + \frac{\partial \phi_y}{\partial x}\right) \\
\gamma_{yz} &= \frac{\partial u_y}{\partial z} + \frac{\partial u_z}{\partial y} = \frac{\partial f}{\partial z}\phi_y \\
\gamma_{xz} &= \frac{\partial u_x}{\partial z} + \frac{\partial u_z}{\partial x} = \frac{\partial f}{\partial z}\phi_x
\end{aligned} \tag{3.4}$$

where ϵ_{xx} and ϵ_{yy} are the normal strains in x- and y- directions, respectively, and γ_{xy}, γ_{yz} and γ_{xz} are the shear strains in x–y, y–z and x–z planes, respectively. Assuming the material constituents of the plate to obey the generalized Hooke's law, the stress–strain relationships can be written in matrix form

$$\begin{Bmatrix} \sigma_{xx} \\ \sigma_{yy} \\ \tau_{xy} \\ \tau_{yz} \\ \tau_{xz} \end{Bmatrix} = \begin{pmatrix} Q_{11} & Q_{12} & 0 & 0 & 0 \\ Q_{21} & Q_{22} & 0 & 0 & 0 \\ 0 & 0 & Q_{66} & 0 & 0 \\ 0 & 0 & 0 & Q_{44} & 0 \\ 0 & 0 & 0 & 0 & Q_{55} \end{pmatrix} \begin{Bmatrix} \epsilon_{xx} \\ \epsilon_{yy} \\ \gamma_{xy} \\ \gamma_{yz} \\ \gamma_{xz} \end{Bmatrix} \tag{3.5}$$

where σ_{xx}, σ_{yy} are the normal stresses; τ_{xy}, τ_{yz} and τ_{xz} are the shear stresses. The reduced stiffness coefficients, Q_{ij} $(i, j = 1, \ 2, \ 4, \ 5, \ 6)$ are then given by

$$Q_{11} = Q_{22} = \frac{E}{1 - \nu^2}, \quad Q_{12} = Q_{21} = \frac{\nu E}{1 - \nu^2} \quad \text{and}$$

$$Q_{44} = Q_{55} = Q_{66} = G = \frac{E}{2(1 + \nu)}$$

Here, E, G and ν are Young's modulus, shear modulus and Poisson's ratio of the material constituent, respectively.

3.4 Mechanical energies

As clearly mentioned in [33], we can now define the strain energy (U) and kinetic energy (T) for the free vibration of isotropic plate by considering the constitutive relations.

$$U = \frac{1}{2}\int_{\Omega}\left[\int_{-h/2}^{h/2}\left(\sigma_{xx}\epsilon_{xx} + \sigma_{yy}\epsilon_{yy} + \tau_{xy}\gamma_{xy} + \tau_{yz}\gamma_{yz} + \tau_{xz}\gamma_{xz}\right)\mathrm{d}z\right]\mathrm{d}x\mathrm{d}y \tag{3.6}$$

$$T = \frac{1}{2}\int_{\Omega}\left[\int_{-h/2}^{h/2}\rho\left\{\left(\frac{\partial u_x}{\partial t}\right)^2 + \left(\frac{\partial u_y}{\partial t}\right)^2 + \left(\frac{\partial u_z}{\partial t}\right)^2\right\}\mathrm{d}z\right]\mathrm{d}x\mathrm{d}y \tag{3.7}$$

Substituting σ_{ii} $(i = x, \ y)$ and τ_{ij} $(i, \ j = x, \ y, \ z)$ in equation (3.6), it becomes

$$U = \frac{1}{2}\int_{\Omega}\left[\int_{-h/2}^{h/2}\left\{Q_{11}\left(\epsilon_{xx}^2 + \epsilon_{yy}^2\right) + 2Q_{12}\epsilon_{xx}\epsilon_{yy} + Q_{66}\gamma_{xy}^2 + Q_{44}\gamma_{yz}^2 + Q_{55}\gamma_{xz}^2\right\}\mathrm{d}z\right]\mathrm{d}x\mathrm{d}y \tag{3.8}$$

Let us first individually expand the terms involved in equation (3.8):

$$
\begin{aligned}
\epsilon_{xx}^2 &= \left(\frac{\partial u}{\partial x} - z\frac{\partial^2 w}{\partial x^2} + f(z)\frac{\partial \phi_x}{\partial x}\right)^2 \\
&= \left(\frac{\partial u}{\partial x}\right)^2 - 2z\frac{\partial u}{\partial x}\frac{\partial^2 w}{\partial x^2} + 2f(z)\frac{\partial u}{\partial x}\frac{\partial \phi_x}{\partial x} \\
&\quad + z^2\left(\frac{\partial^2 w}{\partial x^2}\right)^2 + f^2\left(\frac{\partial \phi_x}{\partial x}\right)^2 - 2zf(z)\frac{\partial \phi_x}{\partial x}\frac{\partial^2 w}{\partial x^2}, \\
\epsilon_{yy}^2 &= \left(\frac{\partial v}{\partial y} - z\frac{\partial^2 w}{\partial y^2} + f(z)\frac{\partial \phi_y}{\partial y}\right)^2 \\
&= \left(\frac{\partial v}{\partial y}\right)^2 - 2z\frac{\partial v}{\partial y}\frac{\partial^2 w}{\partial y^2} + 2f(z)\frac{\partial v}{\partial y}\frac{\partial \phi_y}{\partial y} + z^2\left(\frac{\partial^2 w}{\partial y^2}\right)^2 \\
&\quad + f^2\left(\frac{\partial \phi_y}{\partial y}\right)^2 - 2zf(z)\frac{\partial \phi_y}{\partial y}\frac{\partial^2 w}{\partial y^2}, \\
\epsilon_{xx}\epsilon_{yy} &= \left(\frac{\partial u}{\partial x} - z\frac{\partial^2 w}{\partial x^2} + f(z)\frac{\partial \phi_x}{\partial x}\right)\left(\frac{\partial v}{\partial y} - z\frac{\partial^2 w}{\partial y^2} + f(z)\frac{\partial \phi_y}{\partial y}\right) \\
&= \frac{\partial u}{\partial x}\frac{\partial v}{\partial y} - z\left(\frac{\partial u}{\partial x}\frac{\partial^2 w}{\partial y^2} + \frac{\partial v}{\partial y}\frac{\partial^2 w}{\partial x^2}\right) + f(z)\left(\frac{\partial u}{\partial x}\frac{\partial \phi_y}{\partial y} + \frac{\partial v}{\partial y}\frac{\partial \phi_x}{\partial x}\right) \\
&\quad + z^2\frac{\partial^2 w}{\partial x^2}\frac{\partial^2 w}{\partial y^2} - zf(z)\left(\frac{\partial \phi_x}{\partial x}\frac{\partial^2 w}{\partial y^2} + \frac{\partial \phi_y}{\partial y}\frac{\partial^2 w}{\partial x^2}\right) \\
&\quad + f^2\frac{\partial \phi_x}{\partial x}\frac{\partial \phi_y}{\partial y}, \\
\gamma_{xy}^2 &= \left[\left(\frac{\partial u}{\partial y} + \frac{\partial v}{\partial x}\right) - 2z\frac{\partial^2 w}{\partial x \partial y} + f(z)\left(\frac{\partial \phi_x}{\partial y} + \frac{\partial \phi_y}{\partial x}\right)\right]^2 \\
&= \left(\frac{\partial u}{\partial y} + \frac{\partial v}{\partial x}\right)^2 - 4z\left(\frac{\partial u}{\partial y} + \frac{\partial v}{\partial x}\right)\frac{\partial^2 w}{\partial x \partial y} \\
&\quad + 2f(z)\left(\frac{\partial u}{\partial y} + \frac{\partial v}{\partial x}\right)\left(\frac{\partial \phi_x}{\partial y} + \frac{\partial \phi_y}{\partial x}\right) + 4z^2\left(\frac{\partial^2 w}{\partial x \partial y}\right)^2 \\
&\quad + f^2\left(\frac{\partial \phi_x}{\partial y} + \frac{\partial \phi_y}{\partial x}\right)^2 - 4zf(z)\left(\frac{\partial \phi_x}{\partial y} + \frac{\partial \phi_y}{\partial x}\right)\frac{\partial^2 w}{\partial x \partial y}, \\
\gamma_{yz}^2 &= \left(\frac{\partial f}{\partial z}\right)^2\phi_y^2, \quad \gamma_{xz}^2 = \left(\frac{\partial f}{\partial z}\right)^2\phi_x^2.
\end{aligned}
\tag{3.9}
$$

Substituting the expanded terms from equation (3.9) in equation (3.8), the strain energy (U) takes the form

$$
\begin{aligned}
U = \frac{1}{2} \int_{\Omega} \Bigg[& A_{11}\left\{\left(\frac{\partial u}{\partial x}\right)^2 + \left(\frac{\partial v}{\partial y}\right)^2\right\} - 2B_{11}\left(\frac{\partial u}{\partial x}\frac{\partial^2 w}{\partial x^2} + \frac{\partial v}{\partial y}\frac{\partial^2 w}{\partial y^2}\right) \\
& + 2C_{11}\left(\frac{\partial u}{\partial x}\frac{\partial \phi_x}{\partial x} + \frac{\partial v}{\partial y}\frac{\partial \phi_y}{\partial y}\right) \\
& + D_{11}\left\{\left(\frac{\partial^2 w}{\partial x^2}\right)^2 + \left(\frac{\partial^2 w}{\partial y^2}\right)^2\right\} + E_{11}\left\{\left(\frac{\partial \phi_x}{\partial x}\right)^2 + \left(\frac{\partial \phi_y}{\partial y}\right)^2\right\} \\
& - 2F_{11}\left(\frac{\partial \phi_x}{\partial x}\frac{\partial^2 w}{\partial x^2} + \frac{\partial \phi_y}{\partial y}\frac{\partial^2 w}{\partial y^2}\right) \\
& + 2A_{12}\frac{\partial u}{\partial x}\frac{\partial v}{\partial y} - 2B_{12}\left(\frac{\partial u}{\partial x}\frac{\partial^2 w}{\partial y^2} + \frac{\partial v}{\partial y}\frac{\partial^2 w}{\partial x^2}\right) \\
& + 2C_{12}\left(\frac{\partial u}{\partial x}\frac{\partial \phi_y}{\partial y} + \frac{\partial v}{\partial y}\frac{\partial \phi_x}{\partial x}\right) + 2D_{12}\frac{\partial^2 w}{\partial x^2}\frac{\partial^2 w}{\partial y^2} \\
& + 2E_{12}\frac{\partial \phi_x}{\partial x}\frac{\partial \phi_y}{\partial y} - 2F_{12}\left(\frac{\partial \phi_x}{\partial x}\frac{\partial^2 w}{\partial y^2} + \frac{\partial \phi_y}{\partial y}\frac{\partial^2 w}{\partial x^2}\right) \\
& + A_{66}\left(\frac{\partial u}{\partial y} + \frac{\partial v}{\partial x}\right)^2 - 4B_{66}\left(\frac{\partial u}{\partial y} + \frac{\partial v}{\partial x}\right)\frac{\partial^2 w}{\partial x \partial y} \\
& + 2C_{66}\left(\frac{\partial u}{\partial y} + \frac{\partial v}{\partial x}\right)\left(\frac{\partial \phi_x}{\partial y} + \frac{\partial \phi_y}{\partial x}\right) + 4D_{66}\left(\frac{\partial^2 w}{\partial x \partial y}\right)^2 \\
& + E_{66}\left(\frac{\partial \phi_x}{\partial y} + \frac{\partial \phi_y}{\partial x}\right)^2 \\
& - 4F_{66}\left(\frac{\partial \phi_x}{\partial y} + \frac{\partial \phi_y}{\partial x}\right)\frac{\partial^2 w}{\partial x \partial y} + H_{44}\phi_y^2 + H_{55}\phi_x^2 \Bigg] \mathrm{d}x\mathrm{d}y
\end{aligned}
\tag{3.10}
$$

Taking time derivative of displacement components leads to the kinetic energy (T) as

$$
\begin{aligned}
T = \frac{1}{2} \int_{\Omega} \Bigg\{ & \rho_0\left[\left(\frac{\partial u}{\partial t}\right)^2 + \left(\frac{\partial v}{\partial t}\right)^2 + \left(\frac{\partial w}{\partial t}\right)^2\right] \\
& - 2\rho_1\left(\frac{\partial u}{\partial t}\frac{\partial^2 w}{\partial x \partial t} + \frac{\partial v}{\partial t}\frac{\partial^2 w}{\partial y \partial t}\right) \\
& + \rho_2\left[\left(\frac{\partial^2 w}{\partial x \partial t}\right)^2 + \left(\frac{\partial^2 w}{\partial y \partial t}\right)^2\right] + 2\rho_0^1\left(\frac{\partial u}{\partial t}\frac{\partial \phi_x}{\partial t} + \frac{\partial v}{\partial t}\frac{\partial \phi_y}{\partial t}\right) \\
& + \rho_0^2\left[\left(\frac{\partial \phi_x}{\partial t}\right)^2 + \left(\frac{\partial \phi_y}{\partial t}\right)^2\right] - 2\rho_1^1\left(\frac{\partial \phi_x}{\partial t}\frac{\partial^2 w}{\partial x \partial t} + \frac{\partial \phi_y}{\partial t}\frac{\partial^2 w}{\partial y \partial t}\right) \Bigg\} \mathrm{d}x\mathrm{d}y
\end{aligned}
\tag{3.11}
$$

The extensional, coupling, bending and transverse shear rigidities (as given in equation (3.10)) for the higher-order shear deformation theory can be expressed as,

$$(A_{ij}, B_{ij}, C_{ij}) = \int_{-h/2}^{h/2} Q_{ij}(1, z, f(z))\,\mathrm{d}z; \quad \text{where} \quad i, j = 1,\ 2,\ 6$$
$$(D_{ij}, E_{ij}, F_{ij}) = \int_{-h/2}^{h/2} Q_{ij}(z^2, f^2, zf(z))\,\mathrm{d}z; \quad \text{where} \quad i, j = 1,\ 2,\ 6$$
$$H_{kk} = \int_{-h/2}^{h/2} Q_{kk}\left(\frac{\partial f}{\partial z}\right)^2 \mathrm{d}z; \quad \text{where} \quad k = 4, 5$$

whereas the cross-sectional inertial coefficients in equation (3.11) are written as

$$\rho_i = \int_{-h/2}^{h/2} \rho z^i\,\mathrm{d}z; \quad \text{where} \quad i, j = 0, 1, 2$$
$$\rho_i^j = \int_{-h/2}^{h/2} \rho z^i f^j\,\mathrm{d}z; \quad \text{where} \quad i = 0,\ 1;\ j = 1, 2.$$

Let us assume the displacement components $u(x, y, t)$, $v(x, y, t)$, $w(x, y, t)$, $\phi_x(x, y, t)$ and $\phi_y(x, y, t)$ to be the harmonic type and those may be expressed as

$$\begin{aligned}
u(x, y, t) &= U(x, y)\exp(i\omega t) \\
v(x, y, t) &= V(x, y)\exp(i\omega t) \\
w(x, y, t) &= W(x, y)\exp(i\omega t) \\
\phi_x(x, y, t) &= \frac{1}{a}\Phi_x(x, y)\exp(i\omega t) \\
\phi_y(x, y, t) &= \frac{1}{b}\Phi_y(x, y)\exp(i\omega t)
\end{aligned} \tag{3.12}$$

In equation (3.12), $i = \sqrt{-1}$; $U(x, y), V(x, y), W(x, y), \Phi_x(x, y)$ and $\Phi_y(x, y)$ are the respective amplitudes for these displacement components for free vibration of isotropic plate and the exponential terms indicate the harmonic type variation where ω is the natural frequency.

The mathematical manipulation reveals that the coefficients associated with B_{ij}, C_{ij}, ρ_1 and ρ_0^1 will be zero since $\int_{-h/2}^{h/2} z\,\mathrm{d}z$ and $\int_{-h/2}^{h/2} f(z)\,\mathrm{d}z$ yield the value as zeroes irrespective of the shear deformation theory considered. Substituting the displacement components of equation (3.12) in equations (3.10) and (3.11) gives the maximum strain energy ($U_{\max}$) and the maximum kinetic energy ($T_{\max}$) as stated below

$$\begin{aligned} U_{\max} = \frac{1}{2}\int_{\Omega} \Bigg[& A_{11}\left\{\left(\frac{\partial U}{\partial x}\right)^2 + \left(\frac{\partial V}{\partial y}\right)^2\right\} + 2A_{12}\frac{\partial U}{\partial x}\frac{\partial V}{\partial y} + A_{66}\left(\frac{\partial U}{\partial y} + \frac{\partial V}{\partial x}\right)^2 \\ & + D_{11}\left\{\left(\frac{\partial^2 W}{\partial x^2}\right)^2 + \left(\frac{\partial^2 W}{\partial y^2}\right)^2\right\} + 2D_{12}\frac{\partial^2 W}{\partial x^2}\frac{\partial^2 W}{\partial y^2} + 4D_{66}\left(\frac{\partial^2 W}{\partial x \partial y}\right)^2 \\ & + \frac{E_{11}}{a^2}\left\{\left(\frac{\partial \Phi_x}{\partial x}\right)^2 + \mu^2\left(\frac{\partial \Phi_y}{\partial y}\right)^2\right\} + \frac{2E_{12}}{ab}\frac{\partial \Phi_x}{\partial x}\frac{\partial \Phi_y}{\partial y} \\ & + \frac{E_{66}}{a^2}\left(\frac{\partial \Phi_x}{\partial y} + \mu\frac{\partial \Phi_y}{\partial x}\right)^2 \\ & - \frac{2F_{11}}{a}\left(\frac{\partial \Phi_x}{\partial x}\frac{\partial^2 W}{\partial x^2} + \mu\frac{\partial \Phi_y}{\partial y}\frac{\partial^2 W}{\partial y^2}\right) - \frac{2F_{12}}{a}\left(\frac{\partial \Phi_x}{\partial x}\frac{\partial^2 W}{\partial y^2}\right. \\ & \left. + \mu\frac{\partial \Phi_y}{\partial y}\frac{\partial^2 W}{\partial x^2}\right) \\ & - \frac{4F_{66}}{a}\left(\frac{\partial \Phi_x}{\partial y} + \mu\frac{\partial \Phi_y}{\partial x}\right)\frac{\partial^2 W}{\partial x \partial y} + \frac{H_{44}}{a^2}\left(\Phi_x^2 + \mu^2\Phi_y^2\right)\Bigg] \mathrm{d}x\mathrm{d}y \end{aligned} \tag{3.13}$$

$$\begin{aligned} T_{\max} = \frac{1}{2}\int_{\Omega} \Bigg[& \rho_0(U^2 + V^2 + W^2) + \rho_2\left\{\left(\frac{\partial W}{\partial x}\right)^2 + \left(\frac{\partial W}{\partial y}\right)^2\right\} \\ & + \frac{\rho_0^2}{a^2}\left(\Phi_x^2 + \mu^2\Phi_y^2\right) - \frac{2\rho_1^1}{a}\left(\Phi_x\frac{\partial W}{\partial x} + \mu\Phi_y\frac{\partial W}{\partial y}\right)\Bigg] \mathrm{d}x\mathrm{d}y \end{aligned} \tag{3.14}$$

where $\mu = a/b$, is referred as the aspect ratio of the plate. Now let us normalize x and y involved in equations (3.13) and (3.14) as

$$\xi = \frac{x \sin\alpha - y\cos\alpha}{a \sin\alpha}, \quad \eta = \frac{y}{b\sin\alpha}$$

Further, using these transformation of coordinates in equations (3.13) and (3.14), these expressions can be non-dimensionalized to the form

$$U_{\max} = \frac{D_c}{2a^4}\int_{\Omega}(U_1 + U_2 + U_3 + U_4 + U_5)J \, \mathrm{d}\xi\mathrm{d}\eta \tag{3.15}$$

$$T_{\max} = \frac{\rho_c h}{2}\int_{\Omega}(T_1 + T_2 + T_3 + T_4)J \, \mathrm{d}\xi\mathrm{d}\eta \tag{3.16}$$

where the flexural rigidity $(D_c) = \frac{E_c h^3}{12(1-\nu^2)}$ and the Jacobian $(J) = ab\sin\alpha$. The components inside the expression of $U_{\max}$ and $T_{\max}$ are clearly provided as below.

$$U_1 = 6C_{u1}\delta^2\left\{2\left(\frac{\partial U}{\partial \xi}\right)^2 + 2\left(-\cot\alpha\frac{\partial V}{\partial \xi} + \mu\csc\alpha\frac{\partial V}{\partial \eta}\right)^2 + 4\nu\frac{\partial U}{\partial \xi}\left(-\cot\alpha\frac{\partial V}{\partial \xi} + \mu\csc\alpha\frac{\partial V}{\partial \eta}\right) + (1-\nu)\left(-\cot\alpha\frac{\partial U}{\partial \xi} + \mu\csc\alpha\frac{\partial U}{\partial \eta} + \frac{\partial V}{\partial \xi}\right)^2\right\},$$

$$U_2 = 12C_{u2}\left\{\left(\frac{\partial^2 W}{\partial \xi^2}\right)^2 + \left(-\cot^2\alpha\frac{\partial^2 W}{\partial \xi^2} - 2\mu\csc\alpha\cot\alpha\frac{\partial^2 W}{\partial \xi\partial \eta} + \mu^2\frac{\partial^2 W}{\partial \eta^2}\right)^2 + 2\nu\frac{\partial^2 W}{\partial \xi^2}\left(-\cot^2\alpha\frac{\partial^2 W}{\partial \xi^2} - 2\mu\csc\alpha\cot\alpha\frac{\partial^2 W}{\partial \xi\partial \eta} + \mu^2\frac{\partial^2 W}{\partial \eta^2}\right) + 2(1-\nu)\left(-\cot\alpha\frac{\partial^2 W}{\partial \xi^2} + \mu\csc\alpha\frac{\partial^2 W}{\partial \xi\partial \eta}\right)^2\right\},$$

$$U_3 = 6C_{u3}\left\{2\left(\frac{\partial \Phi_x}{\partial \xi}\right)^2 + 2\mu^2\left(-\cot\alpha\frac{\partial \Phi_y}{\partial \xi} + \mu\csc\alpha\frac{\partial \Phi_y}{\partial \eta}\right)^2 + 4\mu\nu\frac{\partial \Phi_x}{\partial \xi}\left(-\cot\alpha\frac{\partial \Phi_y}{\partial \xi} + \mu\csc\alpha\frac{\partial \Phi_y}{\partial \eta}\right) + (1-\nu)\left(-\cot\alpha\frac{\partial \Phi_x}{\partial \xi} + \mu\csc\alpha\frac{\partial \Phi_x}{\partial \eta} + \mu\frac{\partial \Phi_y}{\partial \xi}\right)^2\right\},$$

$$U_4 = -24C_{u4}\left[\frac{\partial \Phi_x}{\partial \xi}\frac{\partial^2 W}{\partial \xi^2} + \mu\left(-\cot\alpha\frac{\partial \Phi_y}{\partial \xi} + \mu\csc\alpha\frac{\partial \Phi_y}{\partial \eta}\right)\left(-\cot^2\alpha\frac{\partial^2 W}{\partial \xi^2} - 2\mu\csc\alpha\cot\alpha\frac{\partial^2 W}{\partial \xi\partial \eta} + \mu^2\frac{\partial^2 W}{\partial \eta^2}\right) + \nu\left\{\frac{\partial \Phi_x}{\partial \xi}\left(-\cot^2\alpha\frac{\partial^2 W}{\partial \xi^2} - 2\mu\csc\alpha\cot\alpha\frac{\partial^2 W}{\partial \xi\partial \eta} + \mu^2\frac{\partial^2 W}{\partial \eta^2}\right) + \mu\left(-\cot\alpha\frac{\partial \Phi_y}{\partial \xi} + \mu\csc\alpha\frac{\partial \Phi_y}{\partial \eta}\right)\frac{\partial^2 W}{\partial \xi^2}\right\} + (1-\nu)\left(-\cot\alpha\frac{\partial \Phi_x}{\partial \xi} + \mu\csc\alpha\frac{\partial \Phi_x}{\partial \eta} + \mu\frac{\partial \Phi_y}{\partial \xi}\right)\left(-\cot\alpha\frac{\partial^2 W}{\partial \xi^2} + \mu\csc\alpha\frac{\partial^2 W}{\partial \xi\partial \eta}\right)\right],$$

$$U_5 = 6C_{u5}(1-\nu)\delta^2\left(\Phi_x^2 + \mu^2\Phi_y^2\right)$$

$$T_1 = C_{t1}(U^2 + V^2 + W^2),$$
$$T_2 = C_{t2}\left(\frac{1}{\delta^2}\right)\left\{\left(\frac{\partial W}{\partial \xi}\right)^2 + \left(-\cot\alpha\frac{\partial W}{\partial \xi} + \mu \csc\alpha\frac{\partial W}{\partial \eta}\right)^2\right\},$$
$$T_3 = C_{t3}\left(\frac{1}{\delta^2}\right)\left(\Phi_x^2 + \mu^2\Phi_y^2\right),$$
$$T_4 = C_{t4}\left(-\frac{2}{\delta^2}\right)\left(\Phi_x\frac{\partial W}{\partial \xi} - \mu\cot\alpha\Phi_y\frac{\partial W}{\partial \xi} + \mu^2\csc\alpha\Phi_y\frac{\partial W}{\partial \eta}\right)$$

Here in these expressions, $\delta = a/h$ and $\mu = a/b$ are the slenderness and aspect ratios of the FG skew plate and the involved coefficients may be expressed as

$$C_{u1} = \int_{-1/2}^{1/2} \overline{E}\,\mathrm{d}\overline{z}$$
$$C_{u2} = \int_{-1/2}^{1/2} \overline{E}\,\overline{z}^2\mathrm{d}\overline{z}$$
$$C_{u3} = \int_{-1/2}^{1/2} \overline{E}\left\{\overline{z}^{2n+1} - (2n+1)\overline{z}\left(\frac{1}{2}\right)^{2n}\right\}^2 \mathrm{d}\overline{z}$$
$$C_{u4} = \int_{-1/2}^{1/2} \overline{E}\left\{\overline{z}^{2n+2} - (2n+1)\overline{z}^2\left(\frac{1}{2}\right)^{2n}\right\} \mathrm{d}\overline{z}$$
$$C_{u5} = \int_{-1/2}^{1/2} \overline{E}\left[(2n+1)^2\left\{z^{2n} - \left(\frac{1}{2}\right)^{2n}\right\}^2\right] \mathrm{d}\overline{z}$$
$$C_{t1} = \int_{-1/2}^{1/2} \overline{\rho}\,\mathrm{d}\overline{z}$$
$$C_{t2} = \int_{-1/2}^{1/2} \overline{\rho}\,\overline{z}^2\mathrm{d}\overline{z}$$
$$C_{t3} = \int_{1/2}^{1/2} \overline{\rho}\left\{\overline{z}^{2n+1} - (2n+1)\overline{z}\left(\frac{1}{2}\right)^{2n}\right\}^2 \mathrm{d}\overline{z}$$
$$C_{t4} = \int_{-1/2}^{1/2} \overline{\rho}\left\{\overline{z}^{2n+2} - (2n+1)\overline{z}^2\left(\frac{1}{2}\right)^{2n}\right\} \mathrm{d}\overline{z}$$

in which the non-dimensional thickness $-\frac{1}{2} \leqslant \overline{z} \leqslant \frac{1}{2}$, $\overline{E} = \left[\left(1 - \frac{E_m}{E_c}\right)\left(\overline{z} - \frac{1}{2}\right)^k + \frac{E_m}{E_c}\right]$ and $\overline{\rho} = \left[\left(1 - \frac{\rho_m}{\rho_c}\right)\left(\overline{z} - \frac{1}{2}\right)^k + \frac{\rho_m}{\rho_c}\right]$.

3.5 Rayleigh–Ritz approximation

As reported in [33], the amplitudes of vibration are expanded in terms of algebraic polynomial functions by the following series

$$U = \sum_{i=1}^{n} c_i \varphi_i^u, \quad V = \sum_{j=1}^{n} d_j \varphi_j^v, \quad W = \sum_{k=1}^{n} e_k \varphi_k^w,$$
$$\Phi_x = \sum_{l=1}^{n} g_l \varphi_l^1, \quad \Phi_y = \sum_{m=1}^{n} h_m \varphi_m^2$$

where n is the number of polynomials assumed to define the displacements; c_i, d_j, e_k, g_l and h_m are the unknown constant coefficients to be determined and φ_i^u, φ_j^v, φ_k^w, φ_l^1 and φ_m^2 are the admissible functions corresponding to the amplitudes; U, V, W, Φ_x and Φ_y respectively. The admissible functions must satisfy the essential boundary conditions and can be represented as

$$\begin{aligned}
\varphi_i^u(\xi, \eta) &= a_f \psi_i^u(\xi, \eta), \quad i = 0, 1, 2, \ldots, n \\
\varphi_j^v(\xi, \eta) &= a_f \psi_j^v(\xi, \eta), \quad j = 0, 1, 2, \ldots, n \\
\varphi_k^w(\xi, \eta) &= a_f \psi_k^w(\xi, \eta), \quad k = 0, 1, 2, \ldots, n \\
\varphi_l^1(\xi, \eta) &= a_f \psi_l^1(\xi, \eta), \quad l = 0, 1, 2, \ldots, n \\
\varphi_m^2(\xi, \eta) &= a_f \psi_m^2(\xi, \eta), \quad m = 0, 1, 2, \ldots, n
\end{aligned} \tag{3.17}$$

The function $a_f = \xi^p \eta^q (1 - \xi)^r (1 - \eta)^s$ with the exponents p, q, r and s, controls various BCs[1]. The parameter p takes the value 0, 1 or 2 according as the side $\xi = 0$ is free (F), simply supported (S) or clamped (C). Similar interpretations can be given to the parameters q, r and s corresponding to the sides $\eta = 0$, $\xi = 1$ and $\eta = 1$, respectively. Simple algebraic polynomials generated from the Pascal's triangle are considered here to define the admissible functions mentioned in equation (3.17).

Accordingly, Rayleigh Quotient (ω^2) can be obtained by equating $U_{\max}$ and $T_{\max}$. Taking the partial derivatives of the Rayleigh quotient with respect to the constant coefficients involved in the admissible functions, we have

$$\begin{aligned}
\frac{\partial \omega^2}{\partial c_i} &= 0; \quad i = 1, 2, \ldots, n \\
\frac{\partial \omega^2}{\partial d_j} &= 0; \quad j = 1, 2, \ldots, n \\
\frac{\partial \omega^2}{\partial e_k} &= 0; \quad k = 1, 2, \ldots, n \\
\frac{\partial \omega^2}{\partial g_l} &= 0; \quad l = 1, 2, \ldots, n \\
\frac{\partial \omega^2}{\partial h_m} &= 0; \quad m = 1, 2, \ldots, n
\end{aligned}$$

[1] Boundary conditions.

which lead to the generalized eigenvalue problem for the free vibration of the FG skew plate as

$$([\mathbf{K}] - \lambda^2[\mathbf{M}])\{\Delta\} = 0 \tag{3.18}$$

Here, $[\mathbf{K}]$ and $[\mathbf{M}]$ are the stiffness and inertia matrices, respectively, and $\{\Delta\}$ is the column vector of unknown constant coefficients. The eigenvalues $\left(\lambda = \omega a^2\sqrt{\frac{\rho_c h}{D_c}}\right)$ in equation (3.18) are the non-dimensional frequencies for the concerned free vibration problem. In this chapter eigenfrequencies obtained from this eigenvalue problem are clearly addressed in subsequent sections taking different combinations of boundary conditions. A comparison of non-dimensional frequencies with the available literature is also carried out after checking a test of convergence.

3.6 Convergence and comparison studies

In this part, the convergence of first six natural frequencies of isotropic and FG skew plates under different combinations of classical boundary conditions is checked with increase in number of polynomials (n) involved in displacement components. In table 3.1, the natural frequencies of isotropic rhombic plate are evaluated with four different edge supports (CCCC, CCCS, CCCF and SSSS) by assuming power exponent of SDPT (n), slenderness ratio (δ) and skew angle (α) as 1, 100 and 60° respectively. Table 3.2 considers an Al/Al_2O_3 FG parallelogram plate with $n = 1$, $\mu = 0.5$, $k = 1$, $\delta = 100$ and $\alpha = 30°$. It can be viewed that the natural frequencies gradually converge with increase in number of polynomials irrespective of the edge support and the gradation of plate properties, whereas higher mode frequencies do need more polynomials for their corresponding convergence. One can easily notice that present results are in good agreement with the existing literature.

3.7 Numerical results and discussion

In the numerical formulation, one may easily find various parameters which influence the natural frequencies of the Al/Al_2O_3 FG skew plate. Accordingly, the effect of power-law exponent of SDPT (n), aspect ratio (μ), slenderness ratio (δ) and power-law index of gradation (k) on six lowest natural frequencies of FG parallelogram plate are evaluated in this section. In all these tabulations, the skew angles (α) of the skew plate are assumed as 15°, 30°, 45° and 60° with different edge supports viz. SSSS, SCSS, SSSF, SCSC, SCSF and SFSF.

3.7.1 Effect of power-law exponent of SDPT (n)

The natural frequencies of the FG skew plate are evaluated with increase in values of n of SDPT using $\mu = 0.5$, $k = 1$ and $\delta = 10$. This power-law exponent of SDPT assumes its value as 0, 1, 3 and 5 in tables 3.3 and 3.4. It is evident that the aforesaid SDPT acts as classical plate theory while assuming $n = 0$ and accordingly the corresponding results are highest irrespective of the skew angle and edge support

Table 3.1. Convergence and comparison of first six natural frequencies of isotropic rhombic plate with $n = 1$, $\delta = 100$ and $\alpha = 60°$.

Edge support	n	λ_1	λ_2	λ_3	λ_4	λ_5	λ_6
CCCC	6	46.484 5	88.962 7	116.244 1	135.739 5	187.836 9	206.394 1
	10	46.472 4	81.634 5	106.943 8	135.554 2	187.508 4	195.585 8
	15	46.116 5	81.607 4	106.893 2	119.722 8	167.291 7	190.816 8
	19	46.113 3	81.481 5	105.457 8	119.691 2	167.248 8	171.229 0
	21	46.110 3	81.451 3	105.293 0	119.670 3	166.199 3	167.207 5
	Reference [35]	46.150	81.602	105.90	119.72	166.99	
	Reference [36]	46.140	81.691	105.51	119.52	165.80	
	Reference [3]	46.081	81.602	105.11	119.22	165.32	
	Reference [8]	46.166	81.613	105.56	119.98	167.16	
CCCS	6	41.637 6	79.023 0	107.022 8	156.850 5	202.244 6	262.376 2
	10	41.124 1	75.168 6	100.967 9	122.766 7	169.460 0	194.823 2
	15	40.757 2	73.123 9	99.201 7	114.676 6	160.861 8	177.997 6
	19	40.724 2	73.022 1	98.212 1	110.191 9	154.055 9	167.228 2
	21	40.719 8	73.010 1	98.100 2	110.098 3	153.770 3	163.813 7
	Reference [1]	40.969	73.509	98.400	110.74	155.54	
	Reference [8]	40.762	73.132	98.310	110.85	157.51	
CCCF	6	29.416 5	53.790 1	88.769 7	120.821 3	177.824 6	192.070 0
	10	29.350 9	49.425 9	80.131 5	90.677 8	124.219 9	171.138 3
	15	29.056 3	48.450 2	78.354 1	87.424 3	114.684 4	136.422 3
	19	29.001 6	48.081 1	77.190 9	83.990 6	111.603 3	129.230 9
	21	28.997 2	48.080 3	77.130 3	83.937 4	111.586 9	127.277 3
	Reference [8]	29.026	48.139	77.489	84.895	113.70	
SSSS	6	25.759 8	73.730 8	91.714 5	128.322 8	190.638 3	196.559 5
	10	25.756 7	53.554 9	75.029 7	128.304 4	190.581 3	196.495 5
	15	25.303 2	53.547 3	75.012 3	87.358 5	129.985 9	147.998 9
	19	25.301 5	52.992 0	73.170 1	87.346 3	129.957 4	147.309 8
	21	25.301 2	52.611 7	72.623 1	87.335 1	129.930 2	131.720 7
	Reference [37]	24.871	52.704	71.752	83.201		
	Reference [38]	25.069	52.901	72.344	84.780		
	Reference [8]	25.314	52.765	73.006	87.478	130.25	

being assumed. With increase in values of n, the natural frequencies (at each mode) are gradually approaching those evaluated at $n = 0$. In addition, the gradual approach of the frequencies is faster for higher skew angle. This fact was also clearly visible earlier in figure 3.2.

3.7.2 Effect of aspect ratio (μ)

The effect of aspect ratio (μ) on the first six natural frequencies is reported in this section by assuming $k = 1$, $n = 1$ and $\delta = 10$. The increasing values of μ are taken as 0.2, 0.5, 1 and 2 in tables 3.5 and 3.6. It can easily be seen that the non-dimensional

Table 3.2. Convergence and comparison of first six natural frequencies of Al/Al_2O_3 FG parallelogram plate with $n = 1$, $\mu = 0.5$, $k = 1$, $\delta = 100$ and $\alpha = 30°$.

Edge support	n	λ_1	λ_2	λ_3	λ_4	λ_5	λ_6
SFSF	6	31.706	37.298	42.446	169.941	184.794	441.623
	10	30.877	31.683	42.425	58.974	136.671	184.697
	15	28.605	30.834	36.671	58.954	88.443	136.578
	19	28.605	30.062	36.667	54.301	88.415	132.299
	21	26.808	28.562	36.657	44.766	88.363	120.903
	Reference [15]	26.834 9	28.629 7	36.699 3	44.807 9	88.616 4	121.248 6
SCSF	6	30.297	46.736	87.765	187.229	228.304	461.056
	10	28.235	40.561	73.108	118.636	144.073	224.605
	15	26.282	38.361	55.175	108.326	139.030	152.832
	19	25.793	37.667	51.358	95.670	133.422	150.141
	21	25.214	37.631	51.121	74.590	132.295	146.074
	Reference [15]	25.261 7	37.665 2	51.189 9	74.745 3	132.689 6	146.605 8
SCSS	6	43.153	84.080	181.514	193.274	260.582	463.291
	10	41.896	63.272	125.391	147.864	231.457	309.784
	15	39.239	59.320	92.539	143.683	179.222	189.693
	19	38.852	54.393	90.344	138.418	162.864	177.434
	21	38.798	53.497	83.004	130.925	142.382	171.730
	Reference [15]	38.841 6	53.558 9	83.158 9	131.218 8	142.778 7	172.401 7
SCSC	6	44.000	92.324	141.044	191.560	242.461	460.638
	10	43.982	64.731	140.849	146.315	209.955	242.060
	15	40.100	64.700	92.315	146.208	178.745	209.608
	19	40.091	57.398	92.297	138.454	164.798	178.602
	21	40.087	56.339	92.268	126.027	141.337	178.547
	Reference [15]	40.134 2	56.409 7	92.473 2	126.323 8	141.745 6	179.310 2
SSSS	6	40.750	88.871	152.617	186.183	237.424	455.673
	10	40.737	57.397	143.726	152.570	237.193	257.442
	15	37.922	57.375	81.989	143.638	173.560	257.382
	19	37.913	52.552	81.975	137.407	173.421	200.434
	21	37.912	49.850	81.963	116.877	138.835	173.409
	Reference [15]	37.951 5	49.901 2	82.090 9	117.078 6	139.226 6	174.097 7

frequencies follow an ascending pattern with increase in aspect ratio regardless of the edge support and skew angle assumed.

3.7.3 Effect of slenderness ratio (δ)

In this section, the eigenfrequencies have been evaluated for increase in values of slenderness ratio with $\mu = 0.5$, $n = 1$ and $k = 1$. The slenderness ratios take the values as 5, 10, 20 and 100 in tables 3.7 and 3.8. One may easily observe that the greater is the slenderness ratio, the higher are the corresponding natural frequencies of the skew plate.

Table 3.3. Effect of power-law exponent of SDPT on natural frequencies of Al/Al_2O_3 skew plates with $\mu = 0.5$, $k = 1$ and $\delta = 10$.

Edge support	α	n	λ_1	λ_2	λ_3	λ_4	λ_5	λ_6
SSSS	15°	0	130.633	162.746	215.622	239.094	268.812	294.248
		1	111.150	137.588	215.622	230.274	239.094	294.248
		3	111.436	137.973	215.622	230.620	239.094	294.248
		5	111.877	138.570	215.622	231.495	239.094	294.248
	30°	0	37.281	48.789	79.303	111.522	118.504	130.594
		1	35.058	45.625	73.385	102.091	113.101	118.504
		3	35.127	45.733	73.555	102.368	113.580	118.504
		5	35.187	45.825	73.710	102.636	114.042	118.504
	45°	0	19.323	27.112	42.781	61.084	68.168	82.030
		1	18.699	26.055	40.535	57.218	62.540	73.833
		3	18.722	26.096	40.612	57.357	62.734	74.091
		5	18.740	26.128	40.674	57.471	62.896	74.309
	60°	0	13.240	19.887	31.496	44.357	48.094	54.338
		1	12.961	19.309	30.074	41.770	45.193	50.466
		3	12.972	19.332	30.127	41.867	45.303	50.600
		5	12.981	19.351	30.168	41.945	45.389	50.708
SCSS	15°	0	132.362	173.341	216.128	242.151	270.224	283.593
		1	111.749	144.184	212.933	216.128	242.151	283.593
		3	112.043	144.619	213.591	216.128	242.151	283.593
		5	112.493	145.298	214.763	216.128	242.151	283.593
	30°	0	38.148	52.313	80.495	119.014	124.341	134.059
		1	35.732	48.575	72.484	110.073	117.335	119.014
		3	35.806	48.701	72.721	110.481	117.763	119.014
		5	35.870	48.808	72.928	110.880	118.184	119.014
	45°	0	19.957	29.292	44.511	65.982	71.806	82.066
		1	19.273	28.002	41.726	60.755	66.002	73.820
		3	19.298	28.052	41.824	60.936	66.196	74.082
		5	19.318	28.091	41.903	61.087	66.359	74.303
	60°	0	13.779	21.606	33.572	45.336	52.882	54.996
		1	13.465	20.879	31.893	42.580	49.263	51.088
		3	13.477	20.908	31.957	42.682	49.394	51.227
		5	13.487	20.931	32.006	42.765	49.498	51.339
SSSF	15°	0	62.893	118.857	135.613	185.615	192.495	217.806
		1	48.735	99.688	111.600	137.880	185.615	217.806
		3	48.744	99.922	111.935	138.495	185.615	217.806
		5	48.934	100.376	112.451	139.329	185.615	217.806
	30°	0	25.563	36.234	46.160	72.428	103.149	120.807
		1	23.227	34.342	43.260	66.767	103.012	103.149
		3	23.277	34.404	43.353	66.939	103.149	103.439
		5	23.327	34.459	43.433	67.090	103.149	103.851

(*Continued*)

Table 3.3. (*Continued*)

Edge support	α	n	λ_1	λ_2	λ_3	λ_4	λ_5	λ_6
	45°	0	15.188	20.153	28.813	42.328	64.783	70.277
		1	14.537	19.519	27.605	40.242	59.188	63.536
		3	14.558	19.543	27.648	40.319	59.372	63.741
		5	14.575	19.562	27.682	40.380	59.530	63.916
	60°	0	10.905	14.843	22.572	34.137	43.862	48.238
		1	10.652	14.492	21.787	32.581	41.166	44.845
		3	10.661	14.506	21.816	32.641	41.264	44.960
		5	10.669	14.517	21.839	32.688	41.344	45.054

Table 3.4. Effect of power-law exponent of SDPT on natural frequencies of Al/Al_2O_3 skew plates with $\mu = 0.5$, $k = 1$ and $\delta = 10$.

Edge support	α	n	λ_1	λ_2	λ_3	λ_4	λ_5	λ_6
SCSC	15°	0	136.496	179.531	216.712	240.616	287.892	300.390
		1	113.213	146.089	216.712	225.414	240.616	287.892
		3	113.533	146.662	216.712	226.204	240.616	287.892
		5	114.014	147.434	216.712	227.605	240.616	287.892
	30°	0	39.409	55.095	89.397	119.453	120.554	133.127
		1	36.687	50.660	78.649	105.440	114.483	119.453
		3	36.770	50.812	78.954	105.877	115.002	119.453
		5	36.842	50.939	79.223	106.278	115.494	119.453
	45°	0	20.796	31.502	49.088	65.434	71.011	84.993
		1	20.018	29.885	45.388	60.136	64.656	75.753
		3	20.046	29.948	45.516	60.322	64.878	76.039
		5	20.069	29.997	45.618	60.475	65.060	76.281
	60°	0	14.496	23.559	36.825	45.585	53.152	56.615
		1	14.121	22.610	34.609	42.786	49.274	52.216
		3	14.136	22.648	34.692	42.891	49.421	52.369
		5	14.148	22.678	34.756	42.975	49.535	52.492
SCSF	15°	0	56.013	118.576	132.948	183.215	207.097	218.892
		1	44.328	97.435	110.506	141.997	183.215	199.509
		3	44.301	97.586	110.803	142.671	183.215	200.155
		5	44.457	98.033	111.281	143.661	183.215	201.365
	30°	0	24.659	36.901	49.497	71.799	102.807	124.504
		1	22.403	34.910	46.026	64.315	102.807	106.883
		3	22.448	34.975	46.137	64.521	102.807	107.320
		5	22.496	35.032	46.231	64.705	102.807	107.769
	45°	0	15.073	20.880	31.027	45.394	65.203	72.002
		1	14.387	20.206	29.583	42.342	59.597	65.249
		3	14.408	20.232	29.635	42.445	59.783	65.457
		5	14.425	20.253	29.676	42.530	59.941	65.635
	60°	0	10.959	15.651	24.284	37.619	44.548	48.856
		1	10.686	15.242	23.372	35.568	41.821	45.574
		3	10.696	15.258	23.407	35.644	41.920	45.688
		5	10.704	15.271	23.433	35.704	42.000	45.780

SFSF	15°	0	66.526	76.140	117.508	122.536	163.987	187.705
		1	54.508	54.638	100.865	103.561	122.596	149.074
		3	54.615	54.747	101.100	103.861	123.098	149.870
		5	54.864	54.996	101.505	104.329	123.766	150.802
	30°	0	26.202	28.014	35.869	43.432	81.567	101.549
		1	24.543	24.975	33.892	41.199	71.876	96.859
		3	24.595	25.035	33.950	41.279	72.147	97.255
		5	24.641	25.094	34.002	41.346	72.379	97.600
	45°	0	15.258	15.745	21.428	29.165	56.140	63.483
		1	14.590	15.220	20.552	28.132	51.525	58.258
		3	14.611	15.240	20.580	28.173	51.668	58.436
		5	14.628	15.256	20.602	28.205	51.787	58.587
	60°	0	10.526	11.783	16.918	24.934	42.747	42.850
		1	10.290	11.529	16.366	24.106	39.878	40.316
		3	10.298	11.539	16.385	24.139	39.973	40.411
		5	10.305	11.547	16.400	24.165	40.051	40.486

Table 3.5. Effect of aspect ratio on natural frequencies of Al/Al_2O_3 skew plates with $k = 1$, $n = 1$ and $\delta = 10$.

Edge support	α	μ	λ_1	λ_2	λ_3	λ_4	λ_5	λ_6
SSSS	15°	0.2	102.027	104.629	124.190	147.198	207.944	210.713
		0.5	111.150	137.588	215.622	230.274	239.094	294.248
		1.0	163.736	222.474	261.243	328.957	403.079	411.019
		2.0	328.808	401.643	431.244	478.188	574.678	588.496
	30°	0.2	31.306	32.647	38.287	44.962	101.975	107.383
		0.5	35.058	45.625	73.385	102.091	113.101	118.504
		1.0	53.256	80.448	143.328	147.357	152.392	205.514
		2.0	120.164	152.629	237.008	242.882	290.138	320.171
	45°	0.2	16.277	17.386	20.148	23.674	50.654	58.785
		0.5	18.699	26.055	40.535	57.218	62.540	73.833
		1.0	28.675	50.314	80.206	82.981	119.362	119.906
		2.0	67.941	92.194	140.424	177.943	189.637	201.177
	60°	0.2	11.052	12.059	13.998	16.635	32.530	40.650
		0.5	12.961	19.309	30.074	41.770	45.193	50.466
		1.0	20.153	40.389	54.349	64.281	90.656	93.538
		2.0	48.351	70.084	105.810	140.536	151.766	152.629
SCSS	15°	0.2	101.982	105.255	118.794	154.259	208.013	211.059
		0.5	111.749	144.184	212.933	216.128	242.151	283.593
		1.0	178.588	243.988	263.226	334.698	375.754	410.215
		2.0	371.544	437.659	439.013	490.069	576.224	608.203

(Continued)

Table 3.5. (*Continued*)

Edge support	α	μ	λ_1	λ_2	λ_3	λ_4	λ_5	λ_6
	30°	0.2	31.345	32.906	37.280	47.224	76.339	107.459
		0.5	35.732	48.575	72.484	110.073	117.335	119.014
		1.0	60.728	90.633	140.957	153.794	157.752	209.718
		2.0	150.254	176.758	240.207	242.175	294.158	354.215
	45°	0.2	16.315	17.557	20.014	24.858	37.690	58.844
		0.5	19.273	28.002	41.726	60.755	66.002	73.820
		1.0	33.416	56.177	85.530	87.528	120.341	128.425
		2.0	88.317	107.933	147.053	180.055	212.588	224.645
	60°	0.2	11.087	12.195	14.114	17.439	24.367	40.701
		0.5	13.465	20.879	31.893	42.580	49.263	51.088
		1.0	23.727	44.450	59.943	68.052	93.740	102.793
		2.0	63.732	81.516	112.976	154.271	159.645	167.362
SSSF	15°	0.2	65.726	101.741	103.053	115.752	131.392	193.154
		0.5	48.735	99.688	111.600	137.880	185.615	217.806
		1.0	42.072	95.864	143.639	185.840	188.756	262.812
		2.0	78.965	104.781	174.928	241.375	262.665	314.654
	30°	0.2	27.166	31.303	32.862	37.498	52.583	94.014
		0.5	23.227	34.342	43.260	66.767	103.012	103.149
		1.0	21.830	46.147	67.188	110.440	111.015	122.007
		2.0	29.332	62.213	105.381	146.998	171.546	191.346
	45°	0.2	15.276	16.405	17.950	20.426	33.975	48.426
		0.5	14.537	19.519	27.605	40.242	59.188	63.536
		1.0	14.766	29.942	49.023	62.235	78.875	84.885
		2.0	18.588	45.863	79.830	99.957	118.265	129.527
	60°	0.2	10.606	11.305	12.694	14.857	24.776	34.459
		0.5	10.652	14.492	21.787	32.581	41.166	44.845
		1.0	11.477	24.384	40.581	45.841	63.826	69.799
		2.0	14.849	39.267	70.529	71.633	103.266	105.810

Table 3.6. Effect of aspect ratio on natural frequencies of Al/Al_2O_3 skew plates with $k = 1$, $n = 1$ and $\delta = 10$.

Edge support	α	μ	λ_1	λ_2	λ_3	λ_4	λ_5	λ_6
SCSC	15°	0.2	102.012	104.525	120.205	142.156	208.072	211.093
		0.5	113.213	146.089	216.712	225.414	240.616	287.892
		1.0	197.689	252.741	265.868	330.365	394.946	416.053
		2.0	430.318	443.424	472.420	487.023	587.457	663.106
	30°	0.2	31.383	32.882	38.036	44.407	80.452	104.364
		0.5	36.687	50.660	78.649	105.441	114.483	119.453
		1.0	70.912	97.815	152.154	154.927	167.178	208.936
		2.0	182.807	199.998	242.880	264.657	294.745	343.694

	45°	0.2	16.351	17.646	20.433	23.944	40.008	51.555
		0.5	20.018	29.885	45.388	60.136	64.656	75.753
		1.0	39.787	60.991	92.634	94.906	121.118	128.041
		2.0	110.790	124.687	161.274	181.782	211.936	238.053
	60°	0.2	11.121	12.312	14.417	17.178	25.842	33.316
		0.5	14.121	22.610	34.609	42.786	49.274	52.216
		1.0	28.428	47.973	66.485	73.100	98.059	102.266
		2.0	81.104	94.529	122.952	155.602	164.723	182.077
SCSF	15°	0.2	59.946	101.689	102.842	111.607	126.453	191.326
		0.5	44.328	97.435	110.506	141.997	183.215	199.509
		1.0	40.079	92.160	149.298	185.659	194.366	261.891
		2.0	98.234	147.942	192.625	244.573	263.445	320.892
	30°	0.2	26.062	31.353	33.192	36.343	56.503	74.909
		0.5	22.403	34.910	46.026	64.315	102.807	106.883
		1.0	21.583	49.947	74.785	110.944	111.979	118.467
		2.0	43.658	71.262	114.217	148.594	183.544	193.375
	45°	0.2	15.046	16.442	18.076	20.686	31.081	47.676
		0.5	14.387	20.206	29.583	42.342	59.597	65.249
		1.0	15.236	33.915	53.534	64.177	81.385	85.425
		2.0	27.009	50.764	86.966	117.956	119.425	134.484
	60°	0.2	10.568	11.350	12.809	15.465	20.787	37.573
		0.5	10.686	15.242	23.372	35.568	41.821	45.574
		1.0	12.163	28.065	42.051	49.345	69.101	73.501
		2.0	20.971	42.816	75.644	87.236	106.922	112.584
SFSF	15°	0.2	75.767	76.095	101.734	102.664	119.052	125.961
		0.5	54.508	54.638	100.865	103.561	122.596	149.074
		1.0	44.295	44.308	95.419	100.992	137.363	170.835
		2.0	34.276	35.098	85.544	110.817	117.072	162.093
	30°	0.2	28.276	28.437	31.343	32.781	41.498	50.180
		0.5	24.543	24.975	33.892	41.199	71.876	96.859
		1.0	21.586	23.278	42.463	59.935	92.810	105.809
		2.0	17.501	24.086	56.996	69.558	81.224	100.928
	45°	0.2	15.501	15.515	16.716	18.285	25.338	33.273
		0.5	14.590	15.220	20.552	28.132	51.525	58.258
		1.0	13.322	16.773	31.313	47.261	61.339	67.341
		2.0	11.822	21.155	48.144	54.141	54.775	79.353
	60°	0.2	10.599	10.762	11.774	13.365	19.495	26.720
		0.5	10.290	11.529	16.366	24.106	39.878	40.316
		1.0	9.833	14.272	28.327	38.381	48.100	50.239
		2.0	9.295	20.470	39.506	45.941	48.230	76.908

Table 3.7. Effect of slenderness ratio on natural frequencies of Al/Al_2O_3 skew plates with $\mu = 0.5$, $n = 1$ and $k = 1$.

Edge support	α	δ	λ_1	λ_2	λ_3	λ_4	λ_5	λ_6
SSSS	15°	5	82.202	100.411	107.811	119.547	143.670	147.124
		10	111.150	137.588	215.622	230.274	239.094	294.248
		20	128.668	161.573	272.372	378.256	425.212	431.244
		100	138.250	174.004	298.147	421.973	511.677	650.447
	30°	5	30.041	38.157	59.252	60.721	72.535	80.043
		10	35.058	45.625	73.385	102.091	113.101	118.504
		20	37.070	48.679	79.310	112.447	130.549	160.135
		100	37.912	49.850	81.963	116.877	138.835	173.409
	45°	5	16.985	23.048	35.106	44.486	47.409	50.294
		10	18.699	26.055	40.535	57.218	62.540	73.833
		20	19.280	27.088	42.728	61.240	68.137	81.618
		100	19.496	27.456	43.608	62.797	70.393	85.085
	60°	5	12.088	17.521	26.452	35.134	37.941	38.157
		10	12.961	19.309	30.074	41.770	45.193	50.466
		20	13.230	19.879	31.441	44.341	48.092	54.197
		100	13.323	20.076	31.962	45.302	49.189	55.720
SCSS	15°	5	82.445	102.956	108.064	121.076	141.797	143.527
		10	111.749	144.184	212.933	216.128	242.151	283.593
		20	129.755	171.351	263.581	412.964	432.257	438.160
		100	140.058	185.779	296.618	475.658	522.498	640.211
	30°	5	30.492	40.110	57.647	59.507	73.268	81.599
		10	35.732	48.575	72.484	110.073	117.335	119.014
		20	37.882	52.113	79.741	124.575	134.099	158.691
		100	38.798	53.497	83.004	130.925	142.382	171.730
	45°	5	17.435	24.512	35.324	44.719	48.811	53.623
		10	19.273	28.002	41.726	60.755	66.002	73.820
		20	19.904	29.233	44.324	65.947	71.740	81.644
		100	20.139	29.677	45.352	68.005	74.109	85.105
	60°	5	12.505	18.757	27.647	35.638	38.385	40.880
		10	13.465	20.879	31.893	42.580	49.263	51.088
		20	13.763	21.575	33.480	45.298	52.767	54.876
		100	13.867	21.818	34.075	46.321	54.133	56.396
SSSF	15°	5	29.952	55.813	78.577	89.612	92.808	108.903
		10	48.735	99.688	111.600	137.880	185.615	217.806
		20	60.824	118.191	134.928	198.693	287.075	371.230
		100	69.976	126.559	146.342	239.834	338.073	487.498
	30°	5	19.212	29.456	35.950	51.575	52.591	63.298
		10	23.227	34.342	43.260	66.767	103.012	103.149
		20	25.162	36.218	46.250	72.738	122.186	125.533
		100	26.131	36.941	47.479	75.244	131.625	134.288
	45°	5	13.061	17.613	24.298	34.441	37.939	47.063
		10	14.537	19.519	27.605	40.242	59.188	63.536
		20	15.099	20.152	28.835	42.496	64.654	69.999
		100	15.324	20.377	29.314	43.351	66.866	72.889

	60°	5	9.937	13.366	19.574	28.166	31.705	34.423
		10	10.652	14.492	21.787	32.581	41.166	44.845
		20	10.883	14.842	22.570	34.222	43.819	48.071
		100	10.967	14.962	22.863	34.828	44.821	49.391

Table 3.8. Effect of slenderness ratio on natural frequencies of Al/Al_2O_3 skew plates with $\mu = 0.5$, $n = 1$ and $k = 1$.

Edge support	α	δ	λ_1	λ_2	λ_3	λ_4	λ_5	λ_6
SCSC	15°	5	82.867	102.049	108.356	120.308	143.946	145.993
		10	113.213	146.089	216.712	225.414	240.616	287.892
		20	132.698	176.012	287.107	388.260	429.005	433.425
		100	144.440	192.040	330.635	452.113	518.295	666.110
	30°	5	31.061	41.119	59.727	60.948	73.376	79.429
		10	36.687	50.660	78.649	105.440	114.483	119.453
		20	39.073	54.741	87.957	119.627	132.693	163.873
		100	40.087	56.339	92.268	126.027	141.337	178.547
	45°	5	17.979	25.747	37.662	44.910	48.392	51.240
		10	20.018	29.885	45.388	60.136	64.656	75.753
		20	20.726	31.381	48.702	65.284	70.743	84.323
		100	20.988	31.923	50.040	67.327	73.202	88.184
	60°	5	13.020	20.001	29.486	35.762	38.569	40.354
		10	14.121	22.610	34.609	42.786	49.274	52.216
		20	14.469	23.487	36.621	45.541	52.946	56.373
		100	14.590	23.796	37.390	46.575	54.356	58.075
SCSF	15°	5	28.188	52.192	77.643	85.746	91.608	109.446
		10	44.328	97.435	110.506	141.997	183.215	199.509
		20	54.695	117.501	132.710	205.484	288.623	366.431
		100	62.420	126.989	144.451	241.993	386.945	483.403
	30°	5	18.541	29.842	37.517	50.099	51.404	63.958
		10	22.403	34.910	46.026	64.315	102.807	106.883
		20	24.274	36.871	49.629	71.339	124.048	135.434
		100	25.214	37.631	51.121	74.590	132.295	146.074
	45°	5	12.900	18.167	25.768	35.143	37.874	47.382
		10	14.387	20.206	29.583	42.342	59.597	65.249
		20	14.970	20.883	31.020	45.392	65.098	71.711
		100	15.210	21.123	31.574	46.621	67.338	74.517
	60°	5	9.950	13.981	20.807	30.229	31.691	34.973
		10	10.686	15.242	23.372	35.568	41.821	45.574
		20	10.931	15.643	24.275	37.633	44.527	48.745
		100	11.021	15.782	24.607	38.413	45.557	50.012

(*Continued*)

Table 3.8. (*Continued*)

Edge support	α	δ	λ_1	λ_2	λ_3	λ_4	λ_5	λ_6
SFSF	15°	5	31.810	31.816	63.076	66.380	79.340	86.764
		10	54.508	54.638	100.865	103.561	122.596	149.074
		20	68.035	70.938	117.027	123.137	168.462	220.281
		100	74.567	85.475	124.666	131.862	200.557	269.592
	30°	5	20.294	20.625	29.079	34.383	50.622	50.775
		10	24.543	24.975	33.892	41.199	71.876	96.859
		20	26.188	27.284	35.810	43.797	83.092	113.453
		100	26.808	28.562	36.657	44.766	88.363	120.903
	45°	5	13.166	13.679	18.417	24.635	36.380	41.297
		10	14.590	15.220	20.552	28.132	51.525	58.258
		20	15.151	15.726	21.377	29.343	56.259	63.441
		100	15.381	15.903	21.723	29.775	58.324	65.501
	60°	5	9.634	10.710	14.908	21.357	29.877	33.448
		10	10.290	11.529	16.366	24.106	39.878	40.316
		20	10.503	11.777	16.892	25.034	42.601	42.818
		100	10.581	11.861	17.094	25.361	43.724	43.751

3.7.4 Effect of power-law index of gradation (k)

The effect of power-law index of gradation pattern (k) on the natural frequencies can be found in section 3.7.4 with $\mu = 0.5$, $n = 1$ and $\delta = 10$. The values of k are assumed here as 0.1, 0.2, 1 and 2 in tables 3.9 and 3.10. It is worth reporting that the eigenfrequencies of the skew plate follow a descending pattern with increase in values of k irrespective of the boundary support. On the other hand, it is worth mentioning that the said frequencies are gradually decreasing with increase in skew angles of this plate regardless of the edge support. In addition, one can easily assume any FG (metal/ceramic) skew plate as well as any combination of classical BCs to address the corresponding results just by following the present formulation.

3.8 Concluding remarks

In view of the above discussion, the first six natural frequencies of FG skew plate have been evaluated in terms of various physical parameters such as aspect ratio, slenderness ratio, power-law indices of SDPT and gradation pattern along with the skew angles (which mainly decides the configuration of skew plate). One can easily draw the following conclusions based on this chapter.

1. The eigenfrequencies of FG skew plate follow an ascending pattern with increase in both aspect ratio and slenderness ratio regardless of the edge condition assumed. On the other hand, these frequencies gradually decrease with increase in power-law index (k) of gradation pattern.
2. Increasing values of power-law exponent of SDPT (n) have a peculiar effect on the said frequencies. It is evident that the aforesaid SDPT acts as classical

Table 3.9. Effect of power-law index of gradation on natural frequencies of Al/Al_2O_3 skew plates with $\mu = 0.5$, $n = 1$ and $\delta = 10$.

Edge support	α	k	λ_1	λ_2	λ_3	λ_4	λ_5	λ_6
SSSS	15°	0.1	129.317	160.130	252.698	267.904	280.206	344.844
		0.2	125.781	155.784	246.915	260.572	273.793	336.951
		1.0	111.150	137.588	215.622	230.274	239.094	294.248
		2.0	103.690	128.094	194.828	214.933	216.036	265.871
	30°	0.1	40.696	52.978	85.224	118.621	131.587	138.880
		0.2	39.528	51.466	82.799	115.281	127.987	135.702
		1.0	35.058	45.625	73.385	102.091	113.101	118.504
		2.0	33.168	43.085	69.256	96.031	105.524	107.076
	45°	0.1	21.691	30.232	47.047	66.440	72.659	85.808
		0.2	21.058	29.355	45.692	64.544	70.609	83.405
		1.0	18.699	26.055	40.535	57.218	62.540	73.833
		2.0	17.769	24.718	38.389	54.033	58.861	69.351
	60°	0.1	15.030	22.398	34.899	48.495	52.471	58.606
		0.2	14.589	21.744	33.890	47.106	50.970	56.938
		1.0	12.961	19.309	30.074	41.770	45.193	50.466
		2.0	12.339	18.352	28.513	39.481	42.704	47.623
SCSS	15°	0.1	130.022	167.865	248.095	253.291	283.789	332.357
		0.2	126.472	163.345	241.531	247.494	277.294	324.750
		1.0	111.749	144.184	212.933	216.128	242.151	283.593
		2.0	104.213	133.945	195.286	196.932	218.799	256.244
	30°	0.1	41.482	56.413	84.233	128.013	136.452	139.478
		0.2	40.293	54.809	81.869	124.480	132.683	136.286
		1.0	35.732	48.575	72.484	110.073	117.335	119.014
		2.0	33.790	45.819	68.118	102.929	107.537	109.807
	45°	0.1	22.358	32.495	48.444	70.579	76.672	85.796
		0.2	21.707	31.556	47.057	68.584	74.504	83.395
		1.0	19.273	28.002	41.726	60.755	66.002	73.820
		2.0	18.308	26.542	39.438	57.208	62.166	69.324
	60°	0.1	15.615	24.222	37.016	49.438	57.207	59.331
		0.2	15.158	23.517	35.949	48.025	55.578	57.643
		1.0	13.465	20.879	31.893	42.580	49.263	51.088
		2.0	12.815	19.828	30.204	40.229	46.494	48.199
SSSF	15°	0.1	56.740	116.091	129.955	160.848	217.531	255.258
		0.2	55.211	112.979	126.469	156.708	212.553	249.416
		1.0	48.735	99.688	111.600	137.880	185.615	217.806
		2.0	45.362	92.501	103.658	126.969	167.715	196.802
	30°	0.1	26.975	39.862	50.226	77.553	119.870	120.886
		0.2	26.208	38.714	48.789	75.355	116.603	118.119
		1.0	23.227	34.342	43.260	66.767	103.012	103.149
		2.0	21.918	32.512	40.886	62.945	93.202	96.108

(*Continued*)

Table 3.9. (*Continued*)

Edge support	α	k	λ_1	λ_2	λ_3	λ_4	λ_5	λ_6
	45°	0.1	16.866	22.643	32.031	46.707	68.770	73.830
		0.2	16.376	21.983	31.103	45.362	66.833	71.755
		1.0	14.537	19.519	27.605	40.242	59.188	63.536
		2.0	13.797	18.546	26.190	38.109	55.676	59.742
	60°	0.1	12.353	16.807	25.275	37.812	47.797	52.076
		0.2	11.991	16.315	24.539	36.721	46.431	50.592
		1.0	10.652	14.492	21.787	32.581	41.166	44.845
		2.0	10.139	13.789	20.694	30.872	38.892	42.334

Table 3.10. Effect of power-law index of gradation on natural frequencies of Al/Al_2O_3 skew plates with $\mu = 0.5$, $n = 1$ and $\delta = 10$.

Edge support	α	k	λ_1	λ_2	λ_3	λ_4	λ_5	λ_6
SCSC	15°	0.1	131.752	170.164	253.976	262.811	281.989	337.394
		0.2	128.170	165.631	248.163	255.965	275.535	329.673
		1.0	113.213	146.089	216.712	225.414	240.616	287.892
		2.0	105.461	135.318	195.813	207.672	217.411	260.128
	30°	0.1	42.596	58.850	91.442	122.638	133.219	139.993
		0.2	41.379	57.186	88.904	119.261	129.590	136.789
		1.0	36.687	50.660	78.649	105.440	114.483	119.453
		2.0	34.662	47.705	73.684	98.551	106.679	107.933
	45°	0.1	23.224	34.688	52.713	69.862	75.131	88.058
		0.2	22.549	33.690	51.215	67.889	73.020	85.604
		1.0	20.018	29.885	45.388	60.136	64.656	75.753
		2.0	19.004	28.288	42.809	56.613	60.775	71.056
	60°	0.1	16.378	26.235	40.179	49.679	57.230	60.651
		0.2	15.899	25.475	39.028	48.259	55.606	58.931
		1.0	14.121	22.610	34.609	42.786	49.274	52.216
		2.0	13.432	21.445	32.720	40.419	46.454	49.218
SCSF	15°	0.1	51.594	113.492	128.639	165.828	214.719	232.673
		0.2	50.194	110.465	125.163	161.666	209.805	226.648
		1.0	44.328	97.435	110.506	141.997	183.215	199.509
		2.0	41.328	90.332	102.820	129.968	165.547	183.774
	30°	0.1	26.018	40.522	53.450	74.744	120.485	124.372
		0.2	25.278	39.357	51.927	72.650	117.727	120.981
		1.0	22.403	34.910	46.026	64.315	102.807	106.883
		2.0	21.141	33.040	43.446	60.434	92.893	99.633
	45°	0.1	16.693	23.440	34.331	49.166	69.245	75.818
		0.2	16.209	22.758	33.339	47.763	67.295	73.686
		1.0	14.387	20.206	29.583	42.342	59.597	65.249
		2.0	13.651	19.195	28.041	39.989	56.066	61.361

	60°	0.1	12.393	17.678	27.116	41.288	48.556	52.920
		0.2	12.030	17.161	26.329	40.102	47.168	51.410
		1.0	10.686	15.242	23.372	35.568	41.821	45.574
		2.0	10.170	14.497	22.185	33.653	39.515	43.036
SFSF	15°	0.1	63.496	63.654	117.372	120.580	142.942	173.995
		0.2	61.804	61.963	114.174	117.336	139.215	169.563
		1.0	54.508	54.638	100.865	103.561	122.596	149.074
		2.0	50.561	50.695	94.006	96.201	113.238	136.948
	30°	0.1	28.497	29.011	39.338	47.825	83.580	112.672
		0.2	27.682	28.189	38.205	46.451	81.266	109.576
		1.0	24.543	24.975	33.892	41.199	71.876	96.859
		2.0	23.190	23.546	32.096	38.985	67.358	90.550
	45°	0.1	16.928	17.657	23.844	32.640	59.850	67.685
		0.2	16.436	17.143	23.151	31.693	58.155	65.775
		1.0	14.590	15.220	20.552	28.132	51.525	58.258
		2.0	13.849	14.456	19.515	26.701	48.569	54.825
	60°	0.1	11.932	13.370	18.984	27.966	46.302	46.808
		0.2	11.582	12.978	18.430	27.153	44.979	45.468
		1.0	10.290	11.529	16.366	24.106	39.878	40.316
		2.0	9.796	10.974	15.556	22.890	37.677	38.101

plate theory while assuming $n = 0$ and accordingly the corresponding results are highest irrespective of the skew angle and edge support being assumed. With increase in values of n, the natural frequencies at each mode are gradually approaching those evaluated at $n = 0$. In addition, the gradual approach of the frequencies is faster for higher skew angle. This fact was also clearly visible earlier in figure 3.2.

3. In contrast, one may easily visualize that these frequencies are gradually decreasing with increase in skew angles of this plate regardless of the edge support. In addition, one can easily assume any FG (metal/ceramic) skew plate as well as any combination of classical BCs to address the corresponding results just by looking into the present formulation.

References

[1] Nair P S and Durvasula S 1973 Vibration of skew plates *J. Sound Vib.* **26** 1–19

[2] Sathyamoorthy M 1977 Shear and rotary inertia effects on large amplitude vibration of skew plates *J. Sound Vib.* 155–63

[3] Mizusawa T, Katija T and Naruoka M 1979 Vibration of skew plates by using B-spline functions *J. Sound Vib.* **62** 301–8

[4] Mizusawa T, Kajita T and Naruoka M 1980 Analysis of skew plate problems with various constraints *J. Sound Vib.* **73** 575–84

[5] Mizusawa T and Katija T 1987 Vibration of skew plates resting on point supports *J. Sound Vib.* **115** 243–51

[6] Liew K M and Lam K Y 1990 Application of two-dimensional orthogonal plate function to flexural vibration of skew plates *J. Sound Vib.* **139** 241–52
[7] Liew K M, Xiang Y, Kitipornchai S and Wang C M 1993 Vibration of thick skew plates based on Mindlin shear deformation plate theory *J. Sound Vib.* **168** 39–69
[8] Singh B and Chakraverty S 1994 Flexural vibration of skew plates using boundary characteristic orthogonal polynomials in two variables *J. Sound Vib.* **173** 157–78
[9] Wang S 1997 Free vibration analysis of skew fibre-reinforced composite laminates based on first-order shear deformation plate theory *Comput. Struct.* **63** 525–38
[10] Žitňan P 1999 Vibration analysis of rectangular and skew plates by the Rayleigh-Ritz method *J. Sound Vib.* **221** 342–9
[11] Krishna Reddy A R and Palaninathan R 1999 Free vibration of skew laminates *Comput. Struct.* **70** 415–23
[12] Wang C M, Ang K K and Yang L 2000 Free vibration of skew sandwich plates with laminated facings *J. Sound Vib.* **235** 317–40
[13] Mizusawa T and Kondo Y 2001 Application of the spline element method to analyze vibration of skew Mindlin plates with varying thickness in one direction *J. Sound Vib.* **241** 485–501
[14] Upadhyay A K and Shukla K K 2013 Geometrically nonlinear static and dynamic analysis of functionally graded skew plates *Commun. Nonlinear Sci. Numer. Simul.* **18** 2252–79
[15] Chakraverty S and Pradhan K K 2017 Flexural vibration of functionally graded thin skew plates resting on elastic foundations *Int. J. Dyn. Control* **6** 97–121
[16] Memar Ardestani M, Zhang L W and Liew K M 2017 Isogeometric analysis of the effect of CNT orientation on the static and vibration behaviors of CNT-reinforced skew composite plates *Comput. Methods Appl. Mech. Eng.* **317** 341–79
[17] Kiani Y, Dimiti R and Tornabene F 2018 Free vibration of FG-CNT reinforced composite skew cylindrical shells using the Chebyshev–Ritz formulation *Composites* **B147** 169–77
[18] Vinyas M, Nischith G, Loja M A R, Ebrahimi F and Duc N D 2019 Numerical analysis of the vibration response of skew magneto-electro-elastic plates based on the higher-order shear deformation theory *Compos. Struct.* **214** 132–42
[19] Zhang L W 2017 On the study of the effect of in-plane forces on the frequency parameters of CNT-reinforced composite skew plates *Compos. Struct.* **160** 824–37
[20] Xue Y, Jin G, Ding H and Chen M 2018 Free vibration analysis of in-plane functionally graded plates using a refined plate theory and isogeometric approach *Compos. Struct.* **192** 193–205
[21] Reissner E 1975 On transverse bending of plates including the effects of transverse shear deformation *Int. J. Solids Struct.* **25** 495–502
[22] Reddy J N 1984 A refined nonlinear theory of plates with transverse shear deformation *Int. J. Solids Struct.* **20** 881–96
[23] Reddy J N and Phan N D 1985 Stability and vibration of isotropic, orthotropic and laminated plates according to a higher-order shear deformation theory *J. Sound Vib.* **98** 157–70
[24] Reddy J N 1984 A simple higher-order theory for laminated composite plates *J. Appl. Mech.* **51** 745–52
[25] Nayak A K, Moy S S J and Shenoi R A 2002 Free vibration analysis of composite sandwich plates based on Reddy's higher-order theory *Composites* **B33** 505–19
[26] Hosseini-Hashemi S and Arsanjani M 2005 Exact characteristic equations for some classical boundary conditions of vibrating moderately thick rectangular plates *Int. J. Solids Struct.* **42** 819–53

[27] Shimpi R P and Patel H G 2006 Free vibrations of plates using two variable refined plate theory *J. Sound Vib.* **296** 979–99

[28] Shimpi R P, Patel H G and Arya H 2007 New first-order shear deformation plate theories *J. Appl. Mech.* **74** 523–33

[29] Aydogdu M 2009 A new shear deformation theory for laminated composite plates *Compos. Struct.* **89** 94–101

[30] Xiang S, Wang K, Ai Y, Sha Y and Shi H 2009 Analysis of isotropic, sandwich and laminated plates by a meshless method and various shear deformation theories *Compos. Struct.* **91** 31–7

[31] Hosseini-Hashemi S, Fadaee M and Taher H R D 2011 Exact solutions for free flexural vibration of Lévy-type rectangular thick plates via third order shear deformation plate theory *Appl. Math. Model.* **35** 708–27

[32] Thai C H, Ferreira A J M, Bordas S P A, Rabczuk T and Nguyen-Xuan H 2014 Isogeometric analysis of laminated composite and sandwich plates using a new inverse trigonometric shear deformation theory *Eur. J. Mech.* **A43** 89–108

[33] Pradhan K K and Chakraverty S 2015 Transverse vibration of isotropic thick rectangular plates based on new inverse trigonometric shear deformation theories *Int. J. Mech. Sci.* **95** 211–31

[34] Pradhan K K and Chakraverty S 2015 Generalized power-law exponent based shear deformation theory for free vibration of functionally graded beams *Appl. Math. Comput.* **268** 1240–58

[35] Durvasula S 1968 Natural frequencies and modes of skew membranes *J. Acoust. Soc. Am.* **44** 1636–46

[36] Durvasula S and Nair P S 1974 Application of partition method to vibration problems of plates *J. Sound Vib.* **37** 429–45

[37] Durvasula S 1969 Natural frequencies and modes of clamped skew plates *Astronaut. Am. Inst. Aeronaut. J.* **7** 1164–7

[38] Liew K M and Lam K Y 1991 Vibration analysis of multispan plates having orthogonal straight edges *J. Sound Vib.* **147** 255–64

[39] Chakraverty S and Pradhan K K 2017 Flexural vibration of functionally graded thin skew plates resting on elastic foundations *Int. J. Dynam. Control* **6** 97–121

Chapter 4

Advanced mechanical modeling of functionally graded carbon nanotubes-reinforced composite materials and structures

Francesco Tornabene, Rossana Dimitri and Salvatore Brischetto

Parts of this chapter have been adapted from [21] with permission from Elsevier.

This chapter presents the theoretical framework to investigate the mechanical behavior of shell structures and members made of functionally graded materials (FGMs), as largely applied in many engineering fields because of their outstanding mechanical properties (in terms of stiffness, strength and stability). Differently from classical composite materials with distinct substrates, a functional variation of the material properties is ensured by FGMs, such as thermal resistance, thermal conductivity and coefficient of thermal expansion. A continuous change in the coefficient of thermal expansion throughout an interfacial joint between two layers limits the internal thermal stress concentrations at high temperatures together with possible interfacial collapses. A metallic and ceramic phase is employed, usually, in FGMs. Common structures for aerospace applications (e.g. re-entry capsules or turbine blades) have the external surfaces in ceramic material in order to contrast the high temperature values and the internal part in metallic material with structural requirements. This means that FGMs feature a gradual variation of the mechanical properties between the ceramic phase and the metallic one. The application of different FGM laws throughout the thickness direction can change significantly the mechanical response of a structural member, while keeping fixed its geometrical assumptions. This effect is here explored for different geometries, boundary conditions, loading conditions, and mechanical input properties. Several Higher-order Shear Deformation Theories (HSDTs) are also applied to investigate the problems at issue related to the vibration response, the dynamic stability and the critical speed evaluation of thick or moderately thick structures with high values for the in-plane and transverse anisotropy, whose equations are solved numerically by means of the Generalized Differential Quadrature (GDQ) method, due to its accuracy, stability and reliability features.

doi:10.1088/978-0-7503-3483-9ch4

© IOP Publishing Ltd 2021

4.1 Introduction

In recent decades, composite materials have been increasingly applied in many branches of industries, such as aerospace, automobile, nuclear, ships and underwater vehicles, etc, due to their high performances over conventional materials, i.e. high stiffness- and strength-to-weight ratios, high fatigue and corrosion resistances, and easy production. Among them, FGMs are successfully applied for many practical applications, in a context where the scientific community reveals an increased attention to nanoscience and nanotechnology as a novel form of manufacturing materials, while controlling and tailoring their structural units at nanoscale levels. FGMs were introduced for the first time in 1984 to obtain materials with very high thermal resistance, as required in aircraft, space vehicles and other engineering structures. These materials are made of a mixture of two different phases, commonly metal and ceramic, such that their equivalent mechanical properties can change uniformly by simply tuning the volume fraction of constituents in a gradual sense.

Since FGMs have some extraordinary properties, namely a high temperature and corrosion resistance, as well as an improved residual stress distribution, they are widely studied in many fields of the applied sciences and applied as structural components in military, medical, or aerospace industries, as well as in power plants or vessels. Thus, due to their special privileges compared to traditional materials, most industries adopt such materials in lieu of the ordinary ones [1–5]. Due to the large use of these materials in many shell and plate structures, various linear and nonlinear mechanical analyses have been carried out in the literature for these structural members, including their vibration, buckling and post-buckling response, even for rotating systems, such as, high-speed centrifugal separators, rotor systems, drilling pipes, aircraft jet engines and turbines. From a physical perspective, the problem of rotating structures is extremely interesting due to the presence of some peculiar forces caused by the Coriolis and centripetal accelerations that can affect their dynamic response, as highlighted in the book by Li *et al* [6], and in [7, 8]. The natural frequencies of a structural member, indeed, do not depend only on its geometry, mechanical properties and boundary conditions, but also on the applied angular velocity. It is well-known that the bifurcation phenomenon appears consequently in the frequency properties and a critical state characterized by a stiffness reduction of the structural system can be established [9–19], as also illustrated in [20, 21].

The mechanical response of a structural element can be also highly affected by a proper choice of materials and constituents. Nevertheless, the shape of these structures plays a key role as well. Many designers, indeed, aim at finding the optimal structural geometry depending on the desired structural requirements. To this end, it is possible to modify the buckling behavior, the bending response, or the vibration characteristics with a simple variation of the shape and mechanical properties in structural members. Among different structural optimization processes, the assignment of a variable thickness together with its mechanical properties allows one to modify the stiffness of the structure by redistributing the materials within the reference domain without any increase of the structural weight. In other words, variable thicknesses can enhance the mechanical behavior of a structure by increasing the stiffness in those parts of the structure with high stress concentrations and reducing the amount of material where it is not strictly required. Among the FG class of composite materials, there exhist two different approaches to evaluate the smooth variation of the mechanical properties throughout the thickness of a structure, namely, the theory of mixtures [22–26] and the Mori–Tanaka scheme [27–30], where the second approach seems to enable a more accurate representation of a material composition. As far as the FG shell structures are concerned, the vibration effects caused by different phenomena can reduce their strength and safety. Therefore, an accurate frequency and mode shape determination represents a key aspect to account for an adequate technical design of these structures.

In addition, the increased development of nanotechnology has induced a large adoption of nano-scale materials, e.g., carbon nanotubes (CNTs), in many engineering systems and devices, as discovered experimentally by Iijima [31] in 1991 during the production of fullerene by arc discharge evaporation. Improving the properties of materials through a CNT-based reinforcement phase is one of the most relevant topics in modern materials science (metamaterials, heterogeneous materials, architectured materials etc) [32–34]. The noteworthy mechanical, electrical, and thermal properties of FG-CNTs make them very attractive for many current and future engineering applications, more than conventional carbon fiber reinforced composites [35–37]. The shell structures reinforced with CNTs, indeed, are lightweight and resistant to corrosion and feature a high specific strength, with an overall simplification in their manufacturing, transportation, and installation processes.

Modern technology has also allowed a combined use of FGMs and CNTs in various structural elements, which is reflected in the introduction of a great number of advanced theoretical and numerical methods to solve even more complicated problems, with a special focus on mesh-free methods [38–49]. These last numerical approaches are not based on a finite mesh but use a regular or scattered distribution of points to evaluate the approximate solution of a differential problem. Among them, the generalized differential quadrature (GDQ) approach represents an efficient numerical tool to solve complex differential equations, and yields accurate results even with a limited number of collocation points dicretizing the domain [50–54]. This limits many common computational difficulties, especially in a context where nonlinear algorithms and high computational effort are required among complex engineering applications. More specifically, this procedure approximates derivatives of a function at a certain point through a liner sum of all the function values in the definition domain, whereby a key aspect is related to the evaluation of the weighting coefficients. In recent decades, several researchers have widely applied the GDQ method in different engineering fields, including the static, fracturing, and dynamic problems for beams, plates and shells, as well as the analysis of waves and the study of viscous fluid flows [55–69].

The above-mentioned GDQ-based numerical tool is proposed in the present chapter together with some higher-order formulations to handle doubly-curved shell structural components made of FG-CNT materials, including rotating structures. In a context where lower-order models could be inadequate to capture the effective mechanical behavior of shell structures with possible complicated geometries and materials, more refined structural models are required, as discussed in the present chapter within a unified formulation, including a wide range of Higher-order Shear Deformation Theories (HSDTs). The key aspect of the proposed formulation, is that the expansion order of the three dimensional displacements is taken as a free parameter and it can include both the thickness functions and the Zig-Zag (ZZ) effect (known also as Murakami's function). A large systematic investigation aims at showing the effect of the constituent volume fractions and material profiles on the overall structural response. A generalization of the power-law distribution presented in literature [70–85] is proposed. A five parameter-based power-law distribution is assumed for the ceramic volume fraction, whereas different material profiles are selected including both symmetric and asymmetric volume fraction profiles. It is worth mentioning that classical volume fraction profiles can be obtained as special subcases of the general distribution laws presented in this chapter.

4.2 Theoretical formulation

4.2.1 Higher-order theory of shell structures

In the present section we describe the basics of an Equivalent Single Layer (ESL) model, which is applied to study the vibration and dynamic behavior of doubly-curved shell components, including possible rotating structural members. The degrees of freedom of the structures, along with their mechanical properties, are evaluated at the middle surface, which represents the reference domain for the governing equations of each analysed problem. As highlighted in [86–88], a key aspect is related to the description of a curved surface. A generally-curved shell member with a constant thickness h can be considered as a three-dimensional structure in the global reference system $O\,x_1\,x_2\,x_3$, with unit vectors $\mathbf{e}_1$, $\mathbf{e}_2$, $\mathbf{e}_3$. At the same time, the shell middle surface is described in the local reference system $O'\,\alpha_1\,\alpha_2\,\zeta$, in which α_1, α_2 are the orthogonal and principal coordinates of the surface itself, with $\alpha_1 \in [\alpha_1^0, \alpha_1^1]$ and $\alpha_2 \in [\alpha_2^0, \alpha_2^1]$ in the finite domain. The orthogonal direction to the middle surface is defined by $\zeta \equiv \alpha_3$, with $\zeta \in [-h/2, h/2]$. For a structural member made of l layers, the overall shell thickness is computed as

$$h = \sum_{k=1}^{l} h_k \tag{4.1}$$

where h_k stands for the thickness of the k-th layer. This quantity can be defined as

$$h_k = \zeta_{k+1} - \zeta_k \tag{4.2}$$

where ζ_k and ζ_{k+1} refer to the lower and upper thickness coordinates of the k-th layer, respectively. The position vector $\mathbf{r}(\alpha_1, \alpha_2)$ is introduced to identify an arbitrary point P' upon the shell middle surface, namely,

$$\mathbf{r}(\alpha_1, \alpha_2) = r_1(\alpha_1, \alpha_2)\mathbf{e}_1 + r_2(\alpha_1, \alpha_2)\mathbf{e}_2 + r_3(\alpha_1, \alpha_2)\mathbf{e}_3 \tag{4.3}$$

whose functions $r_i(\alpha_1, \alpha_2)$, for $i = 1, 2, 3$, define the shape of the middle surface. On the other hand, the position vector $\mathbf{R}(\alpha_1, \alpha_2, \zeta)$ for an arbitrary point P within the three-dimensional element is defined as

$$\mathbf{R} = \mathbf{r} + \zeta \mathbf{n} \tag{4.4}$$

in which the outward unit normal vector $\mathbf{n}(\alpha_1, \alpha_2)$ reads as follows

$$\mathbf{n} = \frac{\mathbf{r}_{,1} \wedge \mathbf{r}_{,2}}{|\mathbf{r}_{,1} \wedge \mathbf{r}_{,2}|} \tag{4.5}$$

and $\mathbf{r}_{,i} = \partial \mathbf{r}/\partial \alpha_i$, for $i = 1, 2$, and '$\wedge$' stands for the cross product. Different shell geometries can be obtained by simply changing the mathematical description of the position vector $\mathbf{r}(\alpha_1, \alpha_2)$. A doubly-curved surface is characterized by two different radii of curvature $R_1(\alpha_1, \alpha_2)$, $R_2(\alpha_1, \alpha_2)$, whose value can generally vary at each point of the domain. By definition, these quantities can be evaluated as

$$R_1 = -\frac{\mathbf{r}_{,1} \cdot \mathbf{r}_{,1}}{\mathbf{r}_{,11} \cdot \mathbf{n}}, \quad R_2 = -\frac{\mathbf{r}_{,2} \cdot \mathbf{r}_{,2}}{\mathbf{r}_{,22} \cdot \mathbf{n}} \tag{4.6}$$

where $\mathbf{r}_{,ii} = \partial^2 \mathbf{r}/\partial \alpha_i^2$ for $i = 1, 2$, and '$\cdot$' stands for the dot product. We also introduce the Lamè parameters $A_1(\alpha_1, \alpha_2)$, $A_2(\alpha_1, \alpha_2)$ to define the first fundamental form of the reference surface, as follows

$$A_1 = \sqrt{\mathbf{r}_{,1} \cdot \mathbf{r}_{,1}}, \quad A_2 = \sqrt{\mathbf{r}_{,2} \cdot \mathbf{r}_{,2}} \tag{4.7}$$

whereas, for an arbitrary point P throughout the thickness, we introduce the following geometric quantities $H_1(\alpha_1, \alpha_2, \zeta)$, $H_2(\alpha_1, \alpha_2, \zeta)$, i.e.

$$H_1 = 1 + \frac{\zeta}{R_1}, \quad H_2 = 1 + \frac{\zeta}{R_2} \tag{4.8}$$

More details about the shell geometries can be found in the reference books [89, 90]. As far as the kinematic model is concerned, the displacement components $U_1(\alpha_1, \alpha_2, \zeta, t)$, $U_2(\alpha_1, \alpha_2, \zeta, t)$, $U_3(\alpha_1, \alpha_2, \zeta, t)$ of an arbitrary point are defined by means of the kinematic quantities of the corresponding point on the reference surface $u_1^{(\tau)}(\alpha_1, \alpha_2, t)$, $u_2^{(\tau)}(\alpha_1, \alpha_2, t)$, $u_3^{(\tau)}(\alpha_1, \alpha_2, t)$, for each order of the kinematic expansion τ, and selected thickness functions $F_\tau^{\alpha_i}(\zeta)$, for $i = 1, 2, 3$, i.e.

$$\begin{aligned} U_1(\alpha_1, \alpha_2, \zeta, t) &= \sum_{\tau=0}^{N+1} F_\tau^{\alpha_1}(\zeta) u_1^{(\tau)}(\alpha_1, \alpha_2, t) \\ U_2(\alpha_1, \alpha_2, \zeta, t) &= \sum_{\tau=0}^{N+1} F_\tau^{\alpha_2}(\zeta) u_2^{(\tau)}(\alpha_1, \alpha_2, t) \\ U_3(\alpha_1, \alpha_2, \zeta, t) &= \sum_{\tau=0}^{N+1} F_\tau^{\alpha_3}(\zeta) u_3^{(\tau)}(\alpha_1, \alpha_2, t) \end{aligned} \tag{4.9}$$

where N stands for the maximum order of kinematic expansion. Based on relations (4.9), different models can be obtained by varying the value of N and by selecting different thickness functions, as follows

$$F_{\tau}^{\alpha_i} = \zeta^{\tau} \quad \text{for } \tau = 0, 1, \ldots, N$$
$$F_{N+1}^{\alpha_i} = (-1)^k \left(\frac{2}{\zeta_{k+1} - \zeta_k} \zeta - \frac{\zeta_{k+1} + \zeta_k}{\zeta_{k+1} - \zeta_k} \right) \tag{4.10}$$

for $i = 1, 2, 3$. The order of expansion $N + 1$ is associated with Murakami's function, as required to include the so-called zig-zag effect in laminated structures [91–93]. The structural models follow the nomenclature EDZN and EDN, depending on whether the Murakami's function is embedded or not in the theory. Note that the First-order Shear Deformation Theory (FSDT) represents a clear subcase in the proposed compact formulation, whereas its corresponding zig-zag theory is labeled as FSDTZ [93]. Following a compact notation, the degrees of freedom of the problem are defined as

$$\mathbf{u}^{(\tau)}(\alpha_1, \alpha_2, t) = \left[u_1^{(\tau)} \ u_2^{(\tau)} \ u_3^{(\tau)} \right]^T \tag{4.11}$$

for $\tau = 0, 1, 2, \ldots, N, N + 1$, where t represents the time variable. At the same time, it is necessary to evaluate the first-order and second-order time derivatives of the vector in (4.11), for dynamic problems, such that the velocity vector is defined as

$$\dot{\mathbf{u}}^{(\tau)} = \left[\dot{u}_1^{(\tau)} \ \dot{u}_2^{(\tau)} \ \dot{u}_3^{(\tau)} \right]^T \tag{4.12}$$

and the acceleration is determined as

$$\ddot{\mathbf{u}}^{(\tau)} = \left[\ddot{u}_1^{(\tau)} \ \ddot{u}_2^{(\tau)} \ \ddot{u}_3^{(\tau)} \right]^T \tag{4.13}$$

Accordingly, the generalized strain vector $\boldsymbol{\varepsilon}^{(\tau)\alpha_i}(\alpha_1, \alpha_2, t)$ is defined for each order of kinematic expansion,

$$\boldsymbol{\varepsilon}^{(\tau)\alpha_i} = \left[\varepsilon_1^{(\tau)\alpha_i} \ \varepsilon_2^{(\tau)\alpha_i} \ \gamma_1^{(\tau)\alpha_i} \ \gamma_2^{(\tau)\alpha_i} \ \gamma_{13}^{(\tau)\alpha_i} \ \gamma_{23}^{(\tau)\alpha_i} \ \omega_{13}^{(\tau)\alpha_i} \ \omega_{23}^{(\tau)\alpha_i} \ \varepsilon_3^{(\tau)\alpha_i} \right]^T \tag{4.14}$$

for $i = 1, 2, 3$. This is, in turn, related to the generalized displacement vector for each order of kinematic expansion τ as

$$\boldsymbol{\varepsilon}^{(\tau)\alpha_i} = \mathbf{D}_{\Omega}^{\alpha_i} \mathbf{u}^{(\tau)} \tag{4.15}$$

where the kinematic operators $\mathbf{D}_{\Omega}^{\alpha_1}$, $\mathbf{D}_{\Omega}^{\alpha_2}$, $\mathbf{D}_{\Omega}^{\alpha_3}$ are expressed as follows

$$\mathbf{D}_{\Omega}^{\alpha_1} = \begin{bmatrix} \frac{1}{A_1}\frac{\partial}{\partial \alpha_1} & \frac{1}{A_1 A_2}\frac{\partial A_2}{\partial \alpha_1} & -\frac{1}{A_1 A_2}\frac{\partial A_1}{\partial \alpha_2} & \frac{1}{A_2}\frac{\partial}{\partial \alpha_2} & -\frac{1}{R_1} & 0 & 1 & 0 & 0 \\ 0 & 0 & 0 & 0 & 0 & 0 & 0 & 0 & 0 \\ 0 & 0 & 0 & 0 & 0 & 0 & 0 & 0 & 0 \end{bmatrix}^T \tag{4.16}$$

$$\mathbf{D}_{\Omega}^{\alpha_2} = \begin{bmatrix} 0 & 0 & 0 & 0 & 0 & 0 & 0 & 0 & 0 \\ \frac{1}{A_1 A_2}\frac{\partial A_1}{\partial \alpha_2} & \frac{1}{A_2}\frac{\partial}{\partial \alpha_2} & \frac{1}{A_1}\frac{\partial}{\partial \alpha_1} & -\frac{1}{A_1 A_2}\frac{\partial A_2}{\partial \alpha_1} & 0 & -\frac{1}{R_2} & 0 & 1 & 0 \\ 0 & 0 & 0 & 0 & 0 & 0 & 0 & 0 & 0 \end{bmatrix}^T \tag{4.17}$$

$$\mathbf{D}_{\Omega}^{\alpha_3} = \begin{bmatrix} 0 & 0 & 0 & 0 & 0 & 0 & 0 & 0 & 0 \\ 0 & 0 & 0 & 0 & 0 & 0 & 0 & 0 & 0 \\ \frac{1}{R_1} & \frac{1}{R_2} & 0 & 0 & \frac{1}{A_1}\frac{\partial}{\partial \alpha_1} & \frac{1}{A_2}\frac{\partial}{\partial \alpha_2} & 0 & 0 & 1 \end{bmatrix}^T \tag{4.18}$$

On the other hand, the generalized stress resultants can be collected in the corresponding vector $\mathbf{S}^{(\tau)\alpha_i}(\alpha_1, \alpha_2, t)$ defined in the following for each order of kinematic expansion

$$\mathbf{S}^{(\tau)\alpha_i} = \left[N_1^{(\tau)\alpha_i} \quad N_2^{(\tau)\alpha_i} \quad N_{12}^{(\tau)\alpha_i} \quad N_{21}^{(\tau)\alpha_i} \quad T_1^{(\tau)\alpha_i} \quad T_2^{(\tau)\alpha_i} \quad P_1^{(\tau)\alpha_i} \quad P_2^{(\tau)\alpha_i} \quad S_3^{(\tau)\alpha_i} \right]^T \tag{4.19}$$

for $i = 1, 2, 3$. Thus, we introduce vectors $\mathbf{S}_{\alpha_1}^{(\tau)}$, $\mathbf{S}_{\alpha_2}^{(\tau)}$, to define the stress resultants along the external edges, as involved in the boundary conditions

$$\mathbf{S}_{\alpha_1}^{(\tau)} = \left[N_1^{(\tau)\alpha_1} \quad N_{12}^{(\tau)\alpha_1} \quad T_1^{(\tau)\alpha_3} \right]^T \tag{4.20}$$

$$\mathbf{S}_{\alpha_2}^{(\tau)} = \left[N_{21}^{(\tau)\alpha_1} \quad N_2^{(\tau)\alpha_1} \quad T_2^{(\tau)\alpha_3} \right]^T \tag{4.21}$$

for $\tau = 0, 1, 2, \ldots, N, N+1$. In particular, $\mathbf{S}_{\alpha_1}^{(\tau)}$ refers to edges with a constant value of α_1, whereas $\mathbf{S}_{\alpha_2}^{(\tau)}$ refers to the external edges with a constant α_2. By definition, the quantities in (4.19) can be related to the generalized strain components as

$$\mathbf{S}^{(\tau)\alpha_i} = \sum_{\eta=0}^{N+1} \sum_{j=1}^{3} \mathbf{A}^{(\tau\eta)\alpha_i\alpha_j} \boldsymbol{\varepsilon}^{(\eta)\alpha_i} \tag{4.22}$$

where $\mathbf{A}^{(\tau\eta)\alpha_i\alpha_j}$ is the following constitutive matrix

$$\mathbf{A}^{(\tau\eta)\alpha_i\alpha_j} = \begin{bmatrix}
A_{11(20)}^{(\tau\eta)\alpha_i\alpha_j} & A_{12(11)}^{(\tau\eta)\alpha_i\alpha_j} & A_{16(20)}^{(\tau\eta)\alpha_i\alpha_j} & A_{16(11)}^{(\tau\eta)\alpha_i\alpha_j} & 0 & 0 & 0 & 0 & A_{13(10)}^{(\tau\tilde{\eta})\alpha_i\alpha_j} \\
A_{12(11)}^{(\tau\eta)\alpha_i\alpha_j} & A_{22(02)}^{(\tau\eta)\alpha_i\alpha_j} & A_{26(11)}^{(\tau\eta)\alpha_i\alpha_j} & A_{26(02)}^{(\tau\eta)\alpha_i\alpha_j} & 0 & 0 & 0 & 0 & A_{23(01)}^{(\tau\tilde{\eta})\alpha_i\alpha_j} \\
A_{16(20)}^{(\tau\eta)\alpha_i\alpha_j} & A_{26(11)}^{(\tau\eta)\alpha_i\alpha_j} & A_{66(20)}^{(\tau\eta)\alpha_i\alpha_j} & A_{66(11)}^{(\tau\eta)\alpha_i\alpha_j} & 0 & 0 & 0 & 0 & A_{36(10)}^{(\tau\tilde{\eta})\alpha_i\alpha_j} \\
A_{16(11)}^{(\tau\eta)\alpha_i\alpha_j} & A_{26(02)}^{(\tau\eta)\alpha_i\alpha_j} & A_{66(11)}^{(\tau\eta)\alpha_i\alpha_j} & A_{66(02)}^{(\tau\eta)\alpha_i\alpha_j} & 0 & 0 & 0 & 0 & A_{36(01)}^{(\tau\tilde{\eta})\alpha_i\alpha_j} \\
0 & 0 & 0 & 0 & A_{44(20)}^{(\tau\eta)\alpha_i\alpha_j} & A_{45(11)}^{(\tau\eta)\alpha_i\alpha_j} & A_{44(10)}^{(\tau\tilde{\eta})\alpha_i\alpha_j} & A_{45(10)}^{(\tau\tilde{\eta})\alpha_i\alpha_j} & 0 \\
0 & 0 & 0 & 0 & A_{45(11)}^{(\tau\eta)\alpha_i\alpha_j} & A_{55(02)}^{(\tau\eta)\alpha_i\alpha_j} & A_{45(01)}^{(\tau\tilde{\eta})\alpha_i\alpha_j} & A_{55(01)}^{(\tau\tilde{\eta})\alpha_i\alpha_j} & 0 \\
0 & 0 & 0 & 0 & A_{44(10)}^{(\tilde{\tau}\eta)\alpha_i\alpha_j} & A_{45(01)}^{(\tilde{\tau}\eta)\alpha_i\alpha_j} & A_{44(00)}^{(\tilde{\tau}\tilde{\eta})\alpha_i\alpha_j} & A_{45(00)}^{(\tilde{\tau}\tilde{\eta})\alpha_i\alpha_j} & 0 \\
0 & 0 & 0 & 0 & A_{45(10)}^{(\tilde{\tau}\eta)\alpha_i\alpha_j} & A_{55(01)}^{(\tilde{\tau}\eta)\alpha_i\alpha_j} & A_{45(00)}^{(\tilde{\tau}\tilde{\eta})\alpha_i\alpha_j} & A_{55(00)}^{(\tilde{\tau}\tilde{\eta})\alpha_i\alpha_j} & 0 \\
A_{13(10)}^{(\tilde{\tau}\eta)\alpha_i\alpha_j} & A_{23(01)}^{(\tilde{\tau}\eta)\alpha_i\alpha_j} & A_{36(10)}^{(\tilde{\tau}\eta)\alpha_i\alpha_j} & A_{36(01)}^{(\tilde{\tau}\eta)\alpha_i\alpha_j} & 0 & 0 & 0 & 0 & A_{33(00)}^{(\tilde{\tau}\tilde{\eta})\alpha_i\alpha_j}
\end{bmatrix} \tag{4.23}$$

whose terms take the following form

$$A_{nm\ (pq)}^{(\tau\eta)\alpha_i\alpha_j} = \sum_{k=1}^{l} \int_{\zeta_k}^{\zeta_{k+1}} \bar{B}_{nm}^{(k)} F_\eta^{\alpha_j} F_\tau^{\alpha_i} \frac{H_1 H_2}{H_1^p H_2^q} d\zeta \tag{4.24}$$

$$A_{nm\ (pq)}^{(\tilde{\tau}\eta)\alpha_i\alpha_j} = \sum_{k=1}^{l} \int_{\zeta_k}^{\zeta_{k+1}} \bar{B}_{nm}^{(k)} F_\eta^{\alpha_j} \frac{\partial F_\tau^{\alpha_i}}{\partial \zeta} \frac{H_1 H_2}{H_1^p H_2^q} d\zeta \tag{4.25}$$

$$A_{nm\ (pq)}^{(\tau\tilde{\eta})\alpha_i\alpha_j} = \sum_{k=1}^{l} \int_{\zeta_k}^{\zeta_{k+1}} \bar{B}_{nm}^{(k)} \frac{\partial F_\eta^{\alpha_j}}{\partial \zeta} F_\tau^{\alpha_i} \frac{H_1 H_2}{H_1^p H_2^q} d\zeta \tag{4.26}$$

$$A_{nm\ (pq)}^{(\tilde{\tau}\tilde{\eta})\alpha_i\alpha_j} = \sum_{k=1}^{l} \int_{\zeta_k}^{\zeta_{k+1}} \bar{B}_{nm}^{(k)} \frac{\partial F_\eta^{\alpha_j}}{\partial \zeta} \frac{\partial F_\tau^{\alpha_i}}{\partial \zeta} \frac{H_1 H_2}{H_1^p H_2^q} d\zeta \tag{4.27}$$

for $\tau, \eta = 0, 1, 2, \ldots, N, N+1$, $n, m = 1, 2, 3, 4, 5, 6$, $p, q = 0, 1, 2$ and $\alpha_i, \alpha_j = \alpha_1, \alpha_2, \alpha_3$, with

$$\begin{aligned} \bar{B}_{nm}^{(k)} &= \bar{E}_{nm}^{(k)} \quad \text{for} \quad n, m = 1, 2, 3, 6 \\ \bar{B}_{nm}^{(k)} &= \frac{\bar{E}_{nm}^{(k)}}{\chi} \quad \text{for} \quad n, m = 4, 5 \end{aligned} \tag{4.28}$$

where χ is the shear correction factor. In what follows we assume $\chi = 1.2$ for structural models with a kinematic expansion $N \leqslant 2$, including the corresponding zig-zag theories. On the other hand, the elastic constants for the k th layer, are denoted by $\bar{E}_{nm}^{(k)}$, and can be collected in the constitutive matrix $\bar{\mathbf{E}}^{(k)}$ for an isotropic material

$$\bar{\mathbf{E}}^{(k)} = \begin{bmatrix} \bar{E}_{11}^{(k)} & \bar{E}_{12}^{(k)} & 0 & 0 & 0 & \bar{E}_{13}^{(k)} \\ \bar{E}_{12}^{(k)} & \bar{E}_{22}^{(k)} & 0 & 0 & 0 & \bar{E}_{23}^{(k)} \\ 0 & 0 & \bar{E}_{66}^{(k)} & 0 & 0 & 0 \\ 0 & 0 & 0 & \bar{E}_{44}^{(k)} & 0 & 0 \\ 0 & 0 & 0 & 0 & \bar{E}_{55}^{(k)} & 0 \\ \bar{E}_{13}^{(k)} & \bar{E}_{23}^{(k)} & 0 & 0 & 0 & \bar{E}_{33}^{(k)} \end{bmatrix} \tag{4.29}$$

which relates the three-dimensional stress components $\boldsymbol{\sigma}^{(k)}$ and the strain components $\boldsymbol{\varepsilon}^{(k)}$ according to the classical Hooke's laws for a linear elastic medium

$$\boldsymbol{\sigma}^{(k)} = \bar{\mathbf{E}}^{(k)} \boldsymbol{\varepsilon}^{(k)} \tag{4.30}$$

where

$$\boldsymbol{\sigma}^{(k)} = \left[\sigma_1^{(k)} \ \ \sigma_2^{(k)} \ \ \tau_{12}^{(k)} \ \ \tau_{1n}^{(k)} \ \ \tau_{2n}^{(k)} \ \ \sigma_n^{(k)} \right]^T \tag{4.31}$$

$$\boldsymbol{\varepsilon}^{(k)} = \begin{bmatrix} \varepsilon_1^{(k)} & \varepsilon_2^{(k)} & \gamma_{12}^{(k)} & \gamma_{1n}^{(k)} & \gamma_{2n}^{(k)} & \varepsilon_n^{(k)} \end{bmatrix}^T \tag{4.32}$$

The nomenclature in (4.29) is used to have the possibility of introducing the plane stress hypothesis in the models, as required for instance by the FSDT and FSDTZ. In this case, it is

$$\bar{E}_{nm}^{(k)} = Q_{nm}^{(k)} \tag{4.33}$$

in which $Q_{nm}^{(k)}$ are the plane stress-reduced elastic coefficients. In the other cases, the non-reduced elastic coefficients are defined by means of $C_{nm}^{(k)}$ as follows

$$\bar{E}_{nm}^{(k)} = C_{nm}^{(k)} \tag{4.34}$$

Both the coefficients $Q_{nm}^{(k)}$ and $C_{nm}^{(k)}$ can be defined in terms of the Young's modulus $E^{(k)}$ and the Poisson's ratio $\nu^{(k)}$, when the k th layer is made of a FGM, which features an isotropic behavior, with varying mechanical properties throughout the thickness ζ. More specifically, the reduced elastic coefficients $Q_{nm}^{(k)}$ are defined as

$$\begin{aligned}
Q_{11}^{(k)} &= Q_{22}^{(k)} = \frac{E^{(k)}}{1-(\nu^{(k)})^2} \\
Q_{12}^{(k)} &= \frac{\nu^{(k)}E^{(k)}}{1-(\nu^{(k)})^2} \\
Q_{44}^{(k)} &= Q_{55}^{(k)} = Q_{66}^{(k)} = \frac{E^{(k)}}{2(1+\nu^{(k)})} \\
Q_{13}^{(k)} &= Q_{23}^{(k)} = Q_{33}^{(k)} = 0
\end{aligned} \tag{4.35}$$

whereas the non-reduced coefficients $C_{nm}^{(k)}$ are defined as

$$\begin{aligned}
C_{11}^{(k)} &= C_{22}^{(k)} = C_{33}^{(k)} = \frac{E^{(k)}(1-\nu^{(k)})}{(1+\nu^{(k)})(1-2\nu^{(k)})} \\
C_{12}^{(k)} &= C_{13}^{(k)} = C_{23}^{(k)} = \frac{E^{(k)}\nu^{(k)}}{(1+\nu^{(k)})(1-2\nu^{(k)})} \\
C_{44}^{(k)} &= C_{55}^{(k)} = C_{66}^{(k)} = \frac{E^{(k)}}{2(1+\nu^{(k)})}
\end{aligned} \tag{4.36}$$

4.2.2 Mechanical properties for FGMs

For FGMs, the Young's modulus $E^{(k)} = E^{(k)}(\zeta)$ and the Poisson's ratio $\nu^{(k)} = \nu^{(k)}(\zeta)$ vary with continuity along the ζ-direction, and satisfy the following rule of mixtures

$$\begin{aligned}
E^{(k)}(\zeta) &= (E_C^{(k)} - E_M^{(k)})V_C^{(k)}(\zeta) + E_M^{(k)} \\
\nu^{(k)}(\zeta) &= (\nu_C^{(k)} - \nu_M^{(k)})V_C^{(k)}(\zeta) + \nu_M^{(k)}
\end{aligned} \tag{4.37}$$

where $E_C^{(k)}$, $E_M^{(k)}$ are the Young's moduli of the ceramic and metal constituents, respectively, whereas $\nu_C^{(k)}$, $\nu_M^{(k)}$ are the corresponding Poisson's ratios. Similarly, the mass density is derived as

$$\rho^{(k)}(\zeta) = \left(\rho_C^{(k)} - \rho_M^{(k)}\right) V_C^{(k)}(\zeta) + \rho_M^{(k)} \tag{4.38}$$

with $\rho_C^{(k)}$, $\rho_M^{(k)}$ being the densities of the two constituents. In both expressions (4.37) and (4.38), $V_C^{(k)}(\zeta)$ refers to the volume fraction distribution of the ceramic phase, which fulfills the following relation

$$V_C^{(k)}(\zeta) + V_M^{(k)}(\zeta) = 1 \tag{4.39}$$

where $V_M^{(k)}(\zeta)$ refers to the volume fraction distribution of the metal phase. Each variation is here related to the k th layer, since the present approach allows one to consider different properties and constituents for different plies of the structure. In literature there exhist several expressions to characterize the gradual variation of the mechanical properties of an arbitrary FGM ply in the thickness direction. Among them, we propose a five-parameter power-law distribution of the volume fraction $V_C(\zeta)$ of the type

$$V_C^{(k)}(\zeta) = \left(d^{(k)} - a^{(k)}\left(\frac{\zeta_{k+1} - \zeta}{h_k}\right) + b^{(k)}\left(\frac{\zeta_{k+1} - \zeta}{h_k}\right)^{c^{(k)}}\right)^{p^{(k)}} \quad \text{for} \quad E_{\text{top}} > E_{\text{bottom}} \tag{4.40}$$

$$V_C^{(k)}(\zeta) = \left(d^{(k)} - a^{(k)}\left(\frac{\zeta - \zeta_k}{h_k}\right) + b^{(k)}\left(\frac{\zeta - \zeta_k}{h_k}\right)^{c^{(k)}}\right)^{p^{(k)}} \quad \text{for} \quad E_{\text{bottom}} > E_{\text{top}} \tag{4.41}$$

where E_{top}, E_{bottom} stand for the Young's modulus of the constituents at the top and bottom surfaces of the kth FGM layer, respectively; $p^{(k)}$ is the fraction volume index ($0 \leqslant p^{(k)} \leqslant \infty$); and the parameters $a^{(k)}$, $b^{(k)}$, $c^{(k)}$, $d^{(k)}$ control the variation profile along the thickness h_k of the kth lamina. A compact notation is then introduced to define the above-mentioned laws, and its dependence on the selected parameters $a^{(k)}$, $b^{(k)}$, $c^{(k)}$, $d^{(k)}$, $p^{(k)}$ of the distribution, namely

$$FGM^{C}_{M(a^{(k)}/b^{(k)}/c^{(k)}/d^{(k)}/p^{(k)})} \tag{4.42}$$

$$FGM^{M}_{C(a^{(k)}/b^{(k)}/c^{(k)}/d^{(k)}/p^{(k)})} \tag{4.43}$$

4.2.3 Mechanical properties for CNTs

If the kth lamina is reinforced by CNTs, the analytical model accounts for different possible agglomerations of the reinforcing particles, as visible in figure 4.1. The characteristic shape of CNTs, depicted in figure 4.1(a), is one of the main reasons

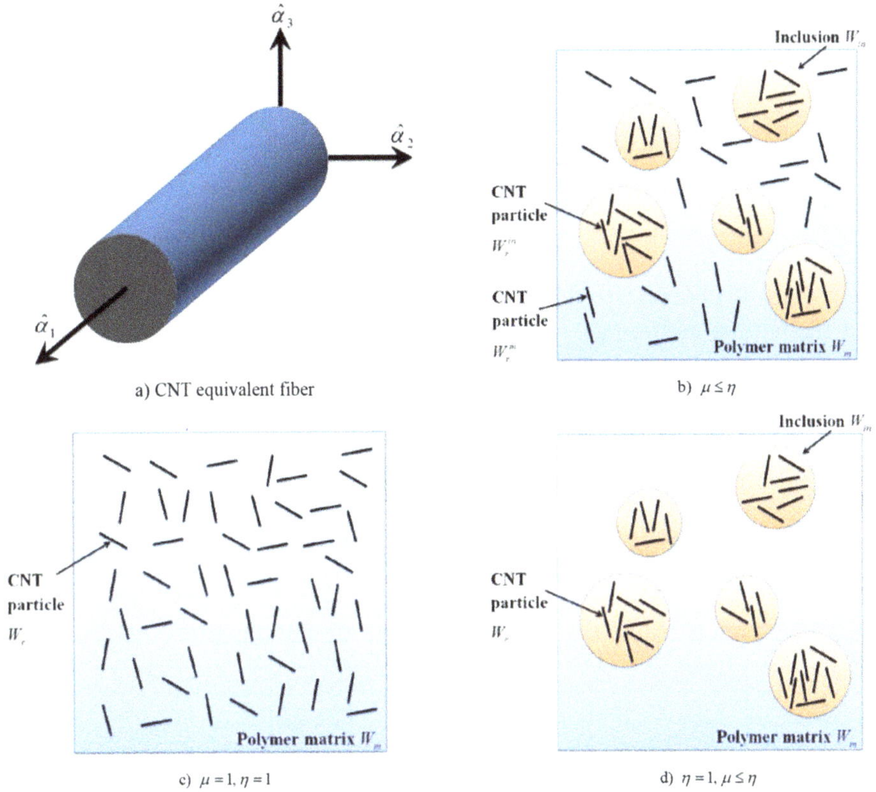

Figure 4.1. CNT reinforced composite layer: (a) effective CNT fiber and local co-ordinate system representation; (b) shell element with inclusion model of CNT agglomeration; (c) example of CNT agglomeration with $\mu = 1, \eta = 1$; example of CNT agglomeration with $\eta = 1, \mu \leqslant \eta$.

which cause the agglomeration of these particles when scattered in a polymer matrix [94].

Thus, in some regions of the CNT-reinforced ply, the concentration of CNTs is higher than the average volume fraction in the ply itself. According to the micromechanical model presented by Shi *et al* [95], these areas can be assumed as spherical shaped inclusions with different elastic properties from the surrounding matrix. Therefore, CNT fibers can be contained both in the matrix and in the spherical inclusions in a different manner (see figure 4.1(b–d)), as will be explained accurately in the following. The total volume W_r of CNTs embedded in the lamina can be defined as

$$W_r = W_r^{in} + W_r^{m} \tag{4.44}$$

where W_r^{in} and W_r^{m} indicate the volume of CNTs scattered in the inclusions and matrix, respectively. Thus, the overall volume of the arbitrary lamina W is given by the following sum

$$W = W_r + W_m \tag{4.45}$$

where W_m refers to the matrix volume. Based on the theory of mixtures, it is convenient to define the mass fraction of both CNTs w_r and matrix w_m as

$$V_r = \frac{W_r}{W}$$
$$V_m = \frac{W_m}{W} \tag{4.46}$$

whose quantities in (4.46) satisfy the relation $V_r + V_m = 1$. The volume fraction of CNTs, V_r, can be characterized by a continuous variation from the bottom to the top side of the lamina assuming

$$V_r(\zeta) = V_r^* V_C^*(\zeta) \tag{4.47}$$

where V_r^* is the CNT volume fraction value, that depends on the mass fraction w_r and on the density of both CNT, ρ_r, and matrix, ρ_m,

$$V_r^* = \left(\frac{\rho_r}{w_r \rho_m} - \frac{\rho_r}{\rho_m} + 1 \right)^{-1} \tag{4.48}$$

whereas $V_C^*(\zeta)$ refers to the CNT distribution along the thickness defined as

$$V_C^*(\zeta) = \left(V_C(\zeta) + f_\rho \right) \tag{4.49}$$

in which f_ρ is an arbitrary constant. A five-parameter exponential law can be taken to define the function $V_C(\zeta)$, as already detailed in equations (4.40) and (4.41). In particular, equation (4.40) has to be employed if the top surface of the layer shows the CNT mechanical properties. In contrast, equation (4.41) supposes that those properties are located on the bottom side of the layer. The reader could find further distributions of the volume fraction along the thickness in [60, 96–98]. The gradual distribution of CNTs along the thickness of a lamina will be denoted univocally, as follows

$$FG - CNT^{\text{top}}_{\text{bottom}(5P)(a^{(k)}/b^{(k)}/c^{(k)}/d^{(k)}/p^{(k)})} \tag{4.50}$$

while referring to the top and bottom external sides of the selected lamina. This means that, for a top surface made of CNTs, and a bottom side made of pure polymer matrix (PM), it will be set top = CNT and bottom = PM. Moreover, the notation ($5P$) specifies that a five-parameter distribution characterized by the parameters ($a^{(k)}/b^{(k)}/c^{(k)}/d^{(k)}/p^{(k)}$) is used to define the profile of the CNT volume fraction along the thickness. All these aspects are summarized graphically in figure 4.2, where some examples of volume fraction distribution of CNTs along the thickness are depicted as a function of the exponent $p^{(k)}$, for an arbitrary layer, using the notation introduced in (4.50) to define the expressions (4.40) and (4.41). By the way, two further parameters are introduced to characterize the CNTs agglomeration, namely,

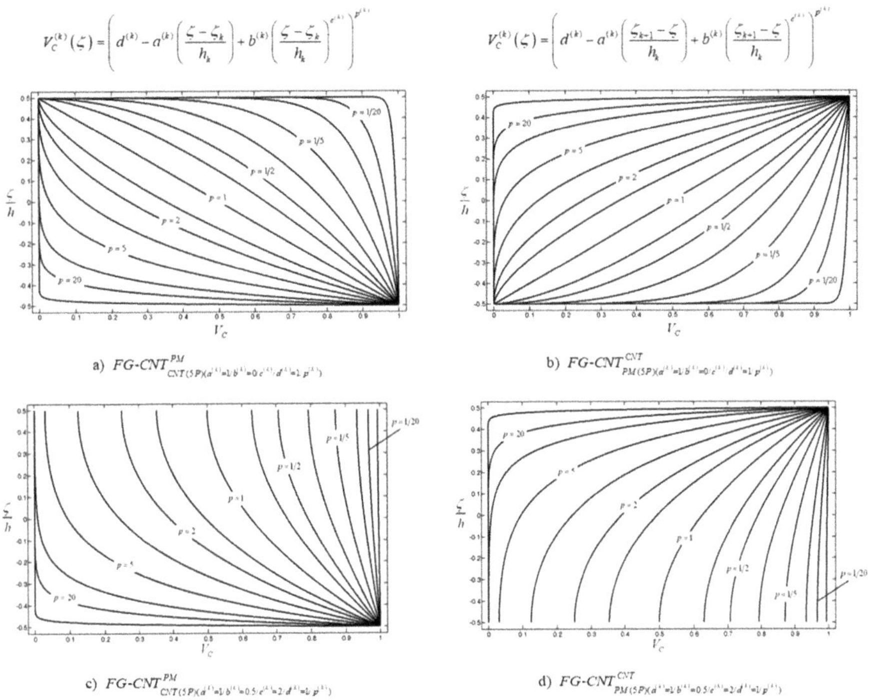

Figure 4.2. Examples of the volume fraction distribution of CNTs, when $p^{(k)}$ varies along the thickness of the kth layer.

$$\mu = \frac{W_{in}}{W}$$
$$\eta = \frac{W_r^{in}}{W_r} \tag{4.51}$$

where W_{in} is the spherical inclusion volume, μ is the volume fraction of the inclusions, and η is the ratio between CNTs inside the inclusions and the total volume of CNTs. When $\mu = 1$, the CNT particles are uniformly scattered within the matrix, where higher agglomerations will get lower values of μ. Note also that the inclusions correspond to the whole domain, for $\mu - 1$ (i.c. $W_{in} - W$), as visiblc in figurc 4.1(c). All thc particlcs arc diffcrcntly allocatcd in the spherical regions, when $\eta = 1$ (figure 4.1(d)). Indeed, a fixed value of $\eta = 1$ involves the equivalent relation $W_r^{in} = W_r$, according to the second relation (4.51). The heterogeneity of a CNT spatial distribution becomes even more pronounced for an increased value of η, while keeping $\eta > \mu$. Considering that the total volume fraction of CNTs V_r can be obtained as the sum of the CNT volume fraction inside the inclusion V_r^{in} with the CNT volume fraction scattered in the matrix V_r^m and by a proper combination of equations (4.46) and (4.51), we get the following relations between the agglomeration parameters

$$\frac{W_r^{in}}{W_{in}} = \frac{V_r\eta}{\mu}$$
$$\frac{W_r^m}{W - W_{in}} = \frac{V_r(1 - \eta)}{1 - \mu} \quad \text{for} \quad \eta \geqslant \mu \tag{4.52}$$

Before defining the overall mechanical properties of the CNT reinforced composite layer, the mechanical properties of both polymer matrix and hybrid inclusions must be evaluated separately. Then, we assume a continuum model in order to define the elastic properties of an equivalent CNT fiber. According to the approach presented by Odegard *et al* [99], the CNT fiber can be reasonably assumed transversely isotropic, with the plane of isotropy orthogonal to the axial direction of the nanotube as synthetically depicted in figure 4.1(a). Assuming the hypothesis of linear elasticity, the constitutive equations of a transversely isotropic material can be written in the following matrix form for the equivalent fiber

$$\begin{bmatrix} \hat{\sigma}_{11} \\ \hat{\sigma}_{22} \\ \hat{\tau}_{12} \\ \hat{\tau}_{13} \\ \hat{\tau}_{23} \\ \hat{\sigma}_{33} \end{bmatrix} = \begin{bmatrix} C_{11}^r & C_{12}^r & 0 & 0 & 0 & C_{12}^r \\ C_{12}^r & C_{22}^r & 0 & 0 & 0 & C_{23}^r \\ 0 & 0 & C_{44}^r & 0 & 0 & 0 \\ 0 & 0 & 0 & C_{44}^r & 0 & 0 \\ 0 & 0 & 0 & 0 & \dfrac{C_{22}^r - C_{23}^r}{2} & 0 \\ C_{12}^r & C_{23}^r & 0 & 0 & 0 & C_{22}^r \end{bmatrix} \begin{bmatrix} \hat{\varepsilon}_{11} \\ \hat{\varepsilon}_{22} \\ \hat{\gamma}_{12} \\ \hat{\gamma}_{13} \\ \hat{\gamma}_{23} \\ \hat{\varepsilon}_{33} \end{bmatrix} \tag{4.53}$$

by means of its elastic constants C_{ij}^r. The same relations (4.53) could be also redefined by means of the Hill's elastic moduli [100–102]

$$\begin{bmatrix} \hat{\sigma}_{11} \\ \hat{\sigma}_{22} \\ \hat{\tau}_{12} \\ \hat{\tau}_{13} \\ \hat{\tau}_{23} \\ \hat{\sigma}_{33} \end{bmatrix} = \begin{bmatrix} n_r & l_r & 0 & 0 & 0 & l_r \\ l_r & k_r + m_r & 0 & 0 & 0 & k_r - m_r \\ 0 & 0 & p_r & 0 & 0 & 0 \\ 0 & 0 & 0 & p_r & 0 & 0 \\ 0 & 0 & 0 & 0 & m_r & 0 \\ l_r & k_r - m_r & 0 & 0 & 0 & k_r + m_r \end{bmatrix} \begin{bmatrix} \hat{\varepsilon}_{11} \\ \hat{\varepsilon}_{22} \\ \hat{\gamma}_{12} \\ \hat{\gamma}_{13} \\ \hat{\gamma}_{23} \\ \hat{\varepsilon}_{33} \end{bmatrix} \tag{4.54}$$

where k_r, l_r, m_r, n_r, p_r are the five independent elastic parameters. By comparing (4.53) with (4.54), it is possible to relate easily the Hill's elastic moduli to the elastic constants C_{ij}^r

$$\begin{aligned} k_r &= \frac{C_{22}^r + C_{23}^r}{2} \\ l_r &= C_{12}^r \\ m_r &= \frac{C_{22}^r - C_{23}^r}{2} \\ n_r &= C_{11}^r \\ p_r &= C_{44}^r \end{aligned} \tag{4.55}$$

The material properties of the fiber could be conveniently described in terms of engineering constants [100–103], as follows

$$\begin{aligned}
E_1^r &= n_r - \frac{l_r^2}{k_r} = C_{11}^r - \frac{2(C_{12}^r)^2}{C_{22}^r + C_{23}^r} \\
E_2^r &= E_3^r = \frac{4m_r(k_r n_r - l_r^2)}{k_r n_r - l_r^2 + m_r n_r} \\
&= \frac{(C_{22}^r - C_{23}^r)(C_{11}^r C_{22}^r + C_{11}^r C_{23}^r - 2(C_{12}^r)^2)}{C_{11}^r C_{22}^r - (C_{12}^r)^2} \\
\nu_{12}^r &= \nu_{13}^r = \frac{l_r}{2k_r} = \frac{C_{12}^r}{C_{22}^r + C_{23}^r} \\
\nu_{23}^r &= \frac{n_r(k_r - m_r) - l_r^2}{n_r(k_r + m_r) - l_r^2} = \frac{C_{11}^r C_{23}^r - (C_{12}^r)^2}{C_{11}^r C_{22}^r - (C_{12}^r)^2} \\
G_{12}^r &= G_{13}^r = p_r = C_{44}^r
\end{aligned} \tag{4.56}$$

The shear modulus in the plane of isotropy G_{23}^r can be evaluated as a function of the other quantities in (4.56)

$$G_{23}^r = \frac{E_2^r}{2(1 + \nu_{23}^r)} \tag{4.57}$$

whereas the other engineering constants can be computed as

$$\begin{aligned}
\frac{\nu_{ij}^r}{E_i^r} &= \frac{\nu_{ji}^r}{E_j^r} \text{ for } i, j = 1, 2, 3 \\
G_{ij}^r &= G_{ji}^r
\end{aligned} \tag{4.58}$$

Assuming $G_{12}^r = G_{13}^r = G_{23}^r$ (i.e. $p_r = m_r$), the inverse relations of (4.56) can be obtained

$$\begin{aligned}
n_r &= \frac{1 - (\nu_{23}^r)^2}{E_2^2 \Delta} \\
l_r &= \frac{\nu_{12}^r(1 + \nu_{23}^r)}{E_1^r E_2^r \Delta} \\
m_r &= p_r = G_{12}^r \\
k_r &= \frac{1 - \nu_{13}^r \nu_{31}^r}{E_1^r E_2^r \Delta} - G_{12}^r
\end{aligned} \tag{4.59}$$

where $\Delta = (1 - 2\nu_{12}^r \nu_{21}^r (1 + \nu_{23}^r) - (\nu_{23}^r)^2)/(E_1^r (E_2^r)^2)$, with a clear reduction to four independent elastic constants for a complete elastic behavior description of a CNT fiber. In table 4.1, the Hill's elastic moduli are listed for several single-walled CNTs (SWCNTs) with various chiral indices, as found in the literature.

According to a classical notation, the microtubules can be indicated as SWCNT(n, m), where n, m stand for the chiral indices which describe the pattern of the carbon atoms. Thus, the so-called armchair tubes, feature the property $n = m$. The elastic properties of the hybrid matrix are estimated through the Mori–Tanaka scheme [27] when the CNTs are oriented randomly in the inclusions [95]. The bulk

Table 4.1. Hill's elastic moduli for several SWCNTs.

Carbon nanotubes	k_r [GPa]	l_r [GPa]	m_r [GPa]	n_r [GPa]	p_r [GPa]	Refs.
SWCNT (5,5)	536	184	132	2143	791	[9, 12]
SWCNT (6,6)	9.9	8.4	4.4	457.6	27	[8]
SWCNT (10,10)	271	88	17	1089	442	[9]
SWCNT (15,15)	181	58	5	726	301	[9]
SWCNT (20,20)	136	43	2	545	227	[9, 12]
SWCNT (50,50)	55	17	0.1	218	92	[9, 12]

moduli K_{in} and K_{out} and the shear moduli G_{in} and G_{out} of the inclusions and the hybrid matrix are defined respectively as

$$\begin{aligned}
K_{\text{in}}(\zeta) &= K_m + \frac{V_r\eta(\delta_r - 3K_m\alpha_r)}{3(\mu - V_r\eta + V_r\eta\alpha_r)} \\
G_{\text{in}}(\zeta) &= G_m + \frac{V_r\eta(\eta_r - 2G_m\beta_r)}{2(\mu - V_r\eta + V_r\eta\beta_r)} \\
K_{\text{out}}(\zeta) &= K_m + \frac{V_r(1-\eta)(\delta_r - 3K_m\alpha_r)}{3(1 - \mu - V_r(1-\eta) + V_r(1-\eta)\alpha_r)} \\
G_{\text{out}}(\zeta) &= G_m + \frac{V_r(1-\eta)(\eta_r - 2G_m\beta_r)}{2(1 - \mu - V_r(1-\eta) + V_r(1-\eta)\beta_r)}
\end{aligned} \tag{4.60}$$

with K_m and G_m being the bulk modulus and the shear modulus of the isotropic matrix, respectively, that can be evaluated as a function of the elastic modulus E_m and Poisson's ratio ν_m, namely,

$$\begin{aligned}
K_m &= \frac{E_m}{3(1 - 2\nu_m)} \\
G_m &= \frac{E_m}{2(1 + \nu_m)}
\end{aligned} \tag{4.61}$$

where several values of E_m and ν_m can be found in the literature. In addition, the quantities α_r, β_r, δ_r and η_r in (4.60) read as follows

$$\begin{aligned}
\alpha_r &= \frac{3(K_m + G_m) + k_r + l_r}{3(G_m + k_r)} \\
\beta_r &= \frac{1}{5}\left(\frac{4G_m + 2k_r + l_r}{3(G_m + k_r)} + \frac{4G_m}{G_m + p_r} + \frac{2(G_m(3K_m + G_m) + G_m(3K_m + 7G_m))}{G_m(3K_m + G_m) + m_r(3K_m + 7G_m)}\right) \\
\delta_r &= \frac{1}{3}\left(n_r + 2l_r + \frac{(2k_r + l_r)(3K_m + G_m - l_r)}{G_m + k_r}\right) \\
\eta_r &= \frac{1}{5}\left(\frac{2}{3}(n_r - l_r) + \frac{8G_m p_r}{G_m + p_r} + \frac{2(k_r - l_r)(2G_m + l_r)}{3(G_m + k_r)}\right. \\
&\quad \left. + \frac{8m_r G_m(3K_m + 4G_m)}{3K_m(m_r + G_m) + G_m(7m_r + G_m)}\right)
\end{aligned} \tag{4.62}$$

Finally, the Mori–Tanaka method provides the effective bulk and shear moduli of the composite layer

$$K(\zeta) = K_{\text{out}}\left(1 + \frac{\mu\left(\dfrac{K_{\text{in}}}{K_{\text{out}}} - 1\right)}{1 + (1 - \mu)\left(\dfrac{K_{\text{in}}}{K_{\text{out}}} - 1\right)\dfrac{1 + \nu_{\text{out}}}{3 - 3\nu_{\text{out}}}}\right)$$
$$G(\zeta) = G_{\text{out}}\left(1 + \frac{\mu\left(\dfrac{G_{\text{in}}}{G_{\text{out}}} - 1\right)}{1 + (1 - \mu)\left(\dfrac{G_{\text{in}}}{G_{\text{out}}} - 1\right)\dfrac{8 - 10\nu_{\text{out}}}{15 - 15\nu_{\text{out}}}}\right) \tag{4.63}$$

where ν_{out} indicates the Poisson's ratio of the hybrid matrix defined as

$$\nu_{\text{out}}(\zeta) = \frac{3K_{\text{out}} - 2G_{\text{out}}}{6K_{\text{out}} + 2G_{\text{out}}} \tag{4.64}$$

Since the composite ply turns out to be isotropic, the Young modulus $E(\zeta)$ and the Poisson's ratio $\nu(\zeta)$ are evaluated through the following expressions

$$E(\zeta) = \frac{9KG}{3K + G}$$
$$\nu(\zeta) = \frac{3K - 2G}{6K + 2G} \tag{4.65}$$

Finally, the density $\rho(\zeta)$ of the composite is given by the mixtures theory

$$\rho(\zeta) = (\rho_r - \rho_m)V_r + \rho_m \tag{4.66}$$

in which ρ_r and ρ_m specify the density of CNTs and the matrix, respectively.

4.2.4 Governing equations of the problem

As highlighted in [20], the governing equations of each problem are obtained by means of the well-known Hamilton's variational principle. Let's start considering a shell element under an arbitrary angular velocity $\mathbf{\Omega}$, of components Ω_1, Ω_2, Ω_3 about the reference system, namely (see figure 4.3),

$$\mathbf{\Omega} = \Omega_1\mathbf{e}_1 + \Omega_2\mathbf{e}_2 + \Omega_3\mathbf{e}_3 \tag{4.67}$$

The rotation is not limited only to the axis of revolution of the doubly-curved shell, but it can have different components about various axes. At this point, we introduce conveniently the angular velocity vector $\boldsymbol{\omega}$, in the local reference system

$$\boldsymbol{\omega} = (\mathbf{\Omega}^T\mathbf{t}_1)\mathbf{t}_1 + (\mathbf{\Omega}^T\mathbf{t}_2)\mathbf{t}_2 + (\mathbf{\Omega}^T\mathbf{n})\mathbf{n} = \omega_1\mathbf{t}_1 + \omega_2\mathbf{t}_2 + \omega_3\mathbf{n} \tag{4.68}$$

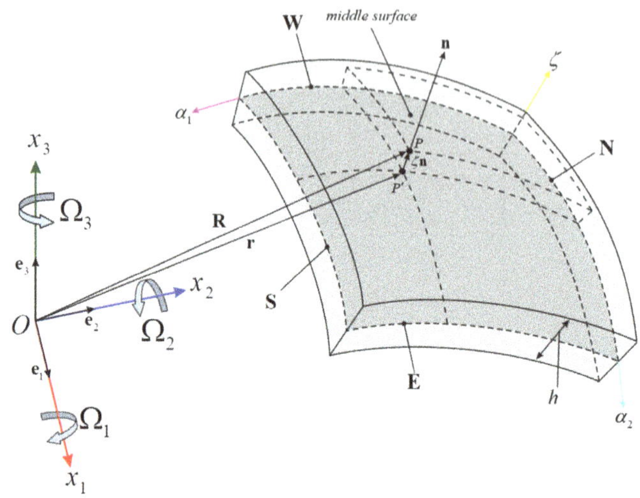

Figure 4.3. Doubly-curved rotating shell element.

where $\mathbf{t}_1$, $\mathbf{t}_2$, $\mathbf{n}$ denote the unit vectors of the corresponding local axes, where $\mathbf{t}_1$, $\mathbf{t}_2$ are evaluated as

$$\mathbf{t}_1 = \frac{\mathbf{r}_{,1}}{|\mathbf{r}_{,1}|}, \quad \mathbf{t}_2 = \frac{\mathbf{r}_{,2}}{|\mathbf{r}_{,2}|} \tag{4.69}$$

and $\mathbf{n}$ is defined by equation (4.5). Thus, we introduce vector $\mathbf{f}_\Omega^{(\tau)}$ to define the load contribution given by the shell rotation, for each order of kinematic expansion $\tau = 0, 1, 2, \dots, N, N+1$, accounting for both the centripetal acceleration $\mathbf{f}_{\Omega Cp}^{(\tau)}$ and the Coriolis effect $\mathbf{f}_{\Omega Cr}^{(\tau)}$ as follows

$$\mathbf{f}_\Omega^{(\tau)} = \mathbf{f}_{\Omega Cp}^{(\tau)} + \mathbf{f}_{\Omega Cr}^{(\tau)} \tag{4.70}$$

Each vector in (4.70) takes the following compact definition

$$\mathbf{f}_{\Omega Cp}^{(\tau)} = \sum_{\eta=0}^{N+1} \mathbf{L}_\Omega^{(\tau\eta)} \mathbf{u}^{(\eta)} + \mathbf{q}_\Omega^{(\tau)} \tag{4.71}$$

$$\mathbf{f}_{\Omega Cr}^{(\tau)} = \sum_{\eta=0}^{N+1} \mathbf{C}_\Omega^{(\tau\eta)} \dot{\mathbf{u}}^{(\eta)} \tag{4.72}$$

where

$$\mathbf{L}_\Omega^{(\tau\eta)} = \tilde{\mathbf{M}}^{(\tau\eta)} \circ \Theta_1 \tag{4.73}$$

$$\mathbf{q}_\Omega^{(\tau)} = \sum_{s=0}^{1} \mathbf{I}_s^{(\tau\eta)} \Theta_1 \mathbf{d}_s \tag{4.74}$$

$$\mathbf{C}_{\Omega}^{(\tau\eta)} = 2\tilde{\mathbf{M}}^{(\tau\eta)} \circ \Theta_2 \tag{4.75}$$

with

$$\mathbf{d}_0 = (\mathbf{r}^T\mathbf{t}_1)\mathbf{t}_1 + (\mathbf{r}^T\mathbf{t}_2)\mathbf{t}_2 + (\mathbf{r}^T\mathbf{n})\mathbf{n} = d_1\mathbf{t}_1 + d_2\mathbf{t}_2 + d_3\mathbf{n} \tag{4.76}$$

$$\mathbf{d}_1 = [0 \quad 0 \quad 1]^T \tag{4.77}$$

and

$$\Theta_1 = \begin{bmatrix} \omega_3^2 + \omega_2^2 & -\omega_1\omega_2 & -\omega_3\omega_1 \\ -\omega_1\omega_2 & \omega_3^2 + \omega_1^2 & -\omega_2\omega_3 \\ -\omega_3\omega_1 & -\omega_2\omega_3 & \omega_2^2 + \omega_1^2 \end{bmatrix} \tag{4.78}$$

$$\Theta_2 = \begin{bmatrix} 0 & \omega_3 & -\omega_2 \\ -\omega_3 & 0 & \omega_1 \\ \omega_2 & -\omega_1 & 0 \end{bmatrix} \tag{4.79}$$

Note that the symbol '∘' introduced in (4.73) and (4.75) refers to the Hadamard product. On the other hand, we assemble the inertial quantities within matrix $\mathbf{I}_s^{(\tau\eta)}$, for $s = 0, 1$

$$\mathbf{I}_s^{(\tau\eta)} = \begin{bmatrix} I_s^{(\tau)\alpha_1} & 0 & 0 \\ 0 & I_s^{(\tau)\alpha_2} & 0 \\ 0 & 0 & I_s^{(\tau)\alpha_3} \end{bmatrix} \tag{4.80}$$

where

$$I_s^{(\tau)\alpha_i} = \sum_{k=1}^{l} \int_{\zeta_k}^{\zeta_{k+1}} \rho^{(k)} \zeta^s F_\tau^{\alpha_i} H_1 H_2 d\zeta \tag{4.81}$$

with $i = 1, 2, 3$ and $s = 0, 1$. Moreover, the mass matrix $\tilde{\mathbf{M}}^{(\tau\eta)}$ is given by

$$\tilde{\mathbf{M}}^{(\tau\eta)} = \begin{bmatrix} I_0^{(\tau\eta)\alpha_1\alpha_1} & I_0^{(\tau\eta)\alpha_1\alpha_2} & I_0^{(\tau\eta)\alpha_1\alpha_3} \\ I_0^{(\tau\eta)\alpha_2\alpha_1} & I_0^{(\tau\eta)\alpha_2\alpha_2} & I_0^{(\tau\eta)\alpha_2\alpha_3} \\ I_0^{(\tau\eta)\alpha_3\alpha_1} & I_0^{(\tau\eta)\alpha_3\alpha_2} & I_0^{(\tau\eta)\alpha_3\alpha_3} \end{bmatrix} \tag{4.82}$$

with

$$I_0^{(\tau\eta)\alpha_i\alpha_j} = \sum_{k=1}^{l} \int_{\zeta_k}^{\zeta_{k+1}} \rho^{(k)} F_\tau^{\alpha_i} F_\eta^{\alpha_j} H_1 H_2 d\zeta \tag{4.83}$$

for $\tau, \eta = 0, 1, 2, \ldots, N, N+1$ and $i, j = 1, 2, 3$. It is worth noticing that matrix $\mathbf{C}_{\Omega}^{(\tau\eta)}$ in equation (4.75), denotes the physical damping effect related to the Coriolis acceleration, with a destabilizing influence in the structural system. As detailed in [6], indeed, the dynamic response of shell structures is affected by the centripetal and Coriolis forces. Their natural frequencies depend on the value of the applied angular velocity, in addition to the mechanical and geometric structural features. Moreover, a bifurcation phenomenon occurs in terms of natural frequencies with a consistent rise of backward and forward waves in the response, caused by the application of the angular velocity. Finally, the vector of inertial forces $\mathbf{f}_I^{(\tau)}$ should be introduced for a fixed order of kinematic expansion τ

$$\mathbf{f}_I^{(\tau)} = -\sum_{\eta=0}^{N+1} \mathbf{M}^{(\tau\eta)}\,\ddot{\mathbf{u}}^{(\eta)} \tag{4.84}$$

where

$$\mathbf{M}^{(\tau\eta)} = \tilde{\mathbf{M}}^{(\tau\eta)} \circ \mathbf{I} \tag{4.85}$$

I being the identity matrix. Thus, the fundamental equations of the structural problem at issue, for each order of kinematic expansion τ, can be arranged in compact form as follows

$$\sum_{\eta=0}^{N+1} \left(\mathbf{L}_\Omega^{(\tau\eta)} + \mathbf{L}^{(\tau\eta)}\right)\mathbf{u}^{(\eta)} + \sum_{\eta=0}^{N+1} \mathbf{C}_\Omega^{(\tau\eta)}\dot{\mathbf{u}}^{(\eta)} + \mathbf{q}_\Omega^{(\tau)} = \sum_{\eta=0}^{N+1} \mathbf{M}^{(\tau\eta)}\ddot{\mathbf{u}}^{(\eta)} \tag{4.86}$$

where the differential operator $\mathbf{L}^{(\tau\eta)}$ assumes the following aspect

$$\mathbf{L}^{(\tau\eta)} = \begin{bmatrix} L_{11}^{(\tau\eta)\alpha_1\alpha_1} & L_{12}^{(\tau\eta)\alpha_1\alpha_2} & L_{13}^{(\tau\eta)\alpha_1\alpha_3} \\ L_{21}^{(\tau\eta)\alpha_2\alpha_1} & L_{22}^{(\tau\eta)\alpha_2\alpha_2} & L_{23}^{(\tau\eta)\alpha_2\alpha_3} \\ L_{31}^{(\tau\eta)\alpha_3\alpha_1} & L_{32}^{(\tau\eta)\alpha_3\alpha_2} & L_{33}^{(\tau\eta)\alpha_3\alpha_3} \end{bmatrix} \tag{4.87}$$

whose explicit expressions can be found in [94]. A dynamic or a static analysis of shell structures in absence of an angular velocity, would represent different subcases of equation (4.87), when vanishing, conveniently, the terms related to $\mathbf{L}_\Omega^{(\tau\eta)}$, $\mathbf{C}_\Omega^{(\tau\eta)}$, $\mathbf{q}_\Omega^{(\tau)}$, $\mathbf{M}^{(\tau\eta)}$. More specifically, the governing equations for a free vibration problem would reduce to the following form

$$\sum_{\eta=0}^{N+1} \mathbf{L}^{(\tau\eta)}\mathbf{u}^{(\eta)} = \sum_{\eta=0}^{N+1} \mathbf{M}^{(\tau\eta)}\ddot{\mathbf{u}}^{(\eta)} \qquad \text{for} \quad \tau = 0, 1, 2, \ldots, N, N+1 \tag{4.88}$$

Furthermore, a proper selection of the kinematic expansion in each principal direction makes it possible to handle complicated models. It must be recalled that the present boundary value problem can be solved only once the boundary conditions are properly enforced. In detail, the following conditions are required along an external edge for a fixed value of α_1

$$\mathbf{u}^{(\tau)} = \bar{\mathbf{u}}^{(\tau)} \quad \text{or} \quad \mathbf{S}_{\alpha_1}^{(\tau)} = \bar{\mathbf{S}}_{\alpha_1}^{(\tau)} \tag{4.89}$$

Similarly, we enforce the following relations along an external edge with a constant α_2

$$\mathbf{u}^{(\tau)} = \bar{\mathbf{u}}^{(\tau)} \quad \text{or} \quad \mathbf{S}_{\alpha_2}^{(\tau)} = \bar{\mathbf{S}}_{\alpha_2}^{(\tau)} \tag{4.90}$$

whose displacement components are collected in the following vector $\bar{\mathbf{u}}^{(\tau)}$, for $\tau = 0, 1, 2, \dots, N, N+1$

$$\bar{\mathbf{u}}^{(\tau)} = \begin{bmatrix} \bar{u}_1^{(\tau)} & \bar{u}_2^{(\tau)} & \bar{u}_3^{(\tau)} \end{bmatrix}^T \tag{4.91}$$

On the other hand, the stress resultants applied along a boundary edge with a fixed value of α_1 or α_2 are collected in the following vectors, respectively

$$\bar{\mathbf{S}}_{\alpha_1}^{(\tau)} = \begin{bmatrix} \bar{N}_1^{(\tau)\alpha_1} & \bar{N}_{12}^{(\tau)\alpha_1} & \bar{T}_1^{(\tau)\alpha_3} \end{bmatrix}^T \tag{4.92}$$

$$\bar{\mathbf{S}}_{\alpha_2}^{(\tau)} = \begin{bmatrix} \bar{N}_{21}^{(\tau)\alpha_1} & \bar{N}_2^{(\tau)\alpha_1} & \bar{T}_2^{(\tau)\alpha_3} \end{bmatrix}^T \tag{4.93}$$

Thus, for a clamped ('C') edge, we enforce that

$$\mathbf{u}^{(\tau)} = \mathbf{0} \tag{4.94}$$

whereas, a free ('F') edge condition is defined by

$$\mathbf{S}_{\alpha_1}^{(\tau)} = \mathbf{0} \quad \text{or} \quad \mathbf{S}_{\alpha_2}^{(\tau)} = \mathbf{0} \tag{4.95}$$

according to the edge coordinates, or a simply-supported ('S') condition is defined by setting

$$N_1^{(\tau)\alpha_1} = 0, \quad u_2^{(\tau)} = u_3^{(\tau)} = 0 \tag{4.96}$$

for $\alpha_1 = \alpha_1^0$ (or $\alpha_1 = \alpha_1^1$), with $\alpha_2^0 \leqslant \alpha_2 \leqslant \alpha_2^1$, and

$$N_2^{(\tau)\alpha_2} = 0, \quad u_1^{(\tau)} = u_3^{(\tau)} = 0 \tag{4.97}$$

for $\alpha_2 = \alpha_2^0$ (or $\alpha_2 = \alpha_2^1$), with $\alpha_1^0 \leqslant \alpha_1 \leqslant \alpha_1^1$. The static of kinematic boundary conditions (4.94)-(4.97) must be enforced for each order of kinematic expansion τ. As shown in figure 4.3, we refer to West-South-East-North (WSEN) to define the sequence of external edges, along with their boundary condition, univocally. If two external sides are overlapped (as occurs in shells of revolution), the following compatibility conditions are enforced along the closing edges for each order of kinematic expansion. In particular, for $\alpha_1^0 = \alpha_1^1$, it is

$$\begin{aligned} \mathbf{u}^{(\tau)}(\alpha_1^0, \alpha_2, t) &= \mathbf{u}^{(\tau)}(\alpha_1^1, \alpha_2, t) \\ \mathbf{S}_{\alpha_1}^{(\tau)}(\alpha_1^0, \alpha_2, t) &= \mathbf{S}_{\alpha_1}^{(\tau)}(\alpha_1^1, \alpha_2, t) \end{aligned} \tag{4.98}$$

On the other hand, for $\alpha_2^0 = \alpha_2^1$, it is

$$\begin{aligned} \mathbf{u}^{(\tau)}(\alpha_1, \alpha_2^0, t) &= \mathbf{u}^{(\tau)}(\alpha_1, \alpha_2^1, t) \\ \mathbf{S}_{\alpha_2}^{(\tau)}(\alpha_1, \alpha_2^0, t) &= \mathbf{S}_{\alpha_2}^{(\tau)}(\alpha_1, \alpha_2^1, t) \end{aligned} \tag{4.99}$$

4.3 A general view on the GDQ-based numerical method

Among different numerical possibilities to treat the problem, in a strong or weak form, in what follows we select the GDQ method to approximate the differential system of equations, along with their boundary conditions. The main aspects of the selected technique are here presented in a one-dimensional domain for the sake of conciseness, where further details could be found in [90, 104]. From a mathematical standpoint, the solution of a function $f(x)$ according to a differential quadrature method, can be approximated by a set of smooth basis functions $\psi_j(x)$ (e.g. the polynomial functions) for $j = 1, 2, \ldots, I_N$, where I_N stands for the total number of grid points, namely,

$$f(x) \cong \sum_{j=1}^{I_N} \lambda_j \psi_j(x) \tag{4.100}$$

The key point is to solve a partial differential system through the approximation (4.100) at the discrete points x_i for $i = 1, 2, \ldots, I_N$. Using the so-called differentiation matrix procedure, the derivative of a given function can be easily found in compact form as

$$\mathbf{f}^{(n)} = \mathbf{D}^{(n)}\mathbf{f} \tag{4.101}$$

where

$$\mathbf{f}^{(n)} = \left[\left.\frac{d^n f(x)}{d^n x}\right|_{x_1}, \left.\frac{d^n f(x)}{d^n x}\right|_{x_2}, \ldots, \left.\frac{d^n f(x)}{d^n x}\right|_{x_{I_N}} \right]^T \tag{4.102}$$

is the vector of the function derivatives at the whole grid points, while

$$\mathbf{f} = [f(x_1), f(x_2), \ldots, f(x_{I_N})]^T \tag{4.103}$$

is the vector of the unknown functions, and $\mathbf{D}^{(n)}$ is the differentiation matrix that must be determined. Thus, the approximation (4.100) can be rewritten in a matrix form as

$$\mathbf{f} = \mathbf{A}\boldsymbol{\lambda} \tag{4.104}$$

where $\boldsymbol{\lambda}$ is the vector of the unknown coefficients λ_j, and $\mathbf{A}$ is the coefficients matrix, with $A_{ij} = \psi_j(x_i)$. Once equation (4.100) is defined, its general n th derivative is related directly to functions $\psi_j(x)$ for linearity reasons, due to the independence of the unknown coefficients λ_j on the variable x, i.e.

$$\frac{d^n f(x)}{dx^n} = \sum_{j=1}^{I_N} \lambda_j \frac{d^n \psi_j(x)}{dx^n}, \qquad \text{for} \quad n = 1, 2, \dots, I_N - 1 \tag{4.105}$$

This last relation can be rewritten in matrix form as

$$\mathbf{f}^{(n)} = \mathbf{A}^{(n)}\boldsymbol{\lambda}, \qquad \text{with } A_{ij}^{(n)} = \left.\frac{d^n \psi_j(x)}{dx^n}\right|_{x_i} \tag{4.106}$$

In order to obtain the differentiation matrix $\mathbf{D}^{(n)}$, the matrix $\mathbf{A}$ must be invertible, whose property depends on the basis functions and on the selected collocation points. The unknown vector is computed from equation (4.104) by a mathematical invertion of matrix $\mathbf{A}$, namely

$$\boldsymbol{\lambda} = \mathbf{A}^{-1}\mathbf{f} \tag{4.107}$$

Thus, the substitution of equation (4.107) into equation (4.106), gets the following relation

$$\mathbf{f}^{(n)} = \mathbf{A}^{(n)}\mathbf{A}^{-1}\mathbf{f} = \mathbf{D}^{(n)}\mathbf{f} \tag{4.108}$$

where the differentiation matrix $\mathbf{D}^{(n)}$ is determined as matrix product between the inverse of the matrix $\mathbf{A}$ of the basis functions at the grid points $\psi_j(x_i)$ and the derivatives of the same function values $\psi_j^{(n)}(x_i)$, collected in $\mathbf{A}^{(n)}$. In the following, all the mathematical steps are generally defined for any order derivative, generally defined as nth

$$\left.\frac{d^n f(x)}{dx^n}\right|_{x_i} = \sum_{j=1}^{I_N} D_{ij}^{(n)} f(x_j) \quad \text{with} \quad D_{ij}^{(n)} = (\mathbf{A}^{(n)}\mathbf{A}^{-1})_{ij} \tag{4.109}$$

This approach is similar to the GDQ method, i.e.

$$\left.\frac{\partial^n f(x)}{\partial x^n}\right|_{x=x_i} \cong \sum_{j=1}^{I_N} \varsigma_{ij}^{n} f(x_j) \tag{4.110}$$

where the coefficients $D_{ij}^{(n)}$ of matrix $\mathbf{D}^{(n)}$ are evaluated as the product between an inverse matrix and its derivative, whereas the $\varsigma_{ij}^{(n)}$ coefficients can be computed by means of the recursive formulae by Shu [50]. Although the two methodologies are appearently similar, the latter does not require any matrix inversion to evaluate the weighting coefficients, and it is selected as the efficient numerical method to solve our mechanical problems. The first key aspect of the proposed methodology relies on the proper choice of the discrete grid distribution in a closed domain $[a, b]$. In other words, the position of I_N discrete points is specified within the reference domain, as follows

$$a = x_1, \quad x_2, \quad \dots, \quad x_k, \quad \dots, \quad x_{I_N-1}, \quad x_{I_N} = b \tag{4.111}$$

In equation (4.110) the arbitrary nodal coordinate $x_k \in [a, b]$ is defined as

$$x_j = \frac{b-a}{d-c}(z_j - c) + a \tag{4.112}$$

where $z_j \in [c, d]$ denotes the position of the kth point in the reference interval $[c, d]$, in which the selected distribution is defined. Thus, considering the interval $[c, d] = [0, 1]$, it is possible to specify z_j as follows

$$z_j = \frac{r_j - r_1}{r_{I_N} - r_1} \tag{4.113}$$

where the values of r_j are defined herein according to the Chebyshev–Gauss–Lobatto grid distribution

$$r_j = \cos\left(\frac{I_N - j}{I_N - 1}\pi\right) \tag{4.114}$$

for $j = 1, 2, \ldots, I_N$ and $r \in [-1, 1]$. According to the above-mentioned numerical approach, the natural frequencies of the most general rotating shell problem are obtained by solving the homogeneous equation associated with (4.86), setting $\mathbf{q}_\Omega^{(\tau)} = \mathbf{0}$, i.e.

$$\sum_{\eta=0}^{N+1} \mathbf{K}^{(\tau\eta)}\mathbf{u}^{(\eta)} + \sum_{\eta=0}^{N+1} \mathbf{C}_\Omega^{(\tau\eta)}\dot{\mathbf{u}}^{(\eta)} - \sum_{\eta=0}^{N+1} \mathbf{M}^{(\tau\eta)}\ddot{\mathbf{u}}^{(\eta)} = \mathbf{0} \tag{4.115}$$

in which $\mathbf{K}^{(\tau\eta)} = \mathbf{L}_\Omega^{(\tau\eta)} + \mathbf{L}^{(\tau\eta)}$, for $\tau = 0, 1, 2, \ldots, N, N+1$. The numerical solution is obtained in the following periodic form, for each order of kinematic expansion

$$\mathbf{u}^{(\tau)} = \bar{\mathbf{U}}^{(\tau)} e^{i\lambda t} \tag{4.116}$$

where i refers to the imaginary unit, whereas $\mathbf{U}^{(\tau)}(\alpha_1, \alpha_2)$ includes the amplitudes of the mode shape, as defined below

$$\bar{\mathbf{U}}^{(\tau)} = \begin{bmatrix} \bar{U}_1^{(\tau)} & \bar{U}_2^{(\tau)} & \bar{U}_3^{(\tau)} \end{bmatrix}^T \tag{4.117}$$

This means that equation (4.115) takes the following form, once the first-order and second-order derivatives of (4.116) are properly computed and substituted

$$\sum_{\eta=0}^{N+1} \mathbf{K}^{(\tau\eta)}\bar{\mathbf{U}}^{(\eta)} + i\lambda \sum_{\eta=0}^{N+1} \mathbf{C}_\Omega^{(\tau\eta)}\bar{\mathbf{U}}^{(\eta)} + \lambda^2 \sum_{\eta=0}^{N+1} \mathbf{M}^{(\tau\eta)}\bar{\mathbf{U}}^{(\eta)} = \mathbf{0} \tag{4.118}$$

The corresponding discrete form is obtained by applying the GDQ method, which provides the following system of equations

$$\mathbf{K}\boldsymbol{\delta} + i\lambda\mathbf{C}\boldsymbol{\delta} + \lambda^2 M\boldsymbol{\delta} = \mathbf{0} \tag{4.119}$$

where $\mathbf{M}$, $\mathbf{C}$ and $\mathbf{K}$ stand for the mass, damping and stiffness matrices of the structure, respectively, of dimensions $N_s \times N_s$, with

$$N_s = 3I_N I_M(N + 2) \tag{4.120}$$

The discrete vector $\boldsymbol{\delta}$, instead, collects the mode shapes for the selected problem. The matrices in equation (4.119) can be conveniently rearranged to account for both contributions of boundary nodes ('b') and internal nodes ('d') in the reference domain, separately. As a consequence, the discrete system takes the following form

$$\begin{bmatrix} \mathbf{K}_{bb} & \mathbf{K}_{bd} \\ \mathbf{K}_{db} & \mathbf{K}_{dd} \end{bmatrix}\begin{bmatrix} \boldsymbol{\delta}_b \\ \boldsymbol{\delta}_d \end{bmatrix} + \lambda\begin{bmatrix} \mathbf{0} & \mathbf{0} \\ \mathbf{0} & i\mathbf{C}_{dd} \end{bmatrix}\begin{bmatrix} \boldsymbol{\delta}_b \\ \boldsymbol{\delta}_d \end{bmatrix} + \lambda^2\begin{bmatrix} \mathbf{0} & \mathbf{0} \\ \mathbf{0} & \mathbf{M}_{dd} \end{bmatrix}\begin{bmatrix} \boldsymbol{\delta}_b \\ \boldsymbol{\delta}_d \end{bmatrix} = \begin{bmatrix} \mathbf{0} \\ \mathbf{0} \end{bmatrix} \tag{4.121}$$

where $\mathbf{M}_{dd}$, $\mathbf{C}_{dd}$, $\mathbf{K}_{dd}$ denote the mass, damping and stiffness matrices evaluated in the internal grid points; $\mathbf{K}_{bb}$, $\mathbf{K}_{bd}$ stand for the stiffness matrices related to the boundary conditions, and involve the discrete points along the external edges of the domain; $\mathbf{K}_{db}$ includes the contribution of the governing equations in the boundary points. It is worth observing that the mass and damping matrices assume a null value along the boundaries, such that boundary points are just related to the stiffness contribution. In such a context, the system (4.121) can be also written as

$$\begin{cases} \mathbf{K}_{bb}\boldsymbol{\delta}_b + \mathbf{K}_{bd}\boldsymbol{\delta}_d = \mathbf{0} \\ \mathbf{K}_{db}\boldsymbol{\delta}_b + \mathbf{K}_{dd}\boldsymbol{\delta}_d + i\lambda\mathbf{C}_{dd}\boldsymbol{\delta}_d + \lambda^2\mathbf{M}_{dd}\boldsymbol{\delta}_d = \mathbf{0} \end{cases} \tag{4.122}$$

that yields to the following relation

$$(\mathbf{K}_{dd} - \mathbf{K}_{db}\mathbf{K}_{bb}^{-1}\mathbf{K}_{bd})\boldsymbol{\delta}_d + i\lambda\mathbf{C}_{dd}\boldsymbol{\delta}_d + \lambda^2\mathbf{M}_{dd}\boldsymbol{\delta}_d = \mathbf{0} \tag{4.123}$$

whose operators have dimesions equal to $N_{s,r} \times N_{s,r}$, with

$$N_{s,r} = 3(I_N - 2)(I_M - 2)(N + 2) \tag{4.124}$$

Thus, a quadratic eigenvalue problem is obtained, which takes the following form

$$\begin{bmatrix} \mathbf{0} & -\mathbf{I} \\ \mathbf{K}_{dd} - \mathbf{K}_{db}\mathbf{K}_{bb}^{-1}\mathbf{K}_{bd} & i\mathbf{C}_{dd} \end{bmatrix}\begin{bmatrix} \boldsymbol{\delta}_d \\ \lambda\boldsymbol{\delta}_d \end{bmatrix} + \lambda\begin{bmatrix} \mathbf{I} & \mathbf{0} \\ \mathbf{0} & \mathbf{M}_{dd} \end{bmatrix}\begin{bmatrix} \boldsymbol{\delta}_d \\ \lambda\boldsymbol{\delta}_d \end{bmatrix} = \mathbf{0} \tag{4.125}$$

where $\mathbf{I}$ denotes the identity matrix. By setting $\mathbf{z}_d = [\boldsymbol{\delta}_d \;\; \lambda\boldsymbol{\delta}_d]^T$, the system (4.125) becomes

$$(\mathbf{H}_1^* + \lambda\mathbf{H}_2^*)\mathbf{z}_d = \mathbf{0} \tag{4.126}$$

in which $\mathbf{H}_1^*$, $\mathbf{H}_2^*$ can be easily defined by a comparison between relations (4.125) and (4.126). This gets a linear eigenvalue problem with complex values of frequencies due to the presence of the damping matrix. In other words, the natural frequencies are found as complex conjugate pairs

$$\lambda = \omega \pm \mathrm{i}\gamma \tag{4.127}$$

in which ω is the real part (circular frequencies of the structure), whereas γ is the imaginary one. The natural frequencies are evaluated as $f = \omega/2\pi$. The discrete solution of the problem is computed as

$$\lambda_k = \omega_k \pm \mathrm{i}\gamma_k \tag{4.128}$$

for $k = 2N_{s,r}$. More details about these quantities and their physical meaning can be found in the recent work by Tornabene *et al* [20, 21]. To compute the critical value of the rotational speed for a selected structure, the governing equation of the problem is solved for increasing values of angular velocity [9–21]. The critical speed of a structural system would represent that value of angular velocity causing a stiffness reduction, with a null natural frequency. As mentioned before, the Coriolis effect causes the instability of a rotating system, as specified in the work by Kim and Lee [105], where the unstable states can depend on the real and imaginary parts of the dynamic solution [106, 107], i.e.

Table 4.2. Comparison between the first ten natural frequency variations of a CCCC square plate made of a single functionally graded lamina of constant thickness $h = 0.05$ m given by GDQ method and FEM. The 2D FEM solution is obtained by commercial software ABAQUS (10 000 S4 elements).

Geometrical properties: $L_x = L_y = 1.0$ m. Lamination scheme: $(FGM^{C}_{M(5P)(a=1/b=0/c/d=1/p=1)})$.
Mechanical properties: $E_C = 168$ GPa, $\nu_C = 0.3$, $\rho_C = 5700$ kg m^{-3} (ceramic); $E_M = 70$ GPa, $\nu_M = 0.3$, $\rho_M = 2707$ kg m^{-3} (metal).

f [Hz]	$\mathrm{FSDT}_{RS}^{\kappa=5/6}$	TSDT_{RS}	$\mathrm{ED2}^{\kappa=5/6}$	ED4	2DFEM
1	436.023	436.066	436.844	437.356	436.237
2	871.761	871.937	873.028	874.883	872.240
3	871.761	871.937	873.028	874.883	872.240
4	1262.896	1263.306	1264.257	1267.974	1263.240
5	1520.858	1521.419	1522.268	1527.438	1522.000
6	1530.371	1530.896	1531.869	1536.948	1531.550
7	1879.047	1879.957	1880.232	1887.842	1879.600
8	1879.047	1879.957	1880.232	1887.842	1879.600
9	2358.708	2360.047	2359.640	2370.928	2361.220
10	2358.708	2360.047	2359.640	2370.928	2361.220

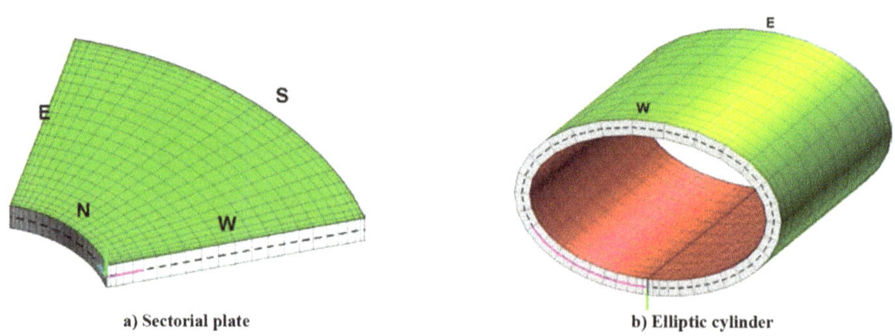

Figure 4.4. Two different panel structures with GDQ discretization and local co-ordinate system representation.

– divergence: $\gamma_k = 0$ and $\omega_k > 0$
– flutter: $\gamma_k \neq 0$ and $\omega_k > 0$

The critical speed is clearly obtained for $\omega_k = 0$. More specifically, the divergence state involves a bifurcation in the natural frequencies (known as backward and forward waves), whereas the flutter state features increasing fluctuations, see also [108].

For a simple free vibration problem of a shell structure, in the absence of angular velocity vector $\boldsymbol{\omega}$, the relation (4.124) reduces to the following form

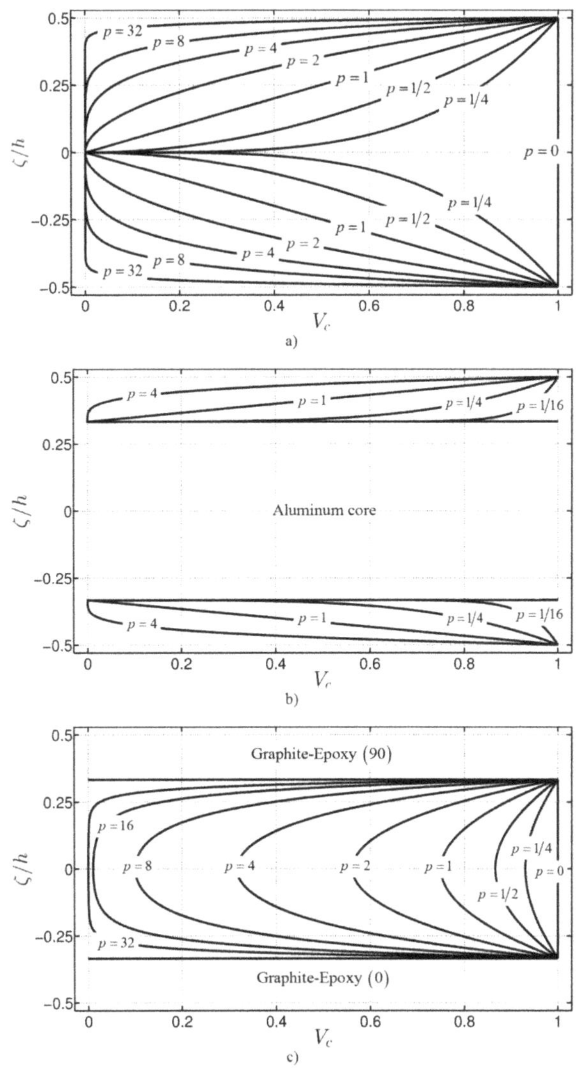

Figure 4.5. Distribution of CNT-reinforcing phase along the thickness.

Table 4.3. First ten natural frequency variations of a CFCC sectorial plate made of one lamina of constant thickness $h = 0.1$ m reinforced by CNTs uniformly distributed along the thickness for different mass fractions w_r varying the parameter of agglomeration μ with $\eta = 1$. The Chebyshev–Gauss–Lobatto grid distributions are employed with $I_N = I_M = 25$.

Geometrical properties: $R_i = 0.5$m, $L = 1.0$m, $\Theta = 60°$.										
$w_r = 0.05$										
f [Hz]	Polymer matrix	$\mu = 0.2$	$\mu = 0.3$	$\mu = 0.4$	$\mu = 0.5$	$\mu = 0.6$	$\mu = 0.7$	$\mu = 0.8$	$\mu = 0.9$	$\mu = 1.0$
1	78.574	94.449	102.564	110.803	119.248	127.983	137.093	146.676	156.843	167.728
2	191.448	230.449	250.392	270.651	291.429	312.932	335.376	359.003	384.092	410.980
3	214.233	256.616	278.252	300.189	322.640	345.816	369.938	395.251	422.034	450.620
4	344.064	414.367	450.321	486.849	524.321	563.107	603.599	646.237	691.527	740.077
5	393.721	472.356	512.523	553.275	595.013	638.136	683.065	730.263	780.268	833.717
6	425.752	510.273	553.423	597.179	641.967	688.209	736.351	786.880	840.357	897.451
7	467.490	560.813	608.486	656.855	706.395	757.580	810.909	866.933	926.287	989.729
8	501.797	605.458	658.496	712.412	767.758	825.088	884.991	948.127	1015.258	1087.302
9	525.423	633.051	688.101	744.037	801.427	860.841	922.883	988.227	1057.653	1132.100
10	611.674	734.547	797.332	861.055	926.348	993.843	1064.203	1138.166	1216.583	1300.473
$w_r = 0.10$										
f [Hz]	Polymer matrix	$\mu = 0.2$	$\mu = 0.3$	$\mu = 0.4$	$\mu = 0.5$	$\mu = 0.6$	$\mu = 0.7$	$\mu = 0.8$	$\mu = 0.9$	$\mu = 1.0$
1	78.574	95.049	104.520	114.867	126.389	139.493	154.777	173.165	196.211	226.787
2	191.448	231.919	255.177	280.595	308.904	341.116	378.707	423.973	480.770	556.254
3	214.233	258.233	283.531	311.157	341.890	376.800	417.445	466.228	527.152	607.566
4	344.064	417.014	458.932	504.746	555.776	613.848	681.630	763.273	865.751	1002.016
5	393.721	475.344	522.270	573.526	630.572	695.410	770.960	861.743	975.307	1125.572
6	425.752	513.493	563.931	619.011	680.288	749.904	830.968	928.288	1049.870	1210.420
7	467.490	564.360	620.056	680.896	748.608	825.572	915.252	1023.013	1157.815	1336.174
8	501.797	609.342	671.121	738.654	813.900	899.573	999.640	1120.289	1271.930	1473.945
9	525.423	637.099	701.267	771.401	849.527	938.445	1042.248	1167.310	1324.341	1533.249
10	611.674	739.205	812.516	892.606	981.763	1083.134	1201.308	1343.403	1521.325	1757.080

$w_r = 0.20$

f [Hz]	Polymer matrix	$\mu = 0.2$	$\mu = 0.3$	$\mu = 0.4$	$\mu = 0.5$	$\mu = 0.6$	$\mu = 0.7$	$\mu = 0.8$	$\mu = 0.9$	$\mu = 1.0$
1	78.574	94.741	104.958	116.618	130.306	146.949	168.152	197.020	240.707	321.540
2	191.448	231.169	256.250	284.873	318.474	359.339	411.416	482.368	589.879	789.351
3	214.233	257.390	284.715	315.897	352.494	396.970	453.568	530.480	646.443	859.252
4	344.064	415.666	460.865	512.442	572.993	646.635	740.492	868.395	1062.281	1422.310
5	393.721	473.797	524.455	582.265	650.122	732.610	837.633	980.479	1196.228	1593.719
6	425.752	511.818	566.287	628.441	701.386	790.037	902.863	1056.211	1287.507	1712.293
7	467.490	562.523	622.651	691.271	771.819	869.735	994.404	1163.970	1420.073	1891.874
8	501.797	607.380	673.956	749.920	839.103	947.586	1085.906	1274.551	1560.938	2094.337
9	525.423	635.042	704.222	783.164	875.841	988.560	1132.237	1328.071	1625.043	2176.930
10	611.674	736.803	815.921	906.210	1012.196	1141.051	1305.158	1528.484	1866.127	2489.590

$w_r = 0.40$

f [Hz]	Polymer matrix	$\mu = 0.2$	$\mu = 0.3$	$\mu = 0.4$	$\mu = 0.5$	$\mu = 0.6$	$\mu = 0.7$	$\mu = 0.8$	$\mu = 0.9$	$\mu = 1.0$
1	78.574	93.259	103.737	115.984	130.804	149.575	174.935	212.840	281.179	487.839
2	191.448	227.553	253.270	283.321	319.682	365.737	427.966	521.012	688.954	1199.015
3	214.233	253.361	281.401	314.184	353.865	404.128	472.006	573.342	755.460	1299.127
4	344.064	409.164	455.504	509.649	575.161	658.135	770.250	937.911	1240.635	2161.260
5	393.721	466.382	518.352	579.104	652.631	745.765	871.565	1059.474	1397.684	2413.579
6	425.752	503.807	559.697	625.032	704.108	804.265	939.526	1141.482	1504.558	2589.719
7	467.490	553.719	615.406	687.518	774.798	885.355	1034.689	1257.753	1659.231	2864.978
8	501.797	597.882	666.118	745.825	842.247	964.363	1129.386	1376.283	1822.665	3186.613
9	525.423	625.109	696.031	778.893	879.147	1006.123	1177.699	1434.311	1897.794	3309.084
10	611.674	725.273	806.428	901.284	1016.082	1161.490	1357.919	1651.422	2180.145	3773.911

$$(\mathbf{K}_{dd} - \mathbf{K}_{db}(\mathbf{K}_{bb})^{-1}\mathbf{K}_{bd})\boldsymbol{\delta}_d = \omega^2\mathbf{M}_{dd}\boldsymbol{\delta}_d \tag{4.129}$$

4.4 Numerical applications

4.4.1 Free vibration problems

The present section presents the numerical results for some free vibration problems of FGM composite doubly-curved shells and panels with a free-form shape, also reinforced by CNTs, thus extending the analysis to shell structures involving possible angular rotations. All the studies are here performed numerically by means of the GDQ method, whose accuracy is consistently checked with respect to predictions from literature and/or results based on a classical finite element method (FEM). The first application considers an FG completely clamped square plate ($L_x = L_y = 1$ m) made of a single ply of constant thickness ($h = 0.05$ m), and

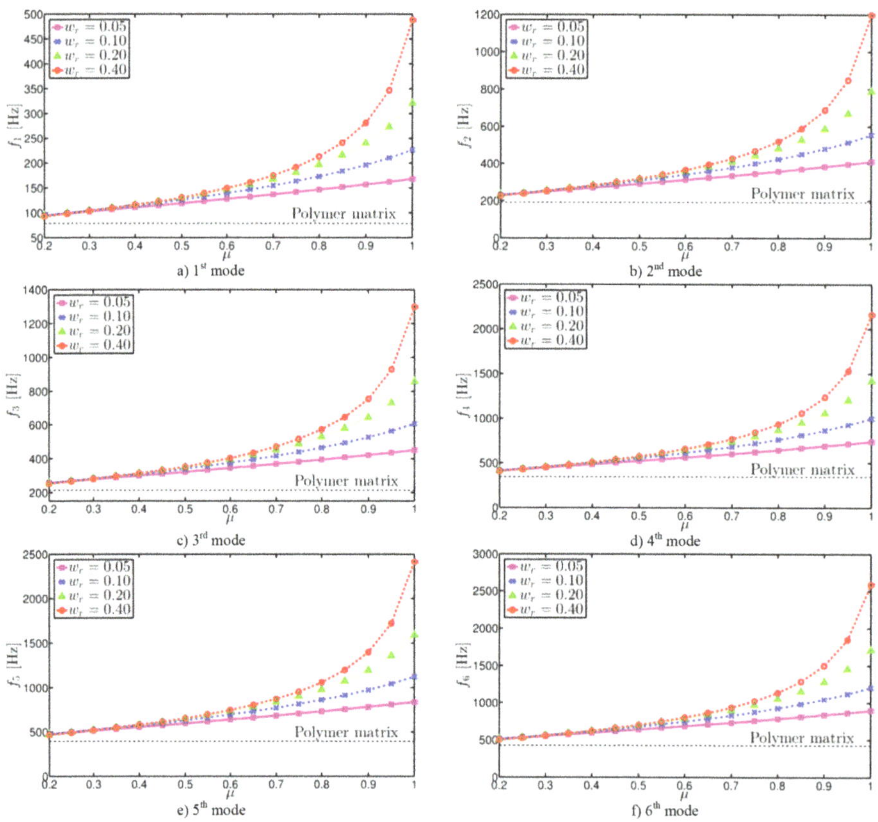

Figure 4.6. Frequency variations of a CFCC sectorial plate made of one lamina of constant thickness $h = 0.1$ m reinforced by CNTs uniformly distributed along the thickness for different mass fractions w_r varying the parameter of agglomeration μ with $\eta = 1$. The Chebyshev–Gauss–Lobatto grid distributions are employed with $I_N = I_M = 25$.

isotropic constituents (ceramic and metal), whose mechanical properties are listed in table 4.2. The volume fraction of the ceramic phase has a linear profile throughout the structural thickness, whose distribution and related parameters are shown in table 4.2, too. The notation $FGM^{C}_{M(5P)(a=1/b=0/c/d=1/p=1)}$ means that the mechanical properties of the metal phase involve the bottom side of the plate. The 2D FEM solution is obtained by the commercial software ABAQUS, employing 10 000 S4 elements. As far as the GDQ solution is concerned, several kinematic models are considered, namely, the $\text{FSDT}^{\kappa=5/6}_{RS}$, TSDT_{RS}, $\text{ED2}^{\kappa=5/6}$ and ED4; where the notation RS stands for the reduced stiffnesses, and $\kappa = 5/6$ stands for the shear correction factor. As visible from the first ten frequencies reported in table 4.2, the GDQ solutions are in good agreement with the FEM-based ones.

After the preliminary validation of the proposed numerical method, we present the results from a large parametric study. First, we analyse the sensitivity of the frequency response for different parameters defining the CNT reinforcing phase. In relation to table 4.1, a SWCNT is considered as reinforcement with chiral index 10,

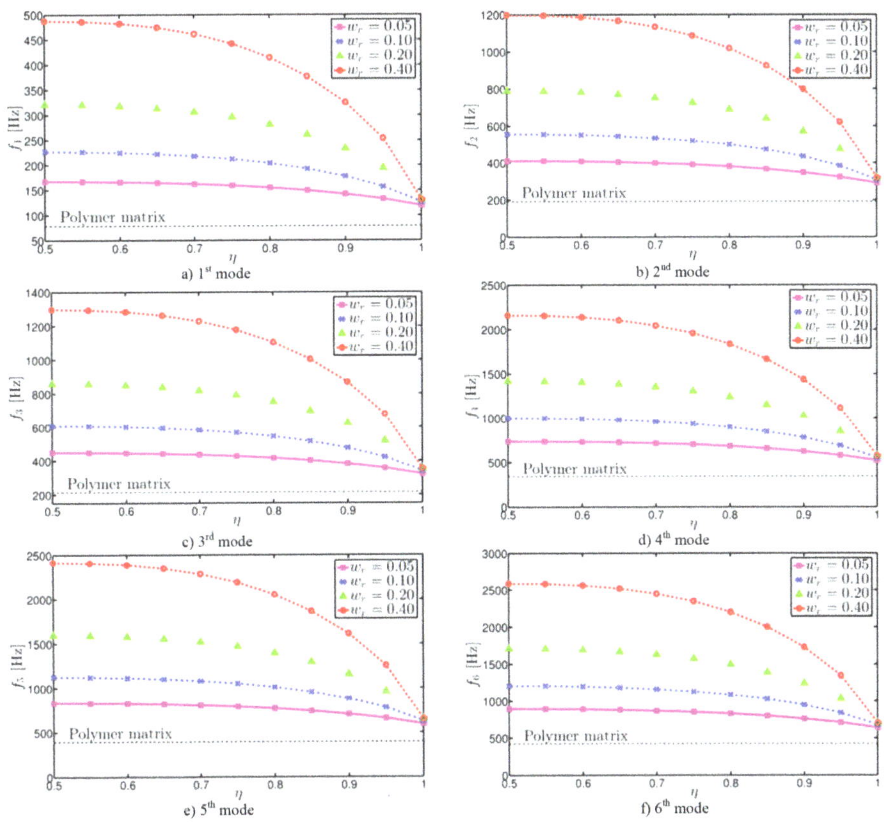

Figure 4.7. Frequency variations of a CFCC sectorial plate made of one lamina of constant thickness $h = 0.1$ m reinforced by CNTs uniformly distributed along the thickness for different mass fractions w_r varying the parameter of agglomeration η with $\mu = 0.5$. The Chebyshev–Gauss–Lobatto grid distributions are employed with $I_N = I_M = 25$.

Table 4.4. First ten natural frequency variations of a CFCC sectorial plate made of one lamina of constant thickness $h = 0.1$ m reinforced by CNTs uniformly distributed along the thickness for different mass fractions w_r varying the parameter of agglomeration η with $\mu = 0.5$. The Chebyshev–Gauss–Lobatto grid distributions are employed with $I_N = I_M = 25$.

Geometrical properties: $R_i = 0.5\text{m}$, $L = 1.0\text{m}$, $\Theta = 60°$.

$w_r = 0.05$

f [Hz]	Polymer matrix	$\eta = 0.5$	$\eta = 0.6$	$\eta = 0.7$	$\eta = 0.8$	$\eta = 0.9$	$\eta = 1.0$
1	78.574	167.728	166.642	162.899	155.422	142.283	119.248
2	191.448	410.980	408.310	399.095	380.679	348.288	291.429
3	214.233	450.620	447.737	437.804	418.005	383.298	322.640
4	344.064	740.077	735.264	718.645	685.424	626.973	524.321
5	393.721	833.717	828.355	809.870	772.993	708.276	595.013
6	425.752	897.451	891.701	871.888	832.386	763.114	641.967
7	467.490	989.729	983.365	961.422	917.648	840.829	706.395
8	501.797	1087.302	1080.196	1055.645	1006.523	919.992	767.758
9	525.423	1132.100	1124.728	1099.272	1048.375	958.800	801.427
10	611.674	1300.473	1292.084	1263.152	1205.401	1103.987	926.348

$w_r = 0.10$

f [Hz]	Polymer matrix	$\eta = 0.5$	$\eta = 0.6$	$\eta = 0.7$	$\eta = 0.8$	$\eta = 0.9$	$\eta = 1.0$
1	78.574	226.787	224.875	218.071	203.946	177.763	126.389
2	191.448	556.254	551.558	534.831	500.080	435.607	308.904
3	214.233	607.567	602.471	584.360	546.838	477.469	341.890
4	344.064	1002.016	993.552	963.398	900.738	784.448	555.776
5	393.721	1125.572	1116.111	1082.457	1012.667	883.488	630.572
6	425.752	1210.420	1200.263	1164.155	1089.331	950.949	680.288
7	467.490	1336.175	1324.943	1284.994	1202.151	1048.812	748.608
8	501.797	1473.945	1461.470	1416.993	1324.489	1152.611	813.900
9	525.423	1533.249	1520.292	1474.120	1378.154	1200.004	849.527
10	611.674	1757.080	1742.291	1689.662	1580.463	1378.197	981.763

$w_r = 0.20$

f [Hz]	Polymer matrix	$\eta = 0.5$	$\eta = 0.6$	$\eta = 0.7$	$\eta = 0.8$	$\eta = 0.9$	$\eta = 1.0$
1	78.574	321.540	318.381	306.677	281.545	233.489	130.306
2	191.448	789.351	781.595	752.833	691.027	572.764	318.474
3	214.233	859.252	850.818	819.638	752.817	625.302	352.494
4	344.064	1422.310	1408.332	1356.490	1245.062	1031.798	572.993
5	393.721	1593.719	1578.070	1520.155	1395.919	1158.619	650.122
6	425.752	1712.294	1695.486	1633.332	1500.103	1245.802	701.386
7	467.490	1891.874	1873.298	1804.549	1657.079	1375.401	771.819
8	501.797	2094.338	2073.748	1997.320	1832.911	1517.978	839.103
9	525.423	2176.931	2155.535	2076.164	1905.528	1578.878	875.841
10	611.674	2489.591	2465.140	2374.592	2180.251	1808.842	1012.196

$w_r = 0.40$

f [Hz]	Polymer matrix	$\eta = 0.5$	$\eta = 0.6$	$\eta = 0.7$	$\eta = 0.8$	$\eta = 0.9$	$\eta = 1.0$
1	78.574	487.840	482.750	461.855	414.831	325.524	130.804
2	191.448	1199.016	1186.514	1135.106	1019.318	799.386	319.682
3	214.233	1299.128	1285.550	1230.072	1105.534	869.099	353.865
4	344.064	2161.262	2138.732	2046.038	1837.210	1440.535	575.161
5	393.721	2413.581	2388.377	2285.158	2053.174	1612.686	652.631
6	425.752	2589.721	2562.659	2452.038	2203.656	1732.076	704.108
7	467.490	2864.981	2835.063	2712.546	2437.200	1914.369	774.798
8	501.797	3186.616	3153.419	3016.601	2708.090	2121.939	842.247
9	525.423	3309.087	3274.598	3132.634	2812.728	2205.031	879.147
10	611.674	3773.914	3734.525	3572.997	3209.714	2519.854	1016.082

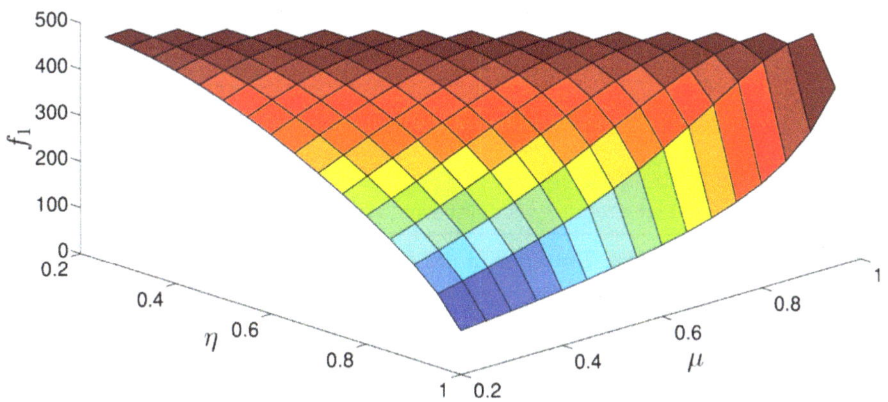

Figure 4.8. First frequency variation of a CFCC sectorial plate made of one lamina of constant thickness $h = 0.1$ m reinforced by CNTs uniformly distributed along the thickness for $w_r = 0.4$, varying both the parameters of agglomeration η, μ. The Chebyshev–Gauss–Lobatto grid distributions is employed with $I_N = I_M = 25$.

for two different shell structures (figure 4.4), i.e. a sectorial plate (figure 4.4(a)), or a complete elliptic cylinder (figure 4.4(b)).

Three different lamination schemes are assumed for each structural element, as shown in figure 4.5, by varying the exponent of the five-parameter distributions (4.40) and (4.41), while selecting an ED4 theory because of its well-known accuracy and stability [109, 110]. More speficially, the first application in figure 4.4(a) relies on a CFCC sectorial plate of inner radius $R_i = 0.5$ m, length $L = 1$ m and opening angle equal to $\Theta = 60°$. The structure is made of a single lamina of constant thickness $h = 0.1$ m reinforced by CNTs, with a uniform distribution along the thickness, accounting for different agglomeration effects. Table 4.3 summarizes the first ten natural frequency results under different mass fractions w_r assumptions,

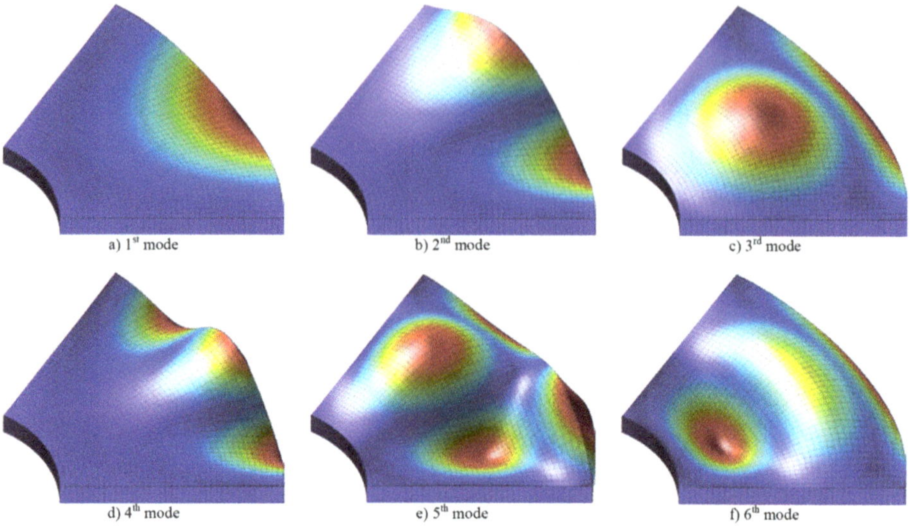

Figure 4.9. First six mode shapes of a CFCC sectorial plate made of one lamina of constant thickness $h = 0.1$ m reinforced by CNTs uniformly distributed along the thickness with $\eta = 1$, $\mu = 0.5$ and $w_r = 0.4$.

Table 4.5. First ten natural frequency variations of a CF elliptic cylinder made of two plies of constant thickness $h_1 = h_2 = 0.15$ m reinforced by CNTs for four different values of the parameter of agglomeration μ with $\eta = 1$ and $w_r = 0.2$, varying the exponents $p^{(1)} = p^{(2)} = p$ of the distributions of CNTs along the thickness. The Chebyshev–Gauss–Lobatto grid distributions are employed with $I_N = 40$ and $I_M = 15$.

Geometrical properties: $L = 6.0$ m, $a = 2.5$ m, $b = 1.5$ m. Lamination scheme: $\left(FG{-}CNT^{PM}_{CNT(5P)(a^{(1)}=1/b^{(1)}=0/c^{(1)}/d^{(1)}=1/p^{(1)})}/FG{-}CNT^{CNT}_{PM(5P)(a^{(2)}=1/b^{(2)}=0/c^{(2)}/d^{(2)}=1/p^{(2)})}\right)$										
$\mu = 0.25$										
f [Hz]	$p = 0$	$p = 1/4$	$p = 1/2$	$p = 1$	$p = 2$	$p = 4$	$p = 8$	$p = 16$	$p = 32$	$p = \infty$
1	18.634	18.675	18.684	18.636	18.210	17.353	16.488	15.830	15.409	14.696
2	19.000	19.012	18.972	18.765	18.426	17.947	17.203	16.386	15.713	14.895
3	19.123	19.163	19.170	19.112	18.879	18.365	17.594	16.764	16.088	15.075
4	27.578	27.595	27.534	27.226	26.402	25.138	23.876	22.930	22.336	21.626
5	43.604	43.663	43.621	43.311	42.389	40.531	38.465	36.897	35.903	34.354
6	43.948	44.057	44.099	43.906	42.579	40.801	38.939	37.266	36.041	34.716
7	43.959	44.069	44.112	44.053	43.693	42.688	40.983	39.012	37.333	34.730
8	44.469	44.497	44.399	44.067	43.710	42.710	41.009	39.038	37.355	34.742
9	46.956	47.017	46.966	46.614	45.584	43.834	41.816	40.033	38.741	36.978
10	53.435	53.505	53.449	53.046	51.831	49.744	47.420	45.469	44.085	42.143
$\mu = 0.50$										
f [Hz]	$p = 0$	$p = 1/4$	$p = 1/2$	$p = 1$	$p = 2$	$p = 4$	$p = 8$	$p = 16$	$p = 32$	$p = \infty$
1	24.337	24.278	24.154	23.524	21.958	19.974	18.143	16.795	15.939	14.696
2	24.936	24.720	24.377	23.803	23.002	21.623	19.857	18.116	16.745	14.895

3	24.985	24.916	24.778	24.390	23.527	22.073	20.253	18.490	17.115	15.075
4	36.188	35.868	35.359	34.097	31.785	28.872	26.210	24.277	23.070	21.626
5	57.005	56.672	56.113	54.670	51.490	46.763	42.389	39.170	37.142	34.354
6	57.307	57.237	57.040	55.221	51.849	47.892	43.758	40.267	37.771	34.716
7	57.318	57.249	57.054	56.405	54.749	51.350	46.893	43.180	39.951	34.730
8	58.576	58.065	57.251	56.424	54.776	51.624	47.506	43.338	39.984	34.742
9	61.409	61.034	60.408	58.800	55.678	51.660	47.550	43.379	40.550	36.978
10	69.792	69.373	68.657	66.762	62.967	57.750	52.652	48.674	45.959	42.143

$\mu = 0.75$

f [Hz]	$p = 0$	$p = 1/4$	$p = 1/2$	$p = 1$	$p = 2$	$p = 4$	$p = 8$	$p = 16$	$p = 32$	$p = \infty$
1	33.837	33.367	32.607	30.410	27.183	23.585	20.443	18.160	16.703	14.696
2	34.750	33.785	32.819	31.681	29.637	26.705	23.453	20.481	18.185	14.895
3	34.848	34.238	33.643	32.418	30.251	27.195	23.854	20.846	18.548	15.075
4	50.567	48.996	47.256	44.017	39.275	34.017	29.461	26.190	24.132	21.626
5	79.345	77.526	75.476	71.520	63.975	55.409	47.898	42.424	38.949	34.354
6	79.539	78.699	76.882	71.654	65.280	57.643	50.343	44.429	40.222	34.716
7	79.548	78.713	77.658	75.314	69.979	61.681	53.835	47.547	43.114	34.730
8	82.198	79.682	77.679	75.345	70.679	63.745	56.150	49.145	43.561	34.742
9	85.510	83.485	81.209	76.831	70.722	63.799	56.215	49.207	43.608	36.978
10	97.037	94.755	92.095	86.801	78.413	68.543	59.662	52.982	48.503	42.143

$\mu = 1.00$

f [Hz]	$p = 0$	$p = 1/4$	$p = 1/2$	$p = 1$	$p = 2$	$p = 4$	$p = 8$	$p = 16$	$p = 32$	$p = \infty$
1	59.822	55.888	51.335	45.114	37.916	30.941	25.212	21.084	18.398	14.696
2	61.486	56.472	53.773	49.514	43.588	36.787	30.476	25.208	21.181	14.895
3	62.355	57.895	55.026	50.543	44.382	37.376	30.908	25.566	21.529	15.075
4	90.461	80.947	74.259	65.142	54.641	44.518	36.236	30.313	26.501	21.626
5	140.059	129.190	120.689	107.180	89.967	73.290	59.515	49.530	43.033	34.354
6	140.059	133.050	122.103	108.387	92.979	76.984	63.385	52.929	45.435	34.716
7	140.693	133.483	127.682	116.181	99.442	82.198	67.620	56.494	48.580	34.730
8	148.603	133.513	127.731	117.871	103.450	86.891	72.345	60.375	50.889	34.742
9	151.803	139.020	129.653	117.935	103.509	86.944	72.428	60.471	50.965	36.978
10	171.570	157.051	145.812	129.496	109.696	90.011	73.723	61.793	53.790	42.143

while varying the agglomeration parameter μ for $\eta = 1$. Based on results in table 4.3, it can be observed that the natural frequencies become greater for an increasing value of μ. The bigger the mass fractions w_r, the more emphasized the effect of this variation is. The same aspect can be also observed in the plots of figure 4.6, where the natural frequencies of the same structure made completely of polymer matrix are plotted in black, too. Some meaningful variations can be observed for $\mu = \eta = 1$, i.e. for a complete agglomeration, when the mass fraction is equal to $w_r = 0.4$. The opposite behavior is observed in the plots of figure 4.7 and in table 4.4, where the response is highly affected by the parameter η, especially for higher mass fractions,

Table 4.6. First ten natural frequency variations of a CF elliptic cylinder made of two plies of constant thickness $h_1 = h_2 = 0.15$ m reinforced by CNTs for three different values of the parameter of agglomeration η with $\mu = 0.5$ and $w_r = 0.2$, varying the exponents $p^{(1)} = p^{(2)} = p$ of the distributions of CNTs along the thickness. The Chebyshev–Gauss–Lobatto grid distributions are employed with $I_N = 40$ and $I_M = 15$.

Geometrical properties: $L = 6.0$ m, $a = 2.5$ m, $b = 1.5$ m. Lamination scheme: $\left(FG\text{-}CNT^{PM}_{CNT(5P)(a^{(1)}=1/b^{(1)}=0/c^{(1)}/d^{(1)}=1/p^{(1)})}/FG\text{-}CNT^{CNT}_{PM(5P)(a^{(2)}=1/b^{(2)}=0/c^{(2)}/d^{(2)}=1/p^{(2)})}\right)$

$\eta = 0.50$

f [Hz]	$p = 0$	$p = 1/4$	$p = 1/2$	$p = 1$	$p = 2$	$p = 4$	$p = 8$	$p = 16$	$p = 32$	$p = \infty$
1	59.844	55.905	51.349	45.124	37.924	30.946	25.215	21.086	18.399	14.696
2	61.509	56.491	53.789	49.527	43.598	36.794	30.481	25.212	21.183	14.895
3	62.379	57.914	55.042	50.557	44.392	37.383	30.913	25.569	21.531	15.075
4	90.496	80.972	74.279	65.157	54.652	44.526	36.241	30.316	26.503	21.626
5	140.110	129.231	120.723	107.206	89.986	73.303	59.523	49.535	43.036	34.354
6	140.110	133.092	122.137	108.414	92.998	76.997	63.394	52.935	45.438	34.716
7	140.745	133.526	127.720	116.210	99.462	82.212	67.630	56.500	48.584	34.730
8	148.661	133.557	127.769	117.902	103.473	86.906	72.356	60.382	50.894	34.742
9	151.860	139.064	129.689	117.966	103.532	86.960	72.438	60.479	50.970	36.978
10	171.633	157.099	145.852	129.526	109.718	90.026	73.733	61.799	53.794	42.143

$\eta = 0.75$

f [Hz]	$p = 0$	$p = 1/4$	$p = 1/2$	$p = 1$	$p = 2$	$p = 4$	$p = 8$	$p = 16$	$p = 32$	$p = \infty$
1	55.136	51.693	47.626	42.048	35.565	29.263	24.087	20.375	17.977	14.696
2	56.669	52.164	49.764	45.973	40.695	34.616	28.903	24.104	20.457	14.895
3	57.448	53.481	50.929	46.937	41.444	35.175	29.323	24.462	20.808	15.075
4	83.342	74.877	68.901	60.724	51.262	42.111	34.631	29.308	25.911	21.626
5	129.105	119.432	111.850	99.833	84.316	69.237	56.787	47.809	42.020	34.354
6	129.106	123.011	113.224	100.855	87.061	72.705	60.400	50.908	44.155	34.716
7	129.659	123.280	118.138	108.126	93.133	77.647	64.459	54.364	47.237	34.730
8	136.859	123.307	118.182	109.448	96.694	81.981	68.788	57.788	49.132	34.742
9	139.891	128.526	120.169	109.507	96.753	82.040	68.871	57.878	49.201	36.978
10	158.131	145.237	135.231	120.654	102.882	85.145	70.429	59.669	52.493	42.143

$\eta = 1.00$

f [Hz]	$p = 0$	$p = 1/4$	$p = 1/2$	$p = 1$	$p = 2$	$p = 4$	$p = 8$	$p = 16$	$p = 32$	$p = \infty$
1	24.337	24.278	24.154	23.524	21.958	19.974	18.143	16.795	15.939	14.696
2	24.936	24.720	24.377	23.803	23.002	21.623	19.857	18.116	16.745	14.895
3	24.985	24.916	24.778	24.390	23.527	22.073	20.253	18.490	17.115	15.075
4	36.188	35.868	35.359	34.097	31.785	28.872	26.210	24.277	23.070	21.626
5	57.005	56.672	56.113	54.670	51.490	46.763	42.389	39.170	37.142	34.354
6	57.307	57.237	57.040	55.221	51.849	47.892	43.758	40.267	37.771	34.716
7	57.318	57.249	57.054	56.405	54.749	51.350	46.893	43.180	39.951	34.730
8	58.576	58.065	57.251	56.424	54.776	51.624	47.506	43.338	39.984	34.742
9	61.409	61.034	60.408	58.800	55.678	51.660	47.550	43.379	40.550	36.978
10	69.792	69.373	68.657	66.762	62.967	57.750	52.652	48.674	45.959	42.143

while keeping fixed $\mu = 0.5$. More in detail, the sensitivity of the vibration response increases when η tends to μ, reaching the highest frequencies for $\mu = \eta$ for each selected mass fraction of CNTs. The sensitivity of solutions gradually reduces for increasing values of η, vanishing completely for $\eta = 1$. The variation of the first frequency f_1 is depicted in the three-dimensional plot of figure 4.8, as a function of both the parameters of agglomeration μ, η, for $w_r = 0.4$, while reproducing the same trend of figures 4.6–4.7. In particular, for $\eta = 1$, the plot reverts to the same curve of figure 4.6(a), whereas for $\mu = 0.5$ the plot corresponds to figure 4.7(a). Note also that fixed values of $\eta = 1$ and $\mu = 0.2$ get the lowest frequency value, which corresponds to an increased deformability of the structure. The first six mode shapes of the selected plate are depicted in figure 4.9 for $\eta = 1$, $\mu = 0.5$ and $w_r = 0.4$, whose

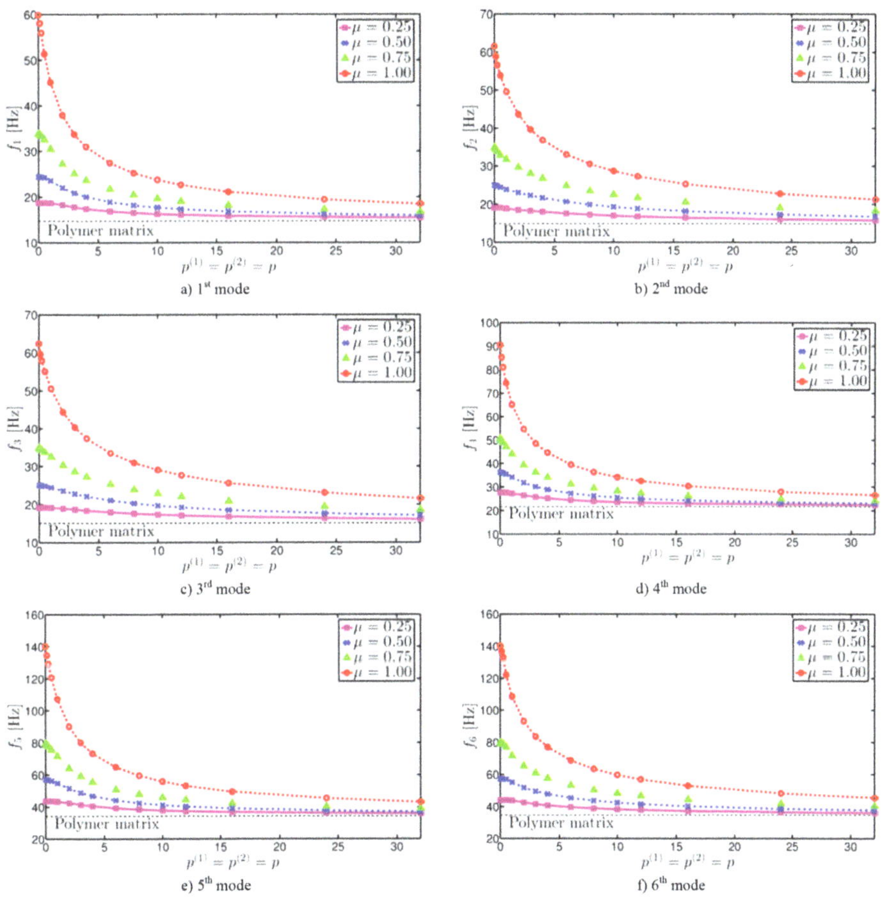

Figure 4.10. Frequency variations of a CF elliptic cylinder made of two plies of constant thickness $h_1 = h_2 = 0.15$ m reinforced by CNTs with lamination scheme $(FG{-}CNT^{PM}_{CNT(5P)(a^{(1)}=1/b^{(1)}=0/c^{(1)}/d^{(1)}=1/p^{(1)})}/FG{-}CNT^{CNT}_{PM(5P)(a^{(2)}=1/b^{(2)}=0/c^{(2)}/d^{(2)}=1/p^{(2)})})$ for four different values of the parameter of agglomeration μ with $\eta = 1$ and $w_r = 0.2$, varying the exponents $p^{(1)} = p^{(2)} = p$ of the distributions of CNTs along the thickness. The Chebyshev–Gauss–Lobatto grid distributions are employed with $I_N = 40$ and $I_M = 15$.

colormaps refer to the three-dimensional displacements, in line with predictions by Shi *et al* [95]. In agreement to the selected boundary conditions, the dark blue color refers to a null value of the displacement field, where the red color stands for the maximum structural displacements.

The second example considers a CF elliptic cylinder [110] of constant thickness $h = 0.3$ m, semi-axes $a = 2.5$ m and $b = 1.5$ m, length $L = 6$ m, as shown in figure 4.4(b). Two parametric studies are here performed, accounting for different exponents $p^{(k)}$ in equations (4.40) and (4.41). The reference lamination scheme is depicted in figure 4.5(a) and defined in tables 4.5–4.6. In particular, table 4.5 summarizes the first ten natural frequencies as a function of $p^{(1)} = p^{(2)} = p$ under

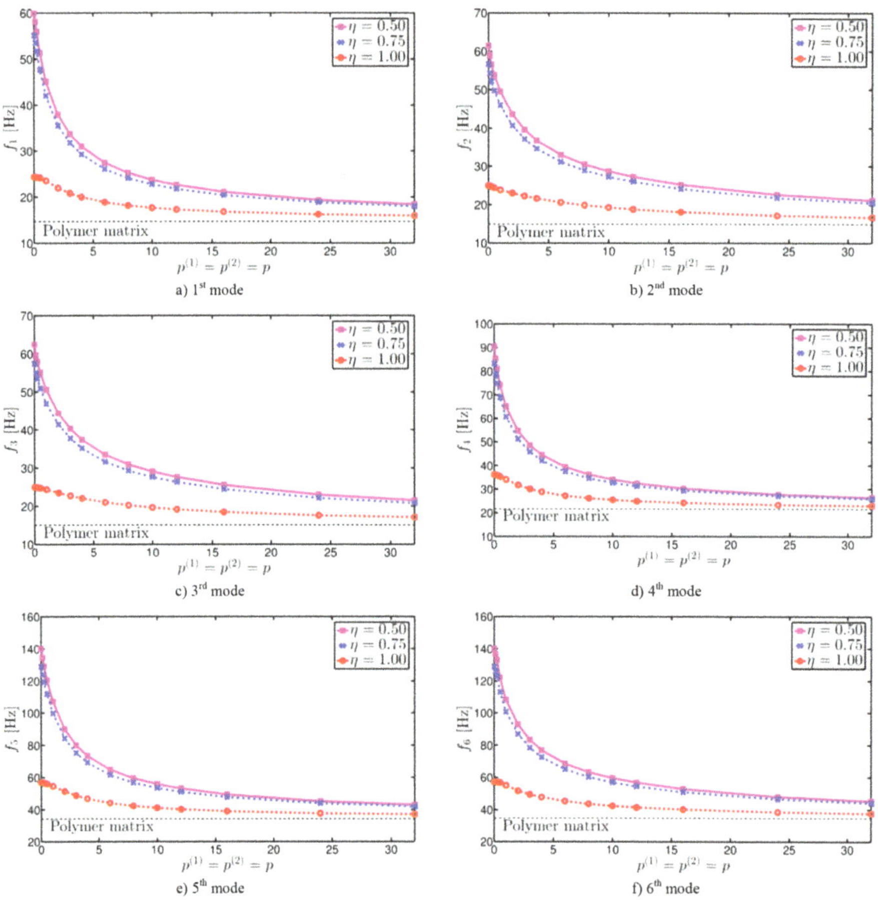

Figure 4.11. Frequency variations of a CF elliptic cylinder made of two plies of constant thickness $h_1 = h_2 = 0.15$ m reinforced by CNTs with lamination scheme $(FG{-}CNT^{PM}_{CNT(5P)(a^{(1)}=1/b^{(1)}=0/c^{(1)}/d^{(1)}=1/p^{(1)})}/FG{-}CNT^{CNT}_{PM(5P)(a^{(2)}=1/b^{(2)}=0/c^{(2)}/d^{(2)}=1/p^{(2)})})$ for three different values of the parameter of agglomeration $\eta = 1$ with $\mu = 0.5$ and $w_r = 0.2$, varying the exponents $p^{(1)} = p^{(2)} = p$ of the distributions of CNTs along the thickness. The Chebyshev–Gauss–Lobatto grid distributions are employed with $I_N = 40$ and $I_M = 15$.

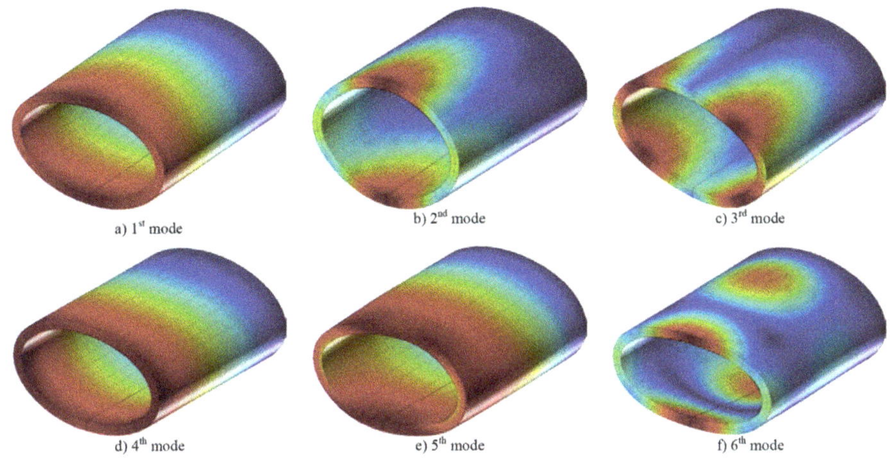

Figure 4.12. First six mode shapes of a CF elliptic made of two plies of constant thickness reinforced by CNTs with lamination scheme $(FG\text{–}CNT^{PM}_{CNT(5P)(a^{(1)}=1/b^{(1)}=0/c^{(1)}/d^{(1)}=1/p^{(1)})}/FG\text{–}CNT^{CNT}_{PM(5P)(a^{(2)}=1/b^{(2)}=0/c^{(2)}/d^{(2)}=1/p^{(2)})})$, assuming $\mu = \eta = 0.5$, $w_r = 0.2$ and $p^{(1)} = p^{(2)} = p = 2$.

Table 4.7. Comparison of the frequency parameter $\tilde{\omega}_{\mathrm{b}}$, $\tilde{\omega}_{\mathrm{f}}$ for a rotating simply-supported isotropic cylindrical shell ($\Omega_2 = 10$ rad/s), with $L/R = 5$ and $h/R = 0.002$. The present solution is achieved by setting $I_N = 35$, $I_M = 21$ for the application of the GDQ method. The mode shape is defined by the corresponding axial wave number m and circumferential wave number n.

Mode shape		Dong *et al* [111]		Liew *et al* [112]		Jafari and Bagheri [113]		Present solution	
m	n	$\tilde{\omega}_{\mathrm{f}}$	$\tilde{\omega}_{\mathrm{b}}$	$\tilde{\omega}_{\mathrm{f}}$	$\tilde{\omega}_{\mathrm{b}}$	$\tilde{\omega}_{\mathrm{f}}$	$\tilde{\omega}_{\mathrm{b}}$	$\tilde{\omega}_{\mathrm{f}}$	$\tilde{\omega}_{\mathrm{b}}$
1	1	0.1846	0.1882	0.1850	0.1864	0.1860	0.1867	0.1846	0.1881
	2	0.0745	0.0775	0.0742	0.0773	0.0756	0.0762	0.0744	0.0774
	3	0.0374	0.0397	0.0374	0.0394	0.0380	0.0384	0.0370	0.0393
	4	0.0241	0.0259	0.0240	0.0255	0.0237	0.0241	0.0229	0.0247
	5	0.0215	0.0230	0.0215	0.0220	0.0200	0.0203	0.0192	0.0207
	6	0.0249	0.0261	0.0248	0.0262	0.0239	0.0241	0.0219	0.0231
	7	0.0313	0.0324	0.0316	0.0321	0.0294	0.0298	0.0281	0.0291
	8	0.0396	0.0406	0.0394	0.0397	0.0368	0.0370	0.0363	0.0372
2	1	0.4573	0.4610	0.4605	0.4607	0.4591	0.4598	0.4573	0.4610
	2	0.2385	0.2414	0.2397	0.2400	0.2401	0.2407	0.2385	0.2414
	3	0.1334	0.1356	0.1341	0.1343	0.1346	0.1350	0.1332	0.1355
	4	0.0828	0.0846	0.0832	0.0834	0.0835	0.0839	0.0825	0.0843
	5	0.0574	0.0589	0.0578	0.0580	0.0574	0.0577	0.0566	0.0581
	6	0.0456	0.0468	0.0460	0.0465	0.0447	0.0449	0.0440	0.0452
	7	0.0426	0.0437	0.0434	0.0436	0.0409	0.0411	0.0403	0.0414
	8	0.0458	0.0468	0.0462	0.0464	0.0435	0.0437	0.0429	0.0439

four different assumptions for the agglomeration parameter μ, while keeping the other quantities equal to $\eta = 1$ and $w_r = 0.2$.

The parametric investigation is repeated for different values of η, whose results are listed in table 4.6 for $\mu = 0.5$ and $w_r = 0.2$. It should be pointed out that the structure is made completely of polymer matrix, when p tends to ∞. Based on the plots in figures 4.10–4.11, the sensitivity of the response increases with μ, reaching the highest variation for $\mu = \eta = 1$, and $p = 0$. Each natural frequency reduced monotonically for increasing values of p, with a similar behavior for all natural frequencies since the overall mechanical behavior is isotropic. The first six mode shapes of the structure under consideration are finally shown in figure 4.12, assuming $\mu = \eta = 0.5$ and $p = 2$, which reflect the the selected boundary condition.

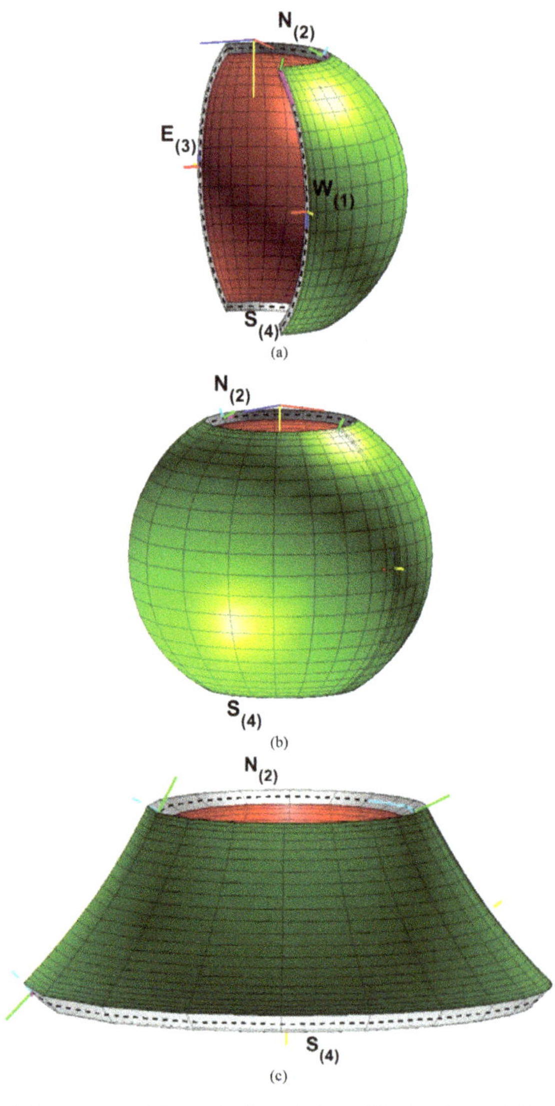

Figure 4.13. Rotating shell structures: (a) panel of revolution with circular meridian; (b) shell of revolution with circular meridian; (c) doubly-curved shell described by a funnel surface.

4.4.2 Critical speed evaluation

The same numerical approach is now applied to study the dynamic response of shell structures subjected to a constant angular speed, while computing their critical speed for different geometries and mechanical properties. The approach is validated, first, by means of a comparative evalation between results and predictions from literature. To this end, a simply-supported cylindrical shell made of one isotropic layer is considered, with radius R, length L and thickness h, under a constant angular speed $\Omega_2 = 10$ rad/s about the revolution axis of the cylinder. The reference solutions

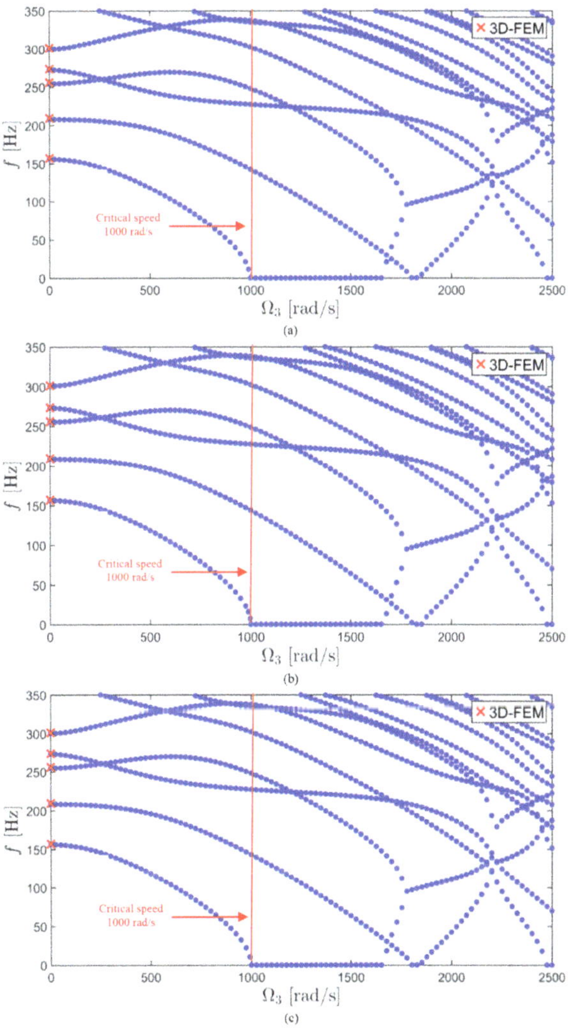

Figure 4.14. Variation of the natural frequencies as a function of the applied angular velocity Ω_3 and critical speed evaluation of a CFCF isotropic panel of revolution with circular meridian varying the structural model: (a) FSDT; (b) ED2; (c) ED3.

resort to [111–113], where results are defined in terms of dimensionless circular frequencies, i.e.

$$\tilde{\omega}_i = 2f\pi R\sqrt{\frac{\rho(1-\nu^2)}{E}} \tag{4.130}$$

for $i = \mathrm{b, f}$, in which 'b' and 'f' stand for the backward and forward waves, respectively. The frequency response is summarized in table 4.7 as a function of the axial and circumferential wave numbers m, n. The shell geometry is defined in a rational form by means of $L/R = 5$ and $h/R = 0.002$. The results are in excellent agreement with the reference ones, which proves the high accuracy of the proposed method to handle also rotating structures.

After this validation step, we compute the critical speed of some shell structures (see figure 4.13) made of FGMs, with mechanical properties

$$\begin{aligned} E_{\mathrm{C}} &= 168\ \mathrm{GPa}, \quad \nu_{\mathrm{C}} = 0.3, \quad \rho_{\mathrm{C}} = 5700\ \mathrm{kg\ m^{-3}} \\ E_{\mathrm{M}} &= 70\ \mathrm{GPa}, \quad \nu_{\mathrm{M}} = 0.3, \quad \rho_{\mathrm{M}} = 2707\ \mathrm{kg\ m^{-3}} \end{aligned} \tag{4.131}$$

for the ceramic (C) and metallic (M) constituents, respectively (e.g. zirconia and aluminum). The three examples include, respectively, a panel of revolution with a circular meridian of radius R, a shell of revolution with circular meridian of radius R, and a doubly-curved shell described by a funnel surface. The first two structures with thickness $h = 0.15$ m, are defined by the following position vector

$$\mathbf{r}(\varphi, \vartheta) = R\sin\varphi\cos\vartheta\,\mathbf{e}_1 - R\sin\varphi\sin\vartheta\,\mathbf{e}_2 + R(1-\cos\varphi)\,\mathbf{e}_3 \tag{4.132}$$

with $\varphi \in [\varphi_0, \varphi_1]$, $\vartheta \in [\vartheta_0, \vartheta_1]$, $\varphi_0 = \pi/6$, $\varphi_1 = 5\pi/6$, $\vartheta_0 = 0$. The value of $\vartheta_1 = \pi$ is set for the panel (figure 4.13(a)), whereas $\vartheta_1 = 2\pi$ is used for the shell (figure 4.13(b)).

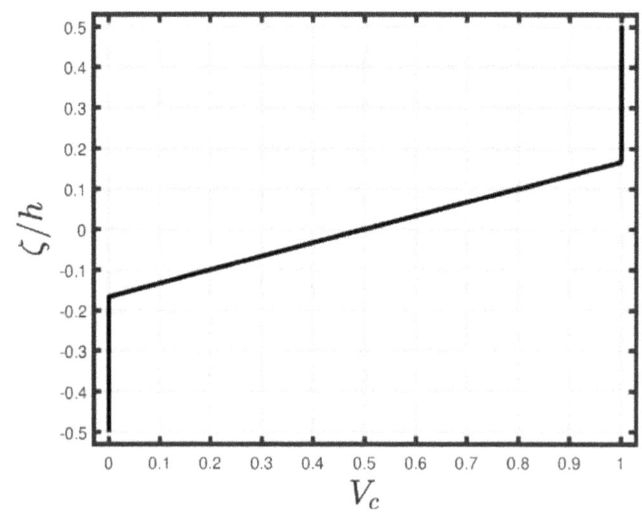

Figure 4.15. Through-the-thickness variation of the volume fraction of the ceramic constituent for the panel of revolution.

The third structure, instead, has a thickness $h = 0.1$ m (figure 4.13(c)), and it is is described as

$$\mathbf{r}(\alpha_1, \alpha_2) = \alpha_1 \cos\alpha_2 \mathbf{e}_1 - \alpha_1 \sin \alpha_2 \mathbf{e}_2 + \log \alpha_1 \mathbf{e}_3 \tag{4.133}$$

with $\alpha_1 \in [\alpha_1^0, \alpha_1^1]$, $\alpha_2 \in [\alpha_2^0, \alpha_2^1]$, $\alpha_1^0 = 0.5$, $\alpha_1^1 = 1$ and $\alpha_2^0 = 0$, $\alpha_2^1 = 2\pi$. The discrete modeling of each selected structure relies on the GDQ method, for a total number of collocation points $I_N = I_M = 25$. The first example starts considering a CFCF panel of revolution made of a single isotropic layer (aluminum), subjected to an angular velocity Ω_3 about the global axis x_3, (i.e. about the axis of revolution).

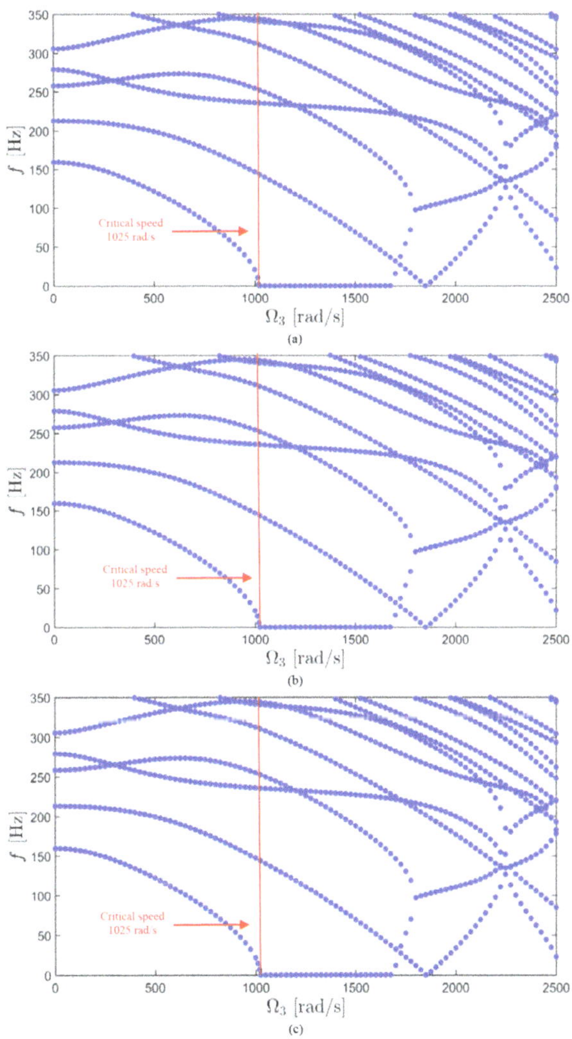

Figure 4.16. Variation of the natural frequencies as a function of the applied angular velocity Ω_3 and critical speed evaluation of a CFCF laminated panel of revolution with circular meridian varying the structural model: (a) FSDT; (b) ED2; (c) ED3.

The variation of the natural frequencies for different angular velocities is shown in figure 4.14, according to three different structural models, namely, a FSDT, ED2 and ED3. As specified in [20], the divergence phenomenon and the bifurcation of the natural frequencies are not particularly meaningful, because of the geometrical properties of the panel, which does not correspond to a complete shell of revolution. On the other hand, it is worth noticing the pronounced flutter instability. The first critical speed of the structure is about $\Omega_{cr} = 1000\ \text{rads}^{-1}$. All the results in figure 4.14 are compared to predictions from a classical three-dimensional finite element method (3D-FEM), only for non-rotating panels (i.e. for $\Omega_3 = 0$), with a very good agreement between them. The same structure is then reconsidered, under the same assumptions for boundary and

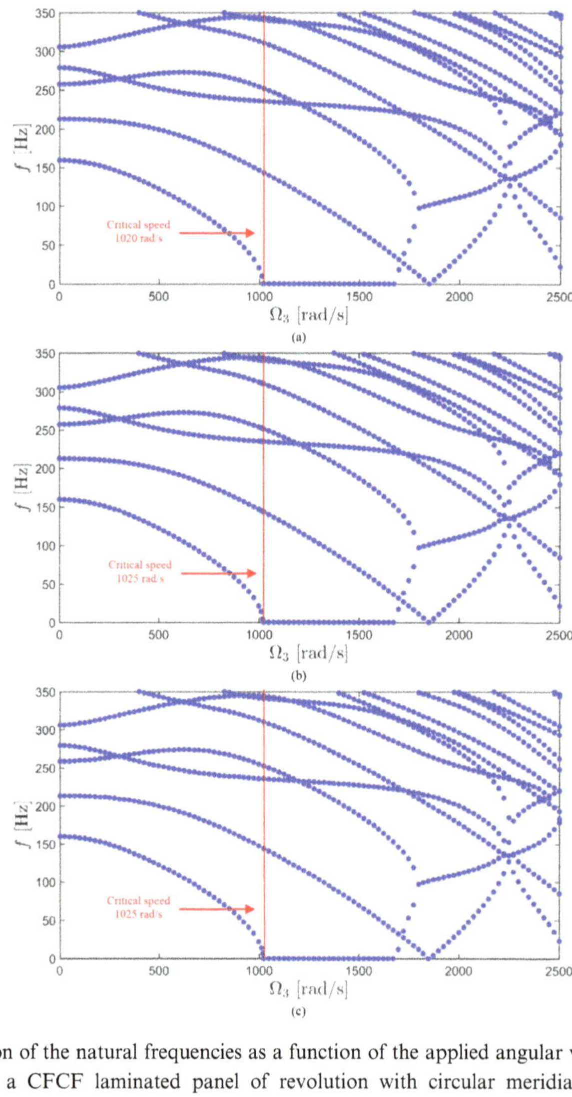

Figure 4.17. Variation of the natural frequencies as a function of the applied angular velocity Ω_3 and critical speed evaluation of a CFCF laminated panel of revolution with circular meridian varying the zig-zag structural model: (a) FSDTZ; (b) EDZ2; (c) EDZ3.

loading conditions, while assuming the lamination scheme (aluminum/FGM_1/zirconia), with $FGM_1 = M^C_{M(a^{(2)}=1/b^{(2)}=0/c^{(2)}/d^{(2)}=1/p^{(2)}=1)}$. The volume fraction distribution of the ceramic constituent V_C along the shell thickness is plotted in figure 4.15, with a linear variation of V_C within the core.

The variation of the natural frequencies versus the applied angular velocity is shown in figures 4.16, 4.17, for three different structural models, up to the third-order of kinematic expansion, by neglecting (i.e. FSDT, ED2 and ED3 theories, in figure 4.16) or including (i.e. FSDTZ, EDZ2 and EDZ3 theories, in figure 4.17) Murakami's function, respectively. Due to the selected lamination scheme made by three different layers, in the present example the use of the FSDTZ, EDZ2 and EDZ3 theories provide a more accurate results than the ones based on the FSDT, ED2 and ED3 approaches. The dynamic response is similar to the one shown in figure 4.14, but the critical speed is about $\Omega_{cr} = 1020 \div 1025$ rad s^{-1}.

As a second application, we consider a CF shell of revolution with circular meridian, as shown in figure 4.13(b). The structure is made of a single FGM layer, with a linear volume fraction distribution $FGM_1 = FGM^C_{M(a^{(1)}=1/b^{(1)}=0/c^{(1)}/d^{(1)}=1/p^{(1)}=1)}$, as depicted in figure 4.18. An increased angular velocity Ω_3 is enforced on the structure, about the axis of revolution x_3, whose effect on the natural frequencies is plotted in figure 4.19 for three different structural models (FSDT, ED2 and ED3). Murakami's function, in this case, is not included since the structure is made of a single layer, without any requirement of interlaminar behavior modeling. As is visible in the plots of figure 4.19, the divergence phenomenon occurs, with a clear bifurcation of the natural frequencies. These properties are completely different if an arbitrary angular velocity is applied to the same geometry. For example, if a rotating speed $\Omega_1 = \Omega_2 = \Omega$ is applied about the axes x_1, x_2, the structural response changes

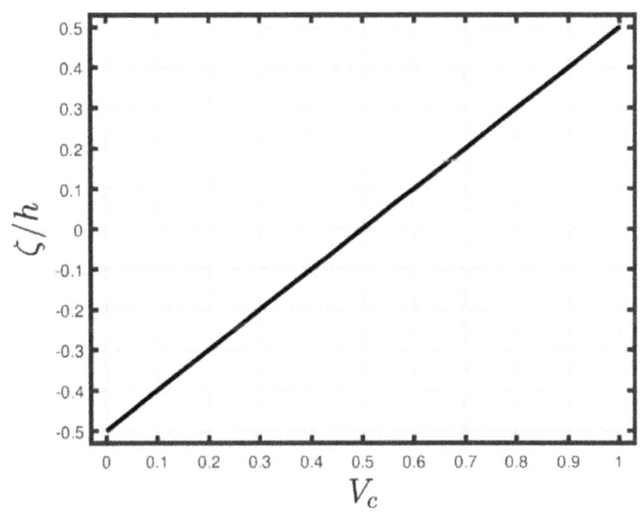

Figure 4.18. Through-the-thickness variation of the volume fraction of the ceramic constituent for the shell of revolution with circular meridian.

completely, as is visible in figure 4.20, despite the application of the same theoretical models.

The last example accounts for a CC doubly-curved shell described by a funnel surface made of a single FGM layer. In this case, a non-linear variation of the mechanical properties is set along the shell thickness, of the type $FGM_2 = FGM^C_{M(a^{(1)}=1/b^{(1)}=1/c^{(1)}=2/d^{(1)}=1/p^{(1)}=5)}$, as shown in figure 4.21. A symmetric distribution is set herein with respect to the shell middle surface (in other words, the same variation can be obtained by switching the position of the ceramic and

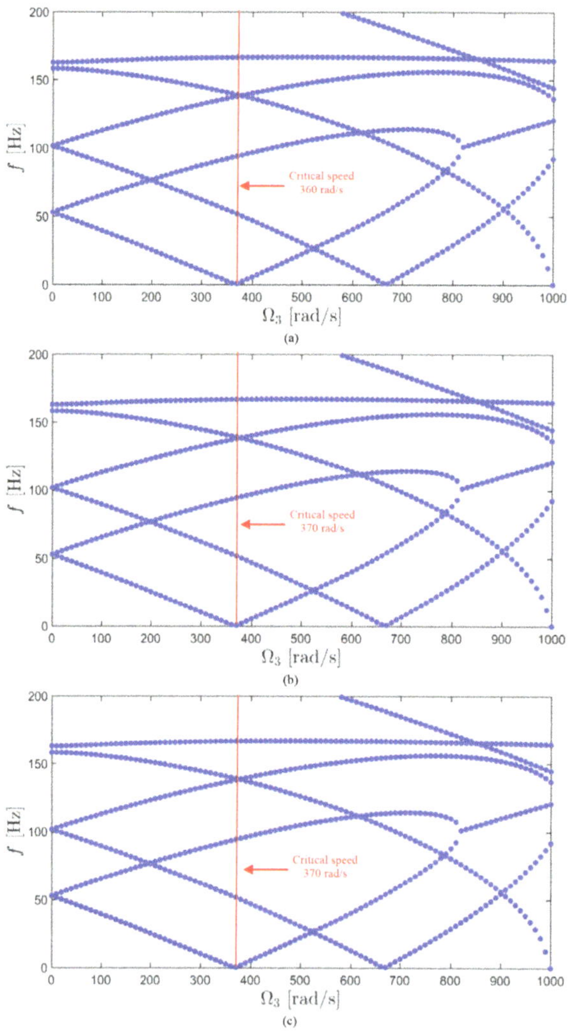

Figure 4.19. Variation of the natural frequencies as a function of the applied angular velocity Ω_3 and critical speed evaluation of a CF shell of revolution made of a single FGM layer with circular meridian varying the structural model: (a) FSDT; (b) ED2; (c) ED3.

metallic constituents), whereby the bottom and top surfaces are completely made of zirconia. Also in this case the structure is subjected to an increased angular velocity Ω_3 about the axis of revolution x_3.

The variation of the natural frequencies is shown in figure 4.22 for three different structural models (FSDT, ED2 and ED3), in line with the previous numerical applications. Even in this case, a pronounced divergence phenomenon is observed together with a bifurcation of the natural frequencies, in line with findings by Tornabene and Bacciocchi [20].

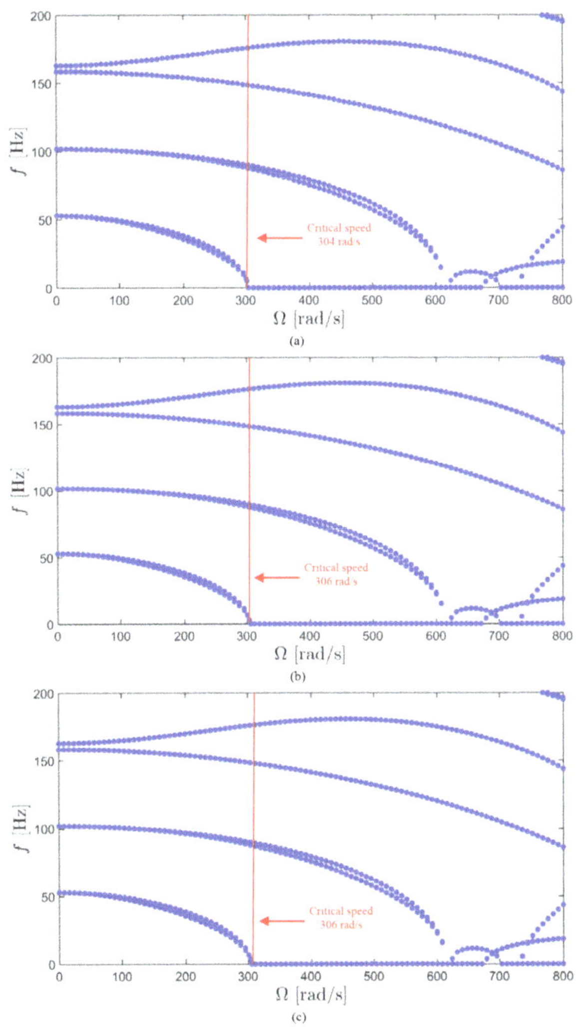

Figure 4.20. Variation of the natural frequencies as a function of the applied angular velocity $\Omega_1 = \Omega_2 = \Omega$ and critical speed evaluation of a CF shell of revolution made of a single FGM layer with circular meridian varying the structural model: (a) FSDT; (b) ED2; (c) ED3.

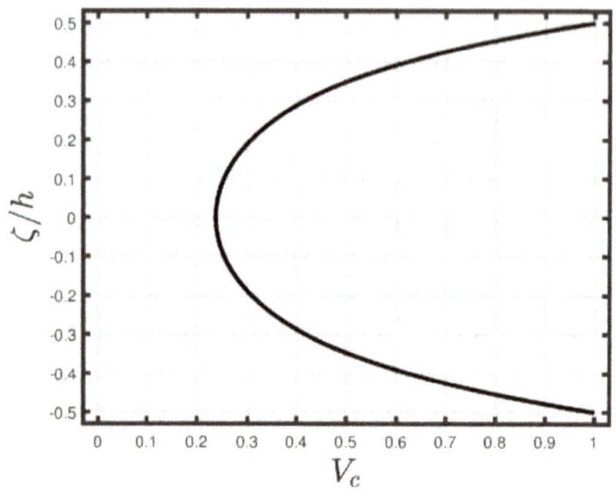

Figure 4.21. Through-the-thickness variation of the volume fraction of the ceramic constituent for the doubly-curved shell described by a funnel surface.

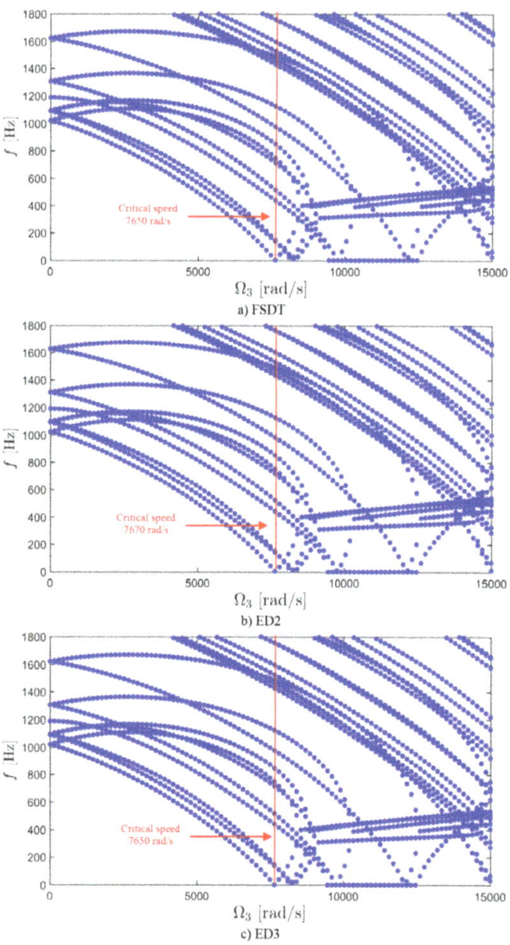

Figure 4.22. Variation of the natural frequencies as a function of the applied angular velocity Ω_3 and critical speed evaluation of a CC doubly-curved shell described by a funnel surface made of a single FGM layer varying the structural model: (a) FSDT; (b) ED2; (c) ED3.

4.5 Conclusions

This chapter provides a theoretical framework based on higher-order formulations combined with a GDQ-based approach to analyse numerically the mechanical behavior of generally-shaped shell structures in a strong form, even for complicated geometries and mechanical properties. The high performance of the proposed approach is explored and checked for different geometries, boundary conditions, loading conditions, and mechanical input properties, including FGMs and CNTs reinforcements within the composite material, with an excellent agreement of results with respect to predictions from literature. Several HSDTs are successfully applied to investigate the vibration response, the dynamic stability and the critical speed of thick or moderately thick structures, whose results are systematically compared to those from a FSDT (classical shell theory). As expected, when thin shell theories are applied to moderately thick shells, the approximation errors become significant, such that HSDTs are strictly required. Due to the increasing use of moderately thick shell structures in several engineering applications, the presented formulation allows one to study and compare several higher-order displacement fields in a unified manner. Thus, the mechanical behavior of doubly-curved structures is defined by the proper selection of several thickness functions that characterize the kinematic expansion along the three principal coordinates. In such a context, more complicated effects can be studied, as summarized in this chapter, where a systematic investigation discusses effects of different reinforcements within a composite material on the overall vibration response, together with possible effects related to an axial rotation geometrical configuration. A five-parameter power law has been properly applied to handle possible variations of the mechanical properties in FGM and CNTs-reinforced layers, whose general expression includes simple models, as commonly used in literature.

References

[1] Birman V and Byrd L W 2007 Modeling and analysis of functionally graded materials and structures *Appl. Mech. Rev.* **60** 195–216

[2] Miyamoto Y, Kaysser W A, Rabin B H, Kawasaki A and Ford R G 2013 *Functionally Graded Materials: Design, Processing and Applications* (New York: Springer)

[3] Noda N 1999 Thermal stresses in functionally graded materials *J. Therm. Stress* **22** 477–512

[4] Shen H S 2016 *Functionally Graded Materials: Nonlinear Analysis of Plates and Shells* (Boca Raton, FL: CRC Press)

[5] Allahkarami F, Tohidi H, Dimitri R and Tornabene F 2020 Dynamic stability of bi-directional functionally graded porous cylindrical shells embedded in an elastic foundation *Appl. Sci.* **10** 1345

[6] Li H, Lam K Y and Ng T Y 2005 *Rotating Shell Dynamics* (Oxford: Elsevier)

[7] Di Taranto R A and Lessen M 1964 Coriolis acceleration effect on the vibration of a rotating thin-walled circular cylinder *J. Appl. Mech.* **31** 700–1

[8] Srinivasan A V and Lauterbach G F 1971 Traveling waves in rotating cylindrical shells *J. Eng. Ind.* **93** 1229–32

[9] Eshleman R L and Eubanks R A 1967 On the critical speeds of a continuous shaft-disk system *J. Eng. Ind.* **89** 1180–88
[10] Eshleman R L and Eubanks R A 1969 On the critical speeds of a continuous rotor *J. Eng. Ind.* **91** 1180–88
[11] Chivens D R and Nelsen H D 1975 The natural frequencies and critical speeds of a rotating, flexible shaft-disk system *J. Eng. Ind.* **97** 881–86
[12] Nelson H D 1980 A finite rotating shaft element using Timoshenko beam theory *J. Mech. Design* **102** 793–803
[13] Dimarogonas A D and Papadopoulus C A 1983 Vibration of cracked shafts in bending *J. Sound Vib.* **91** 583–93
[14] Özgüven H N 1984 On the critical speed of continuous shaft-disk systems *J. Vib. Acoust.* **106** 59–61
[15] Kim C-D and Bert C W 1993 Critical speed analysis of laminated composite, hollow drive shafts *Compos. Eng.* **3** 633–43
[16] Ng T Y and Lam K Y 1999 Vibration and critical speed of a rotating cylindrical shell subjected to axial loading *Appl. Acoust.* **56** 273–82
[17] Daneshjou K, Talebitooti M, Talebitooti R and Googarchin H S 2013 Dynamic analysis and critical speed of rotating laminated conical shells with orthogonal stiffeners using generalized differential quadrature method *Lat. Am. J. Solids Struct.* **10** 349–90
[18] Daneshjou K, Talebitooti M and Talebitooti R 2013 Free vibration and critical speed of moderately thick rotating laminated composite conical shell using generalized differential quadrature method *Appl. Math. Mech. Engl. Ed.* **34** 437–56
[19] Barooti M M, Safarpour H and Ghadiri M 2017 Critical speed and free vibration analysis of spinning 3D single-walled carbon nanotubes resting on elastic foundations *Eur. Phys. J. Plus* **132** 6
[20] Tornabene F and Bacciocchi M 2018 Dynamic stability of doubly-curved multi-layered shells subjected to arbitrarily oriented angular velocities: numerical evaluation of the critical speed *Compos. Struct.* **201** 1031–55
[21] Tornabene F 2019 On the critical speed evaluation of arbitrarily oriented rotating doubly-curved shells made of functionally graded materials *Thin-Walled Struct.* **140** 85–98
[22] Cheng Z-Q and Batra R C 2000 Deflection relationships between the homogeneous plate theory and different functionally graded plate theories *Arch. Mech.* **52** 143–58
[23] Yang J and Shen H-S 2003 Nonlinear bending analysis of shear deformable functionally graded plates subjected to thermo-mechanical loads under various boundary conditions *Compos. Part B Eng.* **34** 103–15
[24] Batra R C 2006 Torsion of a functionally graded cylinder *AIAA J.* **44** 1363–65
[25] Efraim E and Eisenberger M 2007 Exact vibration analysis of variable thickness thick annular isotropic and FGM plates *J. Sound Vib.* **299** 720–38
[26] Tornabene F 2009 Free vibration analysis of functionally graded conical, cylindrical and annular shell structures with a four-parameter power-law distribution *Comput. Methods Appl. Mech. Engrg.* **198** 2911–35
[27] Mori T and Tanaka K 1973 Average stress in matrix and average elastic energy of materials with misfitting inclusions *Acta Metal.* **21** 571–74
[28] Mura T 1982 *Micromechanics of Defects in Solids* (The Hague: Martinus Nijhoff)
[29] Hashin Z 1983 Analysis of composite materials—a survey *J. Appl. Mech.* **50** 481–505

[30] Benveniste Y 1987 A new approach to the application of Mori–Tanaka's theory in composite materials *Mech. Mat.* **6** 147–57

[31] Iijima S 1994 Synthesis of carbon nanotubes *Nature* **354** 56–8

[32] Bouaziz O, Kim H S and Estrin Y 2013 Architecturing of metal-based composites with concurrent nanostructuring: a new paradigm of materials design *Adv. Eng. Mater.* **15** 336–40

[33] Beygelzimer Y, Estrin Y and Kulagin R 2015 Synthesis of hybrid materials by severe plastic deformation: a new paradigm of SPD processing *Adv. Eng. Mater.* **17** 1853–61

[34] Estrin Y, Beygelzimer Y and Kulagin R 2019 Design of architectured materials based on mechanically driven structural and compositional patterning *Adv. Eng. Mater.* **21** 1900487

[35] Ashrafi B, Hubert P and Vengallatore S 2006 Carbon nanotube-reinforced composites as structural materials for micro actuators in micro electromechanical systems *Nanotechnology* **17** 4895–903

[36] Rafiee M, Yang J and Kitipornchai S 2013 Thermal bifurcation buckling of piezoelectric carbon nanotube reinforced composite beams *Comput. Math. Appl.* **66** 1147–60

[37] Araneo R, Bini F, Rinaldi A, Notargiacomo A, Pea M and Celozzi S 2015 Thermal-Electric model for piezoelectric ZnO nanowires *Nanotechnology* **26** 265402

[38] Ng T Y, Lam K Y and Liew K M 2000 Effects of FGM materials on the parametric resonance of plate structures *Comput. Meth. Appl. Mech. Eng.* **190** 953–62

[39] Praveen G N and Reddy J N 1998 Nonlinear transient thermo elastic analysis of functionally graded ceramic–metal plates *Int. J. Solids Struct.* **33** 4457–76

[40] He X Q, Ng T Y, Sivashanker S and Liew K M 2001 Active control of FGM plates with integrated piezoelectric sensors and actuators *Int. J. Solids Struct.* **38** 1641–55

[41] Tornabene F, Viola E and Inman D J 2009 2-D differential quadrature solution for vibration analysis of functionally graded conical, cylindrical shell and annular plate structures *J. Sound Vib.* **328** 259–90

[42] Kiani Y, Dimitri R and Tornabene F 2018 Free vibration study of composite conical panels reinforced with FG-CNTs *Eng. Struct.* **172** 472–82

[43] Sofiyev A H, Tornabene F, Dimitri R and Kuruoglu N 2020 Buckling behavior of FG-CNT reinforced composite conical shells subjected to a combined loading *Nanomaterials* **10** 419

[44] Jena S K, Chakraverty S, Malikan M and Tornabene F 2020 Effects of surface energy and surface residual stresses on vibro-thermal analysis of chiral, zigzag, and armchair types of SWCNTs using refined beam theory *Mech. Based Des. Struct. Mach.*

[45] Ghasemi A R, Mohandes M, Dimitri R and Tornabene F 2019 Agglomeration effects on the vibrations of CNTs/fiber/polymer/metal hybrid laminates cylindrical shell *Compos. Part B: Eng.* **167** 700–16

[46] Mohammadi M, Arefi M, Dimitri R and Tornabene F 2019 Higher-order thermo-elastic analysis of FG-CNTRC cylindrical vessels surrounded by a pasternak foundation *Nanomaterials* **9** 79

[47] Jena S K, Chakraverty S, Malikan M and Tornabene F 2020 Effects of surface energy and surface residual stresses on vibro-thermal analysis of chiral, zigzag, and armchair types of SWCNTs using refined beam theory *Mech. Based Des. Struct. Mach.*

[48] Karami B, Janghorban M, Shahsavari D, Dimitri R and Tornabene F 2019 Nonlocal buckling analysis of composite curved beams reinforced with functionally graded carbon nanotubes *Molecules (Basel)* **24** 2750

[49] Noroozi A R, Malekzadeh P, Dimitri R and Tornabene F 2020 Meshfree radial point interpolation method for the vibration and buckling analysis of FG-GPLRC perforated plates under an in-plane loading *Eng. Struct.* **221** 111000
[50] Shu C 2000 *Differential Quadrature and Its Application in Engineering* (Berlin: Springer)
[51] Bellman R and Casti J 1971 Differential quadrature and long-term integration *J. Math. Anal. Appl.* **34** 235–38
[52] Bert C and Malik M 1996 Differential quadrature method in computational mechanics *Appl. Mech. Rev.* **49** 1–27
[53] Shu C and Du H 1997 Free vibration analysis of composites cylindrical shells by DQM *Compos. Part B Eng.* **28B** 267–74
[54] Liu F-L and Liew K M 1999 Differential quadrature element method: a new approach for free vibration of polar Mindlin plates having discontinuities *Comput. Methods Appl. Mech. Eng.* **179** 407–23
[55] Tornabene F and Viola E 2007 Vibration analysis of spherical structural elements using the GDQ method *Comput. Math. Appl.* **53** 1538–60
[56] Marzani A, Tornabene F and Viola E 2008 Nonconservative stability problems via generalized differential quadrature method *J. Sound Vib.* **315** 176–96
[57] Tornabene F and Viola E 2008 2-D solution for free vibrations of parabolic shells using generalized differential quadrature method *Eur. J. Mech. A-Solid* **27** 1001–25
[58] Tornabene F 2011 2-D GDQ solution for free vibrations of anisotropic doubly-curved shells and panels of revolution *Compos. Struct.* **93** 1854–76
[59] Tornabene F 2011 Free vibrations of anisotropic doubly-curved shells and panels of revolution with a free-form meridian resting on Winkler–Pasternak elastic foundations *Compos. Struct.* **94** 186–206
[60] Tornabene F and Reddy J N 2013 FGM and laminated doubly-curved and degenerate shells resting on nonlinear elastic foundations: a GDQ solution for static analysis with *a posteriori* stress and strain recovery *J. Indian Inst. Sci.* **93** 635–88
[61] Mosallaie Barzoki A A, Ghorbanpour Arani A, Kolahchi R, Mozdianfard M R and Loghman A 2013 Nonlinear buckling response of embedded piezoelectric cylindrical shell reinforced with BNNT under electro–thermo-mechanical loadings using HDQM *Compos. Part B Eng.* **44** 722–27
[62] Khani S, Tabandeh N and Ghomshei M M 2014 Natural frequency analysis of non-uniform smart beams with piezoelectric layers, using differential quadrature method *Compos. Part B Eng.* **58** 303–11
[63] Tornabene F, Fantuzzi N, Ubertini F and Viola E 2015 Strong formulation finite element method based on differential quadrature: a survey *Appl. Mech. Rev.* **67** 020801
[64] Tornabene F, Dimitri R and Viola E 2016 Transient dynamic response of generally-shaped arches based on a GDQ-time-stepping method *Int. J. Mech. Sci.* **114** 277–314
[65] Tornabene F and Brischetto S 2018 3D capability of refined GDQ models for the bending analysis of composite and sandwich plates, spherical and doubly-curved shells *Thin-Walled Struct.* **129** 94–124
[66] Brischetto S and Tornabene F 2018 Advanced GDQ models and 3D stress recovery in multilayered plates, spherical and double-curved panels subjected to transverse shear loads *Compos. Part B: Eng.* **146** 244–69

[67] Dimitri R, Tornabene F and Zavarise G 2018 Analytical and numerical modeling of the mixed-mode delamination process for composite moment-loaded double cantilever beams *Compos. Struct.* **187** 535–53
[68] Tornabene F and Dimitri R 2018 A numerical study of the seismic response of arched and vaulted structures made of isotropic or composite materials *Eng. Struct.* **159** 332–66
[69] Dimitri R, Tornabene F and Reddy J N 2020 Numerical study of the mixed-mode behavior of generally-shaped composite interfaces *Compos. Struct.* **237** 111935
[70] Ng T Y, Lam K Y and Liew K M 2000 Effect of FGM materials on the parametric resonance of plate structures *Comput. Methods Appl. Mech. Eng.* **190** 953–62
[71] Yang J, Kitipornchai S and Liew K M 2003 Large amplitude vibration of thermo-electro-mechanically stressed FGM laminated plates *Comput. Methods Appl. Mech. Eng.* **192** 3861–85
[72] Wu C P and Tsai Y H 2004 Asymptotic DQ solutions of functionally graded annular spherical shells *Eur. J. Mech A-Solids* **23** 283–99
[73] Liew M, He X Q and Kitipornchai S 2004 Finite element method for the feedback control of FGM shells in the frequency domain via piezoelectric sensors and actuators *Comput. Methods Appl. Mech. Eng.* **193** 257–73
[74] Della Croce L and Venini P 2004 Finite elements for functionally graded Reissner–Mindlin plates *Comput. Methods Appl. Mech. Eng.* **193** 705–25
[75] Elishakoff I, Gentilini C and Viola E 2005 Forced vibrations of functionally graded plates in the three-dimensional setting *AIAA J.* **43** 2000–7
[76] Elishakoff I, Gentilini C and Viola E 2005 Three-dimensional analysis of an all-around clamped plate made of functionally graded materials *Acta Mech.* **180** 21–36
[77] Patel B P, Gupta S S, Loknath M S and Kadu C P 2005 Free vibration analysis of functionally graded elliptical cylindrical shells using higher-order theory *Compos. Struct.* **69** 259–70
[78] Abrate S 2006 Free vibration, buckling, and static deflection of functionally graded plates *Compos. Sci. Technol.* **66** 2383–94
[79] Zenkour A M 2006 Generalized shear deformation theory for bending analysis of functionally graded plates *Appl. Math. Model.* **30** 67–84
[80] Pelletier J L and Vel S S 2006 An exact solution for the steady-state thermoelastic response of functionally graded orthotropic cylindrical shells *Int. J. Solids Struct.* **43** 1131–58
[81] Arciniega R A and Reddy J N 2007 Large deformation analysis of functionally graded shells *Int. J. Solids Struct.* **44** 2036–52
[82] Arciniega R A and Reddy J N 2007 Tensor-based finite element formulation for geometrically nonlinear analysis of shell structures *Comput. Methods Appl. Mech. Eng.* **196** 1048–73
[83] Roque C M C, Ferreira A J M and Jorge R M N 2007 A radial basis function for the free vibration analysis of functionally graded plates using refined theory *J. Sound Vib.* **300** 1048–70
[84] Nie G J and Zhong Z 2007 Semi-analytical solution for three dimensional vibration of functionally graded circular plates *Comput. Methods Appl. Mech. Engrg.* **196** 4901–10
[85] Yang J and Shen H S 2007 Free vibration and parametric resonance of shear deformable functionally graded cylindrical panels *J. Sound Vib.* **261** 871–93
[86] Tornabene F 2016 General higher order layer-wise theory for free vibrations of doubly-curved laminated composite shells and panels *Mech. Adv. Mater. Struct.* **23** 1046–67

[87] Tornabene F, Fantuzzi N, Bacciocchi M and Viola E 2018 Mechanical behavior of damaged laminated composites plates and shells: higher-order shear deformation theories *Compos. Struct.* **189** 304–29
[88] Tornabene F, Fantuzzi N and Bacciocchi M 2019 Refined shear deformation theories for laminated composite arches and beams with variable thickness: natural frequency analysis *Eng. Anal. Bound. Elem.* **100** 24–47
[89] Kraus H 1967 *Thin Elastic Shells* (New York: Wiley)
[90] Tornabene F and Bacciocchi M 2018 *Anisotropic Doubly-Curved Shells. Higher-Order Strong and Weak Formulations for Arbitrarily Shaped Shell Structures* (Bologna: Esculapio)
[91] Carrera E 2003 Historical review of zig-zag theories for multi-layered plates and shells *Appl. Mech. Rev.* **56** 287–308
[92] Carrera E 2004 On the use of the Murakami's zig-zag function in the modeling of layered plates and shells *Comput. Struct.* **82** 541–54
[93] Tornabene F, Fantuzzi N and Bacciocchi M 2019 Foam core composite sandwich plates and shells with variable stiffness: effect of curvilinear fiber path on the modal response *J. Sandw. Struct. Mater.* **21** 320–65
[94] Tornabene F, Viola E and Fantuzzi N 2013 General higher-order equivalent single layer theory for free vibrations of doubly-curved laminated composite shells and panels *Compos. Struct.* **104** 94–117
[95] Shi D-L, Huang Y Y, Hwang K-C and Gao H 2004 The effect of nanotube waviness and agglomeration on the elastic property of carbon nanotube-reinforced composites *J. Eng. Mater. – T. ASME* **126** 250–57
[96] Tornabene F and Viola E 2013 Static analysis of functionally graded doubly-curved shells and panels of revolution *Meccanica* **48** 901–30
[97] Tornabene F, Fantuzzi N and Bacciocchi M 2014 Free vibrations of free-form doubly-curved shells made of functionally graded materials using higher-order equivalent single layer theories *Compos. Part B Eng.* **67** 490–509
[98] Tornabene F, Fantuzzi N, Viola E and Batra R C 2015 Stress and strain recovery for functionally graded free-form and doubly-curved sandwich shells using higher-order equivalent single layer theory *Compos. Struct.* **119** 67–89
[99] Odegard G M, Gates T S, Wise K E, Park C and Siochi E J 2003 Constitutive modeling of nanotube-reinforced polymer composites *Compos. Sci. Technol.* **63** 1671–87
[100] Hill R 1965 A self-consistent mechanics of composite materials *J. Mech. Phys. Solids* **13** 213–22
[101] Walpole L J 1969 On the overall elastic moduli of composite materials *J. Mech. Phys. Solids* **17** 235–51
[102] Dong X N, Zhang X, Huang Y Y and Guo X E 2005 A generalized self-consistent estimate for the effective elastic moduli of fiber-reinforced composite materials with multiple transversely isotropic inclusions *Int. J. Mech. Sci.* **47** 922–40
[103] Popov V N, Van Doren V E and Balkanski M 2000 Elastic properties of crystals of single-walled carbon nanotubes *Solid State Commun.* **114** 395–99
[104] Tornabene F and Dimitri R 2020 Generalized differential and integral quadrature: theory and applications *Mathematical Methods in Interdisciplinary Sciences* ed S Chakraverty (New York: Wiley) ch 14
[105] Kim N-I and Lee J 2014 Divergence and flutter behavior of Beck's type of laminated box beams *Int. J. Mech. Sci.* **84** 91–101

[106] Leipholz H 1978 *Stability of Elastic Structures* (New York: Springer)
[107] Leipholz H 1987 Stability theory *An Introduction to the Stability of Dynamic Systems and Rigid Bodies* 2nd edn (New York: Springer)
[108] Bigoni D and Noselli G 2011 Experimental evidence of flutter and divergence instabilities induced by dry friction *J. Mech. Phys. Solids* **59** 2208–26
[109] Tornabene F, Fantuzzi N, Bcciocchi M and Viola E 2016 Effect of agglomeration on the natural frequencies of functionally graded carbon nanotube-reinforced laminated composite doubly-curved shells *Compos. Part B: Eng.* **89** 187–218
[110] Tornabene F, Fantuzzi N, Bacciocchi M and Dimitri R 2015 Dynamic analysis of thick and thin elliptic shell structures made of laminated composite materials *Compos. Struct.* **133** 278–99
[111] Dong Y H, Li Y H, Chen D and Yang J 2018 Vibration characteristics of functionally graded graphene reinforced porous nanocomposite cylindrical shells with spinning motion *Compos. Part B Eng.* **145** 1–13
[112] Liew K M, Ng T Y, Zhao X and Reddy J N 2002 Harmonic reproducing kernel particle method for free vibration analysis of rotating cylindrical shells *Comput. Methods Appl. Mech. Eng.* **191** 4141–57
[113] Jafari A A and Bagheri M 2006 Free vibration of rotating ring stiffened cylindrical shells with non-uniform stiffener distribution *J. Sound Vib.* **296** 353–67

Chapter 5

Vibration of micro/nano structural members: a discrete energy-based formulation

Mukul Saxena and Saikat Sarkar

5.1 Introduction

Continuum mechanics (CM) has played a key role in designing large-scale structures, ranging from bridges and tall buildings to aeroplanes [1, 2]. More recently, with the development of micro- and nano-scale engineering for applications like tissue mechanics [3], flexo-electronics [4] etc, continuum theories have also become crucial for designing such small systems. While molecular dynamics (MD) may be far more accurate than continuum theories, it is practically impossible to solve these problems using MD. The central concept in CM is that it does not see atoms or molecules interacting discretely with one another and conveniently assumes a fictitious continuum hyperspace covering the entire domain of the problem. This assumption makes the notion of a derivative well posed at every material point within the continuum, which allows us to describe the mechanics of the problem in terms of partial differential equation (PDE) based balance laws with well-defined variables like stress and strain. Apart from the full-fledged 3D continuum equations, another interesting aspect of CM is its reduced dimensional models like beams, plates, and shells, which help in solving a large class of problems with drastically reduced computational overhead. However, as the system size reduces, the usual CM equations do not hold sufficiently well. Non-local CM theories, which account for the size effects and microstructural details, may turn out to be quite handy in solving such problems. The journey of non-local CM started long back with the works of the Cosserat brothers [5]. After a long time, following the works of Eringen it found a renewed focus and different non-local CM theories involving higher-order derivatives or additional rotational degrees-of-freedom were proposed [6, 7]. However, these theories face certain criticism: (i) there are too many variants and (ii) the source of such non-locality is not clearly understood. An appropriate unification and physical basis would be necessary before non-local

doi:10.1088/978-0-7503-3483-9ch5 © IOP Publishing Ltd 2021

theories can be more frequently used for solving practical problems. On the other hand, all these theories are derivative-based. Hence they may be inefficient in handling discontinuities.

In this chapter, we address some of these issues. A complete discussion covering all the aspects may be too vast for single chapter. Hence we focus on deriving a molecular-dynamics-like discrete framework for different local and non-local continuum models like beams and plates so that the models can simulate discontinuities. A major aspect of the formulation is that it rationally unifies different local and non-local theories. The approach shown here can be directly applied to derive discrete counterparts for other local and non-local continuum models. In the following sections of the chapter, we consider different reduced dimensional continuum models separately and connect them to their discrete counterparts. Finally some concluding remarks are made.

5.2 Euler–Bernoulli beam theory

The Euler–Bernoulli (EB) [8] beam is used to capture the bending behavior of a structural member under transverse loading. The effect of shear is ignored in this theory. Typically EB theory is applicable to thin slender members. For further discussions, let us consider the dimension of a beam as L_x, L_y, L_z along X, Y and Z axes respectively. A schematic description of the deformation of an EB beam is given in figure 5.1. Here the dotted lines denote the undeformed beam and the blue lines the deformed counterpart. In figure 5.1 the dotted planar cross section ABCD remains planar after deformation, i.e., its deformed counterpart A′B′C′D′ is also planar. Moreover, the deformed neutral axis (N.A.) is normal to the A′B′C′D′ plane just as the undeformed N.A. is normal to the plane ABCD. Overall the theory relies on some crucial assumptions, as mentioned below.

Assumption 1: The beam is slender. This assumption is relatively valid when the length of the beam is much larger than the other two dimensions, i.e., $L >> (b, d)$ where L represents the length of the beam, b and d are the width and the depth of the beam, respectively.

Assumption 2: plane section before deformation remains plane after the deformation.

Assumption 3: The plane normal to the longitudinal axis before deformation remains normal to the deformed axis after the deformation.

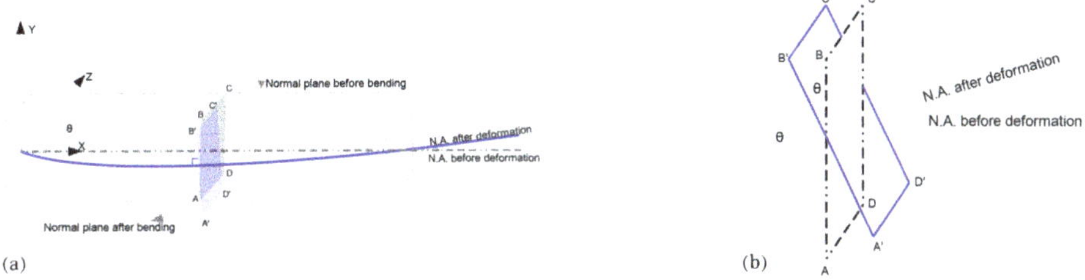

Figure 5.1. Pictorial representation of (a) Euler–Bernoulli beam; (b) normal plane before and after deformation.

The basic idea in deriving a discrete EB theory is to appropriately incorporate the above-mentioned assumptions within a discrete setting. To do so, let us represent the structural member as shown in figure 5.1 with molecules (red colored balls in figure 5.2), which are arranged, say, at a lattice distance $\{l_x, l_y, l_z\}$. Note that in figure 5.2, only four molecules have been considered per cross section, since that appears to be required for a minimal description. The discrete EB theory can also be derived in a similar way as mentioned below, by considering more molecules per cross section. But in that case, appropriately decaying weighting function with increasing distance between two molecules needs to considered while defining the stiffness coefficients. However, our objective here is to discuss a minimal discrete model for clarity in understanding and computational effectiveness. The total kinetic energy for the discrete system of N molecules can be expressed as:

$$T = \sum_{i=1}^{N} \frac{m_i}{2}\left(\dot{u}_i^2 + \dot{v}_i^2 + \dot{w}_i^2\right) \tag{5.1}$$

where u_i, v_i and w_i are displacements along X, Y and Z axes. $\dot{a}$ represents the time derivative of a variable a. m_i is the mass of the ith molecule. The potential energy (U) of the system is represented as,

$$U = \frac{1}{2}\left[\sum_{i}^{N}\sum_{j=1}^{4} K_x\left(\Delta u_{ii_j}\right)^2 + K_y\left(\Delta v_{ii_j}\right)^2 + K_z\left(\Delta w_{ii_j}\right)^2\right] - \sum_{i=1}^{N} q_i^S v_i^S \tag{5.2}$$

where $K := [K_x, K_y, K_z]$ represents the stiffness and Δu_{ii_j}, Δv_{ii_j} and Δw_{ii_j} are the displacements between ith and i_j th molecules along X, Y and Z, respectively. i_j denotes the jth neighbor of the ith molecule. For clarity on this, please refer to figure 5.3 in which the neighborhood molecules of a molecule are shown. We now reduce the dimensionality of the system by defining a set of pseudo molecules (represented by yellow balls in figure 5.2) arranged along the N.A. of the beam. q_i^S is the externally applied force in the direction shown in figure 5.2. For ease of

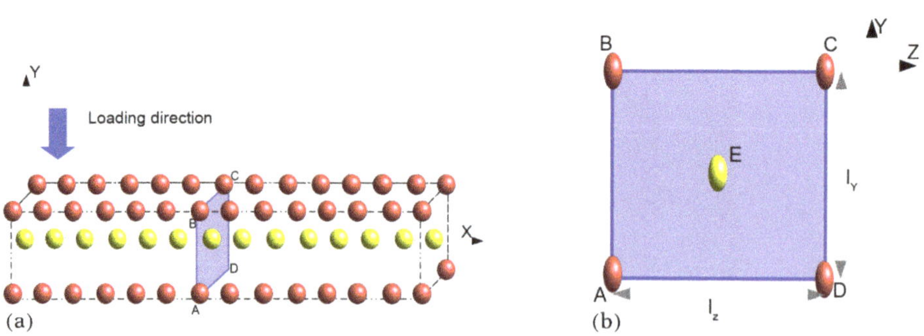

Figure 5.2. Pictorial representation of (a) initial configuration of a beam with N molecules (red) and pseudo molecules (yellow) arranged along the neutral axis (N.A.); (b) cross-section normal to longitudinal axis of beam.

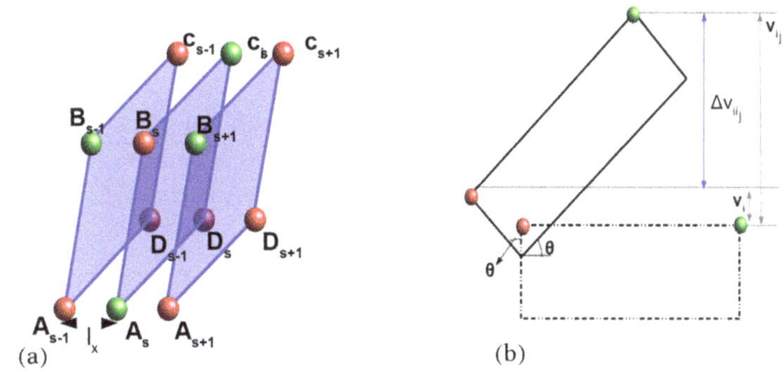

Figure 5.3. Pictorial representation of (a) molecule (suppose B) under the influence of its local neighbors represented by green balls; (b) vertical displacement of molecule i and its neighbor ij (not drawn to scale).

demonstration, let us denote the displacements of the Pth molecule at a cross section S by (u_P^S, v_P^S, w_P^S) along X, Y, Z, respectively. Let $m_i(:=m_A^S + m_B^S + m_C^S + m_D^S)$ be the lumped mass of the pseudo molecule positioned at the center of gravity (C.G.) of the section S. Here i runs over the number of the cross sections. As the beam is slender, the vertical component of the velocity for all the molecules on the plane S can be assumed to equal.

$$\dot{v}_A^S = \dot{v}_B^S = \dot{v}_C^S = \dot{v}_D^S = \dot{v}_i \tag{5.3}$$

$\dot{v}_i$ represents the component of velocity along the Y direction for the pseudo molecule corresponding to plane S. Thus equation (5.1) reduces to:

$$\begin{aligned} T &= \sum_{i=1}^{N} \frac{m_i}{2}\dot{v}_i^2 \\ &= \frac{1}{2}[\dots + m_A^S(\dot{v}_A^S)^2 + m_B^S(\dot{v}_B^S)^2 + m_C^S(\dot{v}_C^S)^2 + m_D^S(\dot{v}_D^S)^2 \dots] \\ &= \frac{1}{2}[\dots + m_i\dot{v}_i^2 + \dots] \end{aligned} \tag{5.4}$$

Taking a continuum limit of the discrete kinetic energy for lattice distance l_x, we obtain the corresponding continuum kinetic energy as,

$$\begin{aligned} T &= \frac{1}{2}\left[\dots + \frac{m_i}{l_x}\dot{v}_i^2 l_x + \dots\right] \\ &\approx \int_0^{L_x} \frac{\nu}{2}\dot{v}^2 dx \end{aligned} \tag{5.5}$$

where ν represents the mass per unit length of the beam and $\dot{v}$ represents the continuum counterparts of velocity in the Y direction.

Now we shall move on to our discussion on the potential energy. Since for EB model, the transverse shear strains can be neglected, for small strains the following is true.

$$\tan\theta \approx \theta = \frac{\partial u}{\partial y} = -\frac{\partial v}{\partial x} \tag{5.6}$$

where u, v, x, y are in usual sense continuum versions of their respective discrete counterparts. The displacements of a molecule (say B) sitting on the plane S can be approximated by the displacement of the pseudo molecule (here E) positioned at the N.A., as given below.

$$u_B^S \approx u_E^S + \left.\frac{\partial u}{\partial y}\right|_S\left(\frac{L_y}{2}\right) + \left.\frac{\partial u}{\partial z}\right|_S\left(\frac{L_z}{2}\right) = u_E^S - \left.\frac{\partial v}{\partial x}\right|_S\left(\frac{L_y}{2}\right) \tag{5.7}$$

$$v_B^S \approx v_E^S + \left.\frac{\partial v}{\partial y}\right|_S\left(\frac{L_y}{2}\right) + \left.\frac{\partial v}{\partial z}\right|_S\left(\frac{L_z}{2}\right) = v_E^S \tag{5.8}$$

$$w_B^S \approx w_E^S + \left.\frac{\partial w}{\partial y}\right|_S\left(\frac{L_y}{2}\right) + \left.\frac{\partial w}{\partial z}\right|_S\left(\frac{L_z}{2}\right) = 0 \tag{5.9}$$

Referring to equation (5.9), the third term of the potential energy as given in equation (5.2) drops out of the analysis. The second term of the potential energy can be expanded as the sum of potentials between interacting neighbors of any molecule i. Since the beam is assumed to be slender, the energy contribution of shear forces can be neglected. Thus for any molecule i and its neighbors from adjacent plane, the term $\left(v_i^S - v_{i_j}^S\right) \approx 0$. Recalling equation (5.2), for the same plane neighbors of any molecule i, the second term of the potential energy as given in equation (5.2) can be expressed as:

$$(v_i - v_{i_j}) \approx (v_E - v_E) = 0 \tag{5.10}$$

Thus the second term of the equation (5.2) vanishes and the equation is reduced to the following.

$$\begin{aligned} U &= \frac{1}{2}\sum_{i}^{N}\sum_{j=1}^{4} K_x(u_i - u_{i_j})^2 - \sum_{i=1}^{N} q_i^S v_i^S \\ &= \frac{1}{2}\Big[\cdots + K_x\left(u_B^S - u_A^S\right)^2 + K_x\left(u_B^S - u_C^S\right)^2 + K_x\left(u_B^S - u_B^{(s-1)}\right)^2 \\ &\quad + K_x\left(u_B^S - u_B^{(s+1)}\right)^2 + \cdots\Big] - \sum_{i=1}^{N} q_i^S v_i^S \end{aligned} \tag{5.11}$$

Recalling equation (5.7):

$$\begin{aligned} u_B^S - u_A^S &\approx u_E^S + \left(\frac{L_y}{2}\right)\left.\frac{\partial u}{\partial y}\right|_S - u_E^S - \left(\frac{L_y}{2}\right)\left.\frac{\partial u}{\partial y}\right|_S = 0 \\ u_B^S - u_C^S &\approx u_E^S + \left(\frac{L_y}{2}\right)\left.\frac{\partial u}{\partial y}\right|_S - u_E^S - \left(\frac{L_y}{2}\right)\left.\frac{\partial u}{\partial y}\right|_S = 0 \end{aligned} \tag{5.12}$$

since, $(u_B^S - u_A^S) = 0$ and $(u_B^S - u_C^S) = 0$. So equation (5.11) gets reduced to:

$$
\begin{aligned}
U &= \frac{1}{2}\left[\cdots + K_x\left(u_B^S - u_B^{S-1}\right)^2 + K_x\left(u_B^S - u_C^{S+1}\right)^2 + \cdots\right] - \sum_{i=1}^{N} q_i^S v_i^S \\
&= \frac{1}{2}\left[\cdots + K_x\left(u_E^S + \left(\frac{L_y}{2}\right)\left(\frac{\partial u}{\partial y}\bigg|_B^S\right) - u_E^{S-1} - \left(\frac{L_y}{2}\right)\frac{\partial u}{\partial y}\bigg|_B^{S-1}\right)^2\right] \\
&\quad + \left[K_x\left(u_E^S + \left(\frac{L_y}{2}\right)\frac{\partial u}{\partial y}\bigg|_B^S - u_E^{S+1} - \left(\frac{L_y}{2}\right)\frac{\partial u}{\partial y}\bigg|_C^{S+1}\right)^2 + \cdots\right] - \sum_{i=1}^{N} q_i^S v_i^S \\
&= \frac{1}{2}\left[\cdots + K_x\left(\left(\frac{L_y}{2}\right)\frac{\partial v}{\partial x}\bigg|_B^S - \left(\frac{L_y}{2}\right)\frac{\partial v}{\partial x}\bigg|_B^{S-1}\right)^2\right. \\
&\quad \left. + K_x\left(\left(\frac{L_y}{2}\right)\frac{\partial v}{\partial x}\bigg|_B^S - \left(\frac{L_y}{2}\right)\frac{\partial v}{\partial x}\bigg|_B^{S+1}\right)^2 + \cdots\right] - \sum_{i=1}^{N} q_i^S v_i^S \\
&= \frac{1}{2}\left(\frac{L_y}{2}\right)^2\left[\cdots + K_x\left(\frac{\partial v}{\partial x}\bigg|_B^S - \frac{\partial v}{\partial x}\bigg|_B^{S-1}\right)^2 + K_x\left(\frac{\partial v}{\partial x}\bigg|_B^S - \frac{\partial v}{\partial x}\bigg|_B^{S+1}\right)^2 + \cdots\right] - \sum_{i=1}^{N} q_i^S v_i^S \\
&= \frac{1}{2}\left(\frac{L_y}{2}\right)^2 (l_x)^2\left[\cdots + K_x\left(\frac{\frac{\partial v}{\partial x}\Big|_B^S - \frac{\partial v}{\partial x}\Big|_B^{S-1}}{l_x}\right)^2 + K_x\left(\frac{\frac{\partial v}{\partial x}\Big|_B^S - \frac{\partial v}{\partial x}\Big|_B^{S+1}}{l_x}\right)^2 + \cdots\right] \\
&\quad + \sum_{i=1}^{N} q_i^S v_i^S
\end{aligned}
\tag{5.13}
$$

Taking a continuum limit of the discrete potential energy for lattice distance l_x, we obtain the corresponding continuum potential energy as,

$$
U = \int_0^{L_x} \frac{1}{2}\left(\frac{L_y}{2}\right)^2 l_x\left[K_x\left(\frac{\partial^2 v}{\partial x^2}\right)^2\right]dx - \int_0^{L_x} qv dx \tag{5.14}
$$

The Lagrangian (L) for the Euler–Bernoulli beam can be approximated as follows,

$$
\begin{aligned}
L &= T - U \\
L &= \int_0^{L_x} \frac{\nu}{2}(\dot{v}^2)dx - \int_0^{L_x} \frac{1}{2}\left(\frac{L_y}{2}\right)^2 (L_x)\left[K_x\left(\frac{\partial^2 v}{\partial x^2}\right)^2\right]dx \\
&\quad + \int_0^{L_x} qv dx \\
L &= \int_0^{L_x} \frac{\nu}{2}(\dot{v}^2)dx - \int_0^{L_x} \frac{1}{2}\left(\frac{L_y}{2}\right)^2 (L_x)[K_x(v_{xx})^2]dx + \int_0^{L_x} qv dx
\end{aligned}
\tag{5.15}
$$

The corresponding Euler–Lagrange equation for $L := L(v, \dot{v}, (v_{xx}))$ is obtained as follows:

$$\frac{\partial L}{\partial v} - \frac{\partial}{\partial t}\left(\frac{\partial L}{\partial \dot{v}}\right) + \frac{\partial^2}{\partial x^2}\left(\frac{\partial L}{\partial v_{xx}}\right) = 0 \tag{5.16}$$

Now,

$$\begin{aligned}
&\frac{\partial L}{\partial v} = q;\\
&\frac{\partial}{\partial t}\left(\frac{\partial L}{\partial \dot{v}}\right) = \nu\frac{\partial \dot{v}}{\partial t} = \nu\frac{\partial^2 v}{\partial t^2};\\
&\frac{\partial^2}{\partial x^2}\left(\frac{\partial L}{\partial (v_{xx})}\right) = -\frac{\partial^2}{\partial x^2}\left[\left(\frac{L_y}{2}\right)^2 (L_x)K_x\left(\frac{\partial^2 v}{\partial x^2}\right)\right]
\end{aligned} \tag{5.17}$$

Putting the values of equation (5.17) in equation (5.16), the governing equation for dynamics of an Euler–Bernoulli beam can be represented as:

$$\frac{\partial^2}{\partial x^2}\left[\left(\frac{L_y}{2}\right)^2 (L_x)K_x\left(\frac{\partial^2 v}{\partial x^2}\right)\right] = -\nu\frac{\partial^2 v}{\partial t^2} + q \tag{5.18}$$

The above equation (5.18) can be further rearranged to:

$$\boxed{\frac{\partial^2}{\partial x^2}\left[EI\left(\frac{\partial^2 v}{\partial x^2}\right)\right] = -\nu\frac{\partial^2 v}{\partial t^2} + q} \tag{5.19}$$

where $E\left(:=\frac{K_x L_x}{A}\right)$ represents the modulus of elasticity of the material and $I\left(:=\left(\frac{L_y}{2}\right)^2 A\right)$ represents the second moment of area.

5.3 Shear deformable beam or Timoshenko beam

As discussed in the previous section of this chapter, the classical theory or EB beam theory neglects the effects of shear deformations. Disregarding the shear deformations makes EB beam theory suitable for thin and slender members but unsuitable for thick and short span members, where shear deformation plays a key role. Timoshenko beam (TB) theory [9] is used for coupling bending and shear deformations. This one-dimensional theory is used to capture the combined effect of bending and shear deformations in a structural member subjected to transverse loading. A schematic description of the deformation of a TB is presented in figure 5.4. Here the dotted lines denote the undeformed orientation of the beam and the red lines represent the deformed counterpart. In figure 5.4 the dotted planar cross section ABCD represents the normal plane before deformation. The plane A′B′C′D′ represents the EB beam deformation, i.e., the plane normal to the deformed neutral axis (N.A.) while the plane A″B″C″D″ represents the further rotation of the plane A′B′C′D′ due to the consideration of shear deformations. So

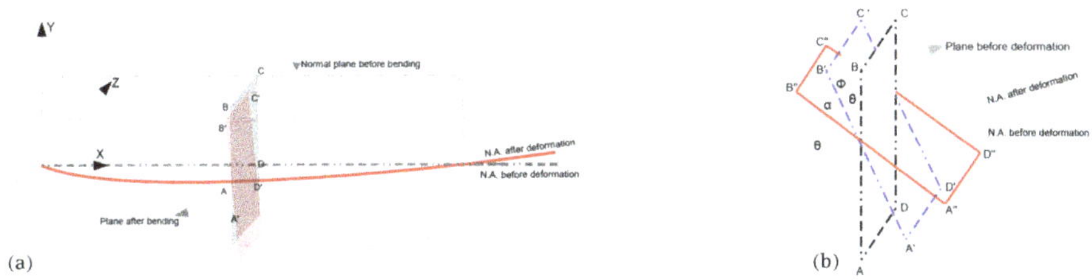

Figure 5.4. Pictorial representation of (a) Timoshenko beam; (b) normal plane before and after deformation.

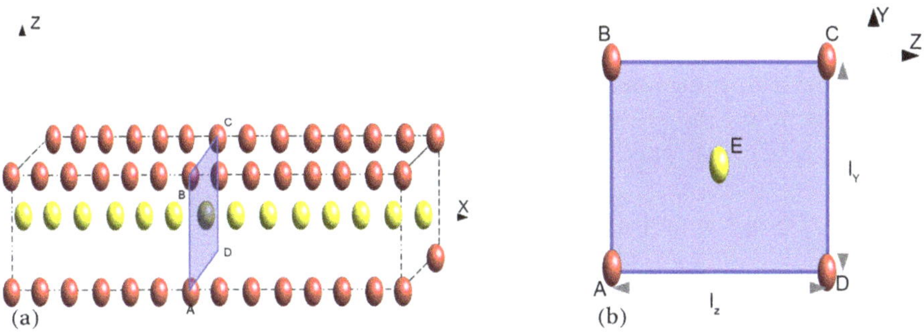

Figure 5.5. Pictorial representation of (a) initial configuration of a beam with N molecules (red) and pseudo molecules (yellow) arranged along the N.A.; (b) cross-section normal to the longitudinal axis of the beam.

the angle of rotation (ϕ) for any plane surface A″B″C″D″ is the sum total of angle of rotation due to bending (θ) and shear deformation angle (α). This can be mathematically represented as,

$$\boxed{\phi = \theta + \alpha} \tag{5.20}$$

The Timoshenko beam theory is built upon the following assumptions as mentioned below:

Assumption 1: plane section before deformation remains plane after the deformation.

Assumption 2: The transverse shear strain is constant for all the points on a given cross section of the beam. Thus the shear stress distribution is uniform throughout the cross section.

The basic idea is to derive a discrete TB theory by incorporating the above-mentioned assumptions. With the similar notations as described in EB theory, let us consider a structural member, as shown in figure 5.5. The discrete system can have more molecules at the cross sections by defining an appropriate decaying function with increasing distances between two molecules. However, our objective is to arrive at a minimal discrete model for clarity in understanding the computational effectiveness. The total kinetic energy for the discrete system of N molecules can be expressed as:

$$T = \sum_{i=1}^{N} \frac{m_i}{2}(\dot{u}_i^2 + \dot{v}_i^2 + \dot{w}_i^2) \tag{5.21}$$

where u_i, v_i, w_i are the displacements along X, Y and Z axes. $\dot{a}$ represents the time derivative of a variable a. m_i is the mass of the ith molecule. The potential energy (U) of the system is represented as:

$$U = \frac{1}{2}\left[\sum_{i}^{N}\sum_{j=1}^{4} K_x\left(\Delta u_{ii_j}\right)^2 + K_y\left(\Delta v_{ii_j}\right)^2 + K_z\left(\Delta w_{ii_j}\right)^2\right] - \sum_{i=1}^{N} q_i^{\ S} v_i^{S} \tag{5.22}$$

where all the notations have the usual meaning as defined in EB beam theory. With the aim to reduce the dimensionality of the system, a set of pseudo molecules (represented by yellow balls in figure 5.5) is arranged along the N.A. of the beam. Let (u_P^S, v_P^S, w_P^S) represent the displacements along X, Y and Z direction, respectively, for the Pth molecule at cross-section S. The displacements of a molecule (say B) sitting on the plane S can be approximated by the displacement of the pseudo molecule (here E) positioned at the N.A., as given below.

$$u_B^S \approx u_E^S + (\phi)|_s\left(\frac{L_y}{2}\right) + \frac{\partial u}{\partial z}\bigg|_s\left(\frac{L_z}{2}\right) = u_E^S + (\phi)|_s\left(\frac{L_y}{2}\right) \tag{5.23}$$

$$v_B^S \approx v_E^S + \frac{\partial v}{\partial y}\bigg|_s\left(\frac{L_y}{2}\right) + \frac{\partial v}{\partial z}\bigg|_s\left(\frac{L_z}{2}\right) = v_E^S \tag{5.24}$$

$$w_B^S \approx w_E^S + \frac{\partial w}{\partial y}\bigg|_s\left(\frac{L_y}{2}\right) + \frac{\partial w}{\partial z}\bigg|_s\left(\frac{L_z}{2}\right) = 0 \tag{5.25}$$

Since it is assumed that the plane section before deformation remains a plane after deformation, the vertical component of velocity of all the molecules on the plane S is assumed to be equal.

$$\begin{aligned}\dot{u}_A^S = \dot{u}_B^S = \dot{u}_C^S = \dot{u}_D^S = \dot{u}_i\\ \dot{v}_A^S = \dot{v}_B^S = \dot{v}_C^S = \dot{v}_D^S = \dot{v}_i\end{aligned} \tag{5.26}$$

Taking the time derivative on both the sides, equation (5.23) can be reformed as:

$$\begin{aligned}\frac{\partial u_B^S}{\partial t} &= \frac{\partial u_E^S}{\partial t} + \frac{\partial}{\partial t}\left(\phi\ |_B^S\ \frac{L_y}{2}\right)\\ \dot{u}_B^S &= \dot{\phi}\ |_B^S\left(\frac{L_y}{2}\right)\end{aligned} \tag{5.27}$$

Thus equation (5.21) can be expressed as:

$$T = \frac{1}{2}\left[\cdots + m_i\dot{\phi}\ |_i^S\left(\frac{L_y}{2}\right) + \cdots\cdots + m_i\dot{v}_i^2 + \cdots\right] \tag{5.28}$$

Taking a continuum limit of the discrete kinetic energy for lattice distance l_x, we obtain the corresponding continuum kinetic energy as,

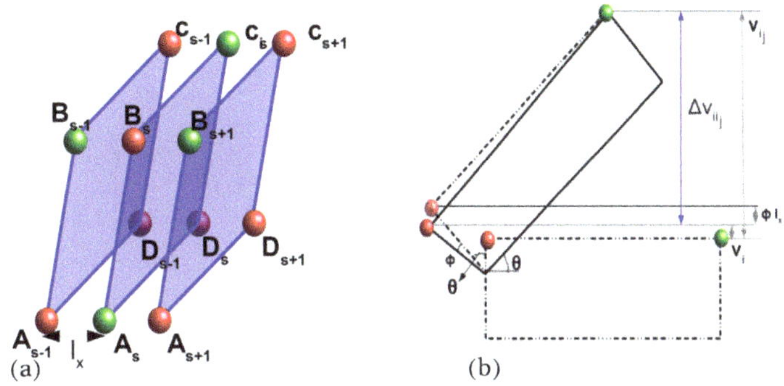

Figure 5.6. Pictorial representation of (a) molecule (suppose B) under the influence of its local neighbors represented by green balls; (b) vertical displacement of molecule i and its neighbor i_j (not drawn to scale).

$$\begin{aligned} T &\approx \int_0^{L_x} \frac{\nu}{2}\left(\dot{\phi}\,|^S\left(\frac{L_y}{2}\right)\right)^2 dx + \int_0^{L_x} \frac{\nu}{2}\dot{v}^2 dx \\ T &\approx \int_0^{L_x} \frac{\rho I}{2}(\dot{\phi}\,|^S)^2 dx + \int_0^{L_x} \frac{\rho A}{2}\dot{v}^2 dx \end{aligned} \tag{5.29}$$

where ν represents the mass per unit length of the beam and $(\dot{\phi}\,|^S, \dot{v})$ represents the continuum counterparts of $\left(\dot{\phi}\,|_i^S, \dot{v}_i\right)$, I represents the second moment of area, ρ represents the mass density of the material and A represents the cross-sectional area of plane S.

Now, the potential energy of the beam can be represented as the sum of the potential energy due to interaction between any ith molecule and its nearest i_j th neighbors. For more clarity on the neighborhood please see figure 5.6 where the green balls represent the neighborhood molecules of molecule B sitting on plane S. Since the deformations along the Z-axis are neglected, so the third term on the right hand side of equation (5.22) drops out of the analysis. The equation thus reduces to:

$$\begin{aligned} U &= \frac{1}{2}\left[\sum_i^N \sum_{j=1}^4 K_x(u_i - u_{i_j})^2 + K_y\left(v_i - v_{i_j} - \phi_i l_x\right)^2\right] - \sum_{i=1}^N q_i^S v_i^S \\ U &= \frac{1}{2}K_x\left[\cdots + \left(u_B^S - u_A^S\right)^2 + \left(u_B^S - u_C^S\right)^2 + \left(u_B^S - u_B^{S-1}\right)^2 + \left(u_B^S - u_B^{S+1}\right)^2 + \cdots\right] \\ &+ \frac{1}{2}K_y\left[\cdots + \left(v_B^S - v_A^S - \phi_B^S l_x\right)^2 + \left(v_B^S - v_C^S - \phi_B^S l_x\right)^2\right. \\ &+ \left.\left(v_B^S - v_B^{S-1} - \phi_B^S l_x\right)^2 + \left(v_B^S - v_B^{S+1} - \phi_B^S l_x\right)^2 + \cdots\right] - \sum_{i=1}^N q_i^S v_i^S \end{aligned} \tag{5.30}$$

$$\because \quad u_i^S - u_{i_j}^S = u_E^S + \phi\,|_i\left(\frac{L_y}{2}\right) - u_E^S - \phi\,|_i\left(\frac{L_y}{2}\right) = 0$$

$$\Longrightarrow u_B^S - u_A^S = 0; \qquad u_B^S - u_C^S = 0$$

$$\because \quad v_i^S - v_{i_j}^S = v_E^S - v_E^S = 0$$

$$\Longrightarrow v_B^S - v_A^S = 0; \qquad v_B^S - v_C^S = 0$$

Thus the above equation reduces to:

$$
\begin{aligned}
U &= \frac{1}{2}K_x\Big[\cdots + \big(u_B^S - u_B^{S-1}\big)^2 + \big(u_B^S - u_B^{S+1}\big)^2 + \cdots\Big] \\
&\quad + \frac{1}{2}K_y\Big[\cdots + \big(v_B^S - v_B^{S-1} - \phi_B^S l_x\big)^2 + \big(v_B^S - v_B^{S+1} - \phi_B^S l_x\big)^2 + \cdots\Big] - \sum_{i=1}^{N} q_i^S v_i^S \\
U &= \frac{1}{2}K_x\Bigg[\cdots + \left(u_E^S + \phi\,|_i^S\left(\frac{L_y}{2}\right) - u_E^{S-1} - \phi\,|_B^{S-1}\left(\frac{L_y}{2}\right)\right)^2 \\
&\quad + \left(u_E^S + \phi\,|_i^S\left(\frac{L_y}{2}\right) - u_E^{S+1} - \phi\,|_B^{S+1}\left(\frac{L_y}{2}\right)\right)^2 + \cdots\Bigg] \\
&\quad + \frac{1}{2}K_y\Big[\cdots + \big(v_B^S - v_B^{S-1} - \phi_B^S l_x\big)^2 + \big(v_B^S - v_B^{S+1} - \phi_B^S l_x\big)^2 + \cdots\Big] - \sum_{i=1}^{N} q_i^S v_i^S \\
U &= \frac{1}{2}K_x\Bigg[\cdots + \left(\phi\,|_i^S\left(\frac{L_y}{2}\right) - \phi\,|_B^{S-1}\left(\frac{L_y}{2}\right)\right)^2 + \left(\phi\,|_i^S\left(\frac{L_y}{2}\right) - \phi\,|_B^{S+1}\left(\frac{L_y}{2}\right)\right)^2 + \cdots\Bigg] \\
&\quad + \frac{1}{2}K_y\Big[\cdots + \big(v_B^S - v_B^{S-1} - \phi_B^S l_x\big)^2 + \big(v_B^S - v_B^{S+1} - \phi_B^S l_x\big)^2 + \cdots\Big] - \sum_{i=1}^{N} q_i^S v_i^S \\
U &= \frac{1}{2}K_x\left(\frac{L_y}{2}\right)^2 (l_x)^2\left[\cdots + \left(\frac{\phi\,|_i^S - \phi\,|_B^{S-1}}{l_x}\right)^2 + \left(\frac{\phi\,|_i^S - \phi\,|_B^{S+1}}{l_x}\right)^2 + \cdots\right] \\
&\quad + \frac{1}{2}K_y(l_x)^2\left[\cdots + \left(\frac{v_B^S - v_B^{S-1} - \phi_B^S l_x}{l_x}\right)^2 + \left(\frac{v_B^S - v_B^{S+1} - \phi_B^S l_x}{l_x}\right)^2 + \cdots\right] - \sum_{i=1}^{N} q_i^S v_i^S
\end{aligned}
\tag{5.31}
$$

Taking a continuum limit of the discrete potential energy for lattice distance l_x, we obtain the corresponding continuum potential energy as,

$$
\begin{aligned}
U &= \int_0^{L_x} \frac{1}{2}\left(\frac{L_y}{2}\right)^2 l_x\left[K_x\left(\frac{\partial\phi}{\partial x}\right)^2\right]dx + \frac{1}{2}\int_0^{L_x} K_y l_x\left(\frac{\partial v}{\partial x} - \phi\right)^2 dx - \int_0^{L_x} qv dx \\
U &= \int_0^{L_x} \frac{1}{2}\left(\frac{L_y}{2}\right)^2 l_x\left[K_x(\phi_x)^2\right]dx + \frac{1}{2}\int_0^{L_x} \left[K_y l_x(v_x - \phi)^2\right] - \int_0^{L_x} qv dx
\end{aligned}
\tag{5.32}
$$

The Lagrangian (L) for the Timoshenko beam can be approximated as,

$$
L = T - U \tag{5.33}
$$

The Lagrangian $L := L_1(v, \dot{v}, (v_x)) + L_2(\phi, \dot{\phi}, (\phi_x))$, the corresponding Euler–Lagrange equations are:

$$
\frac{\partial L}{\partial v} - \frac{\partial}{\partial t}\left(\frac{\partial L}{\partial \dot{v}}\right) + \frac{\partial}{\partial x}\left(\frac{\partial L}{\partial (v_x)}\right) = 0 \tag{5.34}
$$

$$\frac{\partial L}{\partial \phi} - \frac{\partial}{\partial t}\left(\frac{\partial L}{\partial \dot{\phi}}\right) + \frac{\partial}{\partial x}\left(\frac{\partial L}{\partial(\phi_x)}\right) = 0 \tag{5.35}$$

Now,

$$\begin{aligned} &\frac{\partial L}{\partial v} = q; \\ &\frac{\partial}{\partial t}\left(\frac{\partial L}{\partial \dot{v}}\right) = \nu\ddot{v} = \rho A\frac{\partial^2 v}{\partial t^2}; \\ &\frac{\partial}{\partial x}\left(\frac{\partial L}{\partial(v_x)}\right) = \frac{\partial}{\partial x}\left[K_y l_x\left(\frac{\partial v}{\partial x} - \phi\right)\right] \end{aligned} \tag{5.36}$$

$$\begin{aligned} &\frac{\partial L}{\partial \phi} = K_y l_x\left(\frac{\partial v}{\partial x} - \phi\right); \\ &\frac{\partial}{\partial t}\left(\frac{\partial L}{\partial \dot{\phi}}\right) = \rho I\frac{\partial^2 \phi}{\partial t^2}; \\ &\frac{\partial}{\partial x}\left(\frac{\partial L}{\partial(\phi_x)}\right) = \frac{\partial}{\partial x}\left[K_x\left(\frac{L_y}{2}\right)^2 l_x\frac{\partial \phi}{\partial x}\right] = \frac{\partial}{\partial x}\left[EI\frac{\partial \phi}{\partial x}\right] \end{aligned} \tag{5.37}$$

Putting the values of equation (5.36) and equation (5.37) in equation (5.34) and equation (5.35), respectively, the governing equations for dynamics of the Timoshenko beam can be represented as:

$$\boxed{\rho A\frac{\partial^2 v}{\partial t^2} - q = \frac{\partial}{\partial x}\left(K_y l_x\left(\frac{\partial v}{\partial x} - \phi\right)\right)} \tag{5.38}$$

$$\boxed{\rho I\frac{\partial^2 \phi}{\partial t^2} = \frac{\partial}{\partial x}\left(EI\frac{\partial \phi}{\partial x}\right) + K_y l_x\left(\frac{\partial v}{\partial x} - \phi\right)} \tag{5.39}$$

It may be interesting to note that $E\left(:=\frac{K_x l_x}{A}\right)$ represents the Young's modulus of elasticity and $I\left(:=\left(\frac{L_y}{2}\right)^2 A\right)$ represents the second moment of area.

5.4 Kirchhoff–Love theory of thin plates/classical plate theory

Extending the boundaries of EB beam theory from one dimension to two dimensions, Kirchhoff–Love (KL) [10] plate theory is used to capture the bending of a plate subjected to transverse loading. This theory assumes that the plane section normal to the N.A. before bending remains a normal plane after bending. This assumption ignores the effect of shear. Typically, KL plate theory is applicable to thin plates. For further discussion, let us assume the dimensions of a plate as L_x, L_y, L_z along the X, Y and Z axes, respectively. A schematic diagram of a thin plate is given in figure 5.7. The dotted lines represents the plane normal to the N.A.

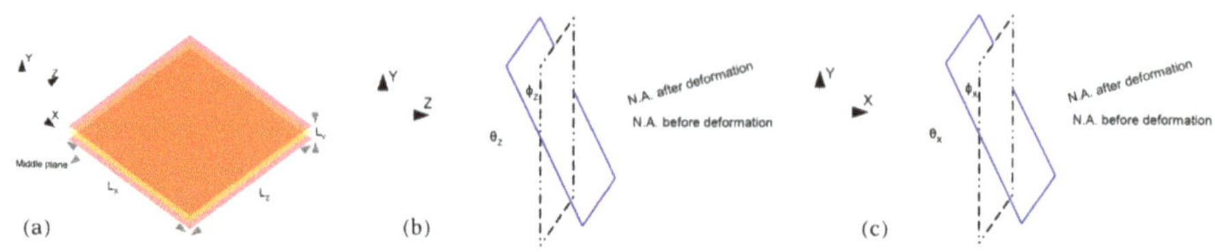

Figure 5.7. Pictorial representation of (a) Kirchhoff–Love plate (thin plate); (b) normal plane before and after deformation along YZ; (c) normal plane before and after deformation along XY.

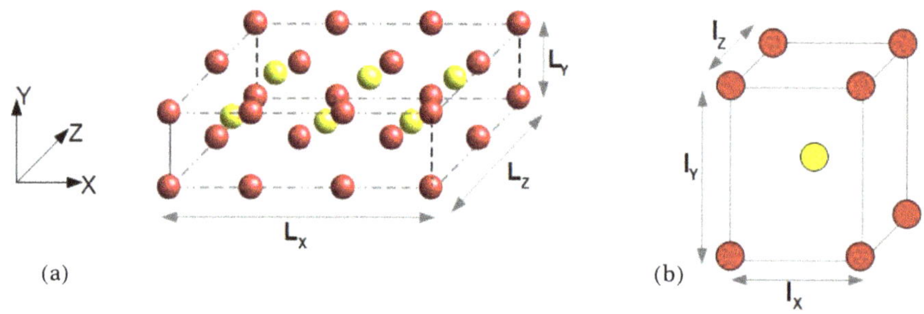

Figure 5.8. Pictorial representation of (a) initial configuration of a plate with N molecules (red) and pseudo molecules (yellow) arranged along the neutral axis (N.A.); (b) cross-section normal to middle plane of the plate.

before deformation. The deformed plane is represented by the blue lines. Since the shear deformations are ignored, as in case of EB beam theory, the plane normal before deformation remains normal plane after deformation. The KL plate theory relies on some crucial assumptions as mentioned below.

Assumption 1: The straight lines normal to the middle plane remain normal throughout the deformation. This assumption is relatively valid for plates undergoing bending unless subjected to significant shear or torsional stresses.

Assumption 2: The plane normal to the middle plane before deformation, remains normal after deformation.

Assumption 3: The thickness of the plate is assumed to remain constant during the deformation.

The basic idea in deriving a discrete KL plate theory is to appropriately incorporate the above assumptions within a discrete setup. To do so, let us represent the structural member as shown in figure 5.8 considering a system of N molecules represented by red balls. The dotted lines represent the initial configuration of molecules arranged at lattice distances $\left[l_x, l_y, l_z\right]$. The red balls represent the molecule positions in a thin plate with dimensions $\left[L_x, L_y, L_z\right]$. The idea is to reduce the dimensionality of the system by defining a set of pseudo molecules (represented by yellow balls) arranged along the middle plane that serves as the neutral membrane of the plate. The total kinetic energy for a thin plate having N molecules and span of L_x and L_z along x and z directions can be expressed as:

$$T = \sum_{i=1}^{N} \frac{m_i}{2}(\dot{u}_i^2 + \dot{v}_i^2 + \dot{w}_i^2) \tag{5.40}$$

where all the variables have their usual meaning as defined in the previous sections of the chapter. Now, the potential energy of the molecules can be represented as the sum of the potential energy due to interaction of the ith molecule with its neighboring molecules i_j.

$$U = \frac{1}{2}\left[\sum_{i}^{N}\sum_{j=1}^{N'} K_x(u_i - u_{i_j})^2 + K_y(v_i - v_{i_j})^2 + K_z(w_i - w_{i_j})^2\right] - \sum_{i=1}^{N} q_i^S v_i^S \tag{5.41}$$

where N' represents the number of neighboring molecules for any molecule i and q_i^S represents the force per unit area acting along the Y-axis. As the plate is assumed to be thin, the shear component of velocity $\{\dot{u}, \dot{w}\}$ is negligible and can be ignored. Considering the assumption that the thickness of the plate remains the same, the vertical component of the velocity for all the molecules on plane S is assumed to be equal.

$$\dot{v}_A^S = \dot{v}_B^S = \dot{v}_C^S = \dot{v}_D^S = \dot{v}_i \tag{5.42}$$

Thus equation (5.40) reduces to:

$$\begin{aligned} T &= \sum_{i=1}^{N} \frac{m_i}{2}\dot{v}_i^2 = \frac{1}{2}[\cdots + m_A^S(\dot{v}_A^S)^2 + m_B^S(\dot{v}_B^S)^2 + m_C^S(\dot{v}_C^S)^2 + m_D^S(\dot{v}_D^S)^2 \ldots] \\ &= \frac{1}{2}[\cdots + m_i\dot{v}_i^2 + \cdots] \end{aligned} \tag{5.43}$$

Taking a continuum limit of the discrete kinetic energy for a lattice distance l_x, we obtain the corresponding continuum kinetic energy as,

$$\begin{aligned} T &= \frac{1}{2}\left[\cdots + \frac{m_i}{l_x}\dot{v}_i^2 l_x + \cdots\right] \\ &\approx \int_0^{L_x}\int_0^{L_z} \frac{\nu}{2}\dot{v}^2 dz dx \end{aligned} \tag{5.44}$$

where ν represents the mass per unit area of the plate and L_x represents the length of the plate along the X-axis.

Now we shall move on to our discussion on the potential energy. Since for KL model, the transverse shear strains can be neglected, for small strains the following is true.

$$\begin{aligned} \gamma_{xy} = 0 &\Longrightarrow \frac{\partial u}{\partial y} = -\frac{\partial v}{\partial x} \\ \gamma_{zy} = 0 &\Longrightarrow \frac{\partial w}{\partial y} = -\frac{\partial v}{\partial z} \end{aligned} \tag{5.45}$$

where u, v, w, x, y, z are in usual sense continuum versions of their respective discrete counterparts. The displacement of any molecule (say B) sitting on a plane S can be approximated about a pseudo molecule (here E) positioned at the N.A., as given below.

$$u_B^S \approx u_E^S + \frac{\partial u}{\partial y}\bigg|_S \left(\frac{L_y}{2}\right) + \frac{\partial u}{\partial z}\bigg|_S \left(\frac{L_y}{2}\right) = u_E^S - \frac{\partial v}{\partial x}\bigg|_S \left(\frac{L_y}{2}\right) \tag{5.46}$$

$$v_B^S \approx v_E^S + \frac{\partial v}{\partial y}\bigg|_S \left(\frac{L_y}{2}\right) + \frac{\partial v}{\partial z}\bigg|_S \left(\frac{L_y}{2}\right) = v_E^S \tag{5.47}$$

$$w_S^B \approx w_S^E + \frac{\partial w}{\partial y}\bigg|_S \left(\frac{L_y}{2}\right) + \frac{\partial w}{\partial z}\bigg|_S \left(\frac{L_y}{2}\right) = w_E^S - \frac{\partial v}{\partial z}\bigg|_S \left(\frac{L_y}{2}\right) \tag{5.48}$$

The second term of equation (5.41) can be expanded as the sum of potentials between interacting neighbors of any ith molecule. Since the plate is assumed to be thin, the energy contribution of shear forces can be neglected. Thus, for any ith molecule and its i_jth nearest neighbors sitting on adjacent planes, the term $(v_i - v_{i_j}) \approx 0$. Recalling the equation (5.47), for the same plane neighbors of any ith molecule, the second term of equation (5.41) can be expressed as:

$$(v_i - v_{i_j}) \approx (v_E - v_E) = 0 \tag{5.49}$$

Thus, equation (5.41) is reduced to:

$$\begin{aligned} U &= \frac{1}{2}\sum_i^N \sum_{j=1}^{N'} \left[K_x(u_i - u_{i_j})^2\right] - \sum_{i=1}^N q_i^S v_i^S \\ &= \frac{1}{2}\Big[\cdots + K_x(u_B^S - u_A^S)^2 + K_x(u_B^S - u_C^S)^2 + K_x(u_B^S - u_B^{(s-1)})^2 \\ &\quad + K_x(u_B^S - u_B^{(s+1)})^2 + \cdots\Big] - \sum_{i=1}^N q_i^S v_i^S \end{aligned} \tag{5.50}$$

Recalling equation (5.46) and equation (5.48):

$$\begin{aligned} u_B^S - u_A^S &\approx u_E^S + \left(\frac{L_y}{2}\right)\frac{\partial u}{\partial y}\bigg|_E - u_E^S - \left(\frac{L_y}{2}\right)\frac{\partial u}{\partial y}\bigg|_E = 0 \\ u_B^S - u_C^S &\approx u_E^S + \left(\frac{L_y}{2}\right)\frac{\partial u}{\partial y}\bigg|_E - u_E^S - \left(\frac{L_y}{2}\right)\frac{\partial u}{\partial y}\bigg|_E = 0 \end{aligned} \tag{5.51}$$

since, $(u_B^S - u_A^S) = (u_B^S - u_C^S) = 0$. So equation (5.50) gets reduced to:

$$U = \frac{1}{2}\Big[\cdots + K_x\left(u_B^{S_x} - u_B^{(S_x-1)}\right)^2 + K_x\left(u_B^{S_z} - u_B^{(S_z+1)}\right)^2 + \cdots\Big] - \sum_{i=1}^N q_i^S v_i^S \tag{5.52}$$

$$U = \frac{1}{2}\left[K_x\left(v_x\,|_B^{S_x}\left(\frac{L_y}{2}\right)^2 - v_x\,|_B^{(S_x-1)}\left(\frac{L_y}{2}\right)^2\right)^2 + K_x\left(v_z\,|_B^{S_z}\left(\frac{L_y}{2}\right)^2 - v_z\,|_B^{(S_z+1)}\left(\frac{L_y}{2}\right)^2\right)^2\right] - \sum_{i=1}^{N} q_i^S v_i^S \tag{5.53}$$

Taking a continuum limit of the discrete potential energy for lattice distances l_x and l_z, we obtain the corresponding continuum potential energy as,

$$U = \int_0^{L_x}\int_0^{L_z} \frac{1}{2}\left(\frac{L_y}{2}\right)^2 l\left[K_x\left(\frac{\partial^2 v}{\partial x^2} + \frac{\partial^2 v}{\partial z^2}\right)^2\right] dzdx - \int_0^{L_x}\int_0^{L_z} qvdzdx \tag{5.54}$$

The Lagrangian for the classical plate can be approximated as:

$$L = T - U \tag{5.55}$$

The Lagrangian $L := L(v, \dot{v}, (v_{xx}), (v_{zz}))$, the corresponding Euler–Lagrange equation is:

$$\frac{\partial L}{\partial v} - \frac{\partial}{\partial t}\left(\frac{\partial L}{\partial \dot{v}}\right) + \frac{\partial^2}{\partial x^2}\left(\frac{\partial L}{\partial (v_{xx})}\right) + \frac{\partial^2}{\partial z^2}\left(\frac{\partial L}{\partial (v_{zz})}\right) = 0 \tag{5.56}$$

$$q - \nu(\ddot{v}) = \left[K_x\frac{\partial^4 v}{\partial x^4}\left(\frac{L_y}{2}\right)^2 l_x + 2K_x\frac{\partial^4 v}{\partial x^2 \partial z^2}\left(\frac{L_y}{2}\right)^2 l_z + K_x\frac{\partial^4 v}{\partial z^4}\left(\frac{L_y}{2}\right)^2 l_x\right] \tag{5.57}$$

For a static analysis of a plate subjected to bending only, i.e., no transverse loading is applied ($q = 0$), the above equation transforms to:

$$\left[\frac{\partial^4 v}{\partial x^4} + 2\frac{\partial^4 v}{\partial x^2 \partial z^2} + \frac{\partial^4 v}{\partial z^4}\right] = 0 \qquad \boxed{\nabla^2\nabla^2 v = 0} \tag{5.58}$$

5.5 Mindlin–Reissner plate theory (MRPT)

As discussed in the previous section of the chapter, Kirchhoff–Love theory is restricted to the analysis of thin plates where thickness is less than 0.1 times the average side. The Mindlin–Reissner (MR) [11] plate theory is used for the thick plates where the normal plane to the mid surface remains a plane but not necessarily orthogonal to the mid-plane after deformation. Thus, the theory is an extension to KL plate theory by taking into consideration the shear deformations occurring through the thickness of the plate. A schematic description of the deformation of an MR plate is presented in figure 5.9. Here the orthogonal plane section before deformation is represented by dotted lines. The plane denoted by blue dotted lines represents a plane normal to the deformed N.A. while the plane denoted by red

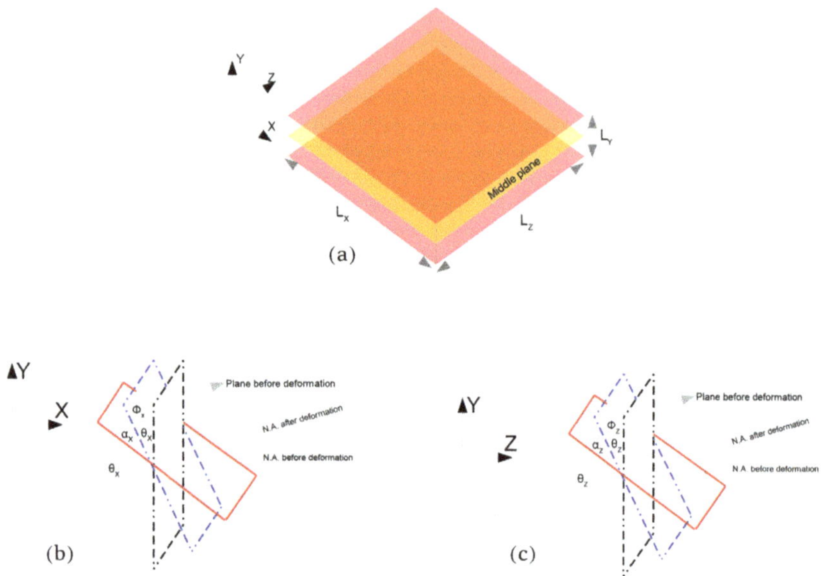

Figure 5.9. Pictorial representation of (a) MR plate (thick plate); (b) normal plane before and after deformation along XY; (c) normal plane before and after deformation along YZ.

dotted lines represents the further rotation of the plane accounting for shear deformation. So, the angle of rotation (ϕ_X) is the sum of rotations due to bending (θ_X) and shear deformation angle (α_X), where $X = x$, z (see figure 5.9).

$$\phi_x = \theta_x + \alpha_x$$
$$\phi_z = \theta_z + \alpha_z$$

The theory relies on the same assumptions as discussed for the Timoshenko beam theory.

With similar notations as defined for KL theory, let us assume a structural member as shown in figure 5.10. The total kinetic energy for a thick plate with N molecules can be expressed as:

$$T = \sum_{i=1}^{N} \frac{m_i}{2}\left(\dot{u}_i^2 + \dot{v}_i^2 + \dot{w}_i^2\right) \tag{5.59}$$

where u_i, v_i, w_i have usual meanings as defined in the previous sections of the chapter. The potential energy of the molecules can be represented as the sum of the potential energy between any molecule i and its nearest neighbors i_j.

$$U = \frac{1}{2}\left[\sum_{i}^{N} \sum_{j=1}^{N'} K_x\left(\Delta u_{ii_j}\right)^2 + K_y\left(\Delta v_{ii_j}\right)^2 + K_z\left(\Delta w_{ii_j}\right)^2\right] - \sum_{i=1}^{N} q_i^S v_i^S \tag{5.60}$$

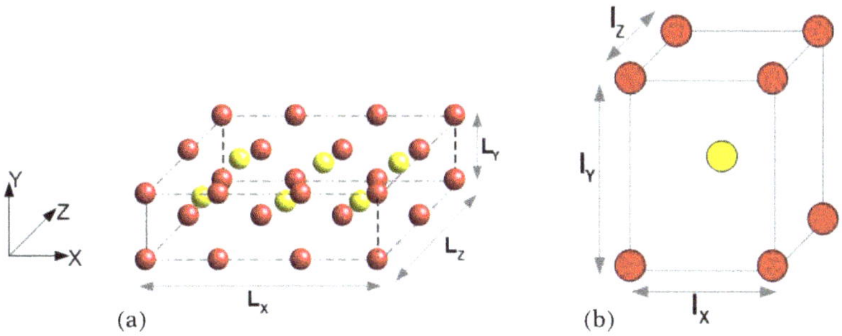

Figure 5.10. Pictorial representation of (a) initial configuration of a plate with N molecules (red) and pseudo molecules (yellow) arranged along the neutral axis (N.A.); (b) cross-section normal to middle plane of the plate.

Let (u_P^S, v_P^S, w_P^S) represent the displacements along X, Y and Z direction, respectively, for the Pth molecule at cross-section S. The displacements of a molecule (say B) sitting on the plane S can be approximated by the displacement of the pseudo molecule (here E) positioned at the N.A., as given below.

$$u_B^S \approx u_E^S + (\phi_x)\big|_S\left(\frac{L_y}{2}\right) + (\phi_z)\big|_S\left(\frac{L_y}{2}\right) \tag{5.61}$$

$$v_B^S \approx v_E^S \tag{5.62}$$

$$w_B^S \approx w_E^S + (\phi_x)\big|_S\left(\frac{L_y}{2}\right) + (\phi_z)\big|_S\left(\frac{L_y}{2}\right) \tag{5.63}$$

Since the plane section remains a plane after deformation, the magnitude of velocity of each molecule on any random plane S is assumed to be equal.

$$\begin{aligned} \dot{u}_A^S &= \dot{u}_B^S = \dot{u}_C^S = \dot{u}_D^S = \dot{u}_i \\ \dot{v}_A^S &= \dot{v}_B^S = \dot{v}_C^S = \dot{v}_D^S = \dot{v}_i \\ \dot{w}_A^S &= \dot{w}_B^S = \dot{w}_C^S = \dot{w}_D^S = \dot{w}_i \end{aligned} \tag{5.64}$$

Taking a continuum limit of the discrete kinetic energy for lattice distance l_x, l_z, we obtain the corresponding continuum kinetic energy as,

$$\begin{aligned} T &= \frac{1}{2}\left[\cdots + \frac{m_i}{l_x l_z}\dot{u}_i^2 l_x l_z + \cdots\cdots + \frac{m_i}{l_x l_z}\dot{v}_i^2 l_x l_z + \cdots + \frac{m_i}{l_x l_z}\dot{w}_i^2 l_x l_z + \cdots\right] \\ &\approx \int_0^{L_x}\int_0^{L_z}\frac{\nu}{2}\dot{u}^2 dz dx + \int_0^{L_x}\int_0^{L_z}\frac{\nu}{2}\dot{v}^2 dz dx + \int_0^{L_x}\int_0^{L_z}\frac{\nu}{2}\dot{w}^2 dz dx \end{aligned} \tag{5.65}$$

where ν represents the mass per unit area of the plate and $\dot{u}$, $\dot{v}$, $\dot{w}$ represents the continuum counterparts of the velocity components along X, Y and Z directions, respectively. Maintaining the requirement that the plane section remains a plane after deformation, the following velocity rules can be written.

$$\begin{aligned} \dot{u}_B^S &= \dot{\phi}_x \big|_S \left(\frac{L_y}{2}\right) \\ \dot{w}_B^S &= \dot{\phi}_z \big|_S \left(\frac{L_y}{2}\right) \end{aligned} \tag{5.66}$$

Thus the equation (5.65) can be presented as:

$$\begin{aligned} T &\approx \int_0^{L_x} \int_0^{L_z} \frac{\nu}{2}\left(\dot{\phi}_x\left(\frac{L_x}{2}\right)\right)^2 dzdx \\ &+ \int_0^{L_x} \int_0^{L_z} \frac{\nu}{2}\dot{v}^2 dzdx + \int_0^{L_x} \int_0^{L_z} \frac{\nu}{2}\left(\dot{\phi}_z\left(\frac{L_z}{2}\right)\right)^2 dzdx \\ T &\approx \int_0^{L_x} \int_0^{L_z} \frac{\rho I}{2}\left(\dot{\phi}_x\right)^2 dzdx + \int_0^{L_x} \int_0^{L_z} \frac{\rho A}{2}\dot{v}^2 dzdx \\ &+ \int_0^{L_x} \int_0^{L_z} \frac{\rho I}{2}\left(\dot{\phi}_z\right)^2 dzdx \end{aligned} \tag{5.67}$$

where I represents the second moment of area, ρ represents the mass density of the material and A represents the cross-sectional area of any plane S.

Recalling equation (5.60), the second term can be expanded as the sum of potentials due to interaction between any molecule i and its neighbor i_j. Since the plane before deformation remains a plane after deformation, so the vertical displacements of all the neighbors lying in the same plane is the same. Thus, $(v_i^S - v_{i_j}^S) = 0$. Similarly, for the adjacent planes at infinitesimally small distances the vertical displacements are approximately the same. Thus, $(v_i^S - v_{i_j}^{S+1}) \approx 0$ and $(v_i^S - v_{i_j}^{S-1}) \approx 0$. The equation thus reduces to:

$$\begin{aligned} U = \frac{1}{2}\Bigg[\sum_i^N \sum_{j=1}^{N'} K_x(u_i - u_{i_j})^2 + K_y\Big(\big(v_i - v_{i_j} - \phi_x l_x\big)^2 \\ + \big(v_i - v_{i_j} - \phi_z l_z\big)^2\Big) + K_z(w_i - w_{i_j})^2\Bigg] - \sum_{i=1}^N q_i^S v_i^S \end{aligned} \tag{5.68}$$

Since, the $u_i^S - u_{i_j}^S = 0$ and $w_i^S - w_{i_j}^S = 0$, the above equation can be written as:

$$
\begin{aligned}
U = {} & \frac{1}{2}K_x\Big[\cdots + \left(u_B^{S_x} - u_B^{S_x-1}\right)^2 + \left(u_B^{S_x} - u_B^{S_x+1}\right)^2 \\
& + \cdots + \left(u_B^{S_z} - u_B^{S_z-1}\right)^2 + \left(u_B^{S_z} - u_B^{S_z+1}\right)^2 + \cdots\Big] \\
& + \frac{1}{2}K_z\Big[\cdots + \left(w_B^{S_x} - w_B^{S_x-1}\right)^2 + \left(w_B^{S_x} - w_B^{S_x+1}\right)^2 \\
& + \cdots + \left(w_B^{S_z} - w_B^{S_z-1}\right)^2 + \left(w_B^{S_z} - w_B^{S_z+1}\right)^2 + \cdots\Big] \\
& + \frac{1}{2}K_y\Big[\cdots + \left(v_B^{S_x} - v_B^{S_x-1} - \phi_{xB}l_x\right)^2 + \left(v_B^{S_x} - v_B^{S_x+1} - \phi_{xB}l_x\right)^2 + \cdots\Big] \\
& + \frac{1}{2}K_y\Big[\cdots + \left(v_B^{S_z} - v_B^{S_z-1} - \phi_{zB}l_z\right)^2 + \left(v_B^{S_z} - v_B^{S_z+1} - \phi_{zB}l_z\right)^2 + \cdots\Big] - \sum_{i=1}^{N} q_i^S v_i^S
\end{aligned}
\tag{5.69}
$$

Replacing the expressions given in equations (5.61)–(5.63) into equation (5.69) and taking continuum limit of the discrete potential energy, we obtain the following continuum version of the potential energy,

$$
\begin{aligned}
U = {} & \int_0^{L_x}\int_0^{L_z} \frac{1}{2}\left(\frac{L_y}{2}\right)^2 l\left[K_x\left(\frac{\partial \phi_x}{\partial x} + \frac{\partial \phi_z}{\partial z}\right)^2\right] dzdx \\
& + \int_0^{L_x}\int_0^{L_z} \frac{1}{2}\left(\frac{L_y}{2}\right)^2 l\left[K_z\left(\frac{\partial \phi_z}{\partial x} + \frac{\partial \phi_x}{\partial z}\right)^2\right] dzdx \\
& + \frac{1}{2}\int_0^{L_x}\int_0^{L_z}\left[K_y l_x^2\left(\frac{\partial v}{\partial x} - \phi_x\right)^2 + K_y l_x^2\left(\frac{\partial v}{\partial z} - \phi_z\right)^2\right] dxdz - \int_0^{L_x}\int_0^{L_z} qv dzdx
\end{aligned}
\tag{5.70}
$$

The Lagrangian (L) for the Mindlin–Reissner plate theory can be approximated as,

$$
L = T - U \tag{5.71}
$$

The Lagrangian $L := L_1(v, \dot{v}, (v_x), (v_z)) + L_2(\phi_x, \dot{\phi}_x, (\phi_x)_x) + L_3(\phi_z, \dot{\phi}_z, ((\phi_z)_x))$, the corresponding Euler–Lagrange equations are:

$$
\frac{\partial L}{\partial v} - \frac{\partial}{\partial t}\left(\frac{\partial L}{\partial \dot{v}}\right) + \frac{\partial}{\partial x}\left(\frac{\partial L}{\partial (v_x)}\right) + \frac{\partial}{\partial z}\left(\frac{\partial L}{\partial (v_z)}\right) = 0 \tag{5.72}
$$

$$
\frac{\partial L}{\partial \phi_x} - \frac{\partial}{\partial t}\left(\frac{\partial L}{\partial \dot{\phi}_x}\right) + \frac{\partial}{\partial x}\left(\frac{\partial L}{\partial ((\phi_x)_x)}\right) = 0 \tag{5.73}
$$

$$
\frac{\partial L}{\partial \phi_z} - \frac{\partial}{\partial t}\left(\frac{\partial L}{\partial \dot{\phi}_z}\right) + \frac{\partial}{\partial x}\left(\frac{\partial L}{\partial ((\phi_z)_x)}\right) = 0 \tag{5.74}
$$

Solving the above equation (5.72), (5.73) and (5.74), the governing equations for the dynamics of Mindlin–Reissner plate theory can be written as:

$$\boxed{q - \rho A\frac{\partial^2 v}{\partial t^2} = \frac{\partial}{\partial x}\left[K_y l_x\left(\frac{\partial v}{\partial x} - \phi_x\right)\right] + \frac{\partial}{\partial z}\left[K_y l_z\left(\frac{\partial v}{\partial z} - \phi_z\right)\right]} \tag{5.75}$$

$$\boxed{\begin{aligned}&\left(\frac{L_y}{2}\right)^2 l_z K_z\frac{\partial^2 \phi_x}{\partial x^2} + \left(\frac{L_y}{2}\right)^2 (l_z K_x + l_x K_z)\frac{\partial^2 \phi_x}{\partial x \partial z}\\ &+ \left(\frac{L_y}{2}\right)^2 l_z K_z\frac{\partial^2 \phi_z}{\partial z^2} + K_y l_x^2\left(\frac{\partial v}{\partial x} - \phi_x\right) - \rho I\frac{\partial^2 \phi_x}{\partial t^2} = 0\end{aligned}} \tag{5.76}$$

$$\boxed{\begin{aligned}&\left(\frac{L_y}{2}\right)^2 l_z K_z\frac{\partial^2 \phi_z}{\partial x^2} + \left(\frac{L_y}{2}\right)^2 (l_z K_x + l_x K_z)\frac{\partial^2 \phi_x}{\partial x \partial z}\\ &+ \left(\frac{L_y}{2}\right)^2 l_z K_z\frac{\partial^2 \phi_z}{\partial z^2} + K_y l_z^2\left(\frac{\partial v}{\partial z} - \phi_z\right) - \rho I\frac{\partial^2 \phi_z}{\partial t^2} = 0\end{aligned}} \tag{5.77}$$

The above equations (5.76) and (5.77) can be rearranged as:

$$\begin{aligned}&A_1\left(\frac{\partial^2 \phi_x}{\partial x^2} + \frac{\partial^2 \phi_x}{\partial z^2}\right) + A_2\frac{\partial}{\partial x}\left(\frac{\partial \phi_x}{\partial x} + \frac{\partial \phi_z}{\partial z}\right) - \rho I\frac{\partial^2 \phi_x}{\partial t^2} - E_x I\frac{\partial^2 \phi_x}{\partial x^2} = 0\\ &A_1\left(\frac{\partial^2 \phi_z}{\partial x^2} + \frac{\partial^2 \phi_z}{\partial z^2}\right) + A_2\frac{\partial}{\partial z}\left(\frac{\partial \phi_x}{\partial x} + \frac{\partial \phi_z}{\partial z}\right) - \rho I\frac{\partial^2 \phi_z}{\partial t^2} - E_z I\frac{\partial^2 \phi_z}{\partial x^2} = 0\end{aligned} \tag{5.78}$$

where $A_1 := \frac{D}{2}(1 - \mu)$ and $A_2 := \frac{D}{2}(1 + \mu)$, $D(:=KA)$ represents the bending rigidity of the plate. For static analysis of homogeneous and isotropic thick plate, the governing equation (5.75) can be represented as:

$$\begin{aligned}&q - KA\nabla^2\left(\frac{\partial \phi_x}{\partial x} + \frac{\partial \phi_z}{\partial z}\right) = 0\\ &\boxed{\frac{q}{D} = \nabla^2\left(\frac{\partial \phi_x}{\partial x} + \frac{\partial \phi_z}{\partial z}\right)}\end{aligned} \tag{5.79}$$

5.6 Non-local theories from discrete to continuum limits

All the theories discussed till now are local theories, i.e., they ignore finer length scale information and long-range information, which may be important for small systems. In this section, we attempt to inlude them. The main idea here is to incorporate the notion of shared mass [12] and long-range interaction while defining the total energy of the system. So, the non-local kinetic energy for N discrete molecules can be written as:

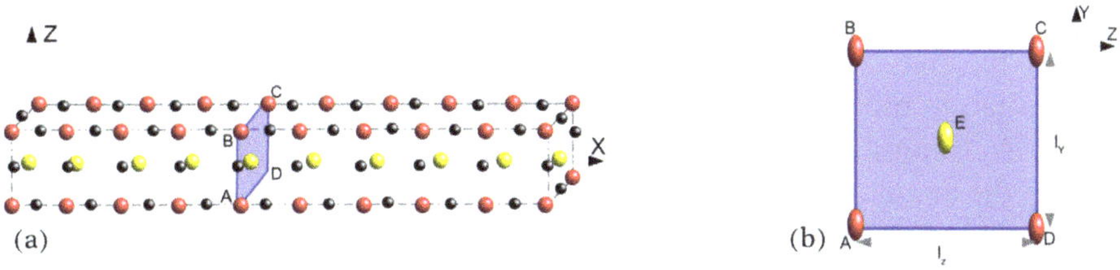

Figure 5.11. Pictorial representation of (a) initial configuration of a beam with N molecules (red), shared masses (black) and pseudo molecules (yellow) arranged along the N.A.; (b) cross-section normal to the longitudinal axis of the beam.

$$T_{NL} = \sum_{i=1}^{N}\left[\frac{1}{2}m_i\dot{v}_i^2 + \sum_{j=1}^{N'}\frac{1}{2}m_b\frac{1}{2}\left(\dot{v}_i - \dot{v}_{i_j}\right)^2 + \sum_{j=1}^{N'}\frac{1}{2}I_b\left(\frac{\dot{v}_i - \dot{v}_{i_j}}{l_x}\right)^2 + \frac{I_o}{2}\dot{\theta}^2\right] \tag{5.80}$$

where $\dot{v}_i$ represents the vertical component of velocity of the ith molecule, I_o represents the mass moment of inertia, m_b represents the common mass shared between adjacent molecules (see figure 5.11), I_b represents the mass moment of inertia for the shared mass m_b and $\dot{\theta}$ represent the angular velocity. $\dot{\theta}$ can be mathematically represented as:

$$\dot{\theta}_i = \sum_{k=-1}^{1}\left(\frac{\dot{v}_{i+k} - \dot{v}_i}{l_x}\right) \tag{5.81}$$

Taking a continuum limit of the discrete kinetic energy we obtain the corresponding continuum kinetic energy as:

$$T_{NL} = \frac{1}{2}\int_{\Omega_x}\left[\nu\left(\frac{\partial v}{\partial t}\right)^2 + 2\left[m_b l_x + \frac{I_b}{l_x}\right]\left(\frac{\partial^2 v}{\partial x \partial t}\right)^2 + I_o l_x\left(\frac{\partial^3 v}{\partial x^2 \partial t}\right)\right]dx \tag{5.82}$$

where ν represents the mass per unit length of the beam.

5.7 Non-local theory for Euler–Bernoulli beam

A schematic description of the deformation of an EB beam is given in figure 5.11. Consider a system of N molecules where the red balls represent the molecules in the beam. The black balls are the shared masses, which arise due to finer length scale information. The yellow balls, as usual denote the pseudo molecules. All the assumptions made in the local EB theory are considered here also. Thus here also, similar to the local EB theory, transverse shear strains can be neglected. The total kinetic energy for the discrete system of N molecules can be expressed using equation (5.82).

$$T = \sum_{i=1}^{N}\left[\frac{1}{2}m_i\dot{v}_i^2 + \sum_{j=1}^{N'}\frac{1}{2}m_b\frac{1}{2}\left(\dot{v}_i - \dot{v}_{i_j}\right)^2 + \sum_{j=1}^{N'}\frac{1}{2}I_b\left(\frac{\dot{v}_i - \dot{v}_{i_j}}{l_x}\right)^2 + \frac{I_o}{2}\dot{\theta}^2\right] \tag{5.83}$$

where $\dot{v}_i$ represents the vertical component of velocity of the ith molecule, I_o represents the mass moment of inertia, m_b represents the common mass shared between adjacent molecules, I_b represents the mass moment of inertia for the shared mass m_b, N' represents the neighborhood domain of each ith molecule and $\dot{\theta}$ represents the angular velocity.

$$\dot{\theta}_i = \sum_{k=-1}^{1}\left(\frac{\dot{v}_{i+k} - \dot{v}_i}{l_x}\right) \tag{5.84}$$

The potential energy (U) of the system is represented as,

$$U = \frac{1}{2}\left[\sum_{i}^{N}\sum_{j=1}^{4} K_x\left(\Delta u_{ii_j}\right)^2 + K_y\left(\Delta v_{ii_j}\right)^2 + K_z\left(\Delta w_{ii_j}\right)^2\right] - \sum_{i=1}^{N} q_i^S v_i^S \tag{5.85}$$

where $K := [K_x, K_y, K_z]$ represents the stiffness and Δu_{ii_j}, Δv_{ii_j} and Δw_{ii_j} are the displacements between ith and i_jth molecules along X, Y and Z, respectively. i_j denotes the jth neighbor of the ith molecule. For clarity on this, please refer to figure 5.3 in which the neighborhood molecules of any molecule are shown. We now reduce the dimensionality of the system by defining a set of pseudo molecules (represented by yellow balls in figure 5.2) arranged along the N.A. of the beam. q_i^S is the externally applied force. For ease of demonstration, let us denote the displacements of the Pth molecule at a cross-section S by (u_P^S, v_P^S, w_P^S) along X, Y, Z, respectively. Let $m_i (:= m_A^S + m_B^S + m_C^S + m_D^S)$ be the lumped mass of the pseudo molecule positioned at the C.G. of section S. Here i runs over the number of the cross-sections. As the beam is slender, the vertical component of the velocity for all the molecules on plane S can be assumed to equal.

$$\dot{v}_A^S = \dot{v}_B^S = \dot{v}_C^S = \dot{v}_D^S = \dot{v}_i \tag{5.86}$$

$\dot{v}_i$ represents the component of velocity along the Y direction for the pseudo molecule corresponding to plane S. Taking a continuum limit of the discrete kinetic energy we obtain the corresponding continuum kinetic energy as:

$$T = \frac{1}{2}\int_0^{L_x}\left[\nu\left(\frac{\partial v}{\partial t}\right)^2 + 2\left[m_b l_x + \frac{I_b}{l_x}\right]\left(\frac{\partial^2 v}{\partial x \partial t}\right)^2 + I_o l_x\left(\frac{\partial^3 v}{\partial x^2 \partial t}\right)\right]dx \tag{5.87}$$

where ν represents the mass per unit length of the beam and $\dot{v}$ represents the continuum counterparts of velocity in the Y direction.

Now we shall move on to our discussion on the potential energy. Since for the EB model, the transverse shear strains can be neglected, for small strains the following is true.

$$\tan\theta \approx \theta = \frac{\partial u}{\partial y} = -\frac{\partial v}{\partial x} \tag{5.88}$$

where u, v, x, y are in the usual sense continuum versions of their respective discrete counterparts. The displacements of a molecule (say B) sitting on plane S can be approximated by the displacement of the pseudo molecule (here E) positioned at the N.A., as given below.

$$u_B^S \approx u_E^S + \left.\frac{\partial u}{\partial y}\right|_S \left(\frac{L_y}{2}\right) + \left.\frac{\partial u}{\partial z}\right|_S \left(\frac{L_z}{2}\right) = u_E^S - \left.\frac{\partial v}{\partial x}\right|_S \left(\frac{L_y}{2}\right) \tag{5.89}$$

$$v_B^S \approx v_E^S + \left.\frac{\partial v}{\partial y}\right|_S \left(\frac{L_y}{2}\right) + \left.\frac{\partial v}{\partial z}\right|_S \left(\frac{L_z}{2}\right) = v_E^S \tag{5.90}$$

$$w_B^S \approx w_E^S + \left.\frac{\partial w}{\partial y}\right|_S \left(\frac{L_y}{2}\right) + \left.\frac{\partial w}{\partial z}\right|_S \left(\frac{L_z}{2}\right) = 0 \tag{5.91}$$

Referring to equation (5.90), the third term of the potential energy as given in equation (5.85) drops out of the analysis. The second term of the potential energy can be expanded as the sum of potentials between interacting neighbors of any molecule i (see figure 5.12). Since the beam is assumed to be slender, the energy contribution of shear forces can be neglected. Thus the second term of equation (5.85) vanishes and the equation is reduced to the following.

$$\begin{aligned} U &= \frac{1}{2}\sum_{i}^{N}\sum_{j=1}^{4} K_x(u_i - u_{i_j})^2 - \sum_{i=1}^{N} q_i^S v_i^S \\ &= \frac{1}{2}\Big[\cdots + K_x\left(u_B^S - u_A^S\right)^2 + K_x\left(u_B^S - u_C^S\right)^2 + K_x\left(u_B^S - u_B^{(s-1)}\right)^2 \\ &\quad + K_x\left(u_B^S - u_B^{(s+1)}\right)^2 + \cdots\Big] - \sum_{i=1}^{N} q_i^S v_i^S \end{aligned} \tag{5.92}$$

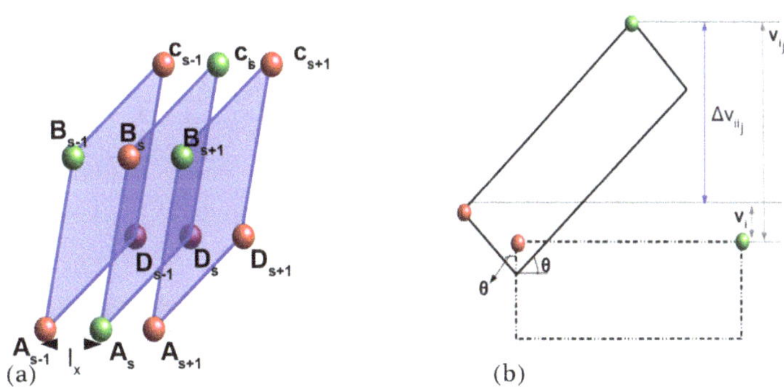

Figure 5.12. Pictorial representation of (a) molecule (suppose B) under the influence of its local neighbors represented by green balls; (b) vertical displacement of molecule i and its neighbor i_j (not drawn to scale).

Recalling equation (5.89):

$$
\begin{aligned}
u_B^S - u_A^S &\approx u_E^S + \left(\frac{L_y}{2}\right)\frac{\partial u}{\partial y}\bigg|_S - u_E^S - \left(\frac{L_y}{2}\right)\frac{\partial u}{\partial y}\bigg|_S = 0 \\
u_B^S - u_C^S &\approx u_E^S + \left(\frac{L_y}{2}\right)\frac{\partial u}{\partial y}\bigg|_S - u_E^S - \left(\frac{L_y}{2}\right)\frac{\partial u}{\partial y}\bigg|_S = 0
\end{aligned} \tag{5.93}
$$

Since, $(u_B^S - u_A^S) = 0$ and $(u_B^S - u_C^S) = 0$. So equation (5.92) gets reduced to:

$$
\begin{aligned}
U &= \frac{1}{2}\left[\cdots + K_x\left(u_B^S - u_B^{S-1}\right)^2 + K_x\left(u_B^S - u_C^{S+1}\right)^2 + \cdots\right] - \sum_{i=1}^{N} q_i^S v_i^S \\
&= \frac{1}{2}\left[\cdots + K_x\left(u_E^S + \left(\frac{L_y}{2}\right)\left(\frac{\partial u}{\partial y}\bigg|_B^S - u_E^{S-1} - \left(\frac{L_y}{2}\right)\frac{\partial u}{\partial y}\bigg|_B^{S-1}\right)^2\right]\right. \\
&\quad + \left[K_x\left(u_E^S + \left(\frac{L_y}{2}\right)\frac{\partial u}{\partial y}\bigg|_B^S - u_E^{S+1} - \left(\frac{L_y}{2}\right)\frac{\partial u}{\partial y}\bigg|_C^{S+1}\right)^2 + \cdots\right] - \sum_{i=1}^{N} q_i^S v_i^S \\
&= \frac{1}{2}\left[\cdots + K_x\left(\left(\frac{L_y}{2}\right)\frac{\partial v}{\partial x}\bigg|_B^S - \left(\frac{L_y}{2}\right)\frac{\partial v}{\partial x}\bigg|_B^{S-1}\right)^2\right. \\
&\quad \left. + K_x\left(\left(\frac{L_y}{2}\right)\frac{\partial v}{\partial x}\bigg|_B^S - \left(\frac{L_y}{2}\right)\frac{\partial v}{\partial x}\bigg|_B^{S+1}\right)^2 + \cdots\right] - \sum_{i=1}^{N} q_i^S v_i^S \\
&= \frac{1}{2}\left(\frac{L_y}{2}\right)^2\left[\cdots + K_x\left(\frac{\partial v}{\partial x}\bigg|_B^S - \frac{\partial v}{\partial x}\bigg|_B^{S-1}\right)^2 + K_x\left(\frac{\partial v}{\partial x}\bigg|_B^S - \frac{\partial v}{\partial x}\bigg|_B^{S+1}\right)^2 + \cdots\right] \\
&\quad - \sum_{i=1}^{N} q_i^S v_i^S \\
&= \frac{1}{2}\left(\frac{L_y}{2}\right)^2 (l_x)^2\left[\cdots + K_x\left(\frac{\frac{\partial v}{\partial x}\Big|_B^S - \frac{\partial v}{\partial x}\Big|_B^{S-1}}{l_x}\right)^2 + K_x\left(\frac{\frac{\partial v}{\partial x}\Big|_B^S - \frac{\partial v}{\partial x}\Big|_B^{S+1}}{l_x}\right)^2 + \cdots\right] \\
&\quad + \sum_{i=1}^{N} q_i^S v_i^S
\end{aligned} \tag{5.94}
$$

Taking a continuum limit of the discrete potential energy for lattice distance l_x, we obtain the corresponding continuum potential energy as,

$$U = \int_0^{L_x} \frac{1}{2}\left(\frac{L_y}{2}\right)^2 l_x \left[K_x\left(\frac{\partial^2 v}{\partial x^2}\right)^2\right] dx - \int_0^{L_x} qv dx \tag{5.95}$$

The Lagrangian (L) for the Euler–Bernoulli beam can be approximated as follows,

$$L = T - U \tag{5.96}$$

The corresponding Euler–Lagrange equation for $L := L(v, \dot{v}, (v_{xx}))$ is obtained as follows:

$$\frac{\partial L}{\partial v} - \frac{\partial}{\partial t}\left(\frac{\partial L}{\partial \dot{v}}\right) + \frac{\partial^2}{\partial x^2}\left(\frac{\partial L}{\partial v_{xx}}\right) = 0 \tag{5.97}$$

Now,

$$\begin{aligned}
\frac{\partial L}{\partial v} &= q; \\
\frac{\partial}{\partial t}\left(\frac{\partial L}{\partial \dot{v}}\right) &= \nu\frac{\partial^2 v}{\partial t^2} - 2\left(m_b l_x + \frac{I_b}{l_x}\right)\left(\frac{\partial^4 v}{\partial t^2 \partial x^2}\right) + I_o l_x \frac{\partial^6 v}{\partial t^2 \partial x^2}; \\
\frac{\partial^2}{\partial x^2}\left(\frac{\partial L}{\partial (v_{xx})}\right) &= -\frac{\partial^2}{\partial x^2}\left[\left(\frac{L_y}{2}\right)^2 (L_x) K_x\left(\frac{\partial^2 v}{\partial x^2}\right)\right]
\end{aligned} \tag{5.98}$$

Putting the values of equation (5.98) in equation (5.97), the governing equation for dynamics of an Euler–Bernoulli beam can be represented as:

$$\frac{\partial^2}{\partial x^2}\left[\left(\frac{L_y}{2}\right)^2 (L_x) K_x\left(\frac{\partial^2 v}{\partial x^2}\right)\right] = -\left[\nu\frac{\partial^2 v}{\partial t^2} - 2\left(m_b l_x + \frac{I_b}{l_x}\right)\left(\frac{\partial^4 v}{\partial t^2 \partial x^2}\right) + I_o l_x \frac{\partial^6 v}{\partial t^2 \partial x^2}\right] + q \tag{5.99}$$

The above equation (5.99) can be further rearranged to:

$$\boxed{\frac{\partial^2}{\partial x^2}\left[EI\left(\frac{\partial^2 v}{\partial x^2}\right)\right] = -\left[\nu\frac{\partial^2 v}{\partial t^2} - 2\left(m_b l_x + \frac{I_b}{l_x}\right)\left(\frac{\partial^4 v}{\partial t^2 \partial x^2}\right) + I_o l_x \frac{\partial^6 v}{\partial t^2 \partial x^2}\right] + q} \tag{5.100}$$

where $E\left(:=\frac{K_x L_x}{A}\right)$ represents the modulus of elasticity of the material and $I\left(:=\left(\frac{L_y}{2}\right)^2 A\right)$ represents the second moment of area. The non-local EBT with multiple non-local parameters ([7, 12]) may be written as:

$$\begin{aligned}
&- m_o\frac{\partial^2 v}{\partial t^2} + \left(\mu_1^{EB} m_o + m_2\right)\frac{\partial^4 v}{\partial t^2 \partial x^2} - \mu_2^{EB} m_2 \frac{\partial^6 v}{\partial x^4 \partial t^2} - \frac{\partial^2}{\partial x^2}\left(EI\frac{\partial^2 v}{\partial x^2}\right) \\
&+ q - \mu_3^{EB}\frac{\partial^2 q}{\partial x^2} = 0
\end{aligned} \tag{5.101}$$

On comparing equation (5.100) with (5.101), we find the correspondence as:

$$\begin{aligned}
&m_o = \nu, \quad \mu_1^{EB} m_o = 2 m_b l_x, \quad m_2 = 2\frac{I_b}{l_x}, \\
&\mu_2^{EB} m_2 = I_o l_x, \quad \mu_3^{EB} = 0
\end{aligned} \tag{5.102}$$

Using the above equation, the two non-zero parameters can be represented as:

$$\mu_1^{EB} = \frac{2m_b l_x}{m_o}$$
$$\mu_2^{EB} = \frac{I_o l_x^2}{2I_b} \tag{5.103}$$

It is interesting to note that as the lattice distance $l_x \to 0$ the shared mass gets lumped on the molecular positions and the non-zero non-local parameters μ_1^{EB} and μ_2^{EB} diminish to zero. This results in the local governing equations for the Euler–Bernoulli beam theory.

$$-m_o\frac{\partial^2 v}{\partial t^2} - \frac{\partial^2}{\partial x^2}\left(EI\frac{\partial^2 v}{\partial x^2}\right) + q = 0 \tag{5.104}$$

5.8 Non-local theory for Timoshenko beam

As discussed in the previous section of this chapter, the classical theory or EB beam theory neglects the effects of shear deformations. Disregarding the shear deformations makes EB beam theory suitable for thin and slender members but unsuitable for thick and short span members, where shear deformation plays a key role. Timoshenko beam (TB) theory is used for coupling bending and shear deformations. This one-dimensional theory is used to capture the combined effect of bending and shear deformations in a structural member subjected to transverse loading. A schematic description of the deformation of a TB is presented in figure 5.13. Here the dotted lines denote the undeformed orientation of the beam. The basic idea is to derive a discrete TB theory by incorporating the above-mentioned assumptions. With the similar notations as described in EB theory, let us consider a structural member as shown in figure 5.5. The discrete system can have more molecules at the cross-sections by defining an appropriate decaying function with increasing distances between two molecules. However, our objective is to arrive at a minimal discrete model for clarity in understanding the computational effectiveness. The total kinetic energy for the discrete system of N molecules can be expressed as:

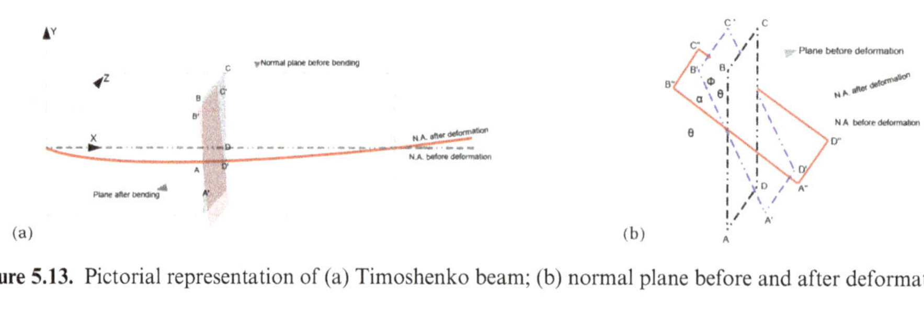

Figure 5.13. Pictorial representation of (a) Timoshenko beam; (b) normal plane before and after deformation.

$$T = \sum_{i=1}^{N}\left[\frac{1}{2}m_i\dot{v}_i^2 + \sum_{j=1}^{N'}\frac{1}{2}m_b\frac{1}{2}\left(\dot{v}_i - \dot{v}_{i_j}\right)^2 + \sum_{j=1}^{N'}\frac{1}{2}I_b\left(\frac{\dot{v}_i - \dot{v}_{i_j}}{l_x}\right)^2 + \frac{I_o}{2}\dot{\theta}^2\right] \tag{5.105}$$

where $\dot{v}_i$ represents the vertical component of velocity of ith molecule, I_o represents the mass moment of inertia, m_b represents the common mass shared (represented by black balls in figure 5.14) between adjacent molecules, I_b represents the mass moment of inertia for the shared mass m_b, N' represents the neighborhood domain of each ith molecule and $\dot{\theta}$ represents the angular velocity.

$$\dot{\theta}_i = \sum_{k=-1}^{1}\left(\frac{\dot{v}_{i+k} - \dot{v}_i}{l_x}\right) \tag{5.106}$$

The potential energy (U) of the system is represented as:

$$U = \frac{1}{2}\left[\sum_{i}^{N}\sum_{j=1}^{4}K_x\left(\Delta u_{ii_j}\right)^2 + K_y\left(\Delta v_{ii_j}\right)^2 + K_z\left(\Delta w_{ii_j}\right)^2\right] - \sum_{i=1}^{N}q_i^S v_i^S \tag{5.107}$$

where all the notations have the usual meaning as defined in EB beam theory. With the aim to reduce the dimensionality of the system, a set of pseudo molecules (represented by yellow balls in figure 5.14) is arranged along the N.A. of the beam. Let (u_P^S, v_P^S, w_P^S) represent the displacements along X, Y and Z direction, respectively for the Pth molecule at cross-section s. The displacements of a molecule (say B) sitting on plane S can be approximated by the displacement of the pseudo molecule (here E) positioned at the N.A., as given below.

$$u_B^S \approx u_E^S + (\phi)|_s\left(\frac{L_y}{2}\right) + \frac{\partial u}{\partial z}\bigg|_s\left(\frac{L_z}{2}\right) = u_E^S + (\phi)|_s\left(\frac{L_y}{2}\right) \tag{5.108}$$

$$v_B^S \approx v_E^S + \frac{\partial v}{\partial y}\bigg|_s\left(\frac{L_y}{2}\right) + \frac{\partial v}{\partial z}\bigg|_s\left(\frac{L_z}{2}\right) = v_E^S \tag{5.109}$$

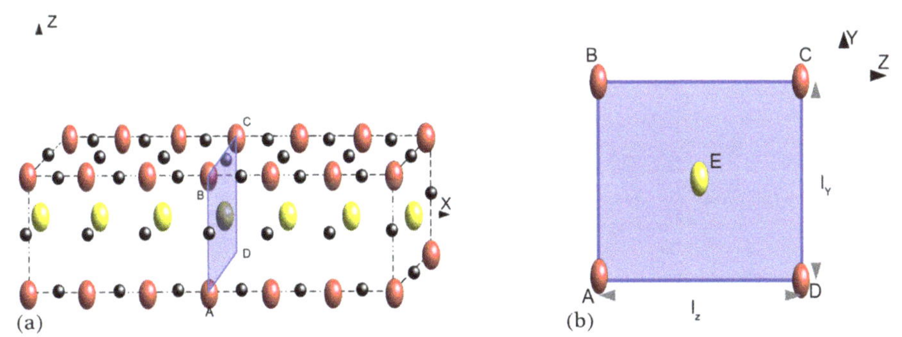

Figure 5.14. Pictorial representation of (a) initial configuration of a beam with N molecules (red), shared masses (black) and pseudo molecules (yellow) arranged along the N.A.; (b) cross-section normal to the longitudinal axis of the beam.

$$w_B^S \approx w_E^S + \frac{\partial w}{\partial y}\bigg|_s \left(\frac{L_y}{2}\right) + \frac{\partial w}{\partial z}\bigg|_s \left(\frac{L_z}{2}\right) = 0 \tag{5.110}$$

Since it is assumed that the plane section before deformation remains a plane after deformation, the vertical component of velocity of all the molecules on plane S is assumed to be equal.

$$\begin{aligned} \dot{u}_A^S = \dot{u}_B^S = \dot{u}_C^S = \dot{u}_D^S = \dot{u}_i \\ \dot{v}_A^S = \dot{v}_B^S = \dot{v}_C^S = \dot{v}_D^S = \dot{v}_i \end{aligned} \tag{5.111}$$

Taking the time derivative on both the sides, equation (5.108) can be reformed as:

$$\begin{aligned} \frac{\partial u_B^S}{\partial t} &= \frac{\partial u_E^S}{\partial t} + \frac{\partial}{\partial t}\left(\phi \,|_B^S \,\frac{L_y}{2}\right) \\ \dot{u}_B^S &= \dot{\phi} \,|_B^S \left(\frac{L_y}{2}\right) \end{aligned} \tag{5.112}$$

Taking a continuum limit of the discrete kinetic energy we obtain the corresponding continuum kinetic energy as:

$$T = \frac{1}{2}\int_0^{L_x}\left[\nu\left(\frac{\partial v}{\partial t}\right)^2 + 2m_b l_x\left(\frac{\partial \dot{v}}{\partial x}\right)^2 + \frac{2I_b}{l_x}\left(\frac{\partial \phi}{\partial t}\right)^2 + I_o l_x\left(\frac{\partial^2 \phi}{\partial x \partial t}\right)^2\right]dx \tag{5.113}$$

where ν represents the mass per unit length of the beam and $\dot{v}$ represents the continuum counterparts of velocity in the Y direction. Now, the potential energy of the beam can be represented as the sum of the potential energy due to interaction between any ith molecule and its nearest i_jth neighbors (see figure 5.15). For more clarity on the neighborhood please see figure 5.6 where the green balls represent the neighborhood molecules of molecule B sitting on plane S. Since the deformations along the z-axis are neglected, so the third term on the right hand side of equation (5.107) drops out of the analysis. The equation thus reduces to:

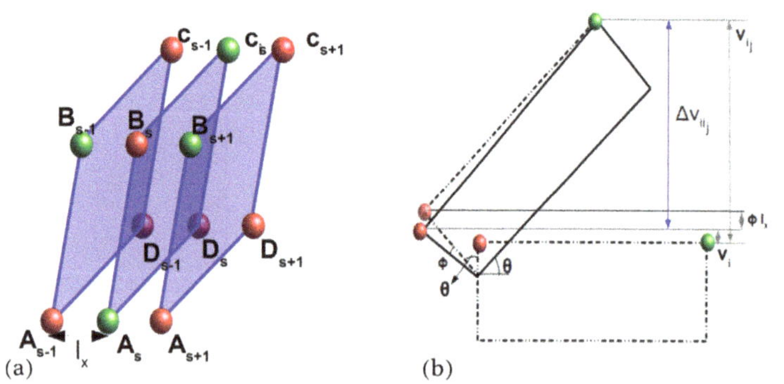

Figure 5.15. Pictorial representation of (a) molecule (suppose B) under the influence of its local neighbors represented by green balls; (b) vertical displacement of molecule i and its neighbor i_j (not drawn to scale).

$$
\begin{aligned}
U &= \frac{1}{2}\left[\sum_{i}^{N}\sum_{j=1}^{4} K_x(u_i - u_{i_j})^2 + K_y\left(v_i - v_{i_j} - \phi_i l_x\right)^2\right] - \sum_{i=1}^{N} q_i^S v_i^S \\
U &= \frac{1}{2}K_x\left[\cdots + \left(u_B^S - u_A^S\right)^2 + \left(u_B^S - u_C^S\right)^2 + \left(u_B^S - u_B^{S-1}\right)^2 + \left(u_B^S - u_B^{S+1}\right)^2 + \cdots\right] \\
&+ \frac{1}{2}K_y\left[\cdots + \left(v_B^S - v_A^S - \phi_B^S l_x\right)^2 + \left(v_B^S - v_C^S - \phi_B^S l_x\right)^2\right. \\
&\left. + \left(v_B^S - v_B^{S-1} - \phi_B^S l_x\right)^2 + \left(v_B^S - v_B^{S+1} - \phi_B^S l_x\right)^2 + \cdots\right] - \sum_{i=1}^{N} q_i^S v_i^S \qquad (5.114) \\
&\because \quad u_i^S - u_{i_j}^S = u_E^S + \phi|_i\left(\frac{L_y}{2}\right) - u_E^S - \phi|_i\left(\frac{L_y}{2}\right) = 0 \\
&\Longrightarrow u_B^S - u_A^S = 0; \qquad u_B^S - u_C^S = 0 \\
&\because \quad v_i^S - v_{i_j}^S = v_E^S - v_E^S = 0 \\
&\Longrightarrow v_B^S - v_A^S = 0; \qquad v_B^S - v_C^S = 0
\end{aligned}
$$

Thus the above equation reduces to:

$$
\begin{aligned}
U &= \frac{1}{2}K_x\left[\cdots + \left(u_B^S - u_B^{S-1}\right)^2 + \left(u_B^S - u_B^{S+1}\right)^2 + \cdots\right] \\
&+ \frac{1}{2}K_y\left[\cdots + \left(v_B^S - v_B^{S-1} - \phi_B^S l_x\right)^2 + \left(v_B^S - v_B^{S+1} - \phi_B^S l_x\right)^2 + \cdots\right] - \sum_{i=1}^{N} q_i^S v_i^S \\
U &= \frac{1}{2}K_x\left[\cdots + \left(u_E^S + \phi|_i^S\left(\frac{L_y}{2}\right) - u_E^{S-1} - \phi|_B^{S-1}\left(\frac{L_y}{2}\right)\right)^2\right. \\
&\left. + \left(u_E^S + \phi|_i^S\left(\frac{L_y}{2}\right) - u_E^{S+1} - \phi|_B^{S+1}\left(\frac{L_y}{2}\right)\right)^2 + \cdots\right] \\
&+ \frac{1}{2}K_y\left[\cdots + \left(v_B^S - v_B^{S-1} - \phi_B^S l_x\right)^2 + \left(v_B^S - v_B^{S+1} - \phi_B^S l_x\right)^2 + \cdots\right] - \sum_{i=1}^{N} q_i^S v_i^S \qquad (5.115) \\
U &= \frac{1}{2}K_x\left[\cdots + \left(\phi|_i^S\left(\frac{L_y}{2}\right) - \phi|_B^{S-1}\left(\frac{L_y}{2}\right)\right)^2 + \left(\phi|_i^S\left(\frac{L_y}{2}\right) - \phi|_B^{S+1}\left(\frac{L_y}{2}\right)\right)^2 + \cdots\right] \\
&+ \frac{1}{2}K_y\left[\cdots + \left(v_B^S - v_B^{S-1} - \phi_B^S l_x\right)^2 + \left(v_B^S - v_B^{S+1} - \phi_B^S l_x\right)^2 + \cdots\right] - \sum_{i=1}^{N} q_i^S v_i^S \\
U &= \frac{1}{2}K_x\left(\frac{L_y}{2}\right)^2 (l_x)^2\left[\cdots + \left(\frac{\phi|_i^S - \phi|_B^{S-1}}{l_x}\right)^2 + \left(\frac{\phi|_i^S - \phi|_B^{S+1}}{l_x}\right)^2 + \cdots\right] \\
&+ \frac{1}{2}K_y(l_x)^2\left[\cdots + \left(\frac{v_B^S - v_B^{S-1} - \phi_B^S l_x}{l_x}\right)^2 + \left(\frac{v_B^S - v_B^{S+1} - \phi_B^S l_x}{l_x}\right)^2 + \cdots\right] - \sum_{i=1}^{N} q_i^S v_i^S
\end{aligned}
$$

Taking a continuum limit of the discrete potential energy for lattice distance l_x, we obtain the corresponding continuum potential energy as,

$$U = \int_0^{L_x} \frac{1}{2}\left(\frac{L_y}{2}\right)^2 l_x\left[K_x\left(\frac{\partial \phi}{\partial x}\right)^2\right]dx + \frac{1}{2}\int_0^{L_x} K_y l_x\left(\frac{\partial v}{\partial x} - \phi\right)^2 dx - \int_0^{L_x} qv dx$$
$$U = \int_0^{L_x} \frac{1}{2}\left(\frac{L_y}{2}\right)^2 l_x\left[K_x(\phi_x)^2\right]dx + \frac{1}{2}\int_0^{L_x} \left[K_y l_x(v_x - \phi)^2\right] - \int_0^{L_x} qv dx \tag{5.116}$$

The Lagrangian (L) for the Timoshenko beam can be represented as,

$$L = T - U$$
$$L = \frac{1}{2}\int_0^{L_x} \left[\nu\left(\frac{\partial v}{\partial t}\right)^2 + 2m_b l_x\left(\frac{\partial \dot{v}}{\partial x}\right)^2 + \frac{2I_b}{l_x}\left(\frac{\partial \phi}{\partial t}\right)^2 + I_o l_x\left(\frac{\partial^2 \phi}{\partial x \partial t}\right)^2\right]dx$$
$$\int_0^{L_x} \frac{1}{2}\left(\frac{L_y}{2}\right)^2 l_x\left[K_x(\phi_x)^2\right]dx + \frac{1}{2}\int_0^{L_x} \left[K_y l_x(v_x - \phi)^2\right] - \int_0^{L_x} qv dx \tag{5.117}$$

The Lagrangian $L := L_1(v, \dot{v}, (v_x)) + L_2(\phi, \dot{\phi}, (\phi_x))$, the corresponding Euler–Lagrange equations are:

$$\frac{\partial L}{\partial v} - \frac{\partial}{\partial t}\left(\frac{\partial L}{\partial \dot{v}}\right) + \frac{\partial}{\partial x}\left(\frac{\partial L}{\partial (v_x)}\right) = 0 \tag{5.118}$$

$$\frac{\partial L}{\partial \phi} - \frac{\partial}{\partial t}\left(\frac{\partial L}{\partial \dot{\phi}}\right) + \frac{\partial}{\partial x}\left(\frac{\partial L}{\partial (\phi_x)}\right) = 0 \tag{5.119}$$

The corresponding Euler–Lagrange equations for the governing equations for dynamics of Timoshenko beam can be represented as:

$$\boxed{\rho A\frac{\partial^2 v}{\partial t^2} - q - \frac{\partial}{\partial x}\left(K_y l_x\left(\frac{\partial v}{\partial x} - \phi\right)\right) - 2m_b l_x\frac{\partial^4 v}{\partial x^2 \partial t^2} = 0} \tag{5.120}$$

$$\boxed{\frac{\partial}{\partial x}\left(EI\frac{\partial \phi}{\partial x}\right) - \frac{2I_b}{l_x}\frac{\partial^2 \phi}{\partial t^2} - K_y l_x\left(\frac{\partial v}{\partial x} - \phi\right) + I_o l_x\frac{\partial^4 \phi}{\partial x^2 \partial t^2} = 0} \tag{5.121}$$

It may be interesting to note that $E\left(:=\frac{K_x l_x}{A}\right)$ represents the Young's modulus of elasticity and $I\left(:=\left(\frac{L_y}{2}\right)^2 A\right)$ represents the second moment of area. The non-local TB equations [7, 12] with non-local parameters can be written as:

$$\frac{\partial}{\partial x}\left[GAk\left(\frac{\partial v}{\partial x} - \phi\right)\right] - \rho A\frac{\partial^2 v}{\partial t^2} + \rho A\mu_1^{TB}\frac{\partial^4 v}{\partial x^2 \partial t^2} + q - \mu_3^{TB}\frac{\partial^2 q}{\partial x^2} = 0$$
$$\frac{\partial}{\partial x}\left(EI\frac{\partial \phi}{\partial x}\right) - GAk\left(\frac{\partial v}{\partial x} - \phi\right) - m_b\frac{\partial^2 \phi}{\partial t^2} + m_b\mu_2^{TB}\frac{\partial^4 \phi}{\partial x^2 \partial t^2} = 0 \tag{5.122}$$

It is interesting to note that the shear modulus G with correction factor k in the traditional approach is equivalent to $\frac{K_y l_x}{A}$ in the present framework. Comparing the equations we get,

$$\rho A \mu_1^{TB} = 2m_b l_x, \quad m_b = \frac{2I_b}{l_x}, \quad m_b \mu_2^{TB} = I_o l_x, \quad \mu_3^{TB} = 0 \tag{5.123}$$

The non-local non-zero parameters can thus be represented as:

$$\begin{aligned} \mu_1^{TB} &= \frac{2m_b l_x}{\rho A} \\ \mu_2^{TB} &= \frac{I_o l_x}{m_b} \end{aligned} \tag{5.124}$$

5.9 Conclusion

In this chapter, we have shown discrete energy counterparts for different reduced dimensional continuum theories. We have shown that a generic molecular dynamics-like discrete energy expression can be reduced by imposition appropriate conditions to arrive at discrete models which are equivalent to different beam and plate theories. A major aspect of our discussion is to unify different local and non-local continuum theories and bring them on the same platform. This a discrete energy-based description is particularly advantageous owing to its derivative-free framework. The approach discussed herein can be directly applied to other local and non-local continuum theories.

References

[1] Malvern L E 1969 *Introduction to the Mechanics of a Continuous Medium* (Upper Saddle River, NJ: Prentice-Hall)

[2] Reddy J N 1993 *An Introduction to the Finite Element Method* (New York: McGraw-Hill) p 27

[3] Cowin S C and Doty S B 2007 *Tissue Mechanics* (Springer: Berlin)

[4] Majdoub M, Sharma P and Cagin T 2008 Enhanced size-dependent piezoelectricity and elasticity in nanostructures due to the flexoelectric effect *Phys. Rev.* **B77** 125424

[5] Cosserat E and Cosserat F 1909 *Théorie des corps déformables*

[6] Eringen A C 2002 *Nonlocal Continuum Field Theories* (Springer: Berlin)

[7] Reddy J 2007 Nonlocal theories for bending, buckling and vibration of beams *Int. J. Eng. Sci.* **45** 288–307

[8] Witmer E 1991-1992 *Elementary Bernoulli-Euler Beam Theory* MIT Unified Engineering Course Notes (Cambridge, MA: MIT Press) pp 114–64

[9] Elishakoff I 2020 *Handbook on Timoshenko-Ehrenfest Beam and Uflyand-Mindlin Plate Theories* (World Scientific: Singapore)

[10] Love A E H 1888 XVI. The small free vibrations and deformation of a thin elastic shell *Phil. Trans. R. Soc.* **A179** 491–546
[11] Mindlin R 1951 Influence of rotatory inertia and shear on flexural motions of isotropic, elastic plates *J. Appl. Mech.* **18** 31–8
[12] Sarkar S and Reddy J 2016 Exploring the source of non-locality in the Euler–Bernoulli and Timoshenko beam models *Int. J. Eng. Sci.* **104** 110–5

IOP Publishing

Modeling and Computation in Vibration Problems, Volume 1
Numerical and semi-analytical methods
S Chakraverty, F Tornabene and J N Reddy

Chapter 6

Effect of thermal environment on nonlinear flutter of laminated composite plates reinforced with graphene nanoplatelets

Hulun Guo, Tianzhi Yang, Krzysztof Kamil Żur, J N Reddy and A J M Ferreira

Parts of this chapter have been adapted from [15] with permission from Elsevier.

Symbols

$E_C^{(k)}$, E_M	Young's moduli of the GPLRC and polymer matrix
$\rho_C^{(k)}$, ρ_{GPL}, ρ_M	Mass density of the GPLRC, GPL and polymer matrix
$\nu_C^{(k)}$, ν_{GPL}, ν_M	Poisson's ratio of the GPLRC, GPL and polymer matrix
$\alpha_C^{(k)}$, α_{GPL}, α_M	Thermal expansion coefficient of the GPLRC, GPL and polymer matrix
a, b, h	Length, width and thickness of plate
Δh	Thickness of each layer
NL	Total number of layers
l_{GPL}, w_{GPL}, h_{GPL}	Average length, width and thickness of the GPLs
$g_{\mathrm{GPL}}^{(k)}$	GPLs weight fraction in kth layer
β	Crack density
u, v, w	Displacements at the mid-plane of the plate along the x, y, and z directions
x, y, z	Cartesian coordinates
$\mathbf{L}^{(k)}$	Compliance matrix
$\mathbf{\Xi}^{(k)}$	Compliance matrix change due to matrix cracks
d	Distance between two adjacent cracks
θ_x, θ_y	Transverse normal rotations
$Q_{ij}^{(k)}$	Stiffness coefficients
ΔT	Temperature change
σ	Stress
ε_{xx}, ε_{yy}, ε_{xy}	Strain
γ_{yz}, γ_{xz}	Strain

 © IOP Publishing Ltd 2021

$\boldsymbol{\varepsilon}_t$	Thermal strain vector
Δp	Aerodynamic pressure
λ	Non-dimensional aerodynamic pressure
U_∞	Free stream velocity
M_∞	Mach number
$u^h(\boldsymbol{x})$	Displacement field
$d_{\max}$	Scaling factor
c_I	Average distance between nodes in the influenced domain
$p_i(\boldsymbol{x})$	Basis function
$w(\boldsymbol{x}-\boldsymbol{x}_I)$	Weight function
$\phi_I(\boldsymbol{x})$	Shape function
α	Penalty parameter
$\boldsymbol{u}_I$	Displacement of node I
$\boldsymbol{u}$	Displacements of all nodes
$\mathbf{M}$	Mass matrix
$\mathbf{K}_p$	Stiffness matrix owing to the plate
$\mathbf{K}_b$	Stiffness matrix owing to the boundary
$\mathbf{K}_a$	Stiffness matrix owing to aerodynamic pressure
$\mathbf{K}_t$	Thermal stiffness matrix
$\mathbf{K}_n(\boldsymbol{u})$	Nonlinear stiffness matrix
ω, $\bar{\omega}$	Frequency and non-dimensional frequency

Functionally graded (FG) composites have been widely studied and used in aerospace engineering due to their high stiffness and strength, and lightweight. Flutter may appear due to the coupling of the inertia, elastic, and aerodynamic forces. Dowell [1] has given a detailed review of panel flutter of plates and shells. Introducing nanofillers into porous materials is a practical way to strengthen their mechanical properties and maintain their potential for lightweight structures simultaneously. Tawfik *et al* [2] studied the effectiveness of a shape memory alloy embedded panel on the flutter boundary, critical buckling temperature, post-buckling deflection and free vibration using a nonlinear finite element analysis (FEA) based on the von Kármán strain-displacement relations. Li and Song [3, 4] studied the aeroelastic flutter behaviors and employed piezoelectric material to suppress the flutter of supersonic composite laminated plate. Asadi and Wang [5] studied the aeroelastic flutter and buckling bounds of FG carbon nanotube reinforced composite (FG-CNTRC) beams subjected to aerodynamic load and axial compression. Most recently, a novel class of multilayer composites called FG multilayer graphene nanoplatelets (GPLs) reinforced composites (GPLRCs) has been synthesized [6, 7]. Subsequently, the vibration, bending, buckling, and post-buckling of GPLRC structures, including nonlinearity and thermal environment, have been studied [8].

Matrix cracking may be initiated as the first damage mode when the plate is subjected to aerodynamic loading. These cracks will reduce the stiffness and strength of laminated composite plate. Laws *et al* [9, 10] have proposed the self-consistent model to evaluate the properties of matrix cracked composite. Fan and Wang [11, 12] used a higher-order shear deformation plate theory with the von Kármán type of kinematic nonlinearity to study the postbuckling and nonlinear vibration of an FG-CNTRC plate in thermal environments. Lei *et al* [13] presented the effect of matrix cracks on nonlinear vibration of FG-CNTRC plate. Pan *et al* [14] employed the first-order shear-deformable plate theory (FSDT) and modified Newton–Raphson method in combination with the arc-length iteration technique to study the nonlinear large deformation of matrix cracked hybrid composite double-curved deep shell containing CNTRC layers. Guo *et al* [15] presented the dynamics of a GPLRC plate with matrix cracks under a supersonic flow. These studies show the matrix cracks have an important influence on the dynamics of FG composites. Thus it is necessary to understand the flutter characteristics of cracked GPLRC structures in structural health monitoring and nondestructive damage evaluation. This is because the predicted flutter data can be used to detect, locate and quantify the extent of the cracks or damage in a GPLRC structure.

High-speed aircraft structural panels are subjected not only to aerodynamic loading, but also to aerodynamic heating. Combined with the nonlinearity from large deformation of the plate, thermal flutter problem will be more prominent and more complicated. Mei *et al* [16] have grouped the vast amount of literature on nonlinear panel flutter. Sohn and Kim [17] investigated the influence of thermal and aerodynamic loads on the static and dynamic stabilities of FG panels. Shiau and Kuo [18] studied the thermal effect on the nonlinear panel flutter of buckled sandwich plates. They found that temperature gradient induced thermal moments and increased the overall stiffness of the plate, and might increase the flutter boundary significantly. Ibrahim *et al* [19, 20] proposed a novel concept in enhancing the thermal buckling and nonlinear flutter behavior of FG plates through embedding shape memory alloy fibers in it. Fazelzadeh *et al* [21] predicted the flutter and thermal buckling behavior of nanocomposite plates reinforced by CNTs. Their results indicated that adding CNTs could enlarge the stable region of the FG-CNTRC plate. Chai *et al* [22] proposed a useful mechanism to suppress the nonlinear flutter and thermal buckling of composite lattice sandwich panels. A review on the aerothermo-elasticity of FG panels can be found in [23]. Numerous studies have demonstrated the importance of thermal effects on the aeroelastic characteristics of FG composites. Thus the nonlinear thermal flutter analysis is necessary for the application of GPLRC in aerospace engineering.

As an efficient computing method, element-free methods are developed to deal with mechanic problems of composite structures. Zhang *et al* [24, 25] proposed the element-free improved moving least-squares Ritz (IMLS-Ritz) method which could avoid forming an ill-conditioned or singular equation system, and need fewer nodes in the domain of influence than other element-free methods. The computational advantages and special features of IMLS-Ritz method have been demonstrated through solving free vibration [26], nonlinear bending [27] and panel flutter [28] of a GPLRC quadrilateral plate. These papers reported the remarkable feature of the element-free IMLS-Ritz method which can be efficiently used for solving dynamic problems of laminated composite plates.

In this chapter we employ the element-free IMLS-Ritz method to obtain the flutter behaviors of a matrix cracked GPLRC plate and nonlinear thermal flutter characteristics of a GPLRC plate.

6.1 Governing equations

The GPLRC plate is considered as a multilayer plate with length a in the x direction, width b in the y direction, and thickness h in the z direction (see figure 6.1). The plate

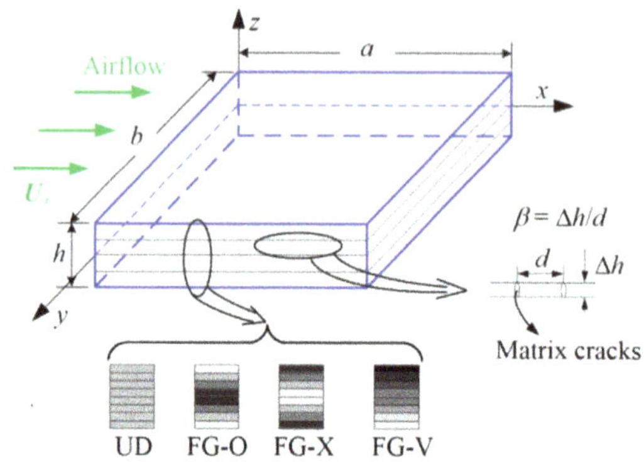

Figure 6.1. Configuration of GPLRC plate.

is subjected to a supersonic airflow along the x direction. Each layer has the same thickness $\Delta h = h/NL$. NL is the total number of layers. Each layer is isotropic and homogenous since GPL nanofillers are randomly oriented and uniformly dispersed. Both the porosity coefficient and GPL volume fraction are graded from layer to layer.

6.1.1 Material properties of GPLRCs

The effective material properties of the kth layer of GPLRC quadrilateral plate can be predicted by the Mori–Tanaka scheme [29].

$$E_C^{(k)} = \frac{3}{8}\frac{1 + \xi_L \eta_L V_{\mathrm{GPL}}^{(k)}}{1 - \eta_L V_{\mathrm{GPL}}^{(k)}} E_M + \frac{5}{8}\frac{1 + \xi_W \eta_W V_{\mathrm{GPL}}^{(k)}}{1 - \eta_W V_{\mathrm{GPL}}^{(k)}} E_M \tag{6.1}$$

where $E_C^{(k)}$ and E_M are Young's moduli of the GPLRC and polymer matrix, respectively. $V_{\mathrm{GPL}}^{(k)}$ is the GPL volume fraction of the kth layer. Parameters η_L and η_W are expressed as follows

$$\eta_L = \frac{E_{\mathrm{GPL}} - E_M}{E_{\mathrm{GPL}} + \xi_L E_M} \tag{6.2}$$

$$\eta_W = \frac{E_{\mathrm{GPL}} - E_M}{E_{\mathrm{GPL}} + \xi_w E_M} \tag{6.3}$$

Parameters ξ_L and ξ_W represent the geometry and size of GPL nanofillers.

$$\xi_L = 2\frac{l_{\mathrm{GPL}}}{h_{\mathrm{GPL}}} \tag{6.4}$$

$$\xi_W = 2\frac{w_{\mathrm{GPL}}}{h_{\mathrm{GPL}}} \tag{6.5}$$

where l_{GPL}, w_{GPL} and h_{GPL} are the average length, width and thickness of the GPLs, respectively.

Similarly, mass density $\rho_C^{(k)}$, Poisson's ratio $\nu_C^{(k)}$ and thermal expansion coefficient $\alpha_C^{(k)}$ of the kth layer of the GPLRC plates can be expressed as follows:

$$\rho_C^{(k)} = \rho_{\mathrm{GPL}} V_{\mathrm{GPL}}^{(k)} + \rho_M(1 - V_{\mathrm{GPL}}^{(k)}) \tag{6.6}$$

$$\nu_C^{(k)} = \nu_{\mathrm{GPL}} V_{\mathrm{GPL}}^{(k)} + \nu_M(1 - V_{\mathrm{GPL}}^{(k)}) \tag{6.7}$$

$$\alpha_C^{(k)} = \alpha_{\mathrm{GPL}} V_{\mathrm{GPL}}^{(k)} + \alpha_M(1 - V_{\mathrm{GPL}}^{(k)}) \tag{6.8}$$

Subscripts 'GPL' and 'M' represent GPLs and matrix material, respectively. $V_{\mathrm{GPL}}^{(k)}$ is defined as follows

$$V_{\mathrm{GPL}}^{(k)} = \frac{g_{\mathrm{GPL}}^{(k)}}{g_{\mathrm{GPL}}^{(k)} + (\rho_{\mathrm{GPL}}/\rho_M)(1 - g_{\mathrm{GPL}}^{(k)})} \tag{6.9}$$

where $g_{\mathrm{GPL}}^{(k)}$ is the GPLs weight fraction. The GPL distributions are dispersed according to four different patterns: UD, FG-O, FG-X and FG-V. The GPL weight fractions of the kth layer for the four GPL distribution patterns are given as

$$g_{\mathrm{GPL}}^{(k)} = \begin{cases} g_{\mathrm{GPL}}^{*} & UD \\ 2g_{\mathrm{GPL}}^{*}\dfrac{NL + 1 - |NL + 1 - 2k|}{NL + 2} & FG - O \\ 2g_{\mathrm{GPL}}^{*}\dfrac{1 + |NL + 1 - 2k|}{NL + 2} & FG - X \\ 2g_{\mathrm{GPL}}^{*}\dfrac{k}{NL + 1} & FG - V \end{cases} \tag{6.10}$$

The material properties and geometry parameters of GPLs and polymer matrix are given as $\rho_{\mathrm{GPL}} = 1.06$ g cm^{-3}, $E_{\mathrm{GPL}} = 1.01$ TPa, $\rho_{\mathrm{M}} = 1.2$ g cm^{-3}, $E_{\mathrm{M}} = 3.0$ GPa, $\nu_{\mathrm{M}} = 0.34$, $\nu_{\mathrm{GPL}} = 0.186$, $\alpha_{\mathrm{M}} = 60 \times 10^{-6}$ K^{-1}, $\alpha_{\mathrm{GPL}} = 5.0 \times 10^{-6}$ K^{-1}, $l_{\mathrm{GPL}} = 2.5$ μm, $w_{\mathrm{GPL}} = 1.5$ μm, $h_{\mathrm{GPL}} = 1.5$ nm, $g_{\mathrm{GPL}} = 1\%$, and $NL = 10$.

6.1.2 Model of matrix cracks

The matrix cracks considered to be elliptic cylindrical cavities are located at each layer. When cracks are introduced, stiffness of composite structures will be reduced. According to [9, 10, 15], the self-consistent estimates for the overall compliance matrix $\mathbf{L}^{(k)}$ of the kth layer of matrix cracked composite can be expressed by

$$\mathbf{L}^{(k)} = \bar{\mathbf{L}}^{(k)} + \frac{1}{4}\pi\beta\boldsymbol{\Xi}^{(k)} \tag{6.11}$$

where $\bar{\mathbf{L}}^{(k)}$ is the compliance matrix of the perfect composite lamina, and β is the crack density parameter. The crack density $\beta = \Delta h/d$, where d is the distance between two adjacent cracks. Therefore, crack density β is the average number of cracks within the side length. When $\beta = 1$, the distance between cracks is equal to layer thickness. As β decreases the distance between cracks increases. When $\beta = 0$, there is no crack. The matrix $\mathbf{\Xi}^{(k)}$ has only three nonzero components which are $\Xi_{22}^{(k)}$, $\Xi_{44}^{(k)}$, and $\Xi_{66}^{(k)}$. It implies that only three compliance coefficients $L_{22}^{(k)}$, $L_{44}^{(k)}$, and $L_{66}^{(k)}$ are affected by the introduction of cracks, the remaining six terms in $\mathbf{L}^{(k)}$ are unchanged. The unchanged components are

$$\begin{aligned} L_{11}^{(k)} &= \bar{L}_{11}^{(k)} = \frac{1}{E_C^{(k)}}, \quad L_{12}^{(k)} = \bar{L}_{12}^{(k)} = -\frac{v_C^{(k)}}{E_C^{(k)}}, \quad L_{13}^{(k)} = \bar{L}_{13}^{(k)} = -\frac{v_C^{(k)}}{E_C^{(k)}}, \\ L_{23}^{(k)} &= \bar{L}_{23}^{(k)} = -\frac{v_C^{(k)}}{E_C^{(k)}}, \quad L_{33}^{(k)} = \bar{L}_{33}^{(k)} = \frac{1}{E_C^{(k)}}, \quad L_{55}^{(k)} = \bar{L}_{55}^{(k)} = \frac{2\left(1 + v_C^{(k)}\right)}{E_C^{(k)}}. \end{aligned} \tag{6.12}$$

The other nonzero components in $\bar{\mathbf{L}}^{(k)}$ are

$$\bar{L}_{22}^{(k)} = \frac{1}{E_C^{(k)}}, \bar{L}_{44}^{(k)} = \frac{2(1 + v_C^{(k)})}{E_C^{(k)}}, \bar{L}_{66}^{(k)} = \frac{2(1 + v_C^{(k)})}{E_C^{(k)}}. \tag{6.13}$$

The component $L_{44}^{(k)}$ is given as [15]

$$L_{44}^{(k)} = \bar{L}_{44}^{(k)} + \frac{1}{2}\beta\left(\beta L_{55}^{(k)} + \sqrt{\beta^2(L_{55}^{(k)})^2 + 4\bar{L}_{44}^{(k)}L_{55}^{(k)}}\right) \tag{6.14}$$

In order to obtain $L_{22}^{(k)}$ and $L_{66}^{(k)}$, we introduce

$$\begin{aligned} X_1 &= \bar{L}_{22}^{(k)}L_{33}^{(k)} - (L_{23}^{(k)})^2, \qquad X_2 = L_{33}^{(k)}\bar{L}_{66}^{(k)}, \\ X_3 &= L_{11}^{(k)}L_{33}^{(k)} - (L_{13}^{(k)})^2, X_4 = 2(L_{12}^{(k)}L_{33}^{(k)} - L_{13}^{(k)}L_{23}^{(k)}). \end{aligned} \tag{6.15}$$

Solving the following equation, one can obtain Y.

$$Y - X_1 - \beta\sqrt{3Y\sqrt{X_3 Y} + (X_2 + X_4)Y - X_1\sqrt{X_3 Y}} = 0 \tag{6.16}$$

Then $L_{22}^{(k)}$ and $L_{66}^{(k)}$ can be obtained.

$$\begin{aligned} L_{22}^{(k)} &= \frac{Y + (L_{23}^{(k)})^2}{L_{33}^{(k)}} \\ L_{66}^{(k)} &= \bar{L}_{66}^{(k)} + (L_{22}^{(k)} - \bar{L}_{22}^{(k)})\sqrt{\frac{X_3}{Y}} \end{aligned} \tag{6.17}$$

The relationships between stiffness coefficients $Q_{ij}^{(k)}$ and compliance coefficients $L_{ij}^{(k)}$ are

$$\begin{aligned} Q_{11}^{(k)} &= \frac{L_{22}^{(k)}}{L_{11}^{(k)}L_{22}^{(k)} - (L_{12}^{(k)})^2}, Q_{12}^{(k)} = \frac{-L_{12}^{(k)}}{L_{11}^{(k)}L_{22}^{(k)} - (L_{12}^{(k)})^2}, Q_{22}^{(k)} = \frac{L_{11}^{(k)}}{L_{11}^{(k)}L_{22}^{(k)} - (L_{12}^{(k)})^2}, \\ Q_{33}^{(k)} &= \frac{1}{L_{66}^{(k)}}, \qquad Q_{44}^{(k)} = \frac{1}{L_{44}^{(k)}}, \qquad Q_{55}^{(k)} = \frac{1}{L_{55}^{(k)}}. \end{aligned} \tag{6.18}$$

6.1.3 First-order shear deformation plate theory

Based on the FSDT [30], the displacement field of plate are written as

$$\begin{aligned} \bar{u} &= u + z\theta_x \\ \bar{v} &= v + z\theta_y \\ \bar{w} &= w \end{aligned} \tag{6.19}$$

where u, v and w are the displacements at the mid-plane of the plate along the x, y, and z directions, and θ_x and θ_y represent the transverse normal rotations about the y- and x-axes, respectively. Considering the von Kármán nonlinearity and thermal effect, the strain-displacement relationship is given by

$$\begin{Bmatrix} \varepsilon_{xx} \\ \varepsilon_{yy} \\ \varepsilon_{xy} \end{Bmatrix} = \boldsymbol{\varepsilon}_0 + z\boldsymbol{\kappa} - \boldsymbol{\varepsilon}_t, \begin{Bmatrix} \gamma_{yz} \\ \gamma_{xz} \end{Bmatrix} = \gamma_0 \tag{6.20}$$

where $\boldsymbol{\varepsilon}_0 = \boldsymbol{\varepsilon}_0^L + \boldsymbol{\varepsilon}_0^N$, $\boldsymbol{\varepsilon}_0^L = \left\{ \frac{\partial u}{\partial x} \quad \frac{\partial v}{\partial y} \quad \frac{\partial u}{\partial y} + \frac{\partial v}{\partial x} \right\}^T$, $\boldsymbol{\varepsilon}_0^N = \left\{ \frac{1}{2}\left(\frac{\partial w}{\partial x}\right)^2 \quad \frac{1}{2}\left(\frac{\partial w}{\partial y}\right)^2 \quad \frac{\partial w}{\partial x}\frac{\partial w}{\partial y} \right\}^T$,
$\boldsymbol{\kappa} = \left\{ \frac{\partial \theta_x}{\partial x} \quad \frac{\partial \theta_y}{\partial y} \quad \frac{\partial \theta_x}{\partial y} + \frac{\partial \theta_y}{\partial x} \right\}^T$, $\gamma_0 = \left\{ \theta_y + \frac{\partial w}{\partial y} \quad \theta_x + \frac{\partial w}{\partial x} \right\}^T$.

The thermal strain vector $\boldsymbol{\varepsilon}_t$ of the kth layer is represented as

$$\boldsymbol{\varepsilon}_t = \begin{Bmatrix} \Delta T \alpha_C^{(k)} \\ \Delta T \alpha_C^{(k)} \\ 0 \end{Bmatrix} \tag{6.21}$$

where ΔT is temperature change. If $\Delta T = 0$, there is no thermal effect.

The stress components of the kth layer can be obtained from the elastic stress–strain constitutive relationship.

$$\begin{Bmatrix} \sigma_{xx}^{(k)} \\ \sigma_{yy}^{(k)} \\ \sigma_{xy}^{(k)} \\ \sigma_{yz}^{(k)} \\ \sigma_{xz}^{(k)} \end{Bmatrix} = \begin{bmatrix} Q_{11}^{(k)} & Q_{12}^{(k)} & 0 & 0 & 0 \\ Q_{21}^{(k)} & Q_{22}^{(k)} & 0 & 0 & 0 \\ 0 & 0 & Q_{33}^{(k)} & 0 & 0 \\ 0 & 0 & 0 & Q_{44}^{(k)} & 0 \\ 0 & 0 & 0 & 0 & Q_{55}^{(k)} \end{bmatrix} \begin{Bmatrix} \varepsilon_{xx}^{(k)} \\ \varepsilon_{yy}^{(k)} \\ \varepsilon_{xy}^{(k)} \\ \varepsilon_{yz}^{(k)} \\ \varepsilon_{xz}^{(k)} \end{Bmatrix} \tag{6.22}$$

where $Q_{ij}^{(k)}$ is defined in equation (6.18).

By introducing $\boldsymbol{\varepsilon} = \{\boldsymbol{\varepsilon}_0^L + \boldsymbol{\varepsilon}_0^N \quad \boldsymbol{\kappa} \quad \gamma_0\}^T$, the strain energy of the GPLRC plate can be written as

$$\begin{aligned} U &= \int_\Omega \sum_{k=1}^{NL} \int_{z_k}^{z_{k+1}} \frac{1}{2}\left(\sigma_{xx}^{(k)}\varepsilon_{xx}^{(k)} + \sigma_{yy}^{(k)}\varepsilon_{yy}^{(k)} + \sigma_{xy}^{(k)}\varepsilon_{xy}^{(k)} + \sigma_{yz}^{(k)}\varepsilon_{yz}^{(k)} + \sigma_{xz}^{(k)}\varepsilon_{xz}^{(k)}\right) dz d\Omega \\ &= \int_\Omega \left(\frac{1}{2}\boldsymbol{\varepsilon}^T \mathbf{S} \boldsymbol{\varepsilon} - \Delta T N \boldsymbol{\varepsilon}\right) d\Omega \end{aligned} \tag{6.23}$$

where $\mathbf{S} = \begin{bmatrix} \mathbf{A} & \mathbf{B} & \mathbf{0} \\ \mathbf{B} & \mathbf{D} & \mathbf{0} \\ \mathbf{0} & \mathbf{0} & \mathbf{A}_s \end{bmatrix}$, $\mathbf{A} = \begin{bmatrix} A_{11} & A_{12} & 0 \\ A_{21} & A_{22} & 0 \\ 0 & 0 & A_{33} \end{bmatrix}$, $\mathbf{B} = \begin{bmatrix} B_{11} & B_{12} & 0 \\ B_{21} & B_{22} & 0 \\ 0 & 0 & B_{33} \end{bmatrix}$,
$\mathbf{D} = \begin{bmatrix} D_{11} & D_{12} & 0 \\ D_{21} & D_{22} & 0 \\ 0 & 0 & D_{33} \end{bmatrix}$, $\mathbf{A}_s = \begin{bmatrix} A_{s,11} & 0 \\ 0 & A_{s,22} \end{bmatrix}$,
$N = \{A_{t,11} + A_{t,12} \;\; A_{t,11} + A_{t,12} \;\; 0 \;\; B_{t,11} + B_{t,12} \;\; B_{t,11} + B_{t,12} \;\; 0 \;\; 0 \;\; 0\}$. The stiffness elements A_{ii}, B_{ij}, D_{ij}, $A_{s,ii}$, $A_{t,ij}$, and $B_{t,ij}$ are expressed as

$$\left(A_{ij}, B_{ij}, D_{ij}\right) = \sum_{k=1}^{NL} \int_{z_k}^{z_{k+1}} Q_{ij}^{(k)}(1, z, z^2)dz, (i, j = 1, 2, 3) \tag{6.24}$$

$$A_{s,ii} = \sum_{k=1}^{NL} \int_{z_k}^{z_{k+1}} k_s Q_{jj}^{(k)} dz, (i, j - 3 = 1, 2) \tag{6.25}$$

$$\left(A_{t,ij}, B_{t,ij}\right) = \sum_{k=1}^{NL} \int_{z_k}^{z_{k+1}} \alpha_c^{(k)} Q_{ij}^{(k)}(1, z)dz, (i, j = 1, 2) \tag{6.26}$$

in which the shear correction factor $k_s = 5/6$.

The kinetic energy of the plate is given by

$$\begin{aligned} \Theta &= \frac{1}{2} \int_\Omega \sum_{k=1}^{NL} \int_{z_k}^{z_{k+1}} \rho_c^{(k)} (\dot{u}_0^2 + \dot{v}_0^2 + \dot{w}_0^2) dz d\Omega \\ &= \frac{1}{2} \int_\Omega \left[I_0(\dot{u}^2 + \dot{v}^2 + \dot{w}^2) + 2I_1\left(\dot{u}\dot{\theta}_x + \dot{v}\dot{\theta}_y\right) + I_2\left(\dot{\theta}_x^2 + \dot{\theta}_y^2\right) \right] d\Omega \end{aligned} \tag{6.27}$$

where $(I_0, I_1, I_2) = \sum_{k=1}^{NL} \int_{z_k}^{z_{k+1}} \rho_c^{(k)} (1, z, z^2)dz$.

Here, it is assumed that the GPLRC plate is subjected to external fluid flow passing over the top surface of the plate. The work done by aerodynamic forces is given by

$$W_a = \int_\Omega \Delta p w d\Omega \tag{6.28}$$

where the aerodynamic pressure Δp of supersonic flow can be determined by the first order piston theory.

$$\Delta p = \frac{\lambda D}{a^3} \left(\frac{\partial w}{\partial x} + \frac{1}{U_\infty} \frac{M_\infty^2 - 2}{M_\infty^2 - 1} \frac{\partial w}{\partial t} \right) \tag{6.29}$$

where λ is the non-dimensional aerodynamic pressure, $D = E_M h^3/[12(1 - \nu_M^2)]$ is a reference bending stiffness, U_∞ is free stream velocity and M_∞ is Mach number. As the aerodynamic damping term in equation (6.29) always stabilizes the flutter boundary [3], to study the flutter boundary of supersonic plate, the aerodynamic load without the aerodynamic damping is used to derive the flutter equation of the supersonic plate.

$$\Delta p = \frac{\lambda D}{a^3}\frac{\partial w}{\partial x} \tag{6.30}$$

Thus the total energy functional of GPLRC plate under supersonic flow can be expressed as

$$\Pi = U - \Theta + W_a \tag{6.31}$$

6.1.4 Solution procedure via the IMLS-Ritz method

The IMLS is employed to construct the shape functions in the element-free method [24, 25]. For a plate, which is discretized by a set of nodes x_i, $i = 1,\dots,m$, in a computational domain Ω, the displacement field can be defined as

$$u^h(\boldsymbol{x}) = \sum_{i=1}^{m} p_i(\boldsymbol{x})a_i(\boldsymbol{x}) = \boldsymbol{p}^T(\boldsymbol{x})\boldsymbol{a}(\boldsymbol{x}) \tag{6.32}$$

where $p_i(\boldsymbol{x})$, $i = 1, 2, \dots, m$ are the basis functions, m is the number of terms in the basis, and the coefficients $a_i(\boldsymbol{x})$ are functions of the spatial coordinates $\boldsymbol{x}$, which are obtained by performing a weighted least-squares fit. This minimization yields the quadratic form

$$J = \sum_{I=1}^{n} w(\boldsymbol{x} - \boldsymbol{x}_I)[u^h(\boldsymbol{x}, \boldsymbol{x}_I) - u(\boldsymbol{x}_I)]^2 \tag{6.33}$$

where $w(\boldsymbol{x}-\boldsymbol{x}_I)$ is a weight function with a domain of influence, and $\boldsymbol{x}_I$ ($I = 1, 2, \dots, n$) are the nodes with domains of influence that cover the point $\boldsymbol{x}$.

In this study, the cubic spline function is chosen as the weight function given by

$$w(\boldsymbol{x} - \boldsymbol{x}_I) = w_I(\boldsymbol{x}) = \begin{cases} \frac{2}{3} - 4Z_I^2 + 4Z_I^3, & 0 \leqslant \left|Z_I\right| \leqslant \frac{1}{2} \\ \frac{4}{3} - 4Z_I + 4Z_I^2 - \frac{4}{3}Z_I^3, & \frac{1}{2} \leqslant \left|Z_I\right| \leqslant 1 \\ 0, & \text{otherwise} \end{cases} \tag{6.34}$$

where $Z_I = \|\boldsymbol{x} - \boldsymbol{x}_I\|/d_I$, $d_I = d_{\max} c_I$ is the size of the support domain of node I, $d_{\max}$ is a scaling factor ranging from 2.0 to 4.0, and c_I is the average distance between nodes in the domain of influence of the point $\boldsymbol{x}$.

For a two-dimensional problem, the weighted orthogonal basis function set can be formed in terms of the polynomial function as

$$p_i = \tilde{p}_i - \sum_{k=1}^{i-1} \frac{(\tilde{p}_i, p_k)}{(p_k, p_k)} p_k, \quad i = 1, 2, 3, \cdots \tag{6.35}$$

where

$$\tilde{\boldsymbol{p}} = (\tilde{p}_i) = (1,\ x,\ y,\ xy,\ x^2,\ y^2,\ \cdots) \tag{6.36}$$

The coefficients $\boldsymbol{a}(\boldsymbol{x})$ are determined by

$$\frac{\partial J}{\partial \boldsymbol{a}} = 0 \tag{6.37}$$

and we have

$$\bar{\mathbf{A}}(\boldsymbol{x})\boldsymbol{a}(\boldsymbol{x}) = \bar{\mathbf{B}}(\boldsymbol{x})\boldsymbol{u} \tag{6.38}$$

where

$$\boldsymbol{u} = (u_1,\ u_2,\ \cdots,\ u_n)^T \tag{6.39}$$

$$\bar{\mathbf{A}}(\boldsymbol{x}) = \bar{\mathbf{P}}^T\mathbf{W}(\boldsymbol{x})\bar{\mathbf{P}} \tag{6.40}$$

$$\bar{\mathbf{B}}(\boldsymbol{x}) = \bar{\mathbf{P}}^T\mathbf{W}(\boldsymbol{x}) \tag{6.41}$$

$$\bar{\mathbf{P}} = \begin{bmatrix} p_1(\boldsymbol{x}_1) & p_2(\boldsymbol{x}_1) & \cdots & p_m(\boldsymbol{x}_1) \\ p_1(\boldsymbol{x}_2) & p_2(\boldsymbol{x}_2) & \cdots & p_m(\boldsymbol{x}_2) \\ \vdots & \vdots & \ddots & \vdots \\ p_1(\boldsymbol{x}_n) & p_2(\boldsymbol{x}_n) & \cdots & p_m(\boldsymbol{x}_n) \end{bmatrix} \tag{6.42}$$

$$\mathbf{W}(\boldsymbol{x}) = \begin{bmatrix} w(\boldsymbol{x} - \boldsymbol{x}_1) & 0 & \cdots & 0 \\ 0 & w(\boldsymbol{x} - \boldsymbol{x}_2) & \cdots & 0 \\ \vdots & \vdots & \ddots & \vdots \\ 0 & 0 & \cdots & w(\boldsymbol{x} - \boldsymbol{x}_n) \end{bmatrix} \tag{6.43}$$

From equation (6.38), coefficients $\boldsymbol{a}(\boldsymbol{x})$ can be obtained.

$$\boldsymbol{a}(\boldsymbol{x}) = \tilde{\mathbf{A}}(\boldsymbol{x})\bar{\mathbf{B}}(\boldsymbol{x})\boldsymbol{u} \tag{6.44}$$

where

$$\tilde{\mathbf{A}}(\boldsymbol{x}) = \begin{bmatrix} \dfrac{1}{(p_1,\ p_1)} & 0 & \cdots & 0 \\ 0 & \dfrac{1}{(p_2,\ p_2)} & \cdots & 0 \\ \vdots & \vdots & \ddots & \vdots \\ 0 & 0 & \cdots & \dfrac{1}{(p_m,\ p_m)} \end{bmatrix} \tag{6.45}$$

in which the inner product is defined as

$$(p_i, p_i) = \sum_{I=1}^{n} w(\boldsymbol{x} - \boldsymbol{x}_I) p_i(\boldsymbol{x}_I) p_i(\boldsymbol{x}_I) \tag{6.46}$$

From equations (6.32) and (6.44), the expression of approximation function $u^h(\boldsymbol{x})$ is

$$u^h(\boldsymbol{x}) = \sum_{I=1}^{n} \phi_I(\boldsymbol{x}) u_I \tag{6.47}$$

where

$$\phi_I(\boldsymbol{x}) = \sum_{j=1}^{m} p_j(\boldsymbol{x}) [\tilde{\mathbf{A}}(\boldsymbol{x}) \bar{\mathbf{B}}(\boldsymbol{x})]_{jI} \tag{6.48}$$

which represents the shape function of the IMLS approximation corresponding to node I.

6.2 Flutter of matrix cracked GPLRC plate

In this section, the flutter characteristics of matrix cracked GPLRC plates under supersonic flow are studied. The IMLS-Ritz method is applied to obtain the vibration equation. The natural frequencies and mode shapes are obtained by solving the eigenvalue problem of the vibration equation. The effects of various parameters on the flutter boundary of matrix cracked GPLRC plates are investigated in detail.

6.2.1 Solution of governing equations

Based on equation (6.49), the approximation displacement of GPLRC plate can be obtained

$$\boldsymbol{u}^h = \begin{Bmatrix} u^h \\ v^h \\ w^h \\ \theta_x^h \\ \theta_y^h \end{Bmatrix} = \sum_{I=1}^{NP} \phi_I \begin{Bmatrix} u_I \\ v_I \\ w_I \\ \theta_{xI} \\ \theta_{yI} \end{Bmatrix} = \sum_{I=1}^{NP} \phi_I \boldsymbol{u}_I \tag{6.49}$$

Simply supported boundary is considered for all four edges of the matrix cracked GPLRC plate. The essential boundary conditions of four boundaries Γ_i $(i = 1,2,3,4)$ of a GPLRC plate are implemented by the penalty method. The variational form of the penalty functional is given by

$$\Pi_b = \frac{\alpha}{2} \sum_{i=1}^{4} \int_{\Gamma_i} (\boldsymbol{u} - \bar{\boldsymbol{u}})^T (\boldsymbol{u} - \bar{\boldsymbol{u}}) d\Gamma_i \tag{6.50}$$

where $\alpha = 10^3 E_M$ is the penalty parameter, $\bar{\boldsymbol{u}}$ is the prescribed displacement on the displacement boundary Γ_i.

Then, the total energy functional for this problem becomes

$$\Pi_t = \Pi + \Pi_b \tag{6.51}$$

Applying the Ritz minimization procedure with

$$\frac{\partial \Pi_t}{\partial \boldsymbol{u}_I} = 0, \boldsymbol{u}_I = \left\{u_I, v_I, w_I, \theta_{xI}, \theta_{yI}\right\}, I = 1, 2, \cdots, NP \tag{6.52}$$

Omitting the nonlinearity and thermal effect, the linear aeroelastic equation of matrix cracked GPLRC plate can be obtained

$$\mathbf{M}\ddot{\boldsymbol{u}} + (\mathbf{K} + \lambda \mathbf{K}_a)\boldsymbol{u} = \mathbf{0} \tag{6.53}$$

where $\mathbf{M} = \int_\Omega \mathbf{\Psi}^T m \mathbf{\Psi} d\Omega$, $\mathbf{K} = \mathbf{K}_p + \mathbf{K}_b$, $\mathbf{K}_p = \int_\Omega \mathbf{\Phi}^T S \mathbf{\Phi} d\Omega$, $\mathbf{K}_b = \sum_{i=1}^{4} \int_{\Gamma_i} \mathbf{\Psi}^T \mathbf{b}_i \mathbf{\Psi} d\Gamma_i$,

$$\mathbf{K}_a = \int_\Omega \mathbf{\Psi}^T \mathbf{P}_a \mathbf{\Psi}_x d\Omega, \qquad \mathbf{m} = \begin{bmatrix} I_0 & 0 & 0 & I_1 & 0 \\ 0 & I_0 & 0 & 0 & I_1 \\ 0 & 0 & I_0 & 0 & 0 \\ I_1 & 0 & 0 & I_2 & 0 \\ 0 & I_1 & 0 & 0 & I_2 \end{bmatrix}, \qquad \mathbf{P}_a = \frac{D}{a^3}\begin{bmatrix} 0 & 0 & 0 & 0 & 0 \\ 0 & 0 & 0 & 0 & 0 \\ 0 & 0 & 1 & 0 & 0 \\ 0 & 0 & 0 & 0 & 0 \\ 0 & 0 & 0 & 0 & 0 \end{bmatrix},$$

$\mathbf{\Psi} = (\boldsymbol{\psi}_1, \boldsymbol{\psi}_2, \cdots, \boldsymbol{\psi}_{NP})$, $\mathbf{\Phi} = (\boldsymbol{\varphi}_1, \boldsymbol{\varphi}_2, \cdots, \boldsymbol{\varphi}_{NP})$, $\mathbf{\Psi}_x = (\boldsymbol{\psi}_{1,x}, \boldsymbol{\psi}_{2,x}, \cdots, \boldsymbol{\psi}_{NP,x})$, $\boldsymbol{\psi}_I = \phi_I \mathbf{I}$,

$$\boldsymbol{\varphi}_I = \begin{bmatrix} \phi_{I,x} & 0 & 0 & 0 & 0 \\ 0 & \phi_{I,y} & 0 & 0 & 0 \\ \phi_{I,y} & \phi_{I,x} & 0 & 0 & 0 \\ 0 & 0 & 0 & \phi_{I,x} & 0 \\ 0 & 0 & 0 & 0 & \phi_{I,y} \\ 0 & 0 & 0 & \phi_{I,y} & \phi_{I,x} \\ 0 & 0 & \phi_{I,y} & 0 & \phi_I \\ 0 & 0 & \phi_{I,x} & \phi_I & 0 \end{bmatrix},$$ $\boldsymbol{\psi}_{I,x} = \phi_{I,x}\mathbf{I}$, in which $\phi_{I,x} = \frac{\partial \phi_I}{\partial x}$, $\phi_{I,y} = \frac{\partial \phi_I}{\partial y}$, and $\mathbf{I}$ is a 5×5 identity matrix.

Thus, all matrices in aeroelastic equation of matrix cracked GPLRC plate equation (6.53) are defined. In the following, matrices $\mathbf{M}$, $\mathbf{K}_p$ and $\mathbf{K}_a$ are evaluated with a 4×4 Gauss integration, and $\mathbf{K}_b$ is evaluated with a 4 points Gauss integration.

The general solution of the homogeneous differential equation corresponding to equation (6.53) can be expressed as

$$\boldsymbol{u}_I = \boldsymbol{q}_I e^{i\omega t} \tag{6.54}$$

Substituting equation (6.54) into equation (6.53) leads to the following eigenvalue problem

$$(\mathbf{K} + \lambda \mathbf{K}_a - \omega^2 \mathbf{M})\boldsymbol{q} = \mathbf{0} \tag{6.55}$$

The natural frequency ω and the corresponding mode shape $\boldsymbol{q}$ under aerodynamic loading can be solved from equation (6.55) [15]. The non-dimensional frequency is defined as

$$\bar{\omega} = \rho_M h a^4 \omega / D \tag{6.56}$$

It is known to all that as the non-dimensional aerodynamic pressure λ increases monotonically from zero, two of these natural frequencies will approach each other and the corresponding modes merge into the same one. The critical aerodynamic pressure λ_c is determined when mode merger occurs first. When λ is greater than λ_c, flutter will begin. Thus λ_c is named the flutter boundary of the plate.

6.2.2 Comparison and convergence studies

In this subsection, comparisons of the natural frequencies obtained by the present method are compared with the results available in open literature. In this section, unless otherwise stated, the geometric parameters of the matrix cracked GPLRC plate in thermal environment is given as, $a = b = 0.45$ m, $h/a = 0.01$.

Firstly, the convergence study was conducted to determine the optimum number of nodes NP and supported size $d_{\max}$. A simply supported isotropic square plate under aerodynamic pressure is studied. The parameters of the square plate are given as, $\rho = 2.7$ g cm^{-3}, $E = 70$ GPa, $\nu = 0.3$. Table 6.1 presents the vibration and flutter solutions of an isotropic square plate obtained by the IMLS-Ritz method and other results available in open literature. It can be seen that the number of nodes NP ranging from 23×23 to 29×29 can furnish converged results when $d_{\max} = 2.2$. When $NP = 23 \times 23$, the results obtained by the present method are in excellent agreement with the values obtained by finite element procedure (Lin *et al* [31]). By varying $d_{\max}$ from 2.1 to 2.4, more accurate results

Table 6.1. Comparison of vibration and flutter solution of square isotropic plates.

			Free vibration		Flutter	
Method	NP	$d_{\max}$	$\bar{\omega}_1^2$	$\bar{\omega}_2^2$	$\bar{\omega}_c^2$	λ_c
Present	23×23	2.2	385.649	2422.871	1853.478	513.45
	25×25	2.2	387.063	2426.652	1852.488	513.22
	27×27	2.2	390.728	2438.263	1851.327	512.98
	29×29	2.2	390.411	2436.701	1850.142	512.74
	29×29	2.1	392.463	2443.975	1856.259	513.59
	29×29	2.2	390.411	2436.701	1850.142	512.74
	29×29	2.3	388.200	2429.960	1843.886	511.92
	29×29	2.4	384.179	2415.286	1830.487	509.41
Lin *et al* [31]			389.637	2435.42	1847.5	512.58

Table 6.2. Comparison study of non-dimensional frequencies $\omega h\sqrt{\rho_M/E_M}$ of GPLRC square plates.

Method	h/a	Mode	UD	FG-O	FG-X	FG-V
NURBS [32]	0.01	1	0.0013	0.0010	0.0014	0.0012
		2	0.0031	0.0026	0.0036	0.0029
	0.02	1	0.0050	0.0042	0.0058	0.0046
		2	0.0126	0.0105	0.0143	0.0115
Present	0.01	1	0.0013	0.0011	0.0014	0.0012
		2	0.0032	0.0026	0.0036	0.0029
	0.02	1	0.0050	0.0041	0.0057	0.0046
		2	0.0125	0.0104	0.0143	0.0114

can be furnished when $d_{\max} = 2.2$. Thus $NP = 29 \times 29$ and $d_{\max} = 2.2$ are adopted for the following parametric studies.

Subsequently, to further validate the present method, the obtained non-dimensional natural frequencies of pure epoxy plate and GPLRC plate with four types of GPLs distributions and different h/a shown in table 6.2 are compared with the results from Thai *et al* [32]. The results by the present solution method are in excellent agreement with the values of NURBS formulation [32].

6.2.3 Numerical results and discussion

In order to well understand the effect of crack density on the flutter characteristics of GPLRC plates, a parameter study of GPLRC plates is presented in the following.

The lowest three natural frequencies of UD plate under five crack density, i.e. $\beta = 0$, $\beta = 0.25$, $\beta = 0.5$, $\beta = 0.75$, and $\beta = 1.0$, with aerodynamic pressure λ increasing from 0 to 2500 are shown in figure 6.2.

When there is no matrix crack, the first order natural frequency is increased and the second order natural frequency is decreased as λ is increasing. These two natural frequencies approach each other at point F ($\lambda_c = 2221.20$). Then mode 1 and mode 2 merge into the same mode. This means flutter is appearing. When there are matrix cracks ($\beta \neq 0$), the natural frequencies will be decreased. For $\beta = 0.25$ and $\beta = 0.5$, mode 2 and mode 3 will exchange at point X, and mode 1 and mode 2 merge into the same mode at point F as λ is increasing. For $\beta = 0.5$ and $\beta = 1.0$, mode 1 and mode 2 will exchange at point X_1, and mode 2 and mode 3 merge into the same mode at point F as λ is increasing. Moreover, the merged mode and mode 1 will exchange at point X_2. The mode exchange and mode merger of UD plate with $\beta = 0.25$ are shown in table 6.3. More importantly, the flutter boundary is dramatically affected by matrix crack density β. In the following, we mainly investigate the effect of different parameters and crack density on the flutter boundary.

The flutter boundary λ_c and the corresponding natural frequency $\bar{\omega}_c$ of GPLRC plates under four types of GPLs distributions, i.e. UD, FG-O, FG-X, and FG-V, with different β are presented in table 6.4. Results show that the

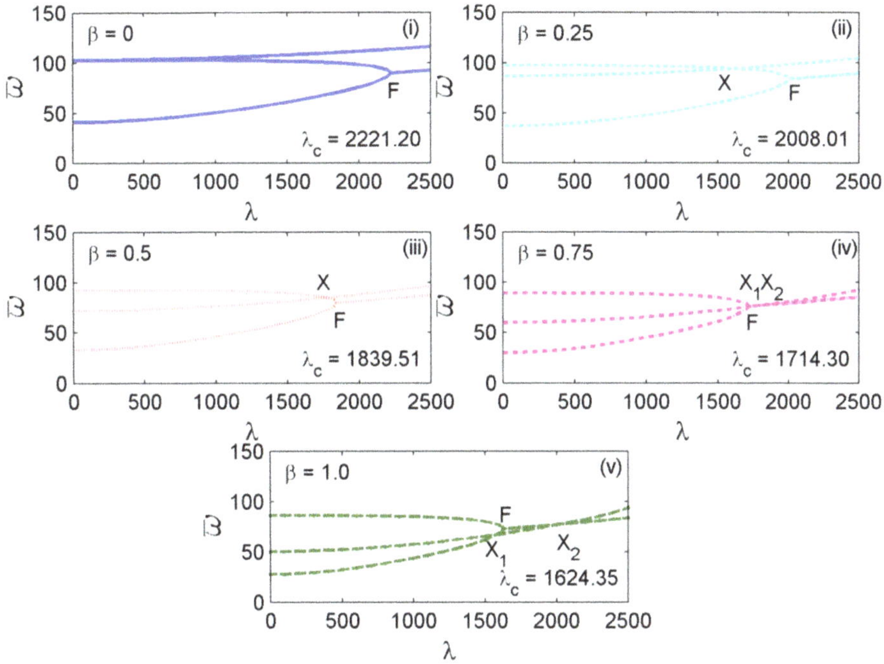

Figure 6.2. The lowest three natural frequencies of UD plate under different aerodynamic pressure λ and crack density β.

Table 6.3. Models of UD plates with different aerodynamic pressure λ ($\beta = 0.25$).

λ	0	1600	1700	2000	2100
Mode 1	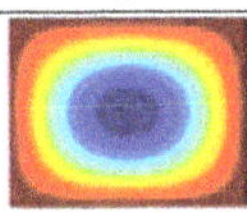 36.88	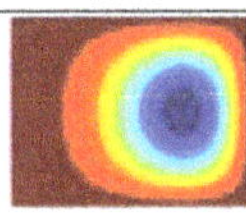 63.12	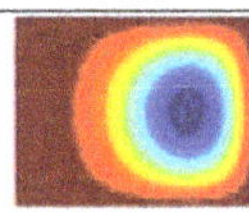 66.42	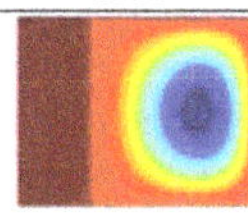 81.75	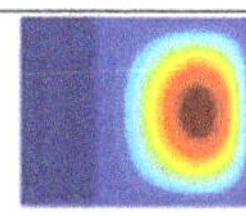 84.98
Mode 2	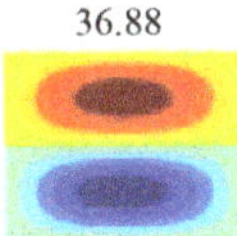 86.67	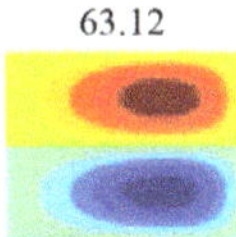 94.10	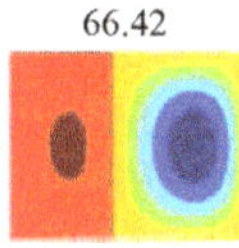 93.66	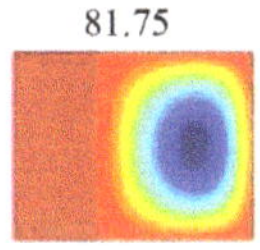 86.13	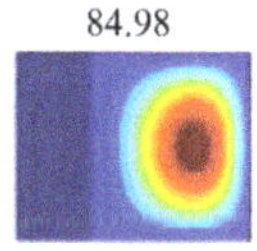 84.98
Mode 3	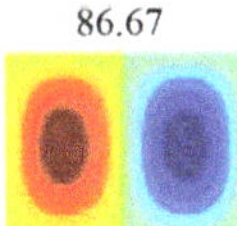 97.31	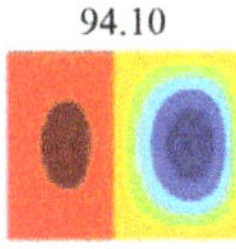 94.49	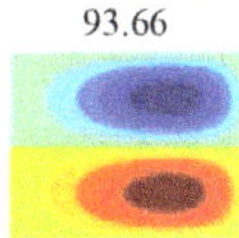 95.04	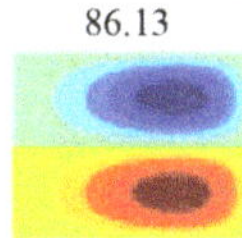 98.19	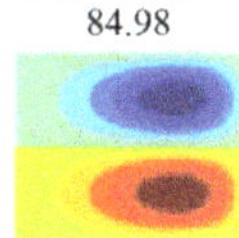 99.35

FG-X plates furnish the highest λ_c and $\bar{\omega}_c$ while FG-O plates attain the lowest λ_c and $\bar{\omega}_c$ when the GPLRC plate has the same β. Moreover, both the flutter boundary λ_c and the corresponding natural frequency $\bar{\omega}_c$ of GPLRC plates for all four types of GPLs distributions are decreased as crack density β is increasing. The reason is that matrix crack leads to the reduction of the stiffness of the

Table 6.4. The flutter boundaries and natural frequencies of four pattern GPLRC plates with different crack density β.

		β				
Pattern		0	0.25	0.5	0.75	1.0
UD	$\bar{\omega}_c$	89.6434	84.0485	79.3809	75.7663	73.0986
	λ_c	2221.20	2008.01	1839.51	1714.30	1624.35
FG-O	$\bar{\omega}_c$	74.6442	69.9869	66.0993	63.0874	60.8636
	λ_c	1540.16	1392.29	1275.37	1188.47	1126.03
FG-X	$\bar{\omega}_c$	102.4445	96.0504	90.7196	86.5929	83.5483
	λ_c	2900.63	2622.46	2402.70	2239.41	2122.11
FG-V	$\bar{\omega}_c$	82.1944	77.0664	72.7868	69.4707	67.0302
	λ_c	1867.51	1688.30	1546.62	1441.29	1365.78

laminated plates. More importantly, the decreasing of λ_c and $\bar{\omega}_c$ is the highest for FG-X plate and lowest for FG-O plate when crack density β increases from 0 to 1.0.

The total layer number NL of GPLRC plates has an important influence on natural frequency and flutter boundary, except for the UD plate [26, 33]. Natural frequency and flutter boundary of UD plate are independent of the total layer number NL [26, 33]. Table 6.5 shows the effect of crack density β on the flutter boundaries of FG-O, FG-X, and FG-V plate with different layer number NL. It shows that increasing NL could increase the flutter boundary of FG-X plates and decrease the flutter boundary of FG-O and FG-V plates. A higher β could decrease the flutter boundary of all three GPL pattern plates.

Figure 6.3 reveals the effects of crack density β on flutter boundary λ_c of UD GPLRC plates with four GPLs weight fraction g_{GPL}. It shows that the larger the g_{GPL} is, the higher the flutter boundary λ_c of GPLRC plates attains. The reason is that natural frequency can be remarkably increased by adding a small amount of GPLs [26, 33]. When the crack density β increases, the flutter boundary will decrease. More important, the decreasing of flutter boundary due to matrix crack density is dramatic as g_{GPL} is increasing.

The effects of GPL geometry and size, in the form of length-to-thickness ratio l_{GPL}/h_{GPL} and length-to-width ratio l_{GPL}/w_{GPL}, on the flutter boundary of UD plates with crack density β are presented in figures 6.4 and 6.5. It can be seen that the flutter boundary increases dramatically when l_{GPL}/h_{GPL} increases from 10 to 2500 and the GPL width w_{GPL} is kept constant. As matrix crack density β increases from 0 to 1.0, the flutter boundaries of UD plates with all l_{GPL}/h_{GPL} are decreased.

Moreover, there is a slight decreasing of the flutter boundary when l_{GPL}/w_{GPL} increases from 1.0 to 25 and the GPL thickness h_{GPL} is kept constant. Matrix crack can also decrease the flutter boundary of the GPLRC plate with different l_{GPL}/w_{GPL}.

The flutter characteristics of square and rectangular GPLRC plates with different crack density β are highlighted in figure 6.6. It can be observed that the

Table 6.5. The flutter boundaries of three pattern GPLRC plates with different layer number NL and crack density β.

		β				
Pattern	NL	0	0.25	0.5	0.75	1.0
FG-O	2	2221.20	2008.01	1839.51	1714.30	1624.35
	4	1795.70	1623.30	1487.03	1385.74	1312.96
	6	1653.78	1494.99	1369.46	1276.15	1209.12
	8	1582.77	1430.80	1310.66	1221.35	1157.19
	10	1540.16	1392.29	1275.37	1188.47	1126.03
	20	1454.92	1315.23	1204.79	1122.69	1063.70
FG-X	2	2221.20	2008.01	1839.51	1714.30	1624.35
	4	2646.01	2392.18	2191.60	2042.61	1935.51
	6	2787.49	2520.13	2308.89	2151.94	2039.19
	8	2858.21	2584.09	2367.52	2206.61	2091.02
	10	2900.63	2622.46	2402.70	2239.41	2122.11
	20	2985.46	2699.19	2473.03	2305.00	2184.29
FG-V	2	2112.12	1909.39	1749.15	1630.06	1544.57
	4	1975.50	1786.03	1636.14	1524.73	1444.81
	6	1917.99	1733.93	1588.42	1480.25	1402.68
	8	1886.90	1705.82	1562.67	1460.01	1379.95
	10	1867.51	1688.30	1546.62	1441.29	1365.78
	20	1827.09	1651.77	1513.16	1410.10	1336.25

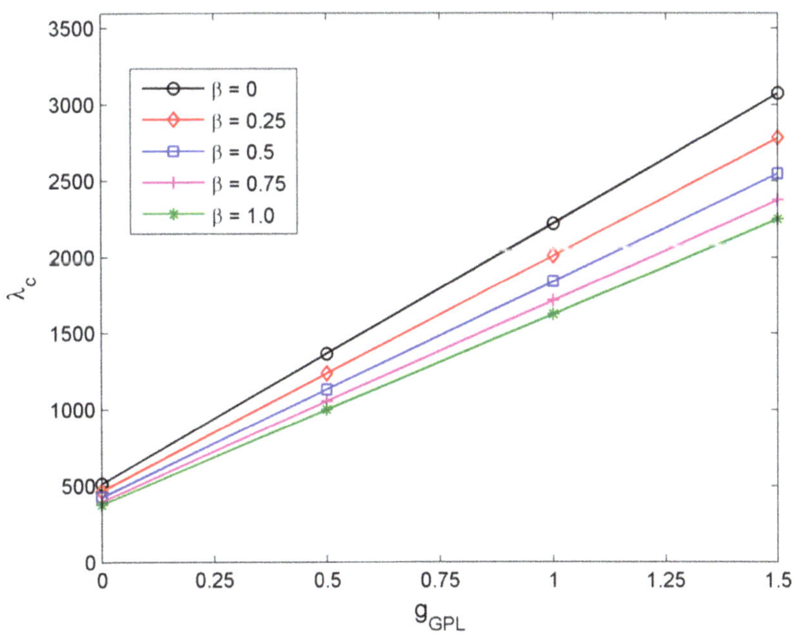

Figure 6.3. The flutter boundaries of UD plates with different GPLs weight fraction g_{GPL} and crack density β.

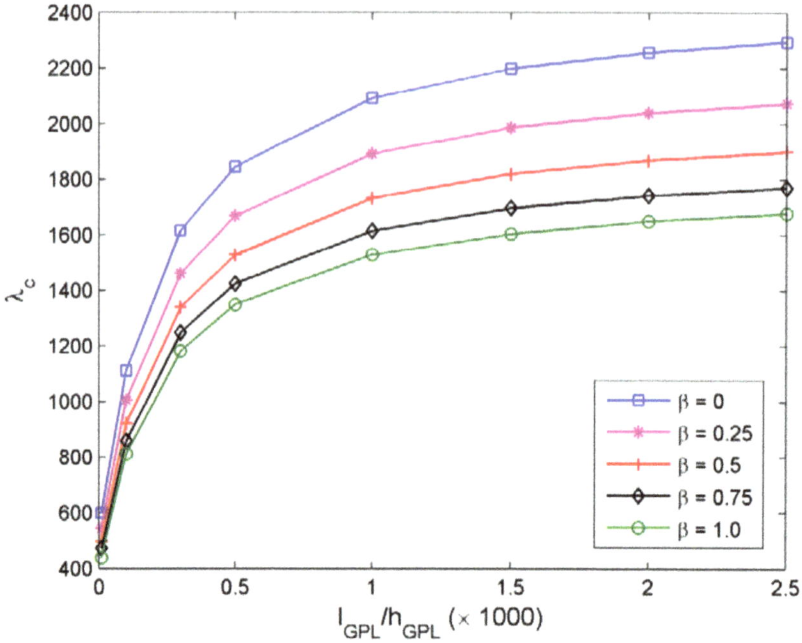

Figure 6.4. The flutter boundaries of UD plates with different l_{GPL}/h_{GPL} and crack density β.

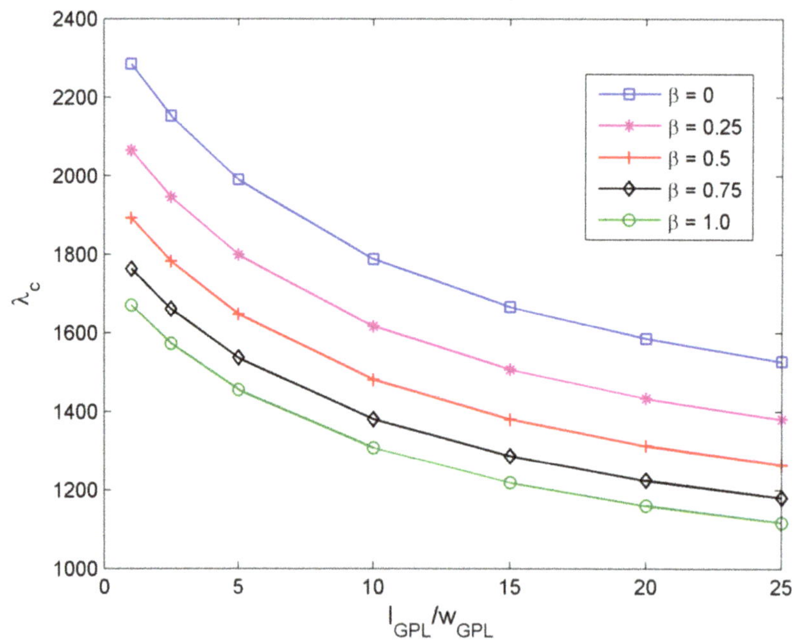

Figure 6.5. The flutter boundaries of UD plates with different l_{GPL}/w_{GPL} and crack density β.

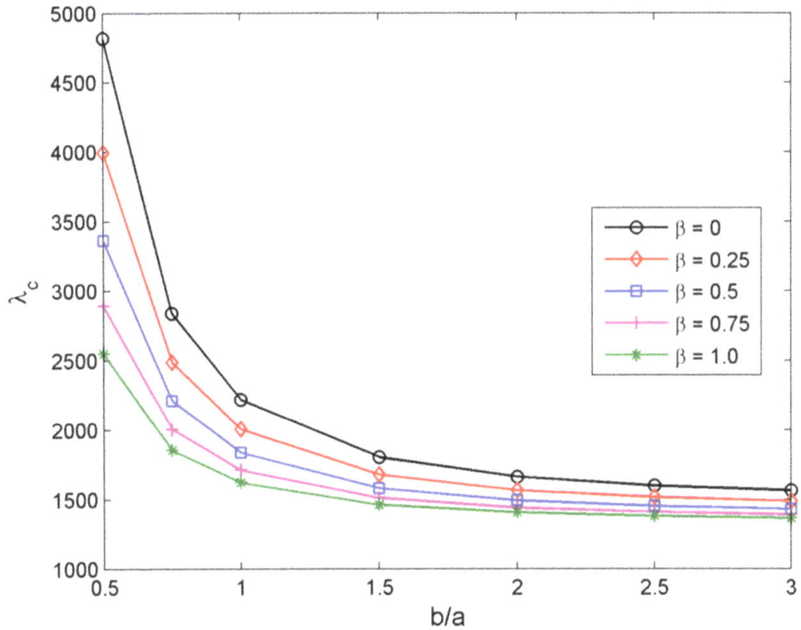

Figure 6.6. The flutter boundaries of UD plates with different length ratio b/a and crack density β.

flutter boundary decreases with the increase in the aspect ratio b/a. Moreover, the flutter boundary of GPLRC plate with different b/a decreases dramatically when the crack density β increases from 0 to 0.5. As the crack density β further increases, the flutter boundary is almost unchanged.

6.3 Nonlinear thermal flutter of GPLRC plate

In the following, thermal buckling and nonlinear flutter behaviors of GPLRC plate under the thermal and aerodynamic loads are analyzed. The IMLS-Ritz method is applied to obtain the nonlinear mechanical model. The Newton–Raphson method is adopted to obtain approximate solutions of the nonlinear governing equations [34]. Thermal buckling and flutter boundaries are defined by eigenvalue analysis. The effects of various parameters on the thermal buckling and flutter boundary of GPLRC plates are investigated in detail.

6.3.1 Discrete solution of governing equations

Similar to section 6.2.1, applying the Ritz minimization procedure with

$$\frac{\partial \Pi}{\partial \boldsymbol{u}_I} = 0, \boldsymbol{u}_I = \{u_I, v_I, w_I, \theta_{xI}, \theta_{yI}\}, I = 1, 2, \cdots, NP \tag{6.57}$$

and omitting the effect of matrix cracks, the nonlinear aerothermoelastic equation of GPLRC plate can be obtained

$$\mathbf{M}\ddot{\boldsymbol{u}} + \left(\mathbf{K}_p + \lambda \mathbf{K}_a - \Delta T \mathbf{K}_t + \mathbf{K}_n(\boldsymbol{u})\right)\boldsymbol{u} = \Delta T \boldsymbol{F}_t \tag{6.58}$$

where $\mathbf{M}$, $\mathbf{K}_p$, $\mathbf{K}_a$, are listed in section 6.2.1, and

$$\mathbf{K}_t = \frac{1}{2}\int_\Omega \left(\mathbf{H}^{(1)T}\boldsymbol{L}^T\boldsymbol{G}^{(1)} + \boldsymbol{G}^{(1)T}\boldsymbol{L}\mathbf{H}^{(1)} + \mathbf{H}^{(2)T}\boldsymbol{L}^T\boldsymbol{G}^{(2)} + \boldsymbol{G}^{(2)T}\boldsymbol{L}\mathbf{H}^{(2)}\right)d\Omega,$$
$$\mathbf{K}_n(\boldsymbol{u}) = \frac{1}{4}\int_\Omega \left(2\mathbf{R}_1 + \boldsymbol{G}^{(1)}\boldsymbol{u}\mathbf{R}_2 + \boldsymbol{G}^{(2)}\boldsymbol{u}\mathbf{R}_3 + \boldsymbol{G}^{(1)T}\boldsymbol{u}^T\mathbf{R}_4 + \boldsymbol{G}^{(2)T}\boldsymbol{u}^T\mathbf{R}_5\right)d\Omega,$$
$$\mathbf{R}_1 = \boldsymbol{G}^{(1)}\boldsymbol{u}\mathbf{P}_1 + \mathbf{P}_1\boldsymbol{u}\boldsymbol{G}^{(1)} + \boldsymbol{G}^{(2)}\boldsymbol{u}\mathbf{P}_2 + \mathbf{P}_2\boldsymbol{u}\boldsymbol{G}^{(2)},$$
$$\mathbf{R}_2 = 2\mathbf{P}_3\boldsymbol{u}\boldsymbol{G}^{(1)} + 3\boldsymbol{G}^{(1)T}\boldsymbol{u}^T\mathbf{P}_3 + \boldsymbol{G}^{(2)T}\boldsymbol{u}^T\mathbf{P}_7 + \mathbf{P}_7\boldsymbol{u}\boldsymbol{G}^{(2)} + \boldsymbol{G}^{(2)}\boldsymbol{u}\mathbf{P}_7,$$
$$\mathbf{R}_3 = 2\mathbf{P}_4\boldsymbol{u}\boldsymbol{G}^{(2)} + 3\boldsymbol{G}^{(2)T}\boldsymbol{u}^T\mathbf{P}_4 + \boldsymbol{G}^{(1)T}\boldsymbol{u}^T\mathbf{P}_7 + \mathbf{P}_7\boldsymbol{u}\boldsymbol{G}^{(1)},$$
$$\mathbf{R}_4 = 2\mathbf{P}_1 + \mathbf{P}_3\boldsymbol{u}\boldsymbol{G}^{(1)} + \mathbf{P}_5\boldsymbol{u}\boldsymbol{G}^{(2)},$$
$$\mathbf{R}_5 = 2\mathbf{P}_2 + \mathbf{P}_4\boldsymbol{u}\boldsymbol{G}^{(2)} + \mathbf{P}_6\boldsymbol{u}\boldsymbol{G}^{(1)},$$
$$\mathbf{P}_1 = \mathbf{H}^{(1)T}S\boldsymbol{\Phi} + \boldsymbol{\Phi}^T\mathbf{S}\mathbf{H}^{(1)},$$
$$\mathbf{P}_2 = \mathbf{H}^{(2)T}S\boldsymbol{\Phi} + \boldsymbol{\Phi}^T\mathbf{S}\mathbf{H}^{(2)},$$
$$\mathbf{P}_3 = \mathbf{H}^{(1)T}\mathbf{S}\mathbf{H}^{(1)},$$
$$\mathbf{P}_4 = \mathbf{H}^{(2)T}\mathbf{S}\mathbf{H}^{(2)},$$
$$\mathbf{P}_5 = \mathbf{H}^{(1)T}\mathbf{S}\mathbf{H}^{(2)},$$
$$\mathbf{P}_6 = \mathbf{H}^{(2)T}\mathbf{S}\mathbf{H}^{(1)},$$
$$\mathbf{P}_7 = \mathbf{H}^{(1)T}\mathbf{S}\mathbf{H}^{(2)} + \mathbf{H}^{(2)T}\mathbf{S}\mathbf{H}^{(1)},$$
$$F_t = \boldsymbol{\Phi}^T N^T,$$
$$\boldsymbol{G}^{(1)} = \left\{g_1^{(1)}, g_2^{(1)}, \cdots, g_{NP}^{(1)}\right\},$$
$$\boldsymbol{G}^{(2)} = \left\{g_1^{(2)}, g_2^{(2)}, \cdots, g_{NP}^{(2)}\right\},$$
$$\mathbf{H}^{(1)} = \left\{\mathbf{h}_1^{(1)}, \mathbf{h}_2^{(1)}, \cdots, \mathbf{h}_{NP}^{(1)}\right\}$$
$$\mathbf{H}^{(2)} = \left\{\mathbf{h}_1^{(2)}, \mathbf{h}_2^{(2)}, \cdots, \mathbf{h}_{NP}^{(2)}\right\},$$
$$\mathbf{h}_I^{(1)} = \begin{bmatrix} 0 & 0 & \phi_{I,x} & 0 & 0 \\ 0 & 0 & 0 & 0 & 0 \\ 0 & 0 & \phi_{I,y} & 0 & 0 \\ 0 & 0 & 0 & 0 & 0 \\ 0 & 0 & 0 & 0 & 0 \\ 0 & 0 & 0 & 0 & 0 \\ 0 & 0 & 0 & 0 & 0 \\ 0 & 0 & 0 & 0 & 0 \end{bmatrix},$$
$$\mathbf{h}_I^{(2)} = \begin{bmatrix} 0 & 0 & 0 & 0 & 0 \\ 0 & 0 & \phi_{I,y} & 0 & 0 \\ 0 & 0 & \phi_{I,x} & 0 & 0 \\ 0 & 0 & 0 & 0 & 0 \\ 0 & 0 & 0 & 0 & 0 \\ 0 & 0 & 0 & 0 & 0 \\ 0 & 0 & 0 & 0 & 0 \\ 0 & 0 & 0 & 0 & 0 \end{bmatrix},$$
$$g_I^{(1)} = \begin{bmatrix} 0 & 0 & \phi_{I,x} & 0 & 0 \end{bmatrix},$$
$$g_I^{(2)} = \begin{bmatrix} 0 & 0 & \phi_{I,y} & 0 & 0 \end{bmatrix}.$$

Since the present shape functions do not have the Kronecker delta property, the full transformation method is employed to impose the essential boundary conditions. According to the full transformation method, we introduce

$$\hat{\boldsymbol{u}} = \boldsymbol{\Lambda}^T \boldsymbol{u} \tag{6.59}$$

where the transformation matrix $\boldsymbol{\Lambda}$ can be expressed as

$$\boldsymbol{\Lambda}_{IJ} = \phi_I(\boldsymbol{x}_J)\boldsymbol{I} \tag{6.60}$$

The simply supported boundary conditions considered in this study are as follows

$$\begin{aligned} v = w = \theta_y = 0 \text{ at } x = 0, \text{ a} \\ u = w = \theta_x = 0 \text{ at } y = 0, b \end{aligned} \tag{6.61}$$

Substituting equation (6.59) into equation (6.58), the nonlinear aerothermoelastic equation is rewritten as

$$\hat{\mathbf{M}}\ddot{\hat{\boldsymbol{u}}} + (\hat{\mathbf{K}} + \lambda\hat{\mathbf{K}}_a - \Delta T\hat{\mathbf{K}}_t + \hat{\mathbf{K}}_n(\hat{\boldsymbol{u}}))\hat{\boldsymbol{u}} = \Delta T\hat{\boldsymbol{F}}_t \tag{6.62}$$

where $\hat{\mathbf{M}} = \boldsymbol{\Lambda}^{-1}\mathbf{M}\boldsymbol{\Lambda}^{-T}$, $\hat{\mathbf{K}} = \boldsymbol{\Lambda}^{-1}\mathbf{K}\boldsymbol{\Lambda}^{-T}$, $\hat{\mathbf{K}}_a = \boldsymbol{\Lambda}^{-1}\mathbf{K}_a\boldsymbol{\Lambda}^{-T}$, $\hat{\mathbf{K}}_t = \boldsymbol{\Lambda}^{-1}\mathbf{K}_t\boldsymbol{\Lambda}^{-T}$, $\hat{\mathbf{K}}_n(\hat{\boldsymbol{u}}) = \boldsymbol{\Lambda}^{-1}\mathbf{K}_n(\boldsymbol{u})\boldsymbol{\Lambda}^{-T}$, $\hat{\boldsymbol{F}}_t = \boldsymbol{\Lambda}^{-1}\boldsymbol{F}_t$.

Thus, all matrices in aeroelastic equation of GPLRC plate equation (6.62) are defined. In the following, matrices $\boldsymbol{\Lambda}$, $\mathbf{M}$, $\mathbf{K}$, $\mathbf{K}_a$, $\mathbf{K}_t$, $\mathbf{K}_n$ and vector $\boldsymbol{F}_t$ are evaluated with a 4 × 4 gauss integration.

Let the total response of the plate be the sum of displacement due to static deformation $\boldsymbol{u}_s$ and that due to dynamic deformation $\boldsymbol{u}_d e^{i\omega t}$.

$$\hat{\boldsymbol{u}} = \boldsymbol{u}_s + \boldsymbol{u}_d e^{i\omega t} \tag{6.63}$$

Substituting equation (6.63) into equation (6.62) and neglecting the higher-order terms related to $\boldsymbol{u}_d$, i.e., if considered for small amplitude vibration, then equation (6.62) can be written into two sets of equations as:

$$(\hat{\mathbf{K}} + \lambda\hat{\mathbf{K}}_a - \Delta T\hat{\mathbf{K}}_t + \hat{\mathbf{K}}_n(\boldsymbol{u}_s))\boldsymbol{u}_s = \Delta T\hat{\boldsymbol{F}}_t \tag{6.64}$$

$$(-\omega^2\hat{\mathbf{M}} + \hat{\mathbf{K}} + \lambda\hat{\mathbf{K}}_a - \Delta T\hat{\mathbf{K}}_t + \hat{\mathbf{K}}_n(\boldsymbol{u}_s))\boldsymbol{u}_d = \mathbf{0} \tag{6.65}$$

Equation (6.64) is a set of nonlinear algebraic equations that yields the static deflection of the thermally buckled plate. The nonlinear equation in equation (6.64) is solved by the Newton–Raphson method [27]. Equation (6.65) is a set of linear equations for the determination of the dynamic behavior of the thermally buckled composite laminated plate. Substituting the static deflection obtained from equation (6.64) into equation (6.65), the natural frequency ω and the corresponding mode shape $\boldsymbol{u}_d$ under can be obtained by solving the eigenvalue problem. As the non-dimensional aerodynamic pressure λ increases monotonically from zero, two of these natural frequencies will approach each other and the corresponding modes merge into the same one. The critical aerodynamic pressure named flutter boundary is determined when mode merger occurs first.

6.3.2 Comparison and convergence studies

In this subsection, comparisons of the flutter boundary, buckling boundary and natural frequency obtained by the present method are compared with the results available in open literature. In this section, the geometric parameters of a GPLRC plate in thermal environment are given as, $a = b = 0.3048$ m, $h/a = 1/240$.

Firstly, the convergence study was conducted to determine the optimum number of nodes NP and supported size d_{max}. A simply supported isotropic square plate with uniform temperature under aerodynamic pressure is studied. The parameters of the square plate are considered as, $\rho = 2.765$ g cm^{-3}, $E = 68.95$ GPa, $\nu = 0.3$, $\alpha = 22.5 \times 10^{-6}$ K^{-1}. The flutter and buckling boundaries at various $\Delta T/\Delta T_{cr}$ (ΔT_{cr} is the buckling temperature of the square isotropic plate) are presented in figures 6.7 and 6.8 by using IMLS-Ritz method with different NP and d_{max}. It can be seen that the number of nodes NP ranging from 11×11 to 17×17 can furnish converged results when $d_{max} = 2.1$. When $NP = 17 \times 17$, the results obtained by the present method are in excellent agreement with the values obtained by Shiau *et al* [35]. By varying d_{max} from 2.1 to 2.4, more accurate results can be furnished when $d_{max} = 2.1$. Thus $NP = 17 \times 17$ and $d_{max} = 2.1$ are adopted for the following parametric studies.

To further validate the present method, the obtained non-dimensional natural frequencies of pure epoxy plate and GPLRC plate with three types of GPLs distributions shown in table 6.6 are compared with the results from Song *et al* [33]. The results by the present solution method are in excellent agreement with the values from analytic method [33].

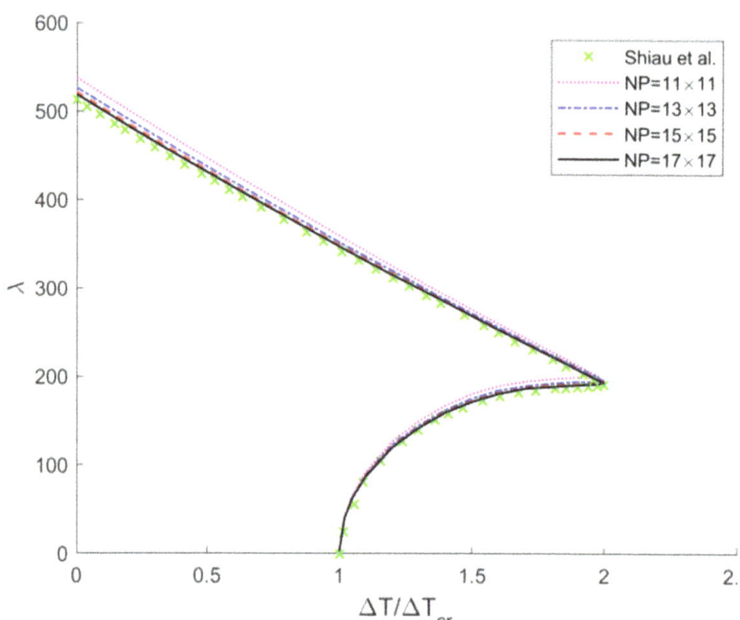

Figure 6.7. Flutter and buckling boundaries of a square isotropic plate with different *NP*.

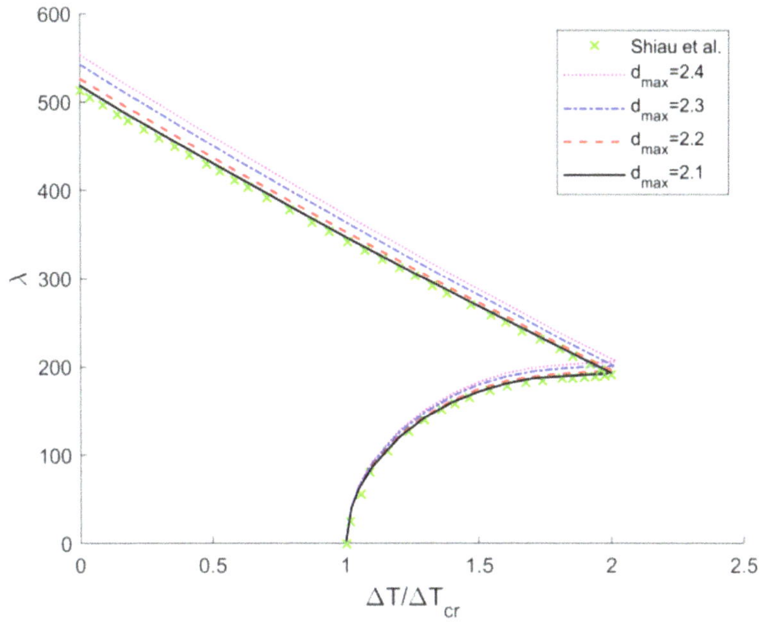

Figure 6.8. Flutter and buckling boundaries of a square isotropic plate with different $d_{\max}$.

Table 6.6. Comparison study of non-dimensional frequencies $\omega h\sqrt{\rho_M/E_M}$ of GPLRC square plates.

Method	Mode	Pure epoxy	UD	FG-O	FG-X
Song *et al* [33]	1	0.0584	0.1216	0.1020	0.1378
	2	0.1391	0.2895	0.2456	0.3249
Present	1	0.0584	0.1216	0.1020	0.1378
	2	0.1391	0.2895	0.2456	0.3250

Subsequently, the critical buckling temperature change ΔT_{cr} of GPLRC plates with three types of GPLs distributions and different geometry parameters obtained by present method are also validated in table 6.7. Direct comparison shows an excellent agreement between the present results and that of Wu *et al* [36].

6.3.3 Numerical results and discussion

In what follows, a detailed parameter study is conducted to reveal the nonlinear thermal buckling and flutter performance of GPLRC plate.

The maximum static deflections and the lowest two non-dimensional natural frequencies $\bar{\omega} = \omega(a/\pi)^2\sqrt{\rho_M h/D}$ of UD plate with different aerodynamic pressure λ and $\Delta T/\Delta T_{cr0}$ (ΔT_{cr0} is the buckling temperature of the pure epoxy plate) are shown in figures 6.9 and 6.10, respectively.

Table 6.7. Comparison study of critical buckling temperature change ΔT_{cr} of GPLRC plates ($g_{GPL} = 0.3\%$).

		$a/b = 1$			$a/b = 1.5$		
Method	Pattern	$a/h = 25$	$a/h = 35$	$a/h = 45$	$a/h = 25$	$a/h = 35$	$a/h = 45$
Wu *et al* [36]	UD	32.539	16.679	10.109	52.564	27.021	16.397
	FG-O	24.817	12.707	7.698	40.147	20.601	12.492
	FG-X	40.261	20.660	12.528	64.949	33.447	20.311
Present	UD	32.546	16.686	10.116	52.573	27.028	16.404
	FG-O	24.824	12.714	7.705	40.155	20.608	12.499
	FG-X	38.978	20.001	12.130	62.890	32.380	19.664

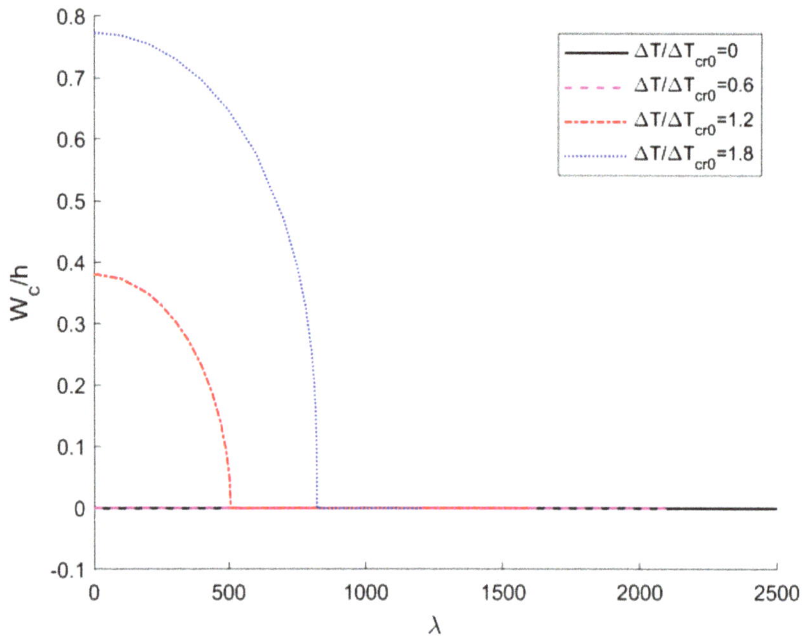

Figure 6.9. The maximum static deflections of UD plate at various λ and $\Delta T/\Delta T_{cr0}$.

For the cases when $\Delta T/\Delta T_{cr0} = 0$ and 0.6, the maximum static deflection is zero. It means the plate is not buckled. The first and second order natural frequencies approach each other as λ increases. When λ is greater than the critical the aerodynamic pressure (flutter boundary), flutter appears. For the cases when $\Delta T/\Delta T_{cr0} = 1.2$ and 1.8, as λ increases, the maximum static deflection decreases and it becomes zero at a critical aerodynamic pressure (buckling boundary). It means the plate is in a buckled condition for the lower dynamic pressure. Meanwhile, the first order natural frequency will drop to zero because of the buckle pattern change. The plate will have a flat condition when the aerodynamic pressure reaches the buckling boundary. When λ still increases, the first order natural frequency will increase from

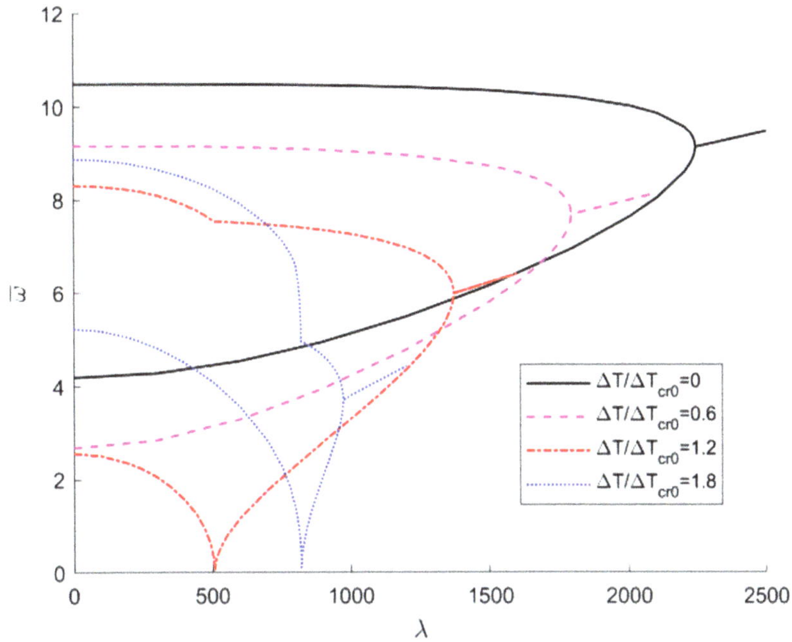

Figure 6.10. The lowest two natural frequencies of UD plate at various λ and $\Delta T/\Delta T_{\mathrm{cr0}}$.

zero. The coalescence occurs between the first and second order natural frequencies at the flutter boundary.

The flutter and buckling boundaries of pure epoxy plate and GPLRC plate with three types of GPLs distributions, i.e. UD, FG-O, and FG-X, are present in figure 6.11. It can be noted that the distribution of GPLs plays an important role in changing the flutter and buckling characteristics of GPLRC plate. The flutter boundaries are increased significantly as GPL is added for all three patterns of GPLRC plate. When $\lambda = 0$, FG-X plate has the highest critical temperature change for thermal buckling, FG-O plate has the lowest one, and UD plate has almost the same value to that of the pure epoxy plate. Similar results for the buckling characteristics of GPLRC plate without aerodynamic loading were reported by Wu *et al* [36]. When there is aerodynamic load, the buckling boundary is increased from zero as the temperature change $\Delta T/\Delta T_{\mathrm{cr0}}$ increases. More importantly, the FG-X plate has the largest stable zone, while the FG-O has the smallest stable zone.

Figures 6.12–6.14 reveal the influences of GPL weight fraction on the flutter and buckling boundaries of UD plate, FG-O plate and FG-X plate, respectively. We can see that both the flutter boundaries and buckling boundaries are increased as GPL weight fraction increases for all three GPLRC plates. When $\lambda = 0$, $\Delta T/\Delta T_{\mathrm{cr0}}$ of UD plate is almost independent of g_{GPL}, $\Delta T/\Delta T_{\mathrm{cr0}}$ of FG-O plate is decreased as g_{GPL} increases, and $\Delta T/\Delta T_{\mathrm{cr0}}$ of FG-X plate is increased as g_{GPL} increases. Thermal buckling appears at $\Delta T/\Delta T_{\mathrm{cr0}}<1$ for FG-O plate, while stable flat condition exists at $\Delta T/\Delta T_{\mathrm{cr0}}>1$ for FG-X plate.

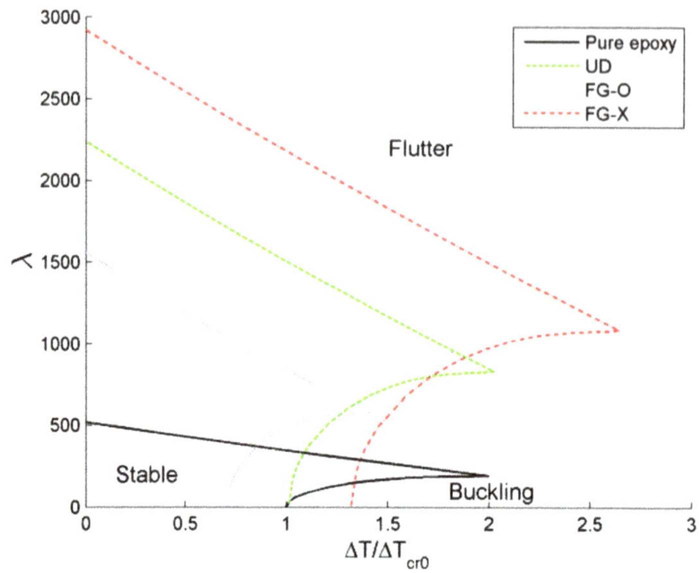

Figure 6.11. Flutter and buckling boundaries of GPLRC plate with different GPLs distributions.

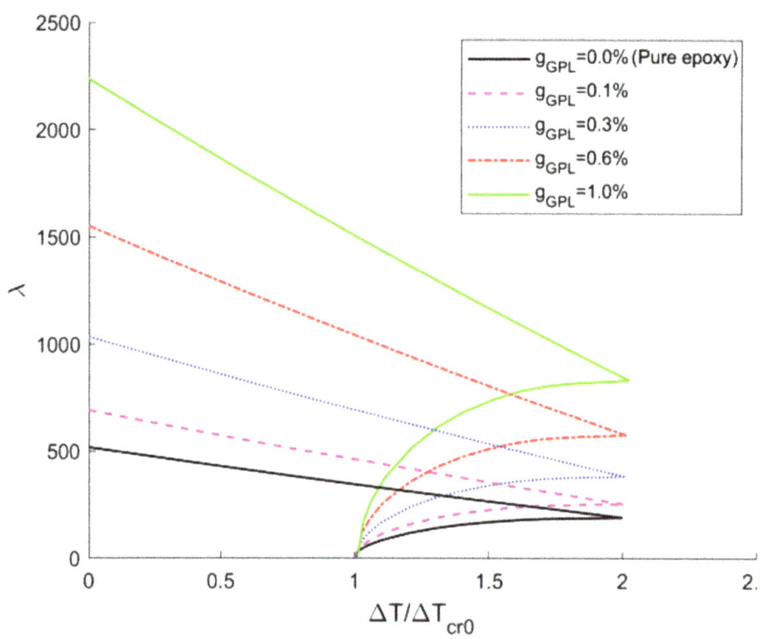

Figure 6.12. Flutter and buckling boundaries of UD plate with different g_{GPL}.

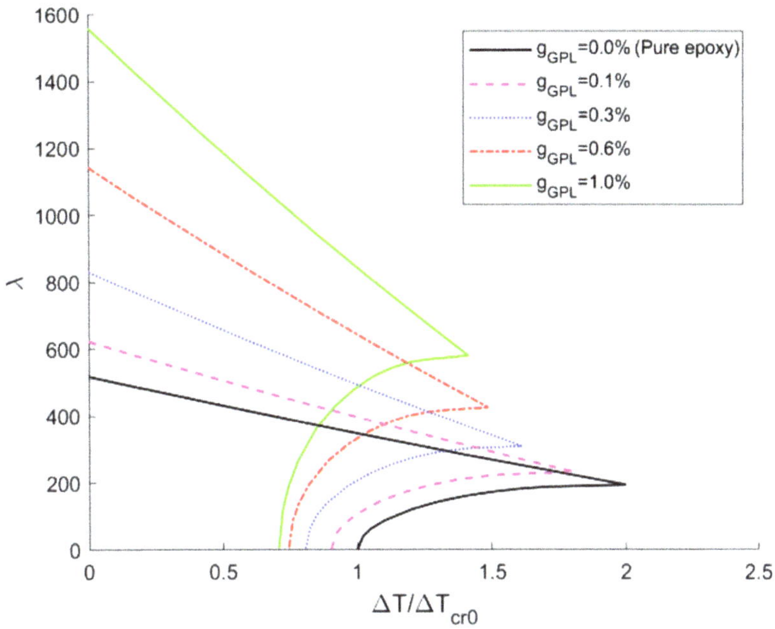

Figure 6.13. Flutter and buckling boundaries of FG-O plate with different g_{GPL}.

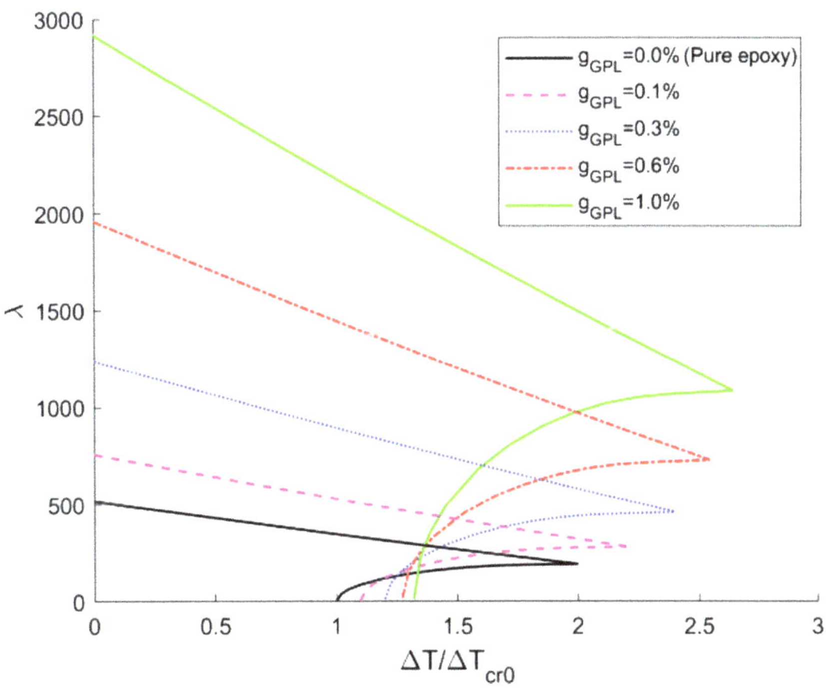

Figure 6.14. Flutter and buckling boundaries of FG-X plate with different g_{GPL}.

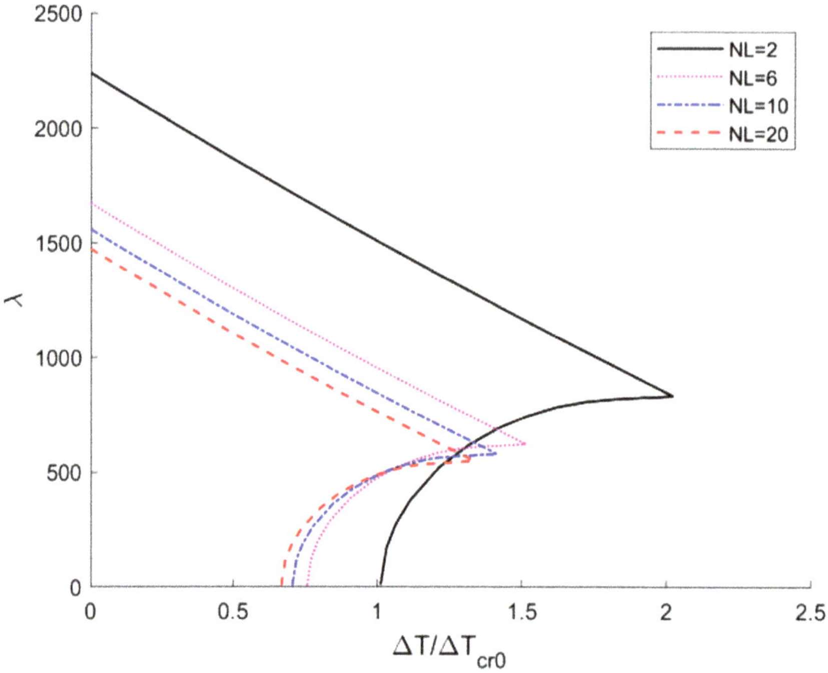

Figure 6.15. Flutter and buckling boundaries of FG-O plate with different *NL*.

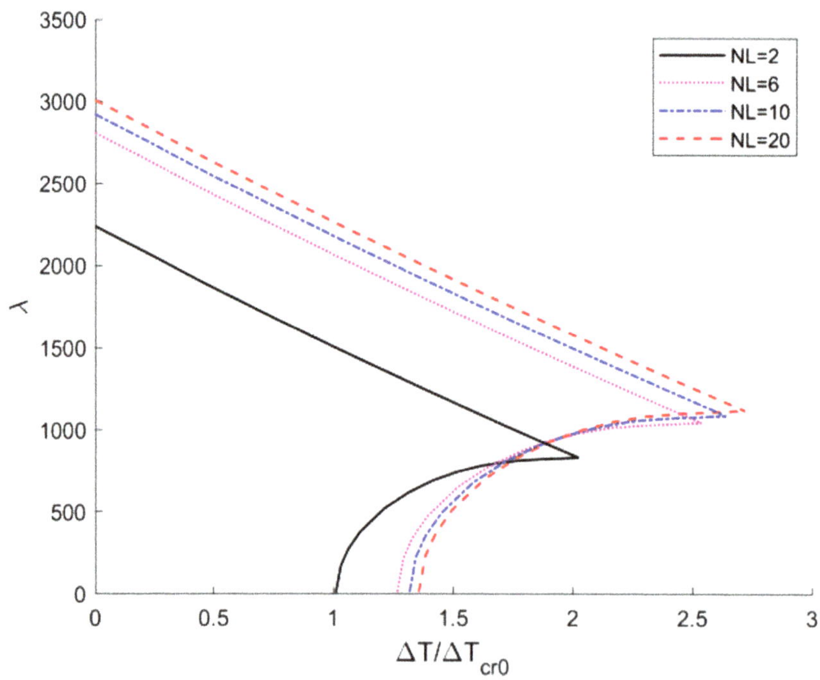

Figure 6.16. Flutter and buckling boundaries of FG-X plate with different *NL*.

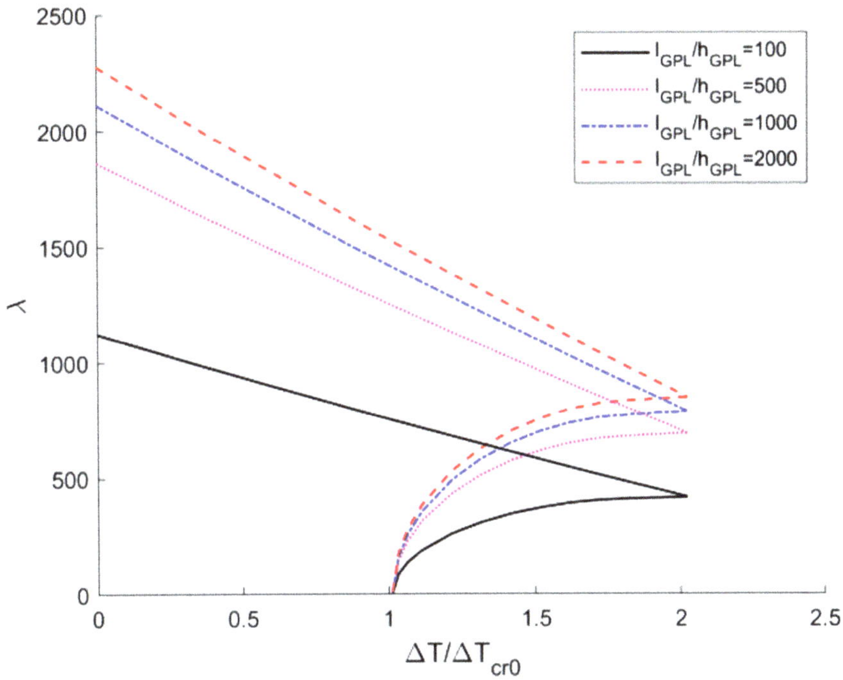

Figure 6.17. Flutter and buckling boundaries of UD plate with different l_{GPL}/h_{GPL}.

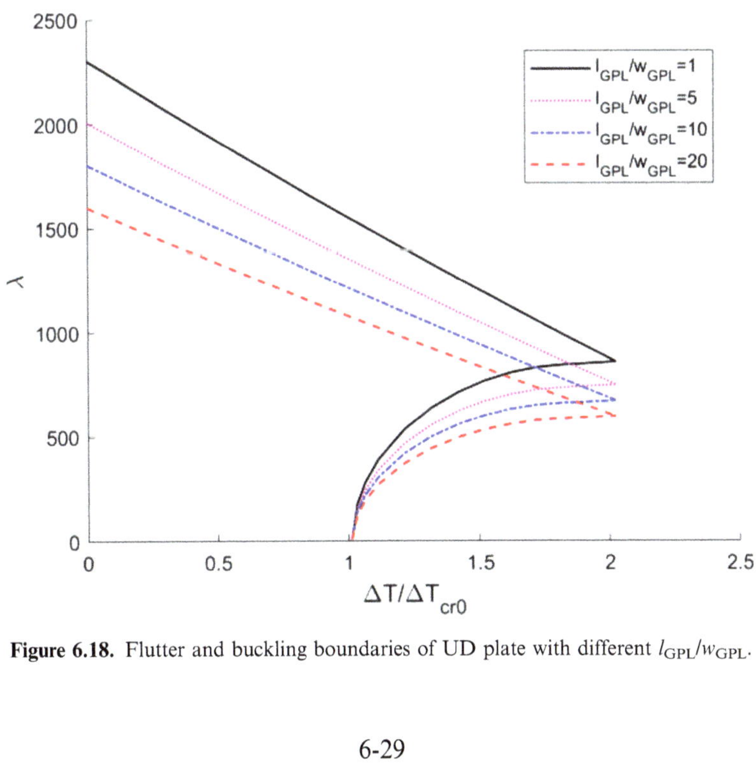

Figure 6.18. Flutter and buckling boundaries of UD plate with different l_{GPL}/w_{GPL}.

The total layer number NL of GPLRC plates has an important influence on natural frequency, except for the UD plate [26, 33]. Natural frequency and flutter boundary of the UD plate are independent of the total layer number NL. Figures 6.15 and 6.16 display the effect of NL on the flutter and buckling boundaries of FG-O and FG-X plates. Increasing NL could increase the flutter and buckling boundaries of the FG-X plate and decrease the flutter and buckling boundaries of the FG-O plate. When NL increases from 10 to 20, the flutter and buckling boundaries of FG-O and FG-X plates are slightly changed. The flutter and buckling boundaries of FG-O and FG-X plates tend to converge to constants.

The effects of GPL geometry and size, in the form of length-to-thickness ratio l_{GPL}/h_{GPL} and length-to-width ratio l_{GPL}/w_{GPL}, on the flutter and buckling boundaries of UD plate are presented in figures 6.17 and 6.18. It can be seen that the flutter and buckling boundaries increase dramatically when l_{GPL}/h_{GPL} increases from 100 to 2000 and the GPL width w_{GPL} is kept constant. Moreover, the flutter and buckling boundaries of UD plate are slightly changed when l_{GPL}/h_{GPL} increases from 1000 to 2000. This means the flutter and buckling boundaries of UD plate tend to converge to constants. Figure 6.18 shows that the flutter and buckling boundaries of UD plate are decreased when l_{GPL}/w_{GPL} increases from 1.0 to 20 and the GPL thickness h_{GPL} is kept constant.

References

[1] Dowell E H 1970 Panel flutter: a review of the aeroelastic stability of plates and shells *AIAA J.* **8** 385–99

[2] Tawfik M, Ro J J and Mei C 2002 Thermal post-buckling and aeroelastic behaviour of shape memory alloy reinforced plates *Smart Mater. Struct.* **11** 297–307

[3] Li F M 2012 Active aeroelastic flutter suppression of a supersonic plate with piezoelectric material *Int. J. Eng. Sci.* **51** 190–203

[4] Song Z G and Li F M 2012 Active aeroelastic flutter analysis and vibration control of supersonic composite laminated plate *Compos. Struct.* **94** 702–13

[5] Asadi H and Wang Q 2017 An investigation on the aeroelastic flutter characteristics of FG-CNTRC beams in the supersonic flow *Compos. Part B-Eng.* **116** 486–99

[6] Stankovich S, Dikin D, Dommett G, Kohlhaas K, Zimney E J, Stach E, Piner R, Nguyen S and Ruoff R 2006 Graphene-based composite materials *Nature* **442** 282–86

[7] Potts J R, Dreyer D R, Bielawski C W and Ruoff R S 2011 Graphene-based polymer nanocomposites *Polymer* **52** 5–25

[8] Zhao S Y, Zhao Z, Yang Z C, Ke L L, Kitipornchai S and Yang J 2020 Functionally graded graphene reinforced composite structures: a review *Eng. Struct.* **210** 110339

[9] Laws N, Dvorak G J and Hejazi M 1983 Stiffness changes in unidirectional composites caused by crack systems *Mech. Mater.* **2** 123–37

[10] Dvorak G, Laws N and Hejazi M 1985 Analysis of progressive matrix cracking in composite laminates I. Thermoelastic properties of a ply with cracks *J. Compos. Mater.* **19** 216–34

[11] Fan Y and Wang H 2016 Nonlinear dynamics of matrix-cracked hybrid laminated plates containing carbon nanotube-reinforced composite layers resting on elastic foundations *Nonlinear Dyn.* **84** 1181–99

[12] Fan Y and Wang H 2016 Thermal postbuckling and vibration of postbuckled matrix cracked hybrid laminated plates containing carbon nanotube reinforced composite layers on elastic foundation *Compos. Struct.* **157** 386–97

[13] Lei Z X, Zhang L W and Liew K M 2018 Modeling large amplitude vibration of matrix cracked hybrid laminated plates containing CNTR-FG layers *Appl. Math. Modell.* **55** 33–48

[14] Pan Z Z, Zhang L W and Liew K M 2019 Modeling geometrically nonlinear large deformation behaviors of matrix cracked hybrid composite deep shells containing CNTRC layers *Comput. Meth. Appl. Mech. Eng.* **355** 753–78

[15] Guo H L, Yang T Z, Żur K K and Reddy J N 2021 On the flutter of matrix cracked laminated composite plates reinforced with graphene nanoplatelets *Thin-Walled Struct.* **158** 107161

[16] Mei C, Abdel-Motagaly K and Chen R 1999 Review of nonlinear panel flutter at supersonic and hypersonic speeds *Appl. Mech. Rev.* **52** 321–32

[17] Sohn K J and Kim J H 2008 Structural stability of functionally graded panels subjected to aero-thermal loads *Compos. Struct.* **82** 317–25

[18] Shiau L C, Kuo S Y and Liu Y P 2012 Aerothermoelastic analysis of composite laminated plates *Compos. Struct.* **94** 1982–90

[19] Ibrahim H H, Tawfik M and Al-Ajmi M 2007 Thermal buckling and nonlinear flutter behavior of functionally graded material panels *J. Aircr.* **44** 1610–18

[20] Ibrahim H H, Tawfik M and Negm H M 2011 Thermal buckling and nonlinear flutter behavior of shape memory alloy hybrid composite plates *J. Vib. Control* **17** 321–33

[21] Fazelzadeh S A, Pouresmaeeli S and Ghavanloo E 2015 Aeroelastic characteristics of functionally graded carbon nanotube-reinforced composite plates under a supersonic flow *Comput. Meth. Appl. Mech. Eng.* **285** 714–29

[22] Chai Y Y, Song Z G and Li F M 2018 Investigations on the aerothermoelastic properties of composite laminated cylindrical shells with elastic boundaries in supersonic airflow based on the Rayleigh-Ritz method *Aerosp. Sci.Technol.* **82–83** 534–44

[23] Marzocca P, Fazelzadeh S A and Hosseini M 2011 A review of nonlinear aero-thermo-elasticity of functionally graded panels *J. Therm. Stresses* **34** 536–68

[24] Liew K M, Lei Z X, Yu J L and Zhang L W 2014 Postbuckling of carbon nanotube-reinforced functionally graded cylindrical panels under axial compression using a meshless approach *Comput. Meth. Appl. Mech. Eng.* **268** 1–17

[25] Zhang L W and Liew K M 2014 An improved moving least-squares Ritz method for two-dimensional elasticity problems *Appl. Math. Comput.* **246** 268–82

[26] Guo H L, Cao S Q, Yang T Z and Chen Y S 2018 Vibration of laminated composite quadrilateral plates reinforced with graphene nanoplatelets using the element-free IMLS-Ritz method *Int. J. Mech. Sci.* **142** 610–21

[27] Guo H L, Cao S Q, Yang T Z and Chen Y S 2018 Geometrically nonlinear analysis of laminated composite quadrilateral plates reinforced with graphene nanoplatelets using the element-free IMLS-Ritz method *Compos. Part B-Eng.* **154** 216–24

[28] Huang K, Guo H L, Qin Z H, Cao S Q and Chen Y S 2020 Flutter analysis of laminated composite quadrilateral plates reinforced with graphene nanoplatelets using the element-free IMLS-Ritz method *Aerosp. Sci.Technol.* **103** 105915

[29] De Villoria R G and Miravete A 2007 Mechanical model to evaluate the effect of the dispersion in nanocomposites *Acta Mater.* **55** 3025–31

[30] Reddy J N 2004 *Mechanics of Laminated Composite Plates and Shells: Theory and Analysis* 2nd edn. (Boca Raton, FL: CRC Press)

[31] Lin K, Lu P and Tarn J 1989 Flutter analysis of composite panels using high-precision finite elements *Comput. Struct.* **33** 561–74

[32] Thai C H, Ferreira A J M, Tran T D and Phung-Van P 2019 Free vibration, buckling and bending analyses of multilayer functionally graded graphene nanoplatelets reinforced composite plates using the NURBS formulation *Compos. Struct.* **220** 749–59

[33] Song M T, Kitipornchai S and Yang J 2017 Free and forced vibrations of functionally graded polymer composite plates reinforced with graphene nanoplatelets *Compos. Struct.* **159** 579–88

[34] Reddy J N 2015 *An Introduction to Nonlinear Finite Element Analysis* 2nd edn (Oxford: Oxford University Press)

[35] Shiau L C, Kuo S Y and Liu Y P 2012 Aerothermoelastic analysis of composite laminated plates *Compos. Struct.* **94** 1982–90

[36] Wu H L, Kitipornchai S and Yang J 2017 Thermal buckling and postbuckling of functionally graded graphene nanocomposite plates *Mater. Des.* **132** 430–41

IOP Publishing

Modeling and Computation in Vibration Problems, Volume 1
Numerical and semi-analytical methods
S Chakraverty, F Tornabene and J N Reddy

Chapter 7

On forced vibrations of piezo-flexomagnetic nano-actuator beams

Mohammad Malikan and Victor A Eremeyev

The effect of excitation frequency on the piezomagnetic Euler–Bernoulli nanobeam taking the flexomagnetic material phenomenon into consideration is investigated in this chapter. The magnetization with strain gradients creates flexomagneticity. We couple simultaneously the piezomagnetic and flexomagnetic properties in an inverse magnetization. Resembling the flexoelectricity, the flexomagneticity is also size-dependent. So, it has further notable influence at a small size. Here, a dynamic loading is transversely acted on the nanobeam. This loading plays the role of the excitation force. Moreover, we combine the equilibrium equations with the nonlocal model of strain gradient elasticity (NSGT). Afterwards, we impose the Hamilton's principle for having the governing equations. The attained linear differential equation is transferred into an algebraic equation by using Galerkin weighted residual method. Then, the gained linear algebraic equation is solved which leads to numerical values of dynamic deflections.

7.1 Flexomagneticity

The flexomagnetic phenomenon, a particular form of coupling between magnetic and mechanical responses, is prominent at the nanoscale. The emergence of flexomagnetic effect is in two cases, direct effect and the inverse one. In response to the strain gradient the magnetic polarization emergences which is called direct flexomagnetic effect. On the other hand, in response to the magnetic polarization gradient the elastic strain emergences of this process are called inverse flexomagneticity effect. As an analogue of the piezomagnetic effect, one can view the flexomagnetic effect [1–4]. Flexomagneticity can be present in all structures, although the piezomagnetic influence can be present only in polar structures having special symmetry [5].

doi:10.1088/978-0-7503-3483-9ch7

© IOP Publishing Ltd 2021

Inevitably, strong strain gradients can exist in nanoscale structures' surfaces, leading to a significant role played by flexomagneticity. Thus, undoubtedly there is a fundamental and basic linkage between the flexomagnetic coupling and unique features appearing in nanoscale.

There are various perspectives of flexomagneticity that stand unexplored [6–11]. For example, the vibration studies of flexomagnetic nanostructures in the framework of forced oscillations, nonlinear behavior of flexomagnetic nanomaterials by considering large amplitudes and post-stability of flexomagnetic nano objects, etc. In this chapter, the tendency is the transition of the effect of flexomagneticity into a nanosize actuator beam excited by a dynamic load to present new results on this new functionality. The effort is needed to theoretically model the nano-actuator for which this is comprehensively performed in regards to the next sections.

7.2 Mathematical modelling

A size-dependent piezo-flexomagnetic nanoscale actuator beam is revealed by figure 7.1 connected to the rectangular coordinates. The beam is joined to a postulated magnetizer to implement the magnetic field influence. A harmonic uniform force is loaded vertically and repeatedly on the beam to simulate the external excitation. The beam is designed squarely.

Let us commence the mathematical modelling of the problem with utilizing the Euler–Bernoulli beam hypothesis as follows [12–16]

$$u_1(x, z, t) = u(x, t) - z\frac{\partial w(x, t)}{\partial x} \tag{7.1}$$

$$u_3(x, z, t) = w(x, t) \tag{7.2}$$

If we take the linear Lagrangian strains as the basis of the formulation, one can obtain

$$\varepsilon_{xx} = \frac{\partial u}{\partial x} - z\frac{\partial^2 w}{\partial x^2} \tag{7.3}$$

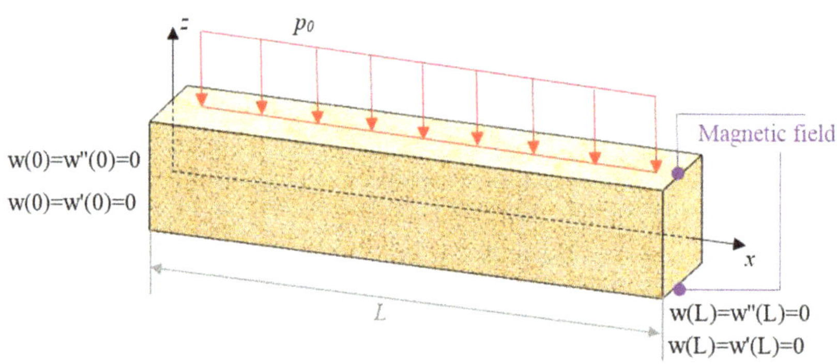

Figure 7.1. Piezo-flexomagnetic nanosize beam subjected to the magnetic potential and transversely acted dynamic load.

$$\eta_{xxz} = \frac{\partial \varepsilon_{xx}}{\partial z} = -\frac{\partial^2 w}{\partial x^2} \tag{7.4}$$

According to the piezomagnetic–flexomagnetic relationships [6–11], we have

$$\sigma_{xx} = C_{11}\varepsilon_{xx} - q_{31}H_z \tag{7.5}$$

$$\xi_{xxz} = g_{31}\eta_{xxz} - f_{31}H_z \tag{7.6}$$

$$B_z = a_{33}H_z + q_{31}\varepsilon_{xx} + f_{31}\eta_{xxz} \tag{7.7}$$

To achieve the non-classical boundary conditions and the governing equations, one may use the Hamilton's principle as

$$\int_{t_1}^{t_2} (\delta K - \delta U + \delta W)dt = 0 \tag{7.8}$$

where

$$\delta U = \int_V (\sigma_{xx}\delta\varepsilon_{xx} + \xi_{xxz}\delta\eta_{xxz} - B_z\delta H_z)dV \tag{7.9}$$

It would be written part by part on the basis of mechanical and magnetic terms as

$$\delta U = \delta\Pi_{U_1}^{\text{Mech}} + \delta\Pi_{U_1}^{\text{Mag}} + \delta\Pi_{U_2}^{\text{Mech}} + \delta\Pi_{U_2}^{\text{Mag}}$$

where

$$\delta\Pi_{U_1}^{\text{Mech}} = -\int_0^L \left(\frac{\partial N_x}{\partial x}\delta u + \frac{\partial^2 M_x}{\partial x^2}\delta w + \frac{\partial^2 T_{xxz}}{\partial x^2}\delta w\right)dx \tag{7.10a}$$

$$\delta\Pi_{U_1}^{\text{Mag}} = -\int_0^L \int_{-h/2}^{h/2} \frac{\partial B_z}{\partial z}\delta\Psi dz dx \tag{7.10b}$$

$$\delta\Pi_{U_2}^{\text{Mech}} = \left(N_x\delta u - M_x\frac{\partial\delta w}{\partial x} - T_{xxz}\frac{\partial\delta w}{\partial x} + \frac{\partial M_x}{\partial x}\delta w + \frac{\partial T_{xxz}}{\partial x}\delta w\right)\Bigg|_0^L \tag{7.11a}$$

$$\delta\Pi_{U_2}^{\text{Mag}} = \int_0^L (B_z\delta\Psi)\Bigg|_{-h/2}^{h/2} dx \tag{7.11b}$$

in which the following integrals are the stress resultants,

$$N_x = \int_{-h/2}^{h/2} \sigma_{xx}dz \tag{7.12}$$

$$M_x = \int_{-h/2}^{h/2} \sigma_{xx}z dz \tag{7.13}$$

$$T_{xxz} = \int_{-h/2}^{h/2} \xi_{xxz} dz \tag{7.14}$$

The performed work by the magnetic field can generate an axial force as follows [17]

$$W = -\frac{1}{2}\int_0^L N_x^0 \left(\frac{\partial w}{\partial x}\right)^2 dx \tag{7.15}$$

$$\delta W = -\int_0^L N_x^0 \left(\frac{\partial \delta w}{\partial x}\frac{\partial w}{\partial x}\right) dx \tag{7.16}$$

The relationship between the magnetic field component and the magnetic potential can be attained as

$$H_z = -\frac{\partial \Psi}{\partial z} \tag{7.17}$$

The below-conducted relations display the magnetic boundary conditions,

$$\Psi\left(+\frac{h}{2}\right) = \psi,\ \Psi\left(-\frac{h}{2}\right) = 0 \tag{7.18a–b}$$

Based on equations (7.7), (7.10b), (7.11b), (7.17) and (7.18), one can acquire the expanded relations of the magnetic potential and the magnetic field component as

$$\Psi = -\frac{q_{31}}{2a_{33}}\left(z^2 - \frac{h^2}{4}\right)\frac{\partial^2 w}{\partial x^2} + \frac{\psi}{h}\left(z + \frac{h}{2}\right) \tag{7.19}$$

$$H_z = z\frac{q_{31}}{a_{33}}\frac{\partial^2 w}{\partial x^2} - \frac{\psi}{h} \tag{7.20}$$

According to equations (7.5)–(7.7), (7.19) and (7.20), one can present the developed relations of the magnetic flux, hyper-stress and axial stress as

$$B_z = q_{31}\frac{\partial u}{\partial x} - f_{31}\frac{\partial^2 w}{\partial x^2} - \frac{a_{33}\psi}{h} \tag{7.21}$$

$$\xi_{xxz} = -\left(g_{31} + \frac{q_{31}f_{31}z}{a_{33}}\right)\frac{\partial^2 w}{\partial x^2} + \frac{f_{31}\psi}{h} \tag{7.22}$$

$$\sigma_{xx} = C_{11}\frac{\partial u}{\partial x} - z\left(C_{11} + \frac{q_{31}^2}{a_{33}}\right)\frac{\partial^2 w}{\partial x^2} + \frac{q_{31}\psi}{h} \tag{7.23}$$

Substituting equations (7.22) and (7.23) into equations (7.12)–(7.14), one can take the expanded relations of the stress and hyper stress resultants as

$$N_x = C_{11}A\frac{\partial u}{\partial x} + q_{31}\psi \tag{7.24}$$

$$M_x = -I_z\left(C_{11} + \frac{q_{31}^2}{a_{33}}\right)\frac{\partial^2 w}{\partial x^2} \tag{7.25}$$

$$T_{xxz} = -g_{31}h\frac{\partial^2 w}{\partial x^2} + f_{31}\psi \tag{7.26}$$

In regard to equation (7.24), one can see the two terms where the second part means the axial force resulting from the magnetic field, then

$$N^{\mathrm{Mag}} = q_{31}\psi \tag{7.27}$$

$$N_x^0 = N^{\mathrm{Mag}} \tag{7.28}$$

Based on the classic kinetic energy relation, one can achieve the inertia terms to make the problem dynamically as

$$K = \frac{1}{2}\int\int_A \rho(z, t)\left[\left(\frac{\partial u_1}{\partial t}\right)^2 + \left(\frac{\partial u_3}{\partial t}\right)^2\right]dAdz \tag{7.29}$$

where

$$K = \frac{1}{2}\int\int_A \rho\left[\left(\frac{\partial u}{\partial t} - z\frac{\partial^2 w}{\partial x \partial t}\right)^2 + \left(\frac{\partial w}{\partial t}\right)^2\right]dAdz \tag{7.30}$$

Therefore,

$$\delta K = \int_A\left[-I_0\frac{\partial^2 u}{\partial t^2}\delta u + I_2\frac{\partial^4 w}{\partial x^2 \partial t^2}\delta w - I_0\frac{\partial^2 w}{\partial t^2}\delta w\right]dA \tag{7.31}$$

in which

$$(I_0, I_2) = \int_{-h/2}^{h/2} \rho(1, z^2)dz \tag{7.32}$$

Heretofore, we presented a massive formulation to receive the final part, that is, the governing equations. So, the following relations

$$\frac{\partial N_x}{\partial x} = I_0\frac{\partial^2 u}{\partial t^2} \tag{7.33}$$

$$\frac{\partial^2 M_x}{\partial x^2} + \frac{\partial^2 T_{xxz}}{\partial x^2} + N_x^0\frac{\partial^2 w}{\partial x^2} = I_0\frac{\partial^2 w}{\partial t^2} - I_2\frac{\partial^4 w}{\partial x^2 \partial t^2} \tag{7.34}$$

Thitherto, all the equations are for a local beam. The considered beam is a size-dependent structure which should be investigated in a small-scale space. Thus far, two functionalities are found for a nanosize material when considered in its real scale, that is, stress nonlocality, which is due to the interaction between atoms and strain gradient effect due to the deformations of atoms themselves. The nonlocal strain gradient theory of elasticity helps to impose these two properties into the model by means of the following relation [18]

$$\left(1 - (e_0a)^2\frac{\partial^2}{\partial x^2}\right)\sigma_{xx}^{\text{Non-local}} = \left(1 - l^2\frac{\partial^2}{\partial x^2}\right)\sigma_{xx}^{\text{Local}} \tag{7.35}$$

where $\mu = (e_0a)^2$, then

$$\left(1 - \mu\frac{\partial^2}{\partial x^2}\right)\sigma_{xx}^{\text{Non-local}} = \left(1 - l^2\frac{\partial^2}{\partial x^2}\right)\left\{C_{11}\frac{\partial u}{\partial x} - z\left(C_{11} + \frac{q_{31}^2}{a_{33}}\right)\frac{\partial^2 w}{\partial x^2} + \frac{q_{31}\Psi}{h}\right\} \tag{7.36}$$

Hence, the above-mentioned relation can proper the nonlocal stress resultants as well,

$$N_x^{\text{Non-local}} = \mu\frac{\partial^2 N_x}{\partial x^2} + \left(1 - l^2\frac{\partial^2}{\partial x^2}\right)\left\{C_{11}\frac{\partial u}{\partial x} + \frac{q_{31}\Psi}{h}\right\} \tag{7.37}$$

$$M_x^{\text{Non-local}} = \mu\frac{\partial^2 M_x}{\partial x^2} - I_z\left(C_{11} + \frac{q_{31}^2}{a_{33}}\right)\left(1 - l^2\frac{\partial^2}{\partial x^2}\right)\left\{\frac{\partial^2 w}{\partial x^2}\right\} \tag{7.38}$$

$$T_{xxz}^{\text{Non-local}} = \mu\frac{\partial^2 T_{xxz}}{\partial x^2} + \left(1 - l^2\frac{\partial^2}{\partial x^2}\right)\left\{-g_{31}h\frac{\partial^2 w}{\partial x^2} + f_{31}\psi\right\} \tag{7.39}$$

By the aid of equations (7.33) and (7.34), equations (7.37–7.39) can be rewritten as

$$N_x^{\text{Non-local}} = \mu I_0\frac{\partial^3 u}{\partial x\partial t^2} + \left(1 - l^2\frac{\partial^2}{\partial x^2}\right)\left\{C_{11}\frac{\partial u}{\partial x} + \frac{q_{31}\Psi}{h}\right\} \tag{7.40}$$

$$M_x^{Nonlocal} = -\mu\left(\frac{\partial^2 T_{xxz}}{\partial x^2} + N_x^0\frac{\partial^2 w}{\partial x^2} - I_0\frac{\partial^2 w}{\partial t^2} + I_2\frac{\partial^4 w}{\partial x^2\partial t^2}\right) - I_z\left(C_{11} + \frac{q_{31}^2}{a_{33}}\right) \times\left(\frac{\partial^2 w}{\partial x^2} - l^2\frac{\partial^4 w}{\partial x^4}\right) \tag{7.41}$$

$$T_{xxz}^{\text{Non-local}} = \left(1 - l^2\frac{\partial^2}{\partial x^2}\right)\left\{-g_{31}h\frac{\partial^2 w}{\partial x^2} + f_{31}\psi\right\} \tag{7.42}$$

Now, there is the possibility to expand equations (7.33) and (7.34) on the basis of equations (7.40)–(7.42) as follows

$$\mu I_0\frac{\partial^4 u}{\partial x^2\partial t^2} - I_0\frac{\partial^2 u}{\partial t^2} + C_{11}\left(1 - l^2\frac{\partial^2}{\partial x^2}\right)\frac{\partial^2 u}{\partial x^2} = 0 \tag{7.43}$$

$$\begin{gathered} - g_{31}h\frac{\partial^4 w}{\partial x^4} + q_{31}\psi\frac{\partial^2 w}{\partial x^2} - p - \mu\left(-g_{31}h\frac{\partial^6 w}{\partial x^6} + q_{31}\psi\frac{\partial^4 w}{\partial x^4} - \frac{\partial^2 p}{\partial x^2}\right) - I_z\left(C_{11} + \frac{q_{31}^2}{a_{33}}\right) \\ \times\left(\frac{\partial^4 w}{\partial x^4} - l^2\frac{\partial^6 w}{\partial x^6}\right) + \mu\left(I_0\frac{\partial^4 w}{\partial x^2\partial t^2} - I_2\frac{\partial^6 w}{\partial x^4\partial t^2}\right) - I_0\frac{\partial^2 w}{\partial t^2} + I_2\frac{\partial^4 w}{\partial x^2\partial t^2} = 0 \end{gathered} \tag{7.44}$$

As can be vividly seen, equation (7.43) is decoupled from equation (7.44), hence, in order to analyze the problem, equation (7.44) is enough only.

7.3 Solution process

Due to having a linear partial differential equation (equation 7.44), it is possible to solve it analytically in order to extract the numerical outcomes more easily [19–21].

$$w(x, t) = \sum_{m=1}^{\infty} X_m(x)\exp(i\omega t) \tag{7.45}$$

This solution means the real part of frequency is taken into the investigation only. Imposing equation (7.45) into equation (7.44) will lead to the following time-independent equation,

$$\Delta_\omega^2 = \frac{\begin{gathered} g_{31}h\frac{\partial^4 X_m}{\partial x^4} - q_{31}\psi\frac{\partial^2 X_m}{\partial x^2} + p + \mu\left(-g_{31}h\frac{\partial^6 X_m}{\partial x^6} + q_{31}\psi\frac{\partial^4 X_m}{\partial x^4} - \frac{\partial^2 p}{\partial x^2}\right) + \\ I_z\left(C_{11} + \frac{q_{31}^2}{a_{33}}\right)\left(\frac{\partial^4 X_m}{\partial x^4} - l^2\frac{\partial^6 X_m}{\partial x^6}\right) \end{gathered}}{\left[-\mu\left(I_0\frac{\partial^2 X_m}{\partial x^2} - I_2\frac{\partial^4 X_m}{\partial x^4}\right) + I_0 - I_2\frac{\partial^2 X_m}{\partial x^2}\right]} \tag{7.46}$$

In order to solve the aforesaid equation, there should be considered some boundary conditions. To do this, with respect to table 7.1, the hinged and fully clamped boundary conditions are applied to equation (7.46).

The described boundary conditions in table 7.1 can be modelled analytically as follows:

Table 7.1. The essential boundary conditions.

Type	Configurations	Conditions
Hinged	S	$w(0, L) = 0$, $w''(0, L) = 0$
Fully clamped	C	$w(0, L) = 0$, $w'(0, L) = 0$

$$X_m(x) = \sin\left(\frac{m\pi}{L}x\right) \tag{7.47}$$

$$X_m(x) = \sin^2\left(\frac{m\pi}{L}x\right) \tag{7.48}$$

where equations (7.47) and (7.48) are expected to, respectively, satisfy hinged and fixed boundary conditions. It is to be noted that in this chapter the mean of the hinged and fixed boundary conditions is the same as the simply-supported and clamped denotations. Furthermore, it should be noted that the following numerical outcomes are demonstrated only for the first mode shape.

To define the transversely acted harmonic load, the following relation is utilized [22].

$$p(x) = \sum_{m=1}^{\infty}\frac{4p_0}{m\pi}\sin\left(\frac{m\pi}{L}x\right)\exp(i\omega t) \tag{7.49}$$

The simplicity of the Galerkin weighted residual method makes this technique more usable. The convenience of this solution technique has been proved to give almost exact results [19, 20]. Moreover, this approach does not require a long solution time or a large volume of procedure. The following integral helps us to attain the numerical answers.

$$\int_0^L X_m(x)\zeta(x)dx \tag{7.50}$$

in which $\zeta(x)$ denotes a residue of equation (7.46).

7.4 Validity

This part of the study compares the present model and technique before running the main results. In doing so, due to nonexistance of a resemblance study, equation (7.46) is diminished to a common nanobeam by neglecting the magnetization. On the other side, the natural frequency is only taken into validation. The comparison equation is as follows:

$$\omega_n^2 = \frac{I_z C_{11}\dfrac{\partial^4 X_m}{\partial x^4}}{\left[-\mu\left(I_0\dfrac{\partial^2 X_m}{\partial x^2} - I_2\dfrac{\partial^4 X_m}{\partial x^4}\right) + I_0 - I_2\dfrac{\partial^2 X_m}{\partial x^2}\right]} \tag{7.51}$$

The variations of the slenderness ratio ($L\,h^{-1}$) and the nonlocal parameter are the basic of the tabular validation shown by table 7.2. The employed elasticity properties of the material and other quantities are dedicated as below

$L = 10$ nm, $h = 1$ nm, $\nu = 0.3$, $l = 0$, $E = 1$TPa, $\rho = 2.7$ kg dm^{-3}

Regarding the used reference, there should be presented the nondimensional relation of the natural frequency, hence

Table 7.2. Non-dimensional natural frequencies of a square nanobeam.

$L\,h^{-1}$	μ	EBT, Navier [23]	EBT [present]
5	0	9.7112	9.7112
	1	9.2647	9.2647
	2	8.8747	8.8747
	3	8.5301	8.5301
	4	8.2228	8.2228
10	0	9.8293	9.8293
	1	9.3774	9.3774
	2	8.9826	8.9826
	3	8.6338	8.6338
	4	8.3228	8.3228
20	0	9.8595	9.8595
	1	9.4062	9.4062
	2	9.0102	9.0102
	3	8.6604	8.6604
	4	8.3483	8.3483

$$\Omega = \omega_n L^2 \sqrt{\frac{\rho A}{EI}} \tag{7.52}$$

The only conflict between our case and the employed reference is the solution type. The Navier-type solution was utilized by [23]. As seen, results are entirely correlated with the reference. No conflicts can be observed even if the values of the slenderness ratio or the nonlocal parameter increase. This validity of outcomes can be a fundamental comparison and basically confirms the utilized solution procedure.

7.5 Discussion and results

This chapter is prepared to analyze a piezo-flexomagnetic nanobeam exposed in a magnetic field and the beam is also excited by an external frequency. The external frequency occurs in a range that can make resonance amplitude. To model a piezo-flexomagnetic nano-actuator beam, table 7.3 exhibits the necessary properties needed for this section by means of [6–11, 24].

7.5.1 Resonance analysis

Figures 7.2–7.4 show the resonance analysis of a piezo-flexomagnetic nanobeam. Defined ratio Δ_ω is the ratio of excitation frequency to the natural frequency of the system. According to the basic vibration texts, we know that a value of 1 for this coefficient is the peak of intensification and oscillations of the system. Hereafter, this coefficient will be known as the frequency coefficient. Figure 7.2 displays the resonance behavior for the two boundary conditions SS and CC. Dynamic deflections for the clamped–clamped boundary condition involve smaller values

Table 7.3. Magneto-mechanical features for the piezo-flexomagnetic nano-actuator beam.

$CoFe_2O_4$
$C_{11} = 286e9$ N m^{-2} $f_{31} = 10^{-10}$ N $Ampere^{-1}$ $q_{31} = 580.3$ N $Ampere^{-1}$ m^{-1} $a_{33} = 1.57 \times 10^{-4}$ N $Ampere^{-2}$
$L = 10$ nm, $\rho = 4.89$ kg dm^{-3}

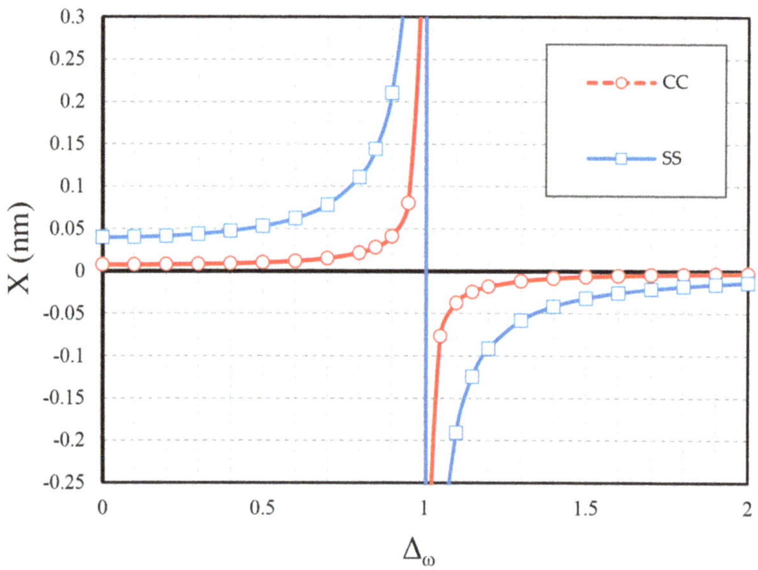

Figure 7.2. Frequency ratio versus dynamic deflection for different cases of edge conditions ($\Psi = 1$ mA, $l = 1$ nm, $e_0a = 0.5$ nm, $P_0 = 0.01$ N m^{-1}).

due to less flexibility at the boundaries. Considering the figure, it can be said that the resonance region seems to be a smaller value in the boundary conditions of the two edges fixed, although the beam goes into the large amplitude of resonance later and with a delay at these boundary conditions. It is worth noting that the mean of resonance region is here $0.9 \leqslant \Delta_\omega \leqslant 1.1$. It is important to pay attention to the fact that in the beam with the boundary conditions of the two ends of the clamped, the mentioned beam jumps to the resonance amplitude and the possibility of damaging the beam in this case is higher. In fact, in beams with boundary conditions of two hinge edges and in the more general case, in beams with more flexible boundary conditions, the beam gradually resonances. Therefore, in beams with CC boundary conditions, measures should be considered to prevent the destruction of the beam material in cases of unpredictable and accidental resonance. Although engineers use

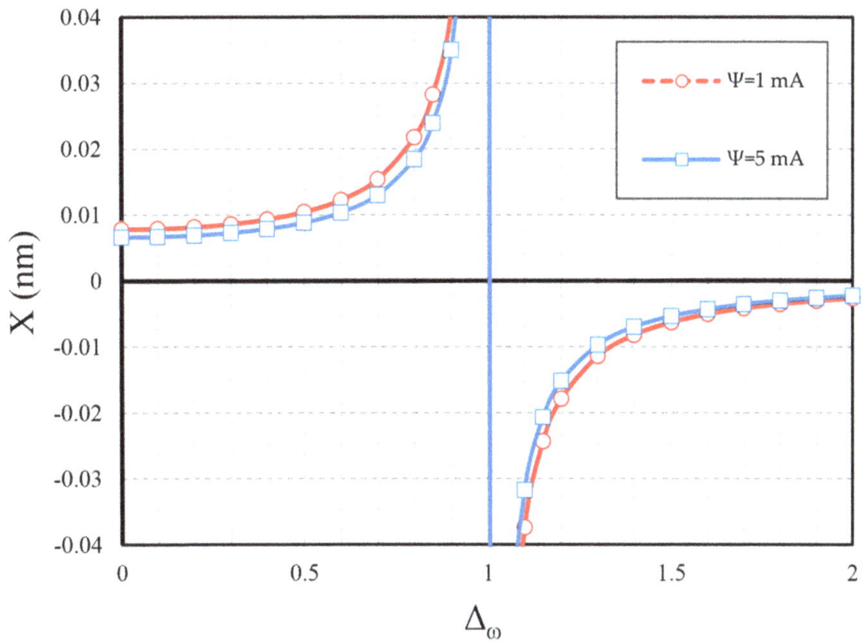

Figure 7.3. Frequency ratio versus dynamic deflection for different cases magnetic potential (l = 1 nm, e_0a = 0.5 nm, P_0 = 0.01 N m^{-1}, CC).

a frequency damper to keep materials and structures removed from the resonance, the CC boundary conditions can be destructive at high excitation frequencies. By examining the diagram further, another point we reach is that after the resonance region, the deflections in both boundary conditions lead to each other, and as a result, in very large cases of excitation and external vibrating forces, after the resonance, the dynamic deflections will be very small. Figures 7.3 and 7.4 will evaluate the effect of an almost stronger magnetic medium and an almost weaker magnetic medium on resonance, respectively. As can be seen by comparing the two figures, the difference in magnetic potential under SS boundary conditions will be more pronounced. It is worthy to mention that before resonance the deflections are in the opposite of the z-axis (see figure 7.1), although they are positively shown by all figures.

Contrarily, after resonance, as the vibration phase changes, the deflections are in line with the z-axis, however, they are negatively represented in all pictured diagrams.

It is also important to note that in stronger magnetic fields, the resonance region will be smaller, and since the higher magnetic potential leads to greater stiffness of the beam material, in this case the beam will jump to resonance, which is not a pleasant event for the material at all. From figure 7.2 this result was obtained for the CC boundary condition even in the case of a weaker magnetic field. Therefore, it can be concluded that whenever the stiffness of the beam and its degree of freedom of movement is lower for any reason, the beam will go to the resonance amplitude with

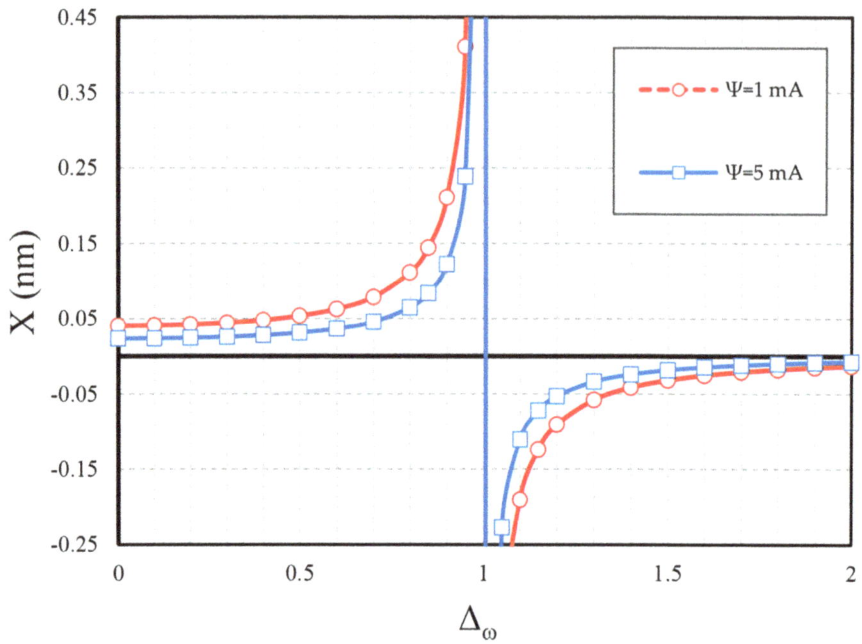

Figure 7.4. Frequency ratio versus dynamic deflection for different cases of magnetic potential (l = 1 nm, e_0a = 0.5 nm, P_0 = 0.01 N m^{-1}, SS).

an improper jump, which is very harmful to the material. As a final point from the last two figures, it can be seen that after the material passes through the resonance region, the effect of the magnetic field will be greatly reduced. This is due to the extreme proximity of the dynamic deflections resulting from both values of magnetic potentials in figures 7.3 and 7.4.

7.5.2 Magnetic field effect

Using this section and figure 7.5, we want to investigate the effect of magnetic field changes in the two boundary conditions CC and SS. As we already know [25], the increase in external magnetic potential leads to more stiffness of the magnetic material, for which one of its reasons can be the contraction and shrinkage of the material. In this way, we came to this conclusion again. As can be seen, when we enter from a magnetic field with a negative potential to a magnetic field with a positive potential, the stiffness of the material increases and the dynamic deflections decrease. But a new result that can be found through this figure is that the effect of the magnetic field and its changes on the CC boundary conditions will be much smaller than the SS boundary conditions. Therefore, although we will see the decreasing results by increasing the external potential in the CC boundary condition, the slope of the decrease in results will not be as noticeable as in SS.

Another important point is that the difference between the dynamic results for the two boundary conditions CC and SS in the negative potential will be greater than

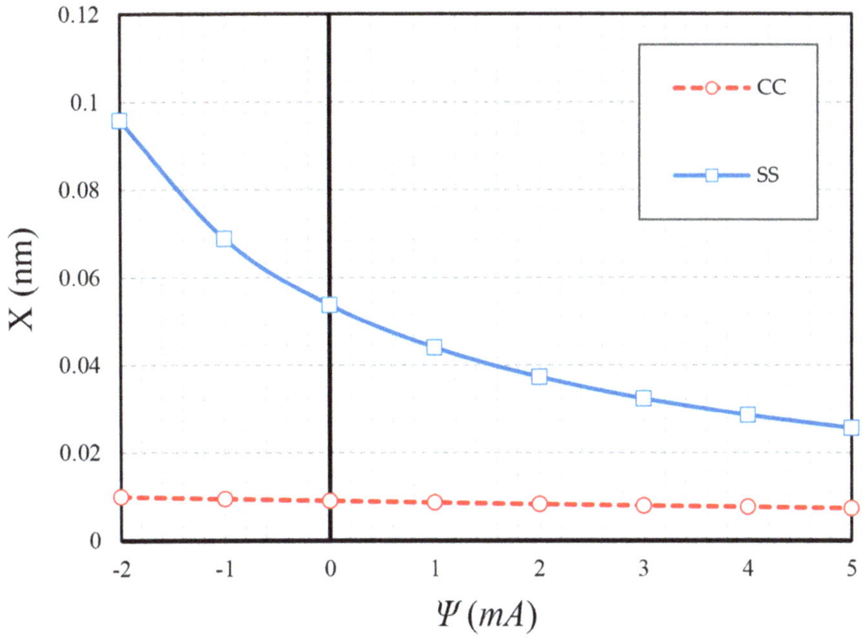

Figure 7.5. Magnetic potential versus dynamic deflection for different cases of edge conditions (l = 1 nm, e_0a = 0.5 nm, Δ_ω = 0.3, P_0 = 0.01 N m^{-1}).

the positive potential. Thus, finally, it can be concluded that in very strong magnetic fields with negative potential, the type of boundary condition will be very important, and vice versa, in very strong magnetic fields with positive potential, the type of boundary condition will not be an important issue.

7.5.3 Dynamic load impact

Figure 7.6 is presented to investigate the effect of the dynamic load transversely applied to the beam, which in turn can be the physical effect of the external frequency. As can be seen from the figure, increasing the amount of dynamic load leads to an increase in dynamic deflections, which will be much more noticeable in the SS boundary conditions. In a way, it can be said that the effect of increasing the dynamic load on the more flexible boundary conditions will be larger and in very large dynamic loads or significant external excitation frequencies, the type of boundary condition is very important.

7.5.4 Small-scale parameters effect

In this part of the present study, with the help of figures 7.7 and 7.8, we examine the effects of small-scale parameters used in NSGT theory. The study is in a state that resonance has not yet occurred. The results are plotted for both CC and SS boundary conditions. Figure 7.7 examines the effect of the nonlocal parameter.

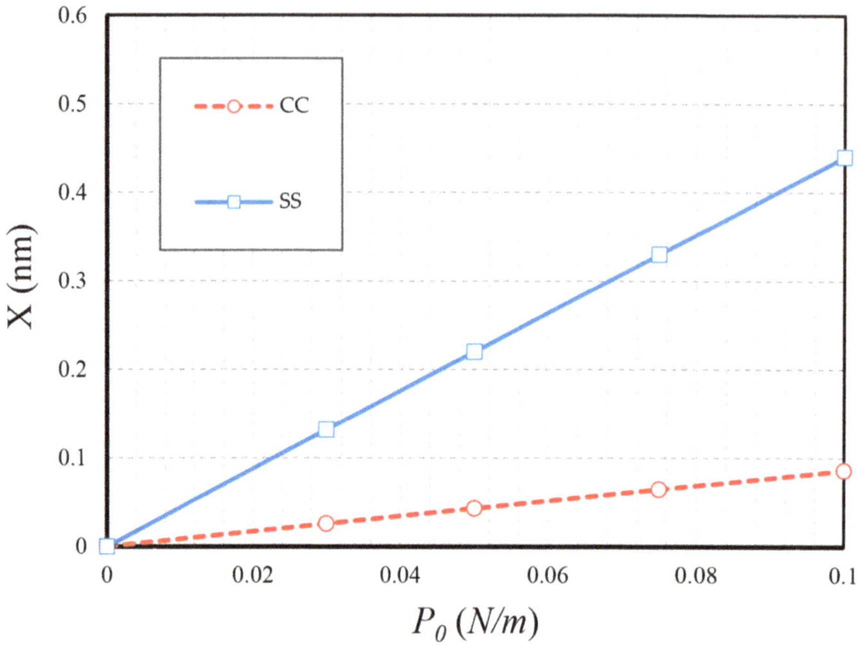

Figure 7.6. Dynamic load versus dynamic deflection for different cases of edge conditions ($\Psi = 1$ mA, $l = 1$ nm, e_0a = 0.5 nm, $\Delta_\omega = 0.3$).

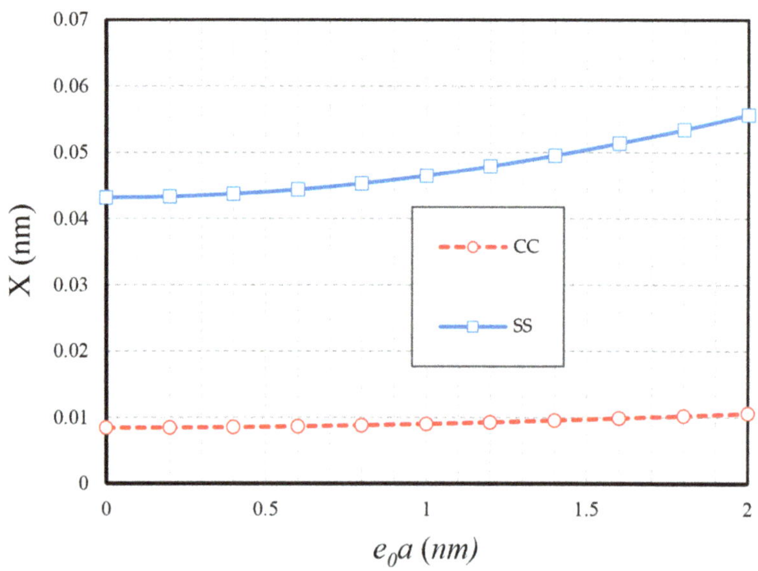

Figure 7.7. Nonlocal parameter versus dynamic deflection for different cases of edge conditions ($\Psi = 1$ mA, $l = 1$ nm, $P_0 = 0.01$ N m^{-1}, $\Delta_\omega = 0.3$).

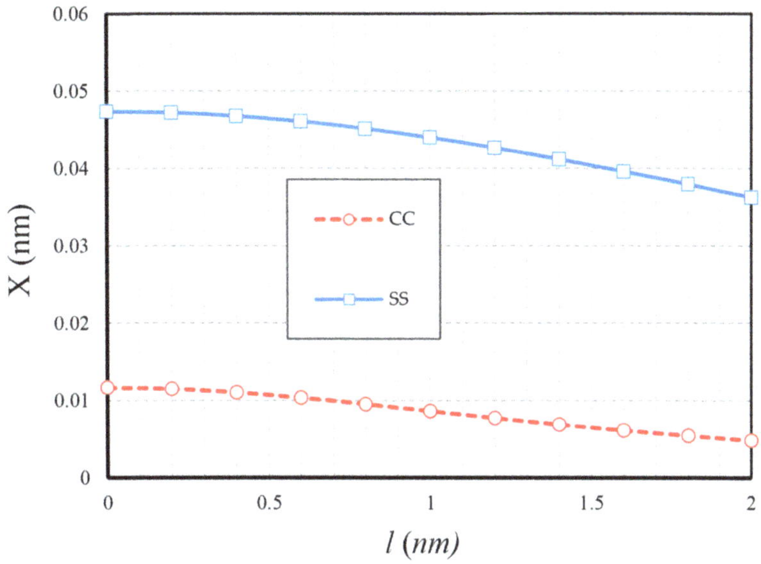

Figure 7.8. SG parameter versus dynamic deflection for different cases of edge conditions (Ψ = 1 mA, e_0a = 0.5 nm, P_0 = 0.01 N m^{-1}, Δ_ω = 0.3).

Using figure 7.8, we will examine the effect of the strain gradient (SG) parameter. As can be clearly seen, as the value of the nonlocal parameter increases, the dynamic deflections will also increase, and this increment will be clearer and more tangible in the SS boundary conditions. On the other hand, it is clear from figure 7.8 that increasing the numerical value of the SG parameter will reduce the dynamic deflections, where the downward trend in both the CC and SS boundary conditions seems to be almost similar. As a result, it can be said that the nonlocal parameter receives a greater effect from the boundary condition type, and this parameter is more sensitive to the boundary condition type than the SG parameter. Increasing the numerical values of the nonlocal parameter will lead to a decrease in the stiffness of the material and increasing the numerical values of the SG parameter will lead to a further stiffness of the beam material [25–27]. Finally, at larger values, the nonlocal parameter will make the boundary condition type more important.

7.6 Conclusion

The provided chapter inspected a new analysis on piezo-flexomagnetic nanoscale actuator beams. Beam modelling was in light of the Euler–Bernoulli beam and nanoscale modelling corresponded to the two functionalities already known for a nanosize structure, that is, nonlocal and strain gradient properties inserted in the relations via NSGT. The response of the beam in opposition to an excitation frequency was measured. In the case of the numerical outputs, the Galerkin weighted residual method reported the numerical values of the dynamic deflections based on the clamped and hinged conditions of two ends. In the wake of pictorial results, there were some notable points that can be outlined as:

- whenever there is a less flexible boundary condition, the structure goes into the resonance amplitude with a jump;
- the less flexible the boundary condition, the smaller the resonant region;
- the stronger the magnetic field, the shorter the resonance region;
- the impact of magnetic field is more pronounced for hinged boundary condition;
- negative magnetic potential made the boundary conditions more vital and vice versa, the positive magnetic potential is capable of making unimportant the boundary conditions if the potential value is chosen adequately big.

References

[1] Fahrner W 2005 *Nanotechnology and Nanoelectronics* (Berlin: Springer)

[2] Kabychenkov A F and Lisovskii F V 2019 Flexomagnetic and flexoantiferromagnetic effects in centrosymmetric antiferromagnetic materials *Tech. Phys.* **64** 980–3

[3] Eliseev E A, Morozovska A N, Glinchuk M D and Blinc R 2009 Spontaneous flexoelectric/flexomagnetic effect in nanoferroics *Phys. Rev.* B **79** 165433

[4] Lukashev P and Sabirianov R F 2010 Flexomagnetic effect in frustrated triangular magnetic structures *Phys. Rev.* B **82** 094417

[5] Yudin P V and Tagantsev A K 2013 Fundamentals of flexoelectricity in solids *Nanotechnology* **24** 432001

[6] Sidhardh S and Ray M C 2018 Flexomagnetic response of nanostructures *J. Appl. Phys.* **124** 244101

[7] Zhang N, Zheng S and Chen D 2019 Size-dependent static bending of flexomagnetic nanobeams *J. Appl. Phys.* **126** 223901

[8] Malikan M and Eremeyev V A 2020 Free vibration of flexomagnetic nanostructured tubes based on stress-driven nonlocal elasticity, Springer Nature, analysis of shells, plates, and beams *Adv. Struct. Mater.* **134**

[9] Malikan M and Eremeyev V A 2020 On the geometrically nonlinear vibration of a piezo-flexomagnetic nanotube *Math. Methods Appl. Sci.* https://doi.org/10.1002/mma.6758

[10] Malikan M, Uglov N S and Eremeyev V A 2020 On instabilities and post-buckling of piezomagnetic and flexomagnetic nanostructures *Int. J. Eng. Sci.* **157** 103395

[11] Malikan M and Eremeyev V A 2020 On nonlinear bending study of a piezo-flexomagnetic nanobeam based on an analytical-numerical solution *Nanomaterials* **10** 10091762

[12] Song X and Li S-R 2007 Thermal buckling and post-buckling of pinned–fixed Euler–Bernoulli beams on an elastic foundation *Mech. Res. Commun.* **34** 164–71

[13] Reddy J N 2010 Nonlocal nonlinear formulations for bending of classical and shear deformation theories of beams and plates *Int. J. Eng. Sci.* **48** 1507–18

[14] Moory-Shirbani M, Sedighi H M, Ouakad H M and Najar F 2018 Experimental and mathematical analysis of a piezoelectrically actuated multilayered imperfect microbeam subjected to applied electric potential *Compos. Struct.* **184** 950–60

[15] Ouakad H M and Sedighi H M 2019 Static response and free vibration of MEMS arches assuming out-of-plane actuation pattern *Int. J. Non Linear Mech.* **110** 44–57

[16] Chen W-R, Chen C-S and Chang H 2020 Thermal buckling analysis of functionally graded Euler–Bernoulli beams with temperature-dependent properties *J. Appl. Comput. Mech.* **6** 457–70

[17] Malikan M and Eremeyev V A 2020 On the dynamics of a visco-piezo-flexoelectric nanobeam *Symmetry* **12** 643

[18] Lim C W, Zhang G and Reddy J N 2015 A Higher-order nonlocal elasticity and strain gradient theory and its applications in wave propagation *J. Mech. Phys. Solids* **78** 298–313

[19] Malikan M and Eremeyev V A 2020 A new hyperbolic-polynomial higher-order elasticity theory for mechanics of thick FGM beams with imperfection in the material composition *Compos. Struct.* **249** 112486

[20] Malikan M, Eremeyev V A and Sedighi H M 2020 Buckling analysis of a non-concentric double-walled carbon nanotube *Acta Mech.* **231** 5007–20

[21] Li L, Han B, Zhang Q-C, Zhao Z-Y and Lu T J 2019 Dynamic response of clamped sandwich beams: analytical modeling *Theor. Appl. Mech. Lett.* **9** 391–6

[22] Malikan M, Nguyen V B and Tornabene F 2018 Damped forced vibration analysis of single-walled carbon nanotubes resting on viscoelastic foundation in thermal environment using nonlocal strain gradient theory *Eng. Sci. Technol. Int. J.* **21** 778–86

[23] Lu L, Guo X and Zhao J 2017 Size-dependent vibration analysis of nanobeams based on the nonlocal strain gradient theory *Int. J. Eng. Sci.* **116** 12–24

[24] Rajput A B, Hazra S and Nath Ghosh N 2013 Synthesis and characterisation of pure single-phase $CoFe_2O_4$ nanopowder via a simple aqueous solution-based EDTA-precursor route *J. Exp. Nanosci.* **8** 629–39

[25] Malikan M, Krasheninnikov M and Eremeyev V A 2020 Torsional stability capacity of a nano-composite shell based on a nonlocal strain gradient shell model under a three-dimensional magnetic field *Int. J. Eng. Sci.* **148** 103210

[26] Malikan M and Eremeyev V A 2020 Post-critical buckling of truncated conical carbon nanotubes considering surface effects embedding in a nonlinear Winkler substrate using the Rayleigh-Ritz method *Mater. Res. Express* **7** 025005

[27] Malikan M, Dimitri R and Tornabene F 2019 Transient response of oscillated carbon nanotubes with an internal and external damping *Compos. Part B: Eng.* **158** 198–205

IOP Publishing

Modeling and Computation in Vibration Problems, Volume 1
Numerical and semi-analytical methods
S Chakraverty, F Tornabene and J N Reddy

Chapter 8

Vibration of size-dependent carbon nanotube-based biosensors in liquid

Fatemeh Sheikhmamoo, Hamid Mohammad Sedighi and Mohammad Shishesaz

Nano-electromechanical systems have made significant advances in various sciences such as mechanics, medicine, and chemistry, due to their unique features and properties, especially in the fields of sensors and actuators. Therefore, it is vital to investigate the behavior of these systems, which are often triggered by electrostatic stimulation. The behavior of nanosized materials can be different from their behavior at the micro-level, so in classical theory, the effects of size are not taken into account; therefore, the modified coupling stress theory for modeling the size-dependent behavior nanotube biosensors are used. The Euler–Bernoulli beam theory has been used for biosensor modeling. The equations and boundary conditions are derived using the Hamiltonian principle. The electrostatic and Casimir forces are the source of the nonlinear equations in this problem. The lateral placement of the biosensor consists of two sections: static and dynamic. In this work, an analytical modified Adomian decomposition method is used to investigate the static response of the system. In the static section, the instability of the system and the pull-in voltage under the influence of surface tension effect and size effect are investigated. The validity of the static results is carried out by comparing the results with the results of previous studies. The main emphasis of this work is investigating the dynamic behavior of the system using the isogeometric method. The isogeometric method uses precise geometry for analysis. The inherent properties of shape functions can be useful in reducing computational cost in one-dimensional problems, compared to the finite element method. The obtained results indicate that the effect size parameter has a direct relation with the system frequency. Therefore, by increasing this parameter, the system frequency increases. It also reduces the frequency with the increase of electrostatic voltage, which increases the dynamic displacement of the system.

doi:10.1088/978-0-7503-3483-9ch8

© IOP Publishing Ltd 2021

8.1 Introduction

Nano-electromechanical systems are a set of sensors and actuators that have attracted the attention of many researchers in recent decades. Since sensors are a powerful tool for identifying biological molecules, they are now used in a variety of fields such as medicine, the chemical industry, the food industry, and manufacturing of pharmaceuticals, health products, and more [1]. One of the most important mechanisms of excitation of these structures is electrostatic excitation. The basis of systems working with this type of excitation is that by exposing the system to electric fields by applying a voltage between the stationary electrode and the moving electrode which is initially placed at a certain distance. If the voltage exceeds its critical value (pull-in voltage), the moveable electrode, which is the nano-beam, becomes unstable and pull-in is onto the ground. This phenomenon is called the phenomenon of pull-in instability. The voltage at which this phenomenon occurs is called the pull-in voltage. Therefore, it is necessary to calculate the pull-in voltage in the design of these types of systems. In recent years, researchers have done a wide range of studies on the instability of nano-systems due to many applications in the design of nanosensors.

Rahaeifard *et al* [2] numerically investigated the deflection and static pull-in voltage of microcantilevers based on the modified couple stress theory. It was shown that the couple stress theory can remove the gap between the experimental observations and the classical theory based simulations for the static pull-in voltage. Yin *et al* [3] investigated the size effect by using the modified couple stress theory on the behavior of electrostatic micro-actuators. With Kong *et al* [4], the static pull-in behavior of an electrostatically actuated Bernoulli–Euler microbeam was investigated based on a modified couple stress theory. The approximate analytical solutions to the pull-in voltage and pull-in displacement of the microbeam were derived by using the Rayleigh–Ritz method. Abdel-Rahman *et al* [5] presented a nonlinear model of a microbeam under electrostatic excitation. This model takes in the mid-plane stretching force. They considered both static deflections and dynamic characteristics of the microbeams. The main difference in nanoscale structures with materials in micro-size is in large surface area/volume ratio of the structure. Therefore, the effects of the surface layer such as surface residual stress and surface elasticity on the behavior of nano-structures cannot be neglected [6]. Koochi *et al* [7] investigated the influence of surface effects (the surface residual stress and surface elasticity) and Casimir force on the pull-in instability of a cantilever nano-actuator using the modified Adomian decomposition (MAD) method. Soroush *et al* [8] applied the MAD method to investigate the effect of the Casimir attraction and van der Waals force on pull-in instability of cantilever nano-actuators. Beni *et al* [9] using modified couple stress theory and MAD solving method, studied the static behavior of electrostatic nano-beams with elastic boundary conditions under the influence of Casimir force. Hosseini-Hashemi *et al* [10] considered the influence of surface on the nonlinear free vibration of an FG Euler–Bernoulli nanobeam using the nonlocal elasticity theory. The dynamic pull-in instability of nonlocal nano-bridges incorporating the surface effect and intermolecular forces was studied by Seddighi *et al* [11]. Lei *et al* [12] investigated the surface effects on the vibrational frequency of double-walled carbon nanotubes using the nonlocal Timoshenko beam model.

Different approaches can be used for free vibration analysis of engineering structures such as finite element method (FEM) and isogeometric analysis (IGA), and so on. The first time in 2005 the concept of IGA was presented by Hughes [13]. Reali (2006) and Cottrell *et al* (2005) studied the isogeometric structural vibration problems and compared their results with the classic finite element method [14, 15]. Weeger *et al* [16] investigated the vibrations of nonlinear Euler–Bernoulli beam utilizing the novel approach of isogeometric finite elements.

Zare *et al* [17] studied the stability of pipes influenced by a fluid using IGA. The problem formulation was developed based on the Euler–Bernoulli beam theory (EBT) and the plug-flow model for fluid forces. Milos *et al* [18] investigated the free vibrations of beam elements using IGA approach. The stiffness and mass matrices have been developed for rotation-free Bernoulli–Euler and Timoshenko beam using the Galerkin method. The natural frequencies of the beam element with the specific boundary conditions were computed using the isogeometric approach. Vinh Loc *et al* [19] investigated the elastostatic and dynamic behavior of Euler–Bernoulli beams using isogeometric approach. They used B-splines as the base functions for geometric representation and analysis. Shojaee *et al* [20] studied the isogeometric free vibration analysis of Kirchhoff plates using the standard Galerkin method. Stoykov *et al* [21] analyzed the dynamic behavior of beams under large displacements and made a comparison between models postulated by the isogeometric analysis and p-FEM. The equation of motion was obtained using the principle of virtual work, considering Timoshenko's theory and von-Karman geometric nonlinearity. The models obtained by IGA were compared with those of the FEM.

A complete search through the previous studies shows up to now, that the effect of size-dependent and surface layer in the presence of dispersion forces has not been the focus of interest to researchers. Therefore, for the first time, this study includes the effects of surface layer and size dependency of the nanotubes on the pull-in behavior of beam-type NEMS via considering the effects of surface stress on Gurtin and Murdoch theory as well as the size-dependent effects on the modified couple stress theory. The nano-biosensor is modeled as a cantilever beam while the MAD method is used to solve for the static deflection. The method of IGA is used to obtain natural frequencies of the bio-sensor for its transverse mode of vibration.

8.1.1 The modified couple stress theory

Figure 8.1 shows a schematic diagram of the cantilever nano-biosensor. Here, L, b, and h correspond to the nanostructured beam length, width, and thickness. The nano-biosensor (with carbon nanotubes (CNTs)) is subjected to the influence of the electrostatic force with fringing field effect and dispersion forces (Casimir force). H_1 is the initial distance between the nano-biosensor and the movable electrodes.

The modified couple stress theory is a continuous nonclassical theory that includes the effects of small scale in the mechanical behavior of nanoscale structures. Here, the coupling stress tensor is symmetric and only contains a length scale parameter. In this theory, the strain energy is a function of both strain tensor (conjugated with stress) and the curvature (conjugated with the couple stress).

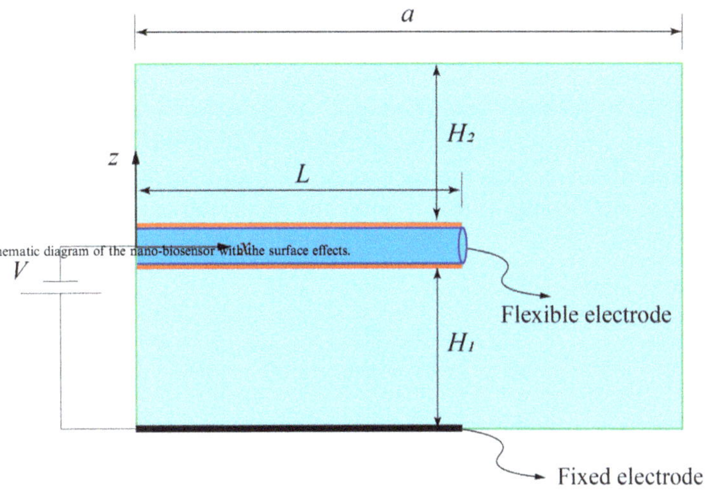

Figure 8.1. Schematic diagram of the nano-biosensor with the surface effects.

According to this theory, the strain energy of a homogeneous elastic material occupying a regional volume V is given by [22]:

$$U_b = \frac{1}{2}\iiint(\sigma_{ij}\varepsilon_{ij} + m_{ij}\chi_{ij})\, dV \tag{8.1}$$

Where, σ_{ij}, ε_{ij}, m_{ij}, and χ_{ij} are the components of stress, strain, deviatoric part of the couple stress tensor, and symmetric curvature tensors, respectively. These tensors can be defined as:

$$\sigma_{ij} = 2\mu\varepsilon_{ij} + \lambda tr(\varepsilon)\delta_{ij} \tag{8.2}$$

$$\varepsilon_{ij} = \frac{1}{2}(u_{i,j} + u_{j,i}) \tag{8.3}$$

$$m_{ij} = 2\mu l^2\chi_{ij} \tag{8.4}$$

$$\chi_{ij} = \frac{1}{2}(\theta_{i,j} + \theta_{j,i}) \tag{8.5}$$

where,

$$\lambda = \frac{Ev}{(1+v)(1-2v)} \qquad \mu = \frac{E}{2(1+v)} \tag{8.6}$$

In the above equations, λ and μ are the Lame's constants, l is the material length scale parameter, u_i *and* θ_i are the displacement components and rotation vectors, respectively. Components of the rotation vectors can be calculated using the following equations.

$$\theta_i = \frac{1}{2}\, curl(u_i) \tag{8.7}$$

Many theories have been proposed that explain how these structures can be treated. The oldest and best-known beam theory is Bernoulli–Euler beam theory (or the classical beam theory). This theory is used for beams with a high length-to-other-dimensions ratio. According to this theory, the displacement field is defined as [23]:

$$\begin{aligned} u_1(x, z, t) &= -z\frac{\partial w(x, t)}{\partial x} \\ u_2(x, z, t) &= 0 \\ u_3(x, z, t) &= w(x, t) \end{aligned} \tag{8.8}$$

Where u_1, u_2, and u_3 are the components of displacement vector of a point with coordinates (x, y, z) on the cross-section of the beam in x, y and z directions, respectively. w is the component of displacement that is defined in equation (8.8). Using equations (8.3) and (8.8), the components of the strain tensor can be expressed as:

$$\begin{aligned} \varepsilon_{xx} &= -z\frac{\partial^2 w}{\partial x^2} \\ \varepsilon_{xy} &= \varepsilon_{xz} = \varepsilon_{yz} = \varepsilon_{yy} = \varepsilon_{zz} = 0 \end{aligned} \tag{8.9}$$

Also, using equations (8.3) and (8.8), components of the strain tensor can be expressed as:

$$\theta_x = \theta_z = 0, \quad \theta_y = -\frac{\partial w}{\partial x} \tag{8.10}$$

Additionally, the non-zero components of the curvature tensor, as well as the coupling stress tensor, are expressed in equations (8.11) and (8.12).

$$\chi_{xy} = -\frac{1}{2}\frac{\partial^2 w}{\partial x^2} \tag{8.11}$$

$$m_{xy} = -\mu l^2\frac{\partial^2 w}{\partial x^2} \tag{8.12}$$

Now, using equation (8.2) in conjunction with the strain components defined in equation (8.9), the components of the classical stress tensor can be represented by equations (8.13)–(8.15).

$$\sigma_{xx} = \frac{E\ (1-\upsilon)}{(1-\ 2\,\upsilon)(1+\upsilon)}\left(-z\frac{\partial^2 w}{\partial x^2}\right) \tag{8.13}$$

$$\sigma_{yy} = \sigma_{zz} = \frac{E\upsilon}{(1-2\upsilon)(1+\upsilon)}\left(-z\frac{\partial^2 w}{\partial x^2}\right) \tag{8.14}$$

$$\sigma_{xz} = \sigma_{yz} = \sigma_{xy} = 0 \tag{8.15}$$

For a slender beam with a large aspect ratio and ($b < 5h$), the Poisson effect is very small and can be neglected [24]. As a result, the above equations are recast in a simplified form as;

$$\begin{aligned} &\sigma_{xx} = E\left(-z\frac{\partial^2 w}{\partial x^2}\right) \\ &\sigma_{yy} = \sigma_{zz} = \sigma_{xy} = \sigma_{xz} = \sigma_{yz} = 0 \end{aligned} \tag{8.16}$$

Therefore, based on the above relations, according to the modified coupling stress theory, the strain energy of the nanobeam is:

$$\begin{aligned} U_b &= \frac{1}{2}\int_0^L \int_A (\sigma_{xx}\varepsilon_{xx} + 2m_{xy}\chi_{xy})\, dAdx \\ &= -\frac{1}{2}\int_0^L (M_x + Y_{xy})\frac{\partial^2 w}{\partial x^2}dx \\ &= \frac{1}{2}\int_0^L (EI + \mu A l^2)\left(\frac{\partial^2 w}{\partial x^2}\right)^2 dx \end{aligned} \tag{8.17}$$

8.1.2 Surface elasticity theory

Considering the effects of the surface stress, based on the theory of Gurtin and Murdoch, the constitutive equations for surface layers are expressed as follows [25].

$$\begin{aligned} &\tau_{ij} = \mu_0(u_{i,j} + u_{j,i}) + (\lambda_0 + \tau_0)u_{k,k}\delta_{ij} + \tau_0(\delta_{ij} - u_{j,i}) \\ &\tau_{ni} = \tau_0\,(u_{n,i}) \end{aligned} \tag{8.18}$$

Here, i and j denote the in-plane Cartesian coordinates indices of the surface and n corresponds to the out-of-plane Cartesian coordinate of the surface. Substituting equation (8.9) in equation (8.18), the non-zero components of the surface stress tensor are obtained as follows.

$$\begin{aligned} &\tau_{xx} = \tau_0 + E_0\left(-z\frac{\partial^2 w}{\partial x^2}\right) \\ &\tau_{zx} = \tau_0\frac{\partial w}{\partial x} \end{aligned} \tag{8.19}$$

where μ_0 and λ_0 are the surface elastic constants, and τ_0 is the residual surface stress. Additionally, E_o is equal to;

$$E_0 = \lambda_0 + 2\mu_0 \tag{8.20}$$

Based on the surface elasticity theory, the strain energy of the surface layer with zero thickness can be expressed as [25]:

$$U_s = \frac{1}{2}\int_0^L \oint_{\partial A} (\tau_{ij}\varepsilon_{ij} + \tau_{ni}\, u_{n,i})\, dA\, dx \tag{8.21}$$

Substituting equations (8.18)—(8.19) into equation (8.21), the strain energy in the surface layer is obtained as follows.

$$\begin{aligned} U_s &= \frac{1}{2}\int_0^L \oint_{\partial A} \left[\left(\tau_0 + E_0\left(-z\frac{\partial^2 w}{\partial x^2}\right)\right)\left(-z\frac{\partial^2 w}{\partial x^2}\right) + \tau_0 n_z^{\,2}\left(\frac{\partial w}{\partial x}\right)^2\right] dA\, dx \\ &= \frac{1}{2}\int_0^L \left[\tau_0 S_0\left(\frac{\partial w}{\partial x}\right)^2 + E_0 I_0\left(\frac{\partial^2 w}{\partial x^2}\right)^2\right] dx \end{aligned} \tag{8.22}$$

For a rectangular cross-section, I_0, S_0 are defined according to equation (8.23) [25].

$$\begin{aligned} I_0 &= \oint z^2 dA = \frac{bh^2}{2} + \frac{h^3}{6} \\ S_0 &= \oint n_z^2\, dA = 2b \end{aligned} \tag{8.23}$$

Accordingly, the strain energy of the nanobeam and its surface layers is obtained as:

$$U_T = \frac{1}{2}\int_0^L \left[(EI + \mu A l^2 + E_0 I_0)\left(\frac{\partial^2 w}{\partial x^2}\right)^2 + (\tau_0 S_0)\left(\frac{\partial w}{\partial x}\right)^2\right] dx \tag{8.24}$$

Also, the kinetic energies of the nanobeam and its surface layers are obtained as follows:

$$T_{\text{beam}} = \frac{1}{2}\int_0^L \int_A \rho\left(\frac{\partial w}{\partial t}\right)^2 dA\, dx \tag{8.25}$$

$$T_{\text{surface layer}} = \frac{1}{2}\int_0^L \oint_{\partial A} \rho_0\left(\frac{\partial w}{\partial t}\right) ds\, dx \tag{8.26}$$

Regarding the external forces per unit length, the work done by these forces can be written as:

$$w_{\text{ext}} = \int_0^L q(x)\, w(x)\, dx \tag{8.27}$$

In the above equation, $q(x)$ is defined as the effective external force and is equal to the sum of following force components;

$$q(x) = q_{\text{elec}} + q_{\text{cas}} + q_{\text{damp}} + q_{\text{pressure}} \tag{8.28}$$

Here, q_{elec} is the electrostatic force, q_{cas} and q_{damp} are the intermolecular Casimir and fluid damping forces, and $q_{pressure}$ is the fluid loading on the nanobeam. These force components can be calculated follows.

Applying a voltage between the moving and fixed electrodes creates a force and also bending of the moving electrode. Considering the fringing field correction, the electrostatic force per unit length exerted on the nano-biosensor is defined by [26]:

$$q_{elec} = \frac{K\varepsilon_0 b V^2}{2(H_1 - w(x, t))^2}\left(1 + 0.65\frac{(H_1 - w(x, t))}{b}\right) \tag{8.29}$$

In this equation, H_1 is the initial distance between the beam and fixed electrode. Additionally, $w(x, t)$ represents the transverse deflection of the beam cross-section, $\varepsilon_0 = 8.854 \times 10^{-12} c^2/Nm^2$ is the named permittivity of vacuum, V is the voltage applied to the nano-biosensor, K and is the dielectric coefficient of the gap medium. by transferring the structures to micro- space and also nano-space, forces such as the Casimir force have effects on the structures. the Casimir force per unit length of the beam can be defined as [27].

$$q_{cas} = \frac{\pi^2\ \bar{h}\ c\ b}{240(H_1 - w(x, t))^4} \tag{8.30}$$

where $\bar{h} = 1.055 \times 10^{-34}$ is the Planck's constant divided by 2π and $c = 2.998 \times 10^8 m\ s^{-1}$ is the speed of light.

It should be noted that the liquid around the nanotubes can affect the vibrational properties of the nanotube. The main force initiating from this medium is attributed to the damping force. This effect is approximated by a damping coefficient C_d (per unit length) multiplied by $\frac{\partial w}{\partial t}$ [28]. As a result, the equivalent damping force is expressed as:

$$q_{damp} = -C_d\frac{\partial w}{\partial t} \tag{8.31}$$

In this equation, the damping coefficient per unit length is expressed as:

$$C_d = \mu_f b^3\left(\frac{1}{H_1^3} + \frac{1}{H_2^3}\right) \tag{8.32}$$

where, μ_f is the fluid viscosity.

Now, utilizing the Hamilton principle, the differential equation for the lateral deflection of the system can be derived by applying equation (8.33) [22]:

$$\int_0^t (\delta T - (\delta U_b + \delta U_s) + \delta w_{ext})dt = 0 \tag{8.33}$$

δT = variation in the total kinetic energy
δU_b = variation in the beam strain energy.
δU_s = variation in the surface layer strain energy.
δW_{ext} = variation in the external work

Substituting equations (8.24)–(8.31) into equation (8.33), the equation of motion for the lateral vibrations of the nano-biosensor is written as:

$$\begin{aligned}\delta w = 0 \rightarrow (\rho A + \rho_0 S_0)\frac{\partial^2 w}{\partial t^2} + C_d\,\frac{\partial w}{\partial t} + (EI + \mu A l^2 + E_0\,I_0)\frac{\partial^4 w}{\partial x^4} - \tau_0 S_0 \frac{\partial^2 w}{\partial x^2} \\ = \frac{\varepsilon_0\,bV^2}{2(\,H_1 - w(x,\,t)\,)^2}\left(1 + 0.65\,\frac{(H_1 - w(x,\,t)\,)}{b}\right) + \frac{\pi^2\bar{h}cb}{240\,(H_1 - w(x,\,t)\,)^4} \\ + b\,(P_2 - P_1)\end{aligned} \tag{8.34}$$

It is observed that the bending deformation of the beam consists of three parts: bending deformation due to the classical theory (associated with EI), $\mu A l^2$ in conjunction with the modified coupling stress theory, and bending deformation of the surface layer (associated with $E_0 I_0$).

Following the complete application of equation (8.33), the boundary conditions of the nano-cantilever beam are obtained as:

$$\begin{aligned} & w(0,\,t) = \frac{\partial w}{\partial t}(0,\,t) = 0 \\ & \frac{\partial^2 w}{\partial x^2}(L,\,t) = 0 \\ & (EI + E_0 I_0 + \mu A l^2)\frac{\partial^3 w}{\partial x^3}(L,\,t) - (\tau_0 S_0)\,\frac{\partial w}{\partial x}(L,\,t) = 0 \end{aligned} \tag{8.35}$$

To normalize the governing equation and its boundary conditions, the following dimensionless quantities are defined.

$$\begin{aligned} & X = \frac{x}{a} \quad W = \frac{w}{H_1} \quad \tau = \frac{t}{t^*} \quad t^* = \sqrt{\frac{\rho A a^4}{EI}} \\ & \eta = \frac{\rho_0 S_0}{\rho A} \quad \varphi = \frac{\tau_0 s_0\, a^2}{EI} \quad \delta_1 = \frac{\mu A l^2}{EI} \quad \delta_2 = \frac{E_0 I_0}{EI} \\ & c = \frac{C_d\, a^4}{EIt^*} \quad \gamma = \frac{0.65 H_1}{b} \quad \beta = \frac{K\varepsilon_0 b L^4 V^2}{2H_1^3 EI} \quad \alpha = \frac{\pi^2\bar{h}cbL^4}{240\,EIH_1^5} \end{aligned} \tag{8.36}$$

Thus, one can recast equations (8.34) and (8.35) in a dimensionless form as:

$$\begin{aligned} (1+\eta)\frac{\partial^2 W}{\partial \tau^2} + c\frac{\partial W}{\partial \tau} + (1+\delta)\frac{\partial^4 W}{\partial X^4} - \varphi\frac{\partial^2 W}{\partial X^2} = \frac{\alpha}{(1-W)^4} \\ + \frac{\beta}{(1-W)^2} + \frac{\gamma\,\beta}{(1-W)} \end{aligned} \tag{8.37}$$

$$\begin{aligned} & W(0,\,\tau) = \frac{\partial W(0,\,\tau)}{\partial X} = 0 \\ & \frac{\partial^2 W(L,\,\tau)}{\partial X^2} = 0, \quad (1+\delta)\frac{\partial^3 W(L,\,\tau)}{\partial X^3} - \varphi\frac{\partial W(L,\,\tau)}{\partial X} = 0 \end{aligned} \tag{8.38}$$

The lateral displacement, $W(X, \tau)$, consists of two parts; a static part as well as a dynamic part. That is:

$$W(X, \tau) = W_s(X) + W_d(X, \tau) \tag{8.39}$$

Where $W_s(X)$ corresponds to the static deflection of the nano-biosensor and $W_d(X, \tau)$ denotes the dynamic deflection about $W_s(X)$. Substituting equation (8.39) in equation (8.34), the nonlinear static deflection of the nanobeam can be obtained by deleting the time-dependent terms equation (8.34). As a result:

$$(1 + \delta)\frac{\partial^4 W_s}{\partial X^4} - \varphi\frac{\partial^2 W_s}{\partial X^2} = \frac{\alpha}{(1 - W_s)^4} + \frac{\beta}{(1 - W_s)^2} + \frac{\gamma\beta}{(1 - W_s)} \tag{8.40}$$

Also, the nonlinear dynamic deflection of the nano-biosensor can be obtained by solving the following equation.

$$\begin{aligned}&(1 + \eta)\frac{\partial^2 W_d}{\partial \tau^2} + c\frac{\partial W_d}{\partial \tau} + (1 + \delta)\frac{\partial^4 W_d}{\partial X^4} - \varphi\frac{\partial^2 W_d}{\partial X^2}\\ &= \frac{\alpha}{(1 - (W_s + W_d))^4} + \frac{\beta}{(1 - (W_s + W_d))^2} + \frac{\beta\gamma}{(1 - (W_s + W_d))}\\ &+ \frac{ba^4}{EIH_1}(P_2 - P_1)\end{aligned} \tag{8.41}$$

Using a Taylor series expansion (TSE) about $W_d(X, \tau) = 0$, the nonlinear electrostatic and the Casimir forces are:

$$\begin{aligned}q_{\text{elec}} &= \frac{\beta}{[1 - W_s(X) - W_d(X, \tau)]^2} + \frac{\beta\gamma}{[1 - W_s(X) - W_d(X, \tau)]}\\ &\simeq \frac{\beta}{[1 - W_s(X)]^2} + \frac{\beta\gamma}{[1 - W_s(X)]} + \left[\frac{2\beta}{[1 - W_s(X)]^3} + \frac{\beta\gamma}{[1 - W_s(X)]^2}\right]W_d(X, \tau)\end{aligned} \tag{8.42.a}$$

$$q_{\text{cas}} = \frac{\alpha}{[1 - W_s(X) - W_d(X, \tau)]^4} \simeq \frac{\alpha}{[1 - W_s(X)]^4} + \frac{4\alpha}{[1 - W_s(X)]^5}W_d(X, \tau) \tag{8.42.b}$$

Substituting equations (8.42.a) and (8.42.b) into equation (8.41) and removing the static part of the resulting equation, the equation for the dynamic deflection of the nano-biosensor can be obtained as follows:

$$\begin{aligned}&(1 + \eta)\frac{\partial^2 W_d}{\partial \tau^2} + c\frac{\partial W_d}{\partial \tau}(1 + \delta) + \frac{\partial^4 W_d}{\partial X^4} - \varphi\frac{\partial^2 W_d}{\partial X^2}\\ &= \frac{\alpha}{(1 - (W_s + W_d))^4} + \frac{\beta\gamma}{(1 - (W_s + W_d))} + \frac{\beta}{(1 - (W_s + W_d))^2}\\ &+ \frac{ba^4}{EIH_1}(P_2 - P_1)\end{aligned} \tag{8.43}$$

8.1.3 Formulation of the fluid pressure on a nano-biosensor

To model the fluid the following assumptions are made.

(a) Fluid flow is a type of potential flow.

(b) The fluid is considered to be incompressible, viscid, and irrotational.

To calculate the hydrodynamic pressure of the fluid flow, we consider the Bernoulli equation as follows [29]:

$$\frac{\partial \phi}{\partial t} + \frac{1}{2}\vec{V}\vec{V} + \frac{P}{\rho_f} - \vec{g}\,\vec{r} = c(t) \tag{8.44}$$

Considering the linear terms, regardless of the gravity effects, the fluid pressures in the upper and lower regions are equal to:

$$P_1 = -\frac{\rho_f}{t^*}\left(\frac{\partial \phi_1}{\partial \tau}\right)_{y_1=1} \tag{8.45.a}$$

$$P_2 = -\frac{\rho_f}{t^*}\left(\frac{\partial \phi_2}{\partial \tau}\right)_{y_2=0} \tag{8.45.b}$$

where ϕ_1 , ϕ_2 denote the velocity potential of the fluid in the lower and upper regions, respectively, and ρ_f is the fluid density. The fluid motion caused by the vibration of the nanotube in each region can be obtained from the Laplace equation using the corresponding velocity potential functions as:

$$\nabla^2\phi_1\ (X, y_1, \tau) = 0 \qquad 0 < X < 1, \quad 0 < y_1 < 1 \tag{8.46.a}$$

$$\nabla^2\phi_2\ (X, y_2, \tau) = 0 \qquad 0 < X < 1, \quad 0 < y_2 < 1 \tag{8.46.b}$$

The velocity potential functions must also satisfy the boundary conditions in regions 1 and 2. These conditions are introduced in the sequel.

8.1.3.1 Fluid boundary conditions in region 1

1. For rigid walls, the fluid velocity is zero, that is:

$$V_x = \frac{\partial \phi_1}{\partial x} = 0 \quad x = 0,\ 1 \tag{8.47.a}$$

$$V_y = \frac{\partial \phi_1}{\partial y_1} = 0 \quad y_1 = 0 \tag{8.47.b}$$

2. For the part of the fluid in contact with the nanobeam, equal velocities exist between the two mediums at their sharing surface. This indicates that [29]:

$$\frac{1}{H_1}\left(\frac{\partial \phi_1}{\partial y_1}\right)_{y_1=1} = \begin{cases} -\dfrac{H_1}{t^*}\dfrac{\partial W_d}{\partial \tau} & 0 < x < \dfrac{L}{a} \\ \dfrac{1}{H_2}\left(\dfrac{\partial \phi_2}{\partial y_2}\right)_{y_2=0} & \dfrac{L}{a} < x < 1 \end{cases} \tag{8.47.c}$$

8.1.3.2 Fluid boundary conditions in region 2

The boundary conditions for the fluid motion in region 2 are similar to those in region 1. These conditions and are stated as [29]:

$$\frac{\partial \phi_2}{\partial x} = 0 \quad x = 0,\ 1 \tag{8.48.a}$$

$$\frac{\partial \phi_2}{\partial y_2} = 0 \quad y_2 = 1 \tag{8.48.b}$$

$$\frac{1}{H_2}\left(\frac{\partial \phi_2}{\partial y_2}\right)_{y_2=0} = \begin{cases} -\dfrac{H_1}{t^*}\dfrac{\partial W_d}{\partial \tau} & 0 < x < \dfrac{L}{a} \\ \dfrac{1}{H_1}\left(\dfrac{\partial \phi_1}{\partial y_1}\right)_{y_1=1} & \dfrac{L}{a} < x < 1 \end{cases} \tag{8.48.c}$$

Now, considering the boundary conditions, velocity potential functions are obtained as follows:

$$\phi_1(x, y_1, t) = \sum_{\infty}^{i=1} A_i(t)\cos\lambda_i x \cosh \beta_i y_1 \tag{8.49.a}$$

$$\phi_2\ (x, y_2, t) = \sum_{\infty}^{i=1} E_i(t)\cos\lambda_i x[\cosh\gamma_i y_2 - \tanh\gamma_i\ \sinh\gamma_i y_2] \tag{8.49.b}$$

where

$$\begin{aligned} \lambda_i &= i\pi \\ \gamma_i &= \frac{i\pi H_2}{a} \\ \beta_i &= \frac{i\pi H_1}{a} \end{aligned} \tag{8.50}$$

$$\frac{1}{H_1}\sum_{\infty}^{i=0} A_i(\tau)\beta_i \cos\lambda_i x \sinh\beta_i = \begin{cases} -\dfrac{H_1}{t^*}\displaystyle\sum_{\infty}^{i=0}\dfrac{\partial W_d}{\partial \tau} & 0 < x < \dfrac{L}{a} \\ -\dfrac{1}{H_2}\displaystyle\sum_{\infty}^{i=0}\gamma_i E_i(\tau)\ (\tanh\gamma_i)\ \cos\lambda_i x & \dfrac{L}{a} < x < 1 \end{cases} \tag{8.51.a}$$

$$-\frac{1}{H_2}\sum_{\infty}^{i=0}\gamma_i\ E_i(\tau)\cos\lambda_i x\ \tanh\gamma_i = \begin{cases} -\dfrac{H_1}{t^*}\displaystyle\sum_{\infty}^{i=0}\frac{\partial W_d}{\partial \tau} & 0 < x < \dfrac{L}{a} \\ \dfrac{1}{H_2}\displaystyle\sum_{\infty}^{i=0}\beta_i A_i(\tau)\ (\sinh\beta_i)\cos\lambda_i x & \dfrac{L}{a} < x < 1 \end{cases} \qquad (8.51.b)$$

$$-\frac{1}{H_2}\sum_{\infty}^{i=0}\gamma_i E_i(\tau)\cos\lambda_i x\ \tanh\gamma_i = \begin{cases} -\dfrac{H_1}{t^*}\displaystyle\sum_{\infty}^{i=0}\frac{\partial W_d}{\partial \tau} & 0 < x < \dfrac{L}{a} \\ \dfrac{1}{H_2}\displaystyle\sum_{\infty}^{i=0}\beta_i A_i(\tau)\ (\sinh\beta_i)\ \cos\lambda_i x & \dfrac{L}{a} < x < 1 \end{cases} \qquad (8.51.c)$$

Using the orthogonal property of trigonometric functions:

$$\int_0^1 \cos\lambda_i x.\cos\lambda_j x\ dx = \begin{cases} \dfrac{1}{2} & i = j \\ 0 & i \neq j \end{cases} \qquad (8.52)$$

$$\frac{1}{2}\sum_{\infty}^{i=0}A_i(\tau)H_2t^*\ \beta_i\sinh\beta_i + \sum_{\infty}^{i=0}H_2H_1^2\frac{\partial W_d}{\partial \tau}\alpha_{ij} + \sum_{\infty}^{i=0}H_1t^*\gamma_iE_i(\tau)\ \mu_{ij}\tanh\gamma_i = 0 \qquad (8.53.a)$$

$$-\frac{1}{2}\sum_{\infty}^{i=0}H_1t^*\gamma_iE_i(\tau)\ \tanh\gamma_i + \sum_{\infty}^{i=0}H_2H_1^2\frac{\partial W_d}{\partial \tau}\alpha_{ij} - \sum_{\infty}^{i=0}A_i(\tau)H_2t^*\beta_i\ \mu_{ij}\sinh\beta_i = 0 \qquad (8.53.b)$$

$$\begin{aligned} \mu_{ij} &= \int_{L/a}^{1} \cos(\lambda_i x)\cos(\lambda_j x)\ dx \\ \alpha_{ij} &= \int_0^{L/a} \cos(\lambda_i x)N_j(x)\ dx \end{aligned} \qquad (8.54)$$

The above equations can be recast in a matrix form as:

$$[U]\{A\} + [B]\{E\} + [D]\left\{\frac{\partial W_d}{\partial \tau}\right\} = 0 \qquad (8.55)$$

$$[R]\{A\} + [P]\{E\} - [D]\left\{\frac{\partial W_d}{\partial \tau}\right\} = 0 \qquad (8.56)$$

The elements of the introduced matrices in equations (8.55) and (8.56) are defined as:

$$\begin{aligned} U_{ij} &= \frac{H_2 t^* \beta_i \sinh\beta_i}{2} \\ B_{ij} &= H_1 t^* \gamma_i \mu_{ij} \tanh\gamma_i \\ D_{ij} &= H_2 H_1^2 \alpha_{ij} \\ R_{ij} &= H_2 t^* \beta_i \mu_{ij} \sinh\beta_i \\ P_{ij} &= \frac{H_1 t^* \gamma_i \tanh\gamma_i}{2} \end{aligned} \tag{8.57}$$

Now, solving equations (8.55) and (8.56), the matrices $\{A\}$ and $\{E\}$ are obtained as shown in equations (8.58.a) and (8.58.b).

$$\{A\} = [-U^{-1}D + U^{-1}B(R\,U^{-1}D + D)(RU^{-1}B - P)^{-1}]\left\{\frac{\partial W_d}{\partial \tau}\right\} \tag{8.58.a}$$

$$\{E\} = -[(R\,U^{-1}D + D)(RU^{-1}B - P)^{-1}]\left\{\frac{\partial W_d}{\partial \tau}\right\} \tag{8.58.b}$$

Additionally, using the potential functions, the parametric pressures can be introduced as follows.

$$P_1 = -\frac{\rho_f}{t^*}\dot{A}\cos\lambda_i x \cosh\beta_i \tag{8.59a}$$

$$P_2 = -\frac{\rho_f}{t^*}\dot{E}\cos\lambda_i x \tag{8.59b}$$

Finally, the vibrational equation is obtained by replacing the pressure terms as shown in equation (8.60).

$$\begin{aligned} &(1 + \eta - A^*)\frac{\partial^2 W_d}{\partial \tau^2} + c\frac{\partial W_d}{\partial \tau} + (1 + \delta)\frac{\partial^4 W_d}{\partial X^4} - \varphi\frac{\partial^2 W_d}{\partial X^2} \\ &- \left(\frac{\gamma\beta}{(1 - W_s)^2} + \frac{2\beta}{(1 - W_s)^3} + \frac{4\alpha}{(1 - W_s)^5}\right)W_d = 0 \end{aligned} \tag{8.60}$$

Here, A^* is equal to:

$$\begin{aligned} A^* = &-\frac{b\,a^4\rho_f}{E\,I\,t^*\,H_1}(\cos\lambda_i x)\big[(RU^{-1}D + D)(RU^{-1}B - P)^{-1} \\ &+ (\cosh\beta_i)(U^{-1}D + U^{-1}B(U^{-1}D + D)(RU^{-1}B - P)^{-1})\big] \end{aligned} \tag{8.61}$$

8.2 Solution

8.2.1 The static pull-in instability of the biosensor using the MAD method

Since the electrostatic and Casimir forces have non-linear nature, solving the static equilibrium equation is complex and time-consuming. Therefore, the MAD method, which is a powerful method for solving nonlinear equations, is used in most cases. To apply MAD for the analysis of the static pull-in instability, we use $y(x) = 1 - w_s(x)$ as the transformation function to recast equation (8.40) into (8.42). Finally, the static equation and boundary conditions can be rewritten as:

$$\frac{\partial^4 y}{\partial x^4} - \psi\frac{\partial^2 y}{\partial x^2} = -\frac{1}{(1+\delta)}\left(\frac{\alpha}{y(x)^4} + \frac{\beta}{y(x)^2} + \frac{\beta\gamma}{y(x)}\right) \tag{8.62}$$

$$\begin{aligned} &y(0) = 1 && y'(0) = 0 \\ &y''\ (L) = 0 && \varphi\, y'(L) - \ (1+\delta)\ \ y'''\ (L) = 0 \end{aligned} \tag{8.63}$$

where,

$$\psi = \frac{\varphi}{1+\delta} \tag{8.64}$$

Now, defining the differential and integral operators as:

$$L^{(n)} = \frac{d^{(n)}}{dx^{(n)}}() \tag{8.65.a}$$

$$L^{(-n)} = \int_0^x \ldots \int_0^x ()dx\ldots dx \tag{8.65.b}$$

equation (8.40) is recast into the following form:

$$L^4\big[y_n(x)\big] = \psi L^2\big[y_n(x)\big] + N(x,\, y(x)) \tag{8.66}$$

The nonlinear terms include the intermolecular Casimir and electrostatic forces as the nonlinear function $N(x, y(x))$ [30] where;

$$N\ (x,\, y(x)) = \sum_{n=0}^{\infty} A_n = \sum_{v=1}^{n} C(v,\, n) h_v(y_0) \tag{8.67}$$

Here, A_n are defined as a series of Adomian polynomials that are derived from the following equations [30, 31].

$$\begin{aligned} &C(v,\, n) = \sum_{pi} \prod_{i=1}^{v} \frac{1}{k_i!} y_{p_i}^{k_i} \quad \sum_{i=1}^{v} k_i\, p_i = n \\ &n > 0 \quad 0 \leqslant i \leqslant n \quad 1 \leqslant p_i \leqslant n - v + 1 \end{aligned} \tag{8.68.a}$$

$$h_v(y_0) = \frac{d^v}{d{y_0}^v}(f(y_0)) \tag{8.68.b}$$

where k_i correspond to the number of repetitions in y_{p_i}. Now, applying four successive integrations on the left side of equation (8.44) we can write:

$$\iiiint \frac{\partial^4 y}{\partial x^4}\, dx = y(x) + \alpha_0 + \alpha_1\, x + \frac{1}{2!} c_1\, x^2 + \frac{1}{3!} c_2\, x^3 \tag{8.69}$$

$$\sum_{\infty}^{n=0} y_n(x) = \alpha_0 + \alpha_1 x + \frac{1}{2!}c_1 x^2 + \frac{1}{3!}c_2 x^3 + \psi L^{-2}\left[\sum_{\infty}^{n=0} y_n(x)\right] + L^{-4}\left[\sum_{\infty}^{n=0} A_n\right] \tag{8.70}$$

Applying the boundary conditions on the left side of the beam (clamped end), we can obtain the constant coefficients α_0 , α_1 as:

$$y(0) = \alpha_0 = 1 \qquad y'(0) = \alpha_1 = 0 \tag{8.71}$$

One of the most important parts of the MAD method is the determination of the Adomian polynomials. For this problem, these polynomials are obtained as follows:

$$A_0 = N(y_0) \quad \text{or} \quad N(y_0) = -\frac{1}{(1+\delta)}\left(\frac{\alpha}{{y_0}^4} + \frac{\beta}{{y_0}^2} + \frac{\beta\gamma}{y_0}\right) \tag{8.72.a}$$

$$\begin{aligned} A_1 &= C(1,\,1)h_1(y_0) = y_1 N'(y_0) \quad \text{or} \\ N'(y_0) &= \frac{1}{(1+\delta)}(4\alpha {y_0}^{-5} + 2\beta {y_0}^{-3} + \beta\gamma {y_0}^{-2}) \end{aligned} \tag{8.72.b}$$

$$\begin{aligned} A_2 &= C(1,\,2)h_1(y_0) + C(2,\,2)h_2(y_0) \\ &= y_2 N'(y_0) + \frac{1}{2!}{y_1}^2 N''(y_0) \\ \text{or } N''(y_0) &= -\frac{1}{(1+\delta)}(20\alpha {y_0}^{-6} + 6\beta {y_0}^{-4} + 2\beta\gamma {y_0}^{-3}) \end{aligned} \tag{8.72.c}$$

$$\begin{aligned} A_3 &= C(1,\,3)h_1(y_0) + C(2,\,3)h_2(y_0) + C(3,\,3)h_3(y_0) \\ &= y_3 N'(y_0) + y_1 y_2 N''(y_0) + \frac{1}{3!}{y_1}^3 N'''(y_0) \\ \text{or } N'''(y_0) &- \frac{1}{(1+\delta)}(120\alpha {y_0}^{\ 7} + 24\beta {y_0}^{-5} + 6\beta\gamma {y_0}^{-4}) \end{aligned} \tag{8.72.d}$$

Additionally, the expressions for y_o, …. y_n are equal to [32].

$$y_0 = \alpha_0 = 1 \tag{8.73.a}$$

$$y_1 = \frac{1}{2!}c_1 x^2 + \frac{1}{3!}c_2 x^3 + \psi L^{-2}[y_0] + L^{-4}[A_0] \tag{8.73.b}$$

$$y_n = \psi L^{-2}[y_{n-1}] + L^{-4}[A_{n-1}] \qquad n \geqslant 2 \tag{8.73.c}$$

On substituting relations (8.52.a–8.52d) into equation (8.55), the static deflection of the nano-biosensor can be approximated as:

$$w_s(x) = 1 - y(x)$$

$$w_s(x) = 1 - \sum_{\infty}^{n=0} y_n(x) = 1 - y_0 - y_1 - y_2 - y_3 - \dots$$

$$\begin{aligned}
w_s(x) = &-\frac{1}{2!}(c_1+\psi)x^2 - \frac{1}{3!}c_2x^3 \\
&-\frac{1}{4!}\left[(c_1\psi+\psi^2) - \frac{1}{1+\delta}(\alpha+\beta+\beta\gamma)\right]x^4 - \frac{1}{5!}c_2\psi x^5 \\
&-\frac{1}{6!}\frac{1}{1+\delta}[(c_1+\psi)(4\alpha+2\beta+\beta\gamma) - \psi(\alpha+\beta+\beta\gamma)]x^6 \\
&-\frac{1}{6!}(c_1\psi^2+\psi^3)x^6 - \frac{1}{7!}\left[\frac{1}{1+\delta}(4\alpha+2\beta+\beta\gamma)c_2 + c_2\psi^2\right]x^7 \\
&-\frac{1}{8!}\frac{1}{1+\delta}[-6(c_1+\psi)^2(10\alpha+3\beta+\beta\gamma) + 2\psi(c_1+\psi)(4\alpha+2\beta+\beta\gamma)]x^8 \\
&-\frac{1}{8!}\frac{1}{1+\delta}\left[-\psi^2(\alpha+\beta+\beta\gamma) - \frac{1}{1+\delta}(4\alpha+2\beta+\beta\gamma)(\alpha+\beta+\beta\gamma)\right]x^8 \\
&-\frac{1}{9!}\frac{c_2}{1+\delta}[2\psi(4\alpha+2\beta+\beta\gamma) - 20(\psi+c_1)(10\alpha+3\beta+\beta\gamma)]x^9 \\
&-\frac{1}{10!}\frac{1}{1+\delta}[-20c_2^2\psi(10\alpha+3\beta+\beta\gamma)]x^{10} \\
&-\frac{1}{10!}\frac{1}{(1+\delta)^2}\left[\begin{pmatrix}(4\alpha+2\beta+\beta\gamma)^2(c_1+\psi) - 2\psi(4\alpha+2\beta+\beta\gamma)(\alpha+\beta+\beta\gamma) \\ +30(\psi+c_1)(10\alpha+3\beta+\beta\gamma)(\alpha+\beta+\beta\gamma)\end{pmatrix}\right]x^{10} \\
&-\frac{1}{11!}\frac{1}{(1+\delta)^2}[(4\alpha+2\beta+\beta\gamma)^2c_2 + 70c_2(10\alpha+3\beta+\beta\gamma)(\alpha+\beta+\beta\gamma)]x^{11} \\
&+\frac{1}{12!}\frac{1}{(1+\delta)^3}[(4\alpha+2\beta+\beta\gamma)^2(\alpha+\beta+\beta\gamma) + 70(10\alpha+3\beta+\beta\gamma)(\alpha+\beta+\beta\gamma)^2]x^{12} + \dots
\end{aligned} \tag{8.74}$$

where constants c_1 and c_2 can be determined by solving equation (8.74) and the boundary conditions at $x = L$.

8.2.2 Dynamic deflection of the biosensor using isogeometric analysis

In this section, the isogeometric approach will be used to solve the dynamic deflection of the nano-biosensor. Hence, we briefly describe the basis functions and curves of B-spline and the Non-Uniform Rational B-Splines (NURBS).

8.2.3 Knot vectors and basic functions

A B-spline curve is defined by a polynomial of degree p, the number of control points n, and a knot vector that is a non-decreasing sequence of parameter value [33]. In other words:

$$\Xi = \{\xi_1, \xi_2, \ldots, \xi_{n+p+1}\} \qquad i = 1, 2, \ldots, n+p+1 \tag{8.75}$$

In equation (8.75), ξ_i are called the knots, and Ξ is the set of coordinates in the parametric space. Mainly, there are two kinds of knot vector, namely periodic and open, that are used and both types may be either uniform or non-uniform. A knot vector is said to be open if its first and last knots are repeated $p + 1$ times and basis functions generated from open knot vectors are interpolatory at the ends of the parametric domain interval.

According to the definition of a knot vector, B-spline base functions are defined by the Cox–DiBor recursive formula as follows [33].

For $p = 0$

$$N_{i,0} = \begin{cases} 1 & \text{if} \quad \xi_i \leqslant \xi \leqslant \xi_{i+1} \\ 0 & \text{otherwise} \end{cases} \tag{8.76}$$

For $p = 1, 2, 3, \ldots$, the B-spline base functions are defined as:

$$N_{i,p}(\xi) = \frac{\xi - \xi_i}{\xi_{i+p} - \xi_i} N_{i,p-1}(\xi) + \frac{\xi_{i+p+1} - \xi}{\xi_{i+p+1} - \xi_{i+1}} N_{i+1,p-1}(\xi) \tag{8.77}$$

where one can point out some of the important features of the B-spline base functions as:

1. Partition of unity: For an open knot vector $\sum_{i=1}^{n} N_{i,p}(\xi) = 1$.
2. Non-negativity: $N_{i,p}(\xi) \geqslant 0$.
3. Continuity: The base functions $N_{i,p}(\xi)$ are p-times continuously differentiable (C^p-continuous) inside a knot span, while at the inner knots of multiplicity $k (k \leqslant p)$, only $p - k$ times.
4. Compact support: $\text{supp}(N_{i,p}) = [\xi_i, \xi_{i+p+1})$, $i = 1, \ldots n$.

The kth derivative of a base function $N_{i,p}(\xi)$ is defined as:

$$\frac{d^k}{d\xi^k} N_{i,p}(\xi) = \frac{p}{\xi_{i+p} - \xi_i} \left(\frac{d^{k-1}}{d\xi^{k-1}} N_{i,p-1}(\xi) \right) + \frac{p}{\xi_{i+p+1} - \xi_{i+1}} \left(\frac{d^{k-1}}{d\xi^{k-1}} N_{i+1,p-1}(\xi) \right) \tag{8.78}$$

Figures 8.2 and 8.3 illustrate the B-spline basis functions and their first derivative, respectively.

8.2.4 B-spline curve

The B-spline curves can be expressed by determining the degree of the basis functions, the control points, and the knot vector. B-spline curves are constructed in the one-dimensional space of linear combinations of base functions and control points as follows [33–35]:

$$X(\xi) = \sum_{i=1}^{n} N_{i,p}(\xi) B_i \quad i = 1, \ldots, n \tag{8.79}$$

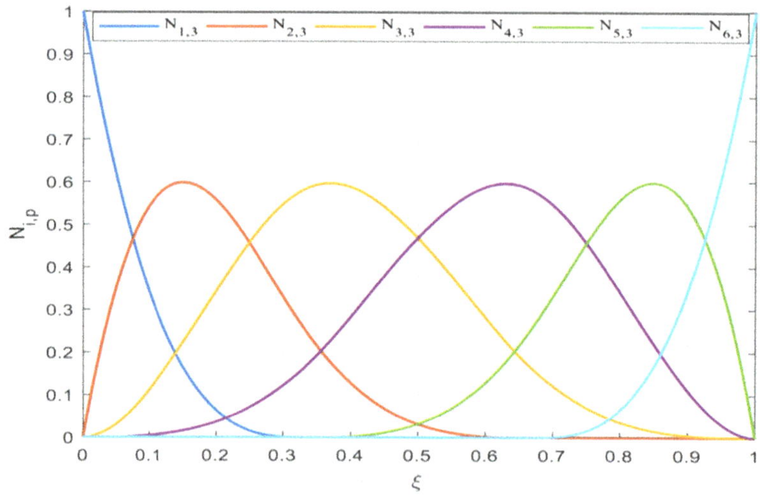

Figure 8.2. B-spline basis functions for p = 3, n = 6, Ξ = {0, 0, 0, 0, 1/3, 2/3, 1, 1, 1, 1}.

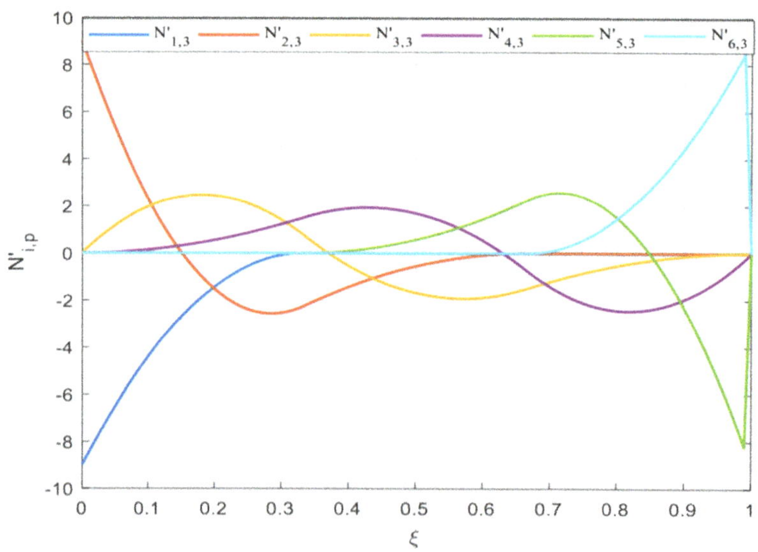

Figure 8.3. The first derivative of B-spline basis functions.

8.2.4.1 Isogeometric weak formulation of the dynamic deflection of biosensor

The formulation of the dynamic deflection of the biosensor was presented in equation (8.60). Using Galerkin's approximation, the weighting functions $W_d(X, \tau)$ are defined as:

$$W_d(X,\tau)=\sum_{i=1}^{n}N_i(X)q_i(\tau) \tag{8.80}$$

where N_i is the ith B-spline basis function of the space coordinate ξ, $q_i(\tau)$ is the ith field variable at each control point that depends only on time. Substituting equation (8.80) into equation (8.60) we obtain the discretized matrix relationship as follows.

$$\sum_{n}^{i=1}M_{ij}\ddot{q}_j(\tau)+\sum_{n}^{i=1}C_{ij}\dot{q}_i(\tau)+\sum_{n}^{i=1}K_{ij}q_i(\tau)=0 \tag{8.81}$$

where,

$$M_{ij}=\int_0^1\Bigg((1+\eta)-\frac{ba^4\rho_f}{EI\overset{*}{t}H_1}(\cos\lambda_i x)\left[\begin{matrix}(RU-1D+D)(RU-1B-P)^{-1}+\\(\cos h\beta_i)(-U^{-1}D+U^{-1}B(RU^{-1}D+D)(RU^{-1}B-P)^{-1})\end{matrix}\right]\Bigg)N_iN_jdX \tag{8.82a}$$

$$C_{ij}=\int_0^1 c\,N_iN_j\,dX \tag{8.82.b}$$

$$K_{ij}=\int_1^0(1+\delta)\,N_i''\;N_i''\;dX+\int_1^0\varphi\,N_i'\;N_j'\,dX-\int_1^0\left(\frac{\beta\gamma}{[1-W_s(X)]^{2}}+\frac{2\beta}{[1-W_s(X)]^{3}}+\frac{4\alpha}{[1-W_s(X)]^{5}}\right)N_iN_jdX \tag{8.82c}$$

The general solution of the vibrational equation above can be written as:

$$q_i(\tau)=\bar{q}e^{i\,\omega\,\tau} \tag{8.83}$$

Substituting (8.83) into (8.81), natural frequencies of beams can be calculated by solving eigenvalue equation

$$(-\,\omega^2\,M_{ij}+K_{ij}-i\omega\,C_{ij})\,\bar{q}\;=\{0\} \tag{8.84}$$

8.3 Results and discussion

8.3.1 Validate of the static analysis

To validate the method of the solution presented in this section, the results based on the modified couple stress theory and the MAD method are compared with the results presented in [7]. As observed in figure 8.4, it can be concluded that there is a good agreement between the results of present work and those of [7]. Figure 8.4 shows the variations in w_s versus x in system location for the applied voltages from the initial value of zero to that of pull-in voltage.

To study the static instability behavior of the biosensor, equation (8.40) has been solved utilizing the analytical method (modified Adomian analysis) and the effects of β, τ_0, γ, α parameters have been investigated. The dimensions and physical characteristics of the model are given in table 8.1.

The influences of the dimensionless Casimir force and the surface layer on the pull-in voltage of the nano-biosensor are shown in figure 8.5. In this figure, one can observe that β_{PI} decreases with the increase in α. Additionally, increasing the Casimir force reduces the dimensional voltage parameter. Moreover, the presence of the surface layer and the Casimir force causes the pull-in voltage to be higher than a case in which the presence of the surface layer is ignored.

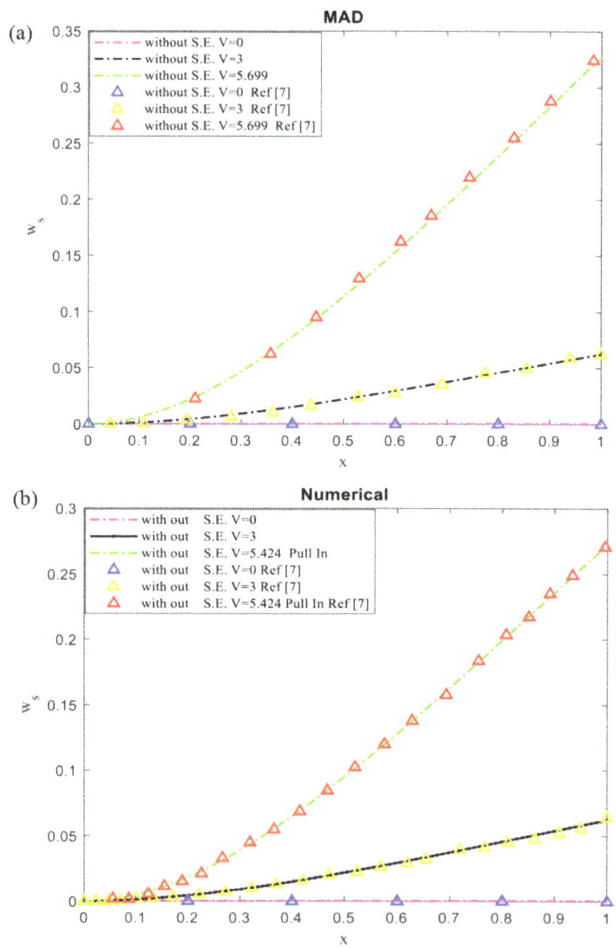

Figure 8.4. Deflection of a cantilever nano-biosensor for different applied voltages, (a) MAD, (b) numerical.

Table 8.1. The data used in the calculations.

Parameters	Value
Nanobeam length, L	1000 nm
Container length, a	1200 nm
Nanobeam width, b	250 nm
Nanobeam thickness, h	50 nm
Nanobeam mass density, ρ	10490 (kg m^3)
Liquid depth of domain 1, H_1	50 nm
Liquid depth of domain 2, H_2	80 nm
Young's modulus, E	76 GPa
Surface stress, τ_0	0.89 (N m^{-1})
Surface modulus, E_0	1.22 (N m^{-1})
Dielectric coefficient of the fluid	80
Fluid density, ρ_f	1000 (kg m^3)

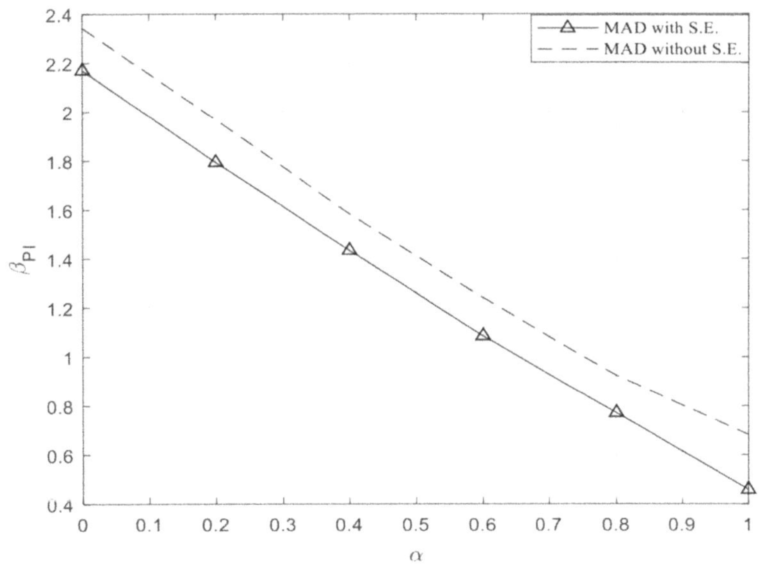

Figure 8.5. The effects of dimensionless Casimir force on β_{PI} with and without the surface effect, ($\delta_1 = 0.5$, $\tau_0 = 0.89$, $E_0 = 1.22$).

Figure 8.6 shows the effect of fringing field on the dimensionless voltage and the final displacement of the nanobeam in the presence of the surface layer and considering the size effect. According to the fringing field effect, this parameter, $\gamma = 0.65\ \frac{H_1}{b}$ is directly related to $\frac{H_1}{b}$. In other words, narrowing the beam and increasing the distance between the two electrodes increase this parameter. It can be seen from figure 8.6. that the effect of the fringing field has a significant effect on the final displacement and pull-in voltage of the nanobeam. In other words, increasing

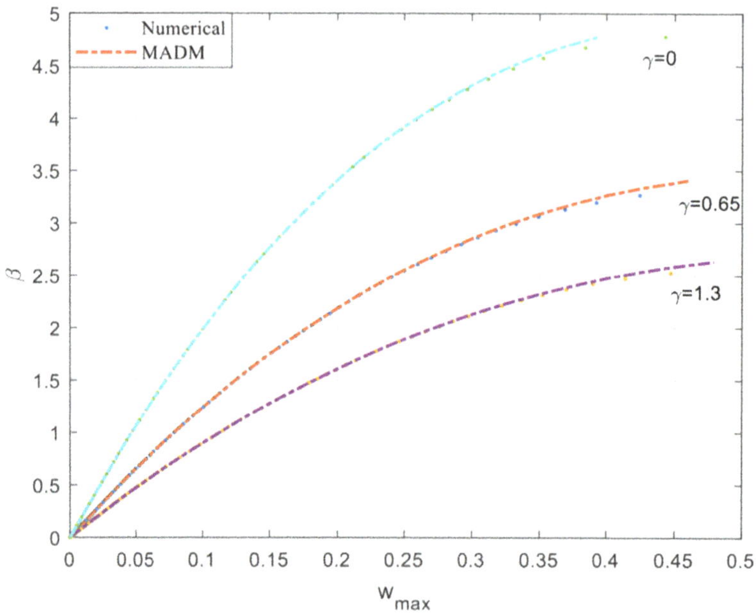

Figure 8.6. Influence of the fringing field effect on the dimensionless voltage for various values γ, ($\delta_1 = 0.5$, τ_0=0.89, $E_0 = 1.22$).

the fringing field parameter reduces the dimensional voltage and increases the displacement of the beam.

Figure 8.7. shows the influence of classical theory and the modified couple stress theory on the pull-in voltage as a function of length. According to these results, increasing the length of the nanobeam causes the system to converge to the pull-in instability at lower voltages. In other words, by increasing the length of the nanobeam, the differences between the classical theory and the modified couple stress decrease.

8.3.2 Validation of the dynamic analysis

To validate and compare the results obtained by IGA method with those of previously published data in [36] table 8.2 is generated. The results in this table show that the proposed method has good accuracy and can be used for the solution of such problems. To investigate the convergence of the system frequency, the corresponding results versus the number of control points were plotted in figure 8.8. In this figure, the vertical axis represents the first frequency of the system and the horizontal axis represents the number of control points. Results in this figure indicate that at control point 13, convergence to a specific value is obtained. Additionally, figure 8.9 is plotted to investigate the effect of beam wetting and compare and validate the results deduced from equations (8.59a) and (8.59b). (related to the pressure term) with those in [37]. In this figure, the results obtained on the first mode of vibration are plotted. As observed, a good agreement exists between these results.

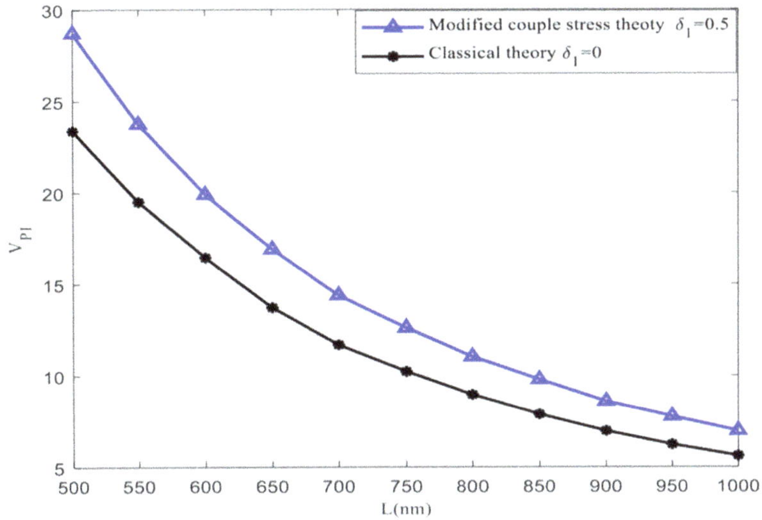

Figure 8.7. Influence of the size effect and length of the nanobeam on the pull-in voltage, ($\tau_0 = 0.89$, $E_0 = 1.22$).

Table 8.2. Validation and comparison of dimensionless frequency result with [36].

Mode of vibration	Reference [36]	Present work	Percentage error
First	3.5160	3.5165	0.014
Second	22.034	22.0219	0.054
Third	61.6909	61.6174	0.110

The surface effects on the system frequency are plotted in figure 8.10. In this figure, the vertical axis represents the frequency of the system and the horizontal axis represents the material length scale parameter. According to these results, the surface effect increases the frequency of the system. It is also observed that the length scale parameter is directly proportional to the frequency of the system. Increasing this parameter increases the strain energy of the system and hence, increasing the magnitude of the stiffness matrix. Therefore, examining the structure at the micro- and nano-levels, it is necessary to take into account the surface effects, which are observed in figure 8.10.

Figure 8.11. illustrates the effects of electrostatic voltage on the system frequency. The results show that as the electrostatic voltage increases, the system frequency decreases. It should be noted that the instability occurs when the system frequency becomes zero. This occurs at $V = 5$. Figure 8.12 shows the effect of fringing field parameter on the frequency of the nano-biosensor for selected values of the nanobeam thickness. As observed, the increase in fringing field parameter reduces the frequency of the nano-biosensor, resulting in a sudden drop to zero and hence,

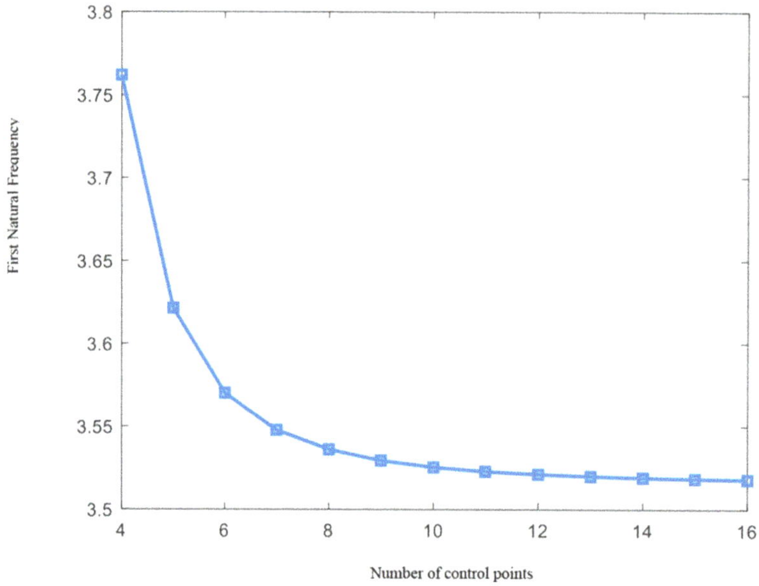

Figure 8.8. The convergence of first system frequency with the increase in the number of control points.

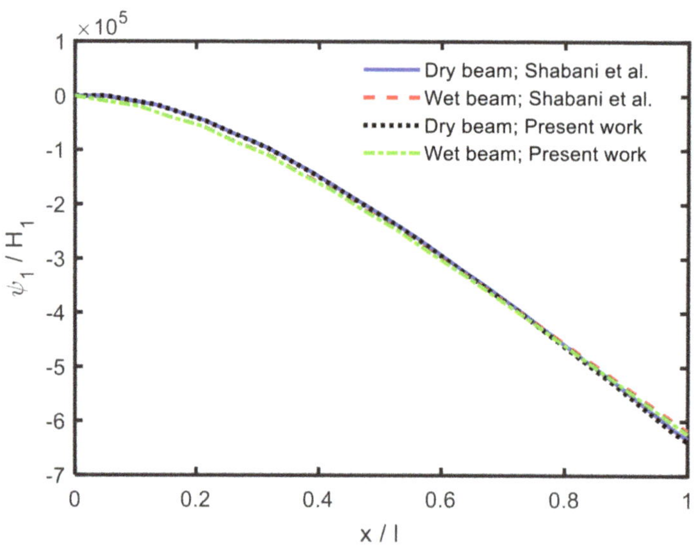

Figure 8.9. Comparison of the first mode of vibration for dry and wet nanobeams.

initiating dynamic instability of the system. The resulting instability decreases with an increase in the thickness of the nanobeam. Thus, increasing the thickness of the nanobeam helps to stabilize the system.

Furthermore, the effect of nonlocal theory on the variation of system frequencies is investigated and the corresponding results are shown in figure 8.13. According to

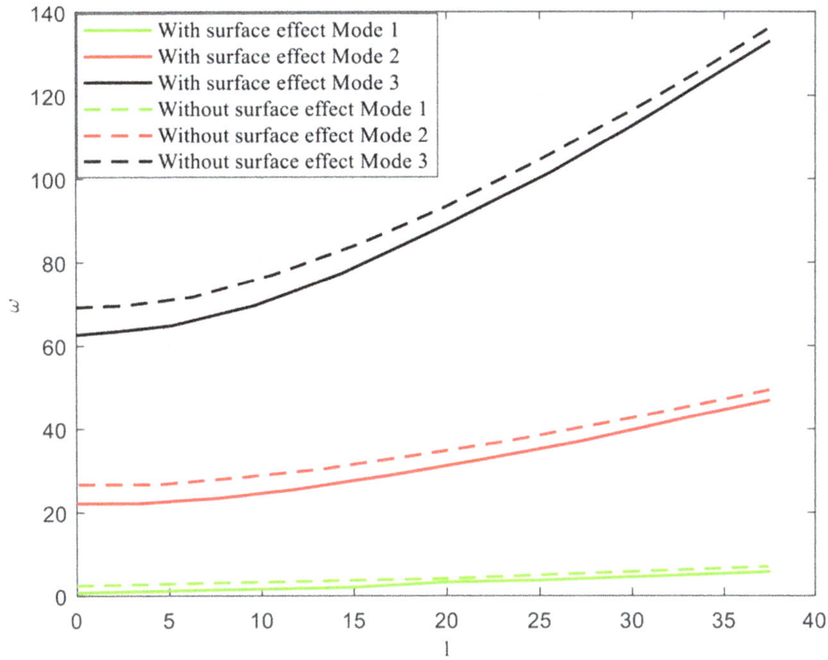

Figure 8.10. The effects of surface on the frequencies of the system.

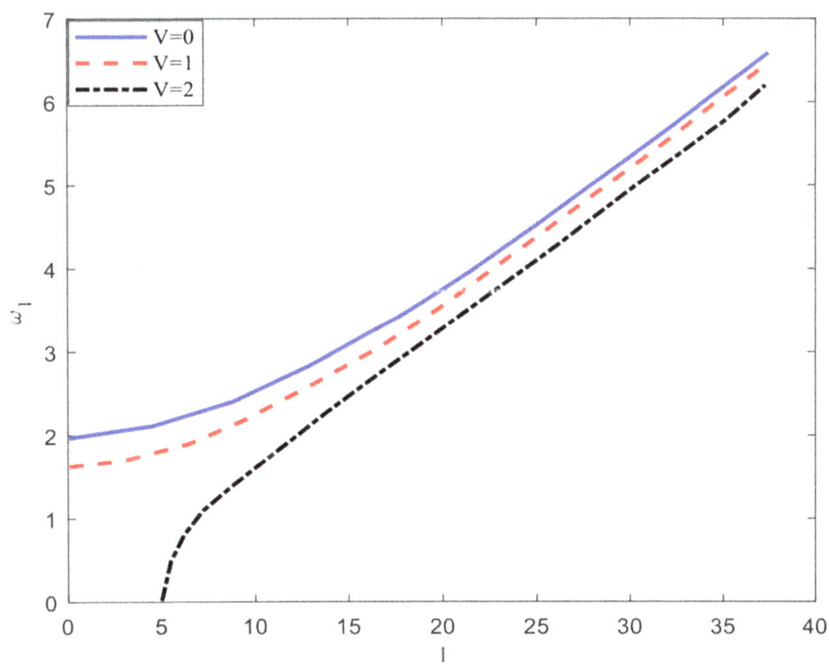

Figure 8.11. The influence of electrostatic voltage on the first frequency of the system.

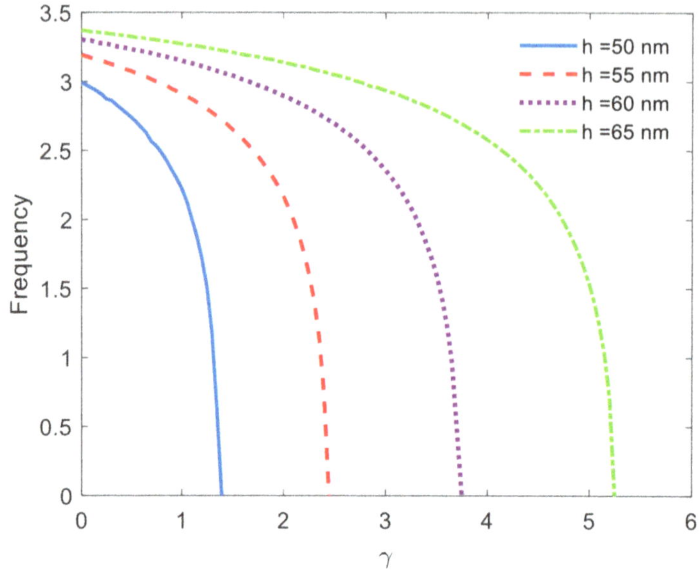

Figure 8.12. The influence of fringing field parameter on the first frequency of the system.

these findings, the modified coupling stress theory predicts higher frequency values than the classical theory. The differences between these theories appear to be considerable showing the necessity of using nonlocal theories for behavioral investigation of such problems. Also, increasing the dimensional voltage results in a reduction in system frequency and hence, pushing the system toward instability.

8.4 Conclusion

In this chapter, the static and vibrational behavior of a cantilever nano-biosensor submerged in a liquid was investigated. The applied forces consisted of Casimir, electrostatic, and fluid pressure. To account for the effects of size and surface strain energy on the behavior of nano-biosensor, the modified couple stress theory was used in conjunction with the Gurtin–Murdoch surface elasticity theory. Due to nonlinearity in the electrostatic and the Casimir forces, the MAD method was adopted to solve the static nonlinear equation. Additionally, IGA was utilized to study the vibrational behavior of the submerged biosensor. The influences of different parameters such as the initial gap between electrodes, size dependency, surface effects, and pull-in voltage on the frequency of the system were examined. The initial gap between the beam and fixed electrode were shown to have an important influence on the frequency of the system such that increasing this gap reduced the system frequency. Further results showed that increasing this gap increases the fringing field parameter. The existence of this parameter has a significant effect on the final displacement and the static pull-in voltage. This means that increasing the fringing field parameter reduces the pull-in voltage and hence increases the final displacement of the nanobeam. Additionally, the presence of the Casimir intermolecular force causes the system to lose its stability faster. In other

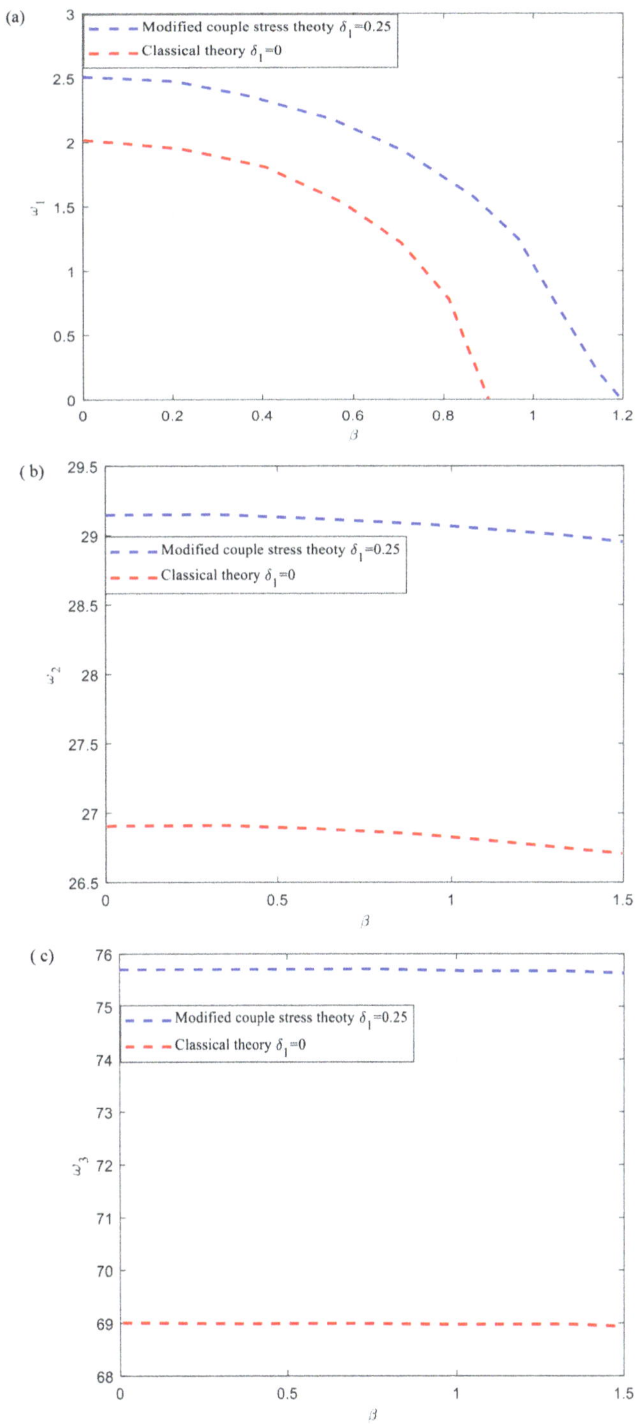

Figure 8.13. The effect of nonlocal theory on the variation of system frequencies, (a) first mode, (b) second mode, (c) third mode.

words, increasing this force component reduces the amount of voltage required to cause instability. This can be interpreted as the highest system performance will occur at the lowest Casimir force. Moreover, increasing the length of the nanobeam takes the system at lower voltages to the state of pull-in and instability. However, increasing the length of the nanobeam reduces the effect of size parameter on the pull-in voltage, as well as the difference between the classical theory and the modified coupling stress. According to the extracted results, the effect size parameter is directly proportional to the frequency of the system. As the electrostatic voltage increases, the system frequency decreases while increasing the dynamic displacement of the system.

References

[1] Younis M I 2011 *MEMS Linear and Nonlinear Statics and Dynamics* (Cham: Springer) p 20

[2] Rahaeifard M, Kahrobaiyan M H, Asghari M and Ahmadian M T 2011 Static pull-in analysis of microcantilevers based on the modified couple stress theory *Sens. Actuators A: Phys.* **171** 370–4

[3] Yin L, Qian Q and Wang L 2011 Size effect on the static behavior of electrostatically actuated microbeams *Acta Mech. Sin.* **27** 445

[4] Kong S L 2013 Size effect on pull-in behaviors of electrostatically actuated cantilever micro-beams *Appl. Mech. Mater.* **300** 889–92

[5] Abdel-Rahman E M, Younis M I and Nayfeh A H 2002 Characterization of the mechanical behavior of an electrically actuated microbeam *J. Micromech. Microeng.* **12** 759

[6] Gurtin M E and Murdoch A I 1975 A continuum theory of elastic material surfaces *Arch. Ration. Mech. Anal.* **57** 291–323

[7] Koochi A, Kazemi A, Khandani F and Abadyan M 2012 Influence of surface effects on size-dependent instability of nano-actuators in the presence of quantum vacuum fluctuations *Phys. Scr.* **85** 035804

[8] Soroush R, Koochi A, Kazemi A S, Noghrehabadi A, Haddadpour H and Abadyan M 2010 Investigating the effect of Casimir and van der Waals attractions on the electrostatic pull-in instability of nano-actuators *Phys. Scr.* **82** 045801

[9] Beni Y T, Koochi A and Abadyan M 2011 Theoretical study of the effect of Casimir force, elastic boundary conditions and size dependency on the pull-in instability of beam-type NEMS *Phys. E: Low-Dimens. Syst. Nanostruct.* **43** 979–88

[10] Hosseini-Hashemi S, Nazemnezhad R and Bedroud M 2014 Surface effects on nonlinear free vibration of functionally graded nanobeams using nonlocal elasticity *Appl. Math. Model.* **38** 3538–53

[11] Sedighi H M, Daneshmand F and Abadyan M 2015 Dynamic instability analysis of electrostatic functionally graded doubly-clamped nano-actuators *Compos. Struct.* **124** 55–64

[12] Lei X W, Natsuki T, Shi J X and Ni Q Q 2012 Surface effects on the vibrational frequency of double-walled carbon nanotubes using the nonlocal Timoshenko beam model *Compos. Part B: Eng.* **43**(1) 64–9

[13] Hughes T J, Cottrell J A and Bazilevs Y 2005 Isogeometric analysis: CAD, finite elements, NURBS, exact geometry and mesh refinement *Compu. Methods Appl. Mech. Eng.* **194** 4135–95

[14] Reali A 2006 An isogeometric analysis approach for the study of structural vibrations *J. Earthq. Eng.* **10** 1–30

[15] Cottrell J A, Reali A, Bazilevs Y and Hughes T J 2005 Isogeometric analysis of structural vibrations *Comp. Methods Appl. Mech. Eng.* **195** 5257–96
[16] Weeger O, Wever U and Simeon B 2013 Isogeometric analysis of nonlinear Euler–Bernoulli beam vibrations *Nonlinear Dyn.* **72** 813–35
[17] Zare A, Eghtesad M and Daneshmand F 2017 Numerical investigation and dynamic behavior of pipes conveying fluid based on isogeometric analysis *Ocean Eng.* **140** 388–400
[18] Jockovic M 2016 Free vibration analysis of beam element using Isogeometric analysis *4th Int. Conf. on Contemporary Achievements in Civil Engineering* 22 269–78
[19] Vinh Loc T, Hoang Chien T and Xuan Hung N 2013 Geometrically nonlinear and dynamic analysis of Euler–Bernoulli beams using isogeometric approach *J. Sci.* **3** 2–9
[20] Shojaee S, Izadpanah E, Valizadeh N and Kiendl J 2012 Free vibration analysis of thin plates by using a NURBS-based isogeometric approach *Finite Elem. Anal. Des.* **61** 23–34
[21] Stoykov S, Hofreither C and Margenov S 2014 Isogeometric analysis for nonlinear dynamics of Timoshenko beams *Int. Conf. on Numerical Methods and Applications* (Cham: Springer) 138–46
[22] Yang F A C M, Chong A C M, Lam D C C and Tong P 2002 Couple stress based strain gradient theory for elasticity *Int. J. Solids Struct.* **39** 2731–43
[23] Reddy J N 2017 *Energy Principles and Variational Methods in Applied Mechanics* (New York: Wiley)
[24] Park S K and Gao X L 2006 Bernoulli–Euler beam model based on a modified couple stress theory *J. Micromech. Microeng.* **16** 2355
[25] Gurtin M E and Murdoch A I 1978 Surface stress in solids *Int. J. Solids Struct.* **14** 431–40
[26] Huang J M, Liew K M, Wong C H, Rajendran S, Tan M J and Liu A Q 2001 Mechanical design and optimization of capacitive micromachined switch *Sens. Actuators A: Phys.* **93** 273–85
[27] Israelachvili J N 1992 *Intermolecular and Surface Forces* (New York: Academic)
[28] Hosaka H, Itao K and Kuroda S 1995 Damping characteristics of beam-shaped micro-oscillators *Sens. Actuators A: Phys.* **49** 87–95
[29] Fox R W, McDonald A T and Pritchard P J 1998 *Introduction to Fluid Mechanics* (New York: Wiley)
[30] Wazwaz A M 1999 A reliable modification of Adomian decomposition method *Appl. Math. Comput.* **102** 77–86
[31] Rach R 1984 A convenient computational form for the Adomian polynomials *J. Math. Anal. Appl.* **102** 415–9
[32] Wazwaz A M 2001 The numerical solution of sixth-order boundary value problems by the modified decomposition method *Appl. Math. Comput.* **118** 311–25
[33] Cottrell J A, Hughes T J and Bazilevs Y 2009 *Isogeometric Analysis: toward Integration of CAD and FEA* (New York: Wiley)
[34] Piegl L and Tiller W 2012 *The NURBS Book* (Cham: Springer)
[35] Rogers D F 2000 *An Introduction to NURBS: With Historical Perspective* (Amsterdam: Elsevier)
[36] Rao S S 1995 *Mechanical Vibrations Laboratory Manual* (New York: Addison-Wesley)
[37] Shabani R, Hatami H, Golzar F G, Tariverdilo S and Rezazadeh G 2013 Coupled vibration of a cantilever micro-beam submerged in a bounded incompressible fluid domain *Acta Mech.* **224** 841–50

IOP Publishing

Modeling and Computation in Vibration Problems, Volume 1
Numerical and semi-analytical methods
S Chakraverty, F Tornabene and J N Reddy

Chapter 9

Continuum 3D and 2D shell models for free vibration analysis of single-walled and double-walled carbon nanotubes

Salvatore Brischetto, Francesco Tornabene and Rossana Dimitri

The present chapter proposes 3D shell solutions based on the exponential matrix methodology for the study of the free vibration response of simply-supported Single- and Double-Walled Carbon NanoTubes (SWCNTs and DWCNTs). A continuum approach based on the three-dimensional theory of elasticity is employed to represent discrete elements such as SWCNTs and DWCNTs. Therefore, SWCNTs are modeled as isotropic cylinders where appropriate equivalent thickness and Young's modulus values are opportunely defined. DWCNTs are modeled as two concentric isotropic cylinders with proper equivalent thickness and Young's modulus values. The two isotropic concentric cylinders are connected by means of interlaminar continuity conditions or van der Waals interactions. These two different choices can be evaluated in terms of natural frequencies and vibration modes. Such studies can be performed by means of a Layer-Wise (LW) approach that allows one to consider separately each cylinder. The effects of the van der Waals interactions and interlaminar continuity are evaluated for different DWCNT lengths, diameters and vibration modes. Proposed analytical 3D shell models are compared with some classical and refined 2D models based on the Generalized Differential Quadrature Method (GDQM) in the case of free vibration study of SWCNTs and DWCNTs with different lengths, diameters and equivalent elastic properties. Moreover, comparisons with well-known continuum beam models from the literature are proposed to analyze the differences between 3D, 2D and 1D approaches depending on the carbon nanotube length and on the thickness of its walls. Proposed results have been presented in [1]–[3] where readers can find all the details here omitted for the sake of brevity.

doi:10.1088/978-0-7503-3483-9ch9 © IOP Publishing Ltd 2021

9.1 3D continuum shell model

The free vibration analysis of the proposed SWCNTs and DWCNTs is conducted by means of a general exact 3D shell model based on the exponential matrix methodology. The exact closed-form solution of the differential equations of equilibrium written in orthogonal mixed curvilinear coordinates is obtained by considering simply-supported boundary conditions and 0°/90° orthotropic angles. The equations consider a geometry for spherical shells without simplifications and they automatically degenerate in equations for cylindrical shell panels, cylindrical closed shells and plates by means of simple geometrical considerations. The method is based on a LW approach. Therefore, the continuity of displacements and transverse shear/normal stresses must be correctly imposed at each interface between the layers of the considered multilayered structures. The novelty introduced by the present 3D shell model is the general formulation for all the geometries (square and rectangular plates, cylindrical and spherical shell panels, and cylindrical closed shells). The system of second order differential equations must be reduced to a system of first order differential equations, and afterwards it can be exactly solved using the exponential matrix methodology and the Navier-type solution.

In the present section of the chapter, the first part is dedicated to equations of motion in orthogonal mixed curvilinear coordinates for free frequency analysis of plates and shells. The second part shows the geometrical relations and three-dimensional constitutive equations for spherical shells using an exact geometrical approach. These relations easily degenerate into the geometrical relations for cylindrical shells and plates. The third part gives the closed form solution of equations of motion and their extension to multilayer cases. Vibration modes related to calculated frequencies are obtained from the same system of equations. DWCNTs and SWCNTs are analyzed as continuum cylinders; therefore, the cylindrical shell equations must be considered as a particular case of the spherical shell ones when one of the two radii of curvature is considered as infinite. All the details about the proposed formulation can be found in [1]–[7].

9.1.1 3D equilibrium equations in orthogonal mixed curvilinear coordinates

A shell structure is a three-dimensional body bounded by two closely spaced curved surfaces where one dimension (in the proposed cases, the thickness dimension) is small if compared with the other dimensions in the two curved plane directions. The thickness was defined in [8] and [9] as the distance between the curved surfaces evaluated through the normal to the middle surface of the shell at the considered point. The middle surface Ω_0 of the shell is the locus of points that are positioned midway between these defined surfaces. The geometry and the reference system employed for the proposed formulation are shown in figure 9.1 where the orthogonal mixed curvilinear coordinates (α, β, z) are clearly indicated. Displacement components are called u, v, and w in α, β and z directions, respectively. The square of an infinitesimal linear segment defined in the layer can be calculated as

$$ds^2 = g_1 d\alpha^2 + g_2 d\beta^2 + g_3 dz^2, \tag{9.1}$$

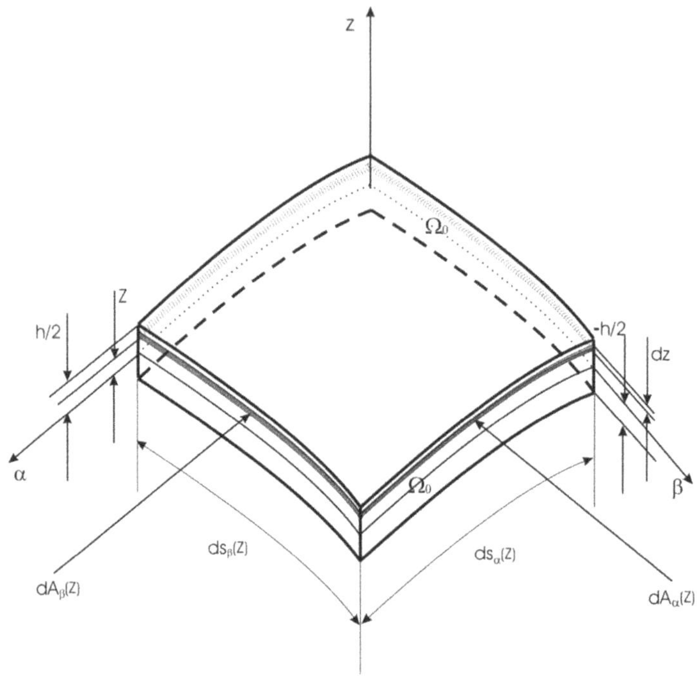

Figure 9.1. Notations, geometry and orthogonal mixed curvilinear coordinates for shells.

where

$$g_1 = H_\alpha^2 = \left[A\left(1 + \frac{z}{R_\alpha}\right)\right]^2, \quad g_2 = H_\beta^2 = \left[B\left(1 + \frac{z}{R_\beta}\right)\right]^2, \quad g_3 = H_z^2 = 1. \quad (9.2)$$

The geometrical parameters g_1, g_2, g_3, A, B, R_α, R_β are linked by using the Lamb's equations as discussed in [9]; in fact, the three-dimensional space where the three independent coordinates α, β and z change is defined as a Euclidean space. R_α and R_β are called the principal radii of curvature along the two curvilinear in-plane coordinates α and β, respectively. A and B are the coefficients of the first fundamental form of the mid-surface Ω_0. The out-of-plane rectilinear coordinate z is defined as the thickness coordinate and it varies from $-h/2$ to $+h/2$ (where h is the global thickness of the whole structure, as shown in figure 9.1). Parameter $g_3 = 1$ because z is a rectilinear coordinate.

The fundamental shell element is defined as the differential element bounded by two surfaces dz apart at a distance z from the middle surface and four ruled surfaces. These last surfaces have generators that are the normals to the middle surface along the parametric curves defined as $\alpha = \alpha_0$, $\alpha = \alpha_0 + d\alpha$, $\beta = \beta_0$ and $\beta = \beta_0 + d\beta$ [9]. The edge lengths of this fundamental shell element are clearly indicated in figure 9.1, they are defined as

$$\begin{aligned} ds_\alpha(z) &= A(1 + z/R_\alpha)d\alpha, \\ ds_\beta(z) &= B(1 + z/R_\beta)d\beta, \end{aligned} \quad (9.3)$$

the differential areas of the edge faces of the fundamental shell element as shown in figure 9.1 are

$$\begin{aligned} dA_\alpha(z) &= A(1 + z/R_\alpha)d\alpha dz, \\ dA_\beta(z) &= B(1 + z/R_\beta)d\beta dz, \end{aligned} \tag{9.4}$$

the associated infinitesimal area is calculated as

$$d\Omega(z) = [A(1 + z/R_\alpha)][B(1 + z/R_\beta)]d\alpha d\beta, \tag{9.5}$$

therefore, the related volume of the fundamental shell element is

$$dV(z) = [A(1 + z/R_\alpha)][B(1 + z/R_\beta)]d\alpha d\beta dz. \tag{9.6}$$

The parametric coefficients for constant radii of curvature shells have the terms A and B equal to 1

$$\begin{aligned} &H_\alpha = \left(1 + \frac{z}{R_\alpha}\right) = \left(1 + \frac{\tilde{z} - h/2}{R_\alpha}\right), \quad H_\beta = \left(1 + \frac{z}{R_\beta}\right) = \left(1 + \frac{\tilde{z} - h/2}{R_\beta}\right), \\ &H_z = 1, \end{aligned} \tag{9.7}$$

H_α and H_β depend on z or $\tilde{z}$ coordinate. The difference between the two types of z coordinate are clearly displayed in figure 9.2.

The three differential equations of equilibrium are written for the particular case of free frequency analysis of multilayered spherical shells including N_L layers. The main hypothesis is constant radii of curvature R_α and R_β along curved in-plane α and β directions, respectively. The most general form for variable radii of curvature is given in [10] and [11]. However, the spherical shell equations are

$$\begin{aligned} &H_\beta(z)\frac{\partial \sigma_{\alpha\alpha}^k}{\partial \alpha} + H_\alpha(z)\frac{\partial \sigma_{\alpha\beta}^k}{\partial \beta} + H_\alpha(z)H_\beta(z)\frac{\partial \sigma_{\alpha z}^k}{\partial z} \\ &+ \left(\frac{2H_\beta(z)}{R_\alpha} + \frac{H_\alpha(z)}{R_\beta}\right)\sigma_{\alpha z}^k = \rho^k H_\alpha(z)H_\beta(z)\ddot{u}^k, \end{aligned} \tag{9.8}$$

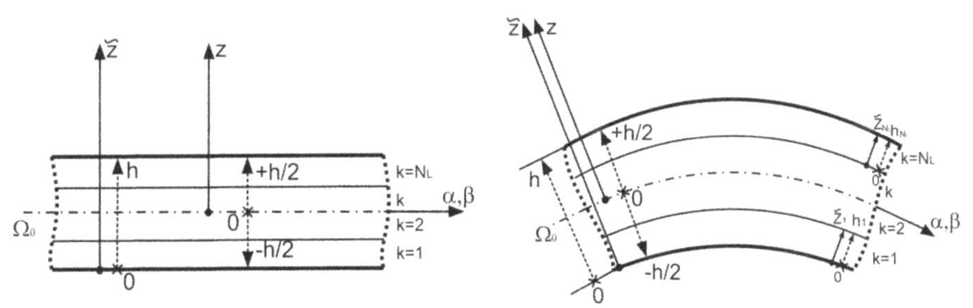

Figure 9.2. Thickness coordinates and orthogonal mixed curvilinear reference systems for plate and shell structures.

$$H_\beta(z)\frac{\partial\sigma^k_{\alpha\beta}}{\partial\alpha} + H_\alpha(z)\frac{\partial\sigma^k_{\beta\beta}}{\partial\beta} + H_\alpha(z)H_\beta(z)\frac{\partial\sigma^k_{\beta z}}{\partial z} + \left(\frac{2H_\alpha(z)}{R_\beta} + \frac{H_\beta(z)}{R_\alpha}\right)\sigma^k_{\beta z} = \rho^k H_\alpha(z)H_\beta(z)\ddot{v}^k, \tag{9.9}$$

$$H_\beta(z)\frac{\partial\sigma^k_{\alpha z}}{\partial\alpha} + H_\alpha(z)\frac{\partial\sigma^k_{\beta z}}{\partial\beta} + H_\alpha(z)H_\beta(z)\frac{\partial\sigma^k_{zz}}{\partial z} - \frac{H_\beta(z)}{R_\alpha}\sigma^k_{\alpha\alpha} - \frac{H_\alpha(z)}{R_\beta}\sigma^k_{\beta\beta} + \left(\frac{H_\beta(z)}{R_\alpha} + \frac{H_\alpha(z)}{R_\beta}\right)\sigma^k_{zz} = \rho^k H_\alpha(z)H_\beta(z)\ddot{w}^k, \tag{9.10}$$

where ρ^k is the mass density, $(\sigma^k_{\alpha\alpha}, \sigma^k_{\beta\beta}, \sigma^k_{zz}, \sigma^k_{\beta z}, \sigma^k_{\alpha z}, \sigma^k_{\alpha\beta})$ are the six 3D stress components and $\ddot{u}^k$, $\ddot{v}^k$ and $\ddot{w}^k$ are the derivatives of the three displacement components with respect to time. The superscript k indicates the dependence on the layer. R_α and R_β are measured on the curved mid-surface Ω_0 of the entire multilayered shell. $H_\alpha(z)$ and $H_\beta(z)$ vary with continuity through the thickness of the multilayered shell and for this reason they depend on the thickness coordinate z. However, from this point on, in order not to weigh down the notation we will imply their dependence on z by always writing them as H_α and H_β.

9.1.2 3D geometrical and constitutive relations

The three-dimensional theory of elasticity in orthogonal mixed curvilinear coordinates gives the following geometrical relations between strains and displacements as already demonstrated in [10] and [12] for the generic k layer of the multilayered shell structure

$$\epsilon^k_{\alpha\alpha} = \frac{1}{(1 + z/R_\alpha)}\left(\frac{1}{A}\frac{\partial u^k}{\partial\alpha} + \frac{v^k}{AB}\frac{\partial A}{\partial\beta} + \frac{w^k}{R_\alpha}\right), \tag{9.11}$$

$$\epsilon^k_{\beta\beta} = \frac{1}{(1 + z/R_\beta)}\left(\frac{u^k}{AB}\frac{\partial B}{\partial\alpha} + \frac{1}{B}\frac{\partial v^k}{\partial\beta} + \frac{w^k}{R_\beta}\right), \tag{9.12}$$

$$\epsilon^k_{zz} = \frac{\partial w^k}{\partial z}, \tag{9.13}$$

$$\gamma^k_{\alpha\beta} = \frac{A(1 + z/R_\alpha)}{B(1 + z/R_\beta)}\frac{\partial}{\partial\beta}\left[\frac{u^k}{A(1 + z/R_\alpha)}\right] + \frac{B(1 + z/R_\beta)}{A(1 + z/R_\alpha)}\frac{\partial}{\partial\alpha}\left[\frac{v^k}{B(1 + z/R_\beta)}\right], \tag{9.14}$$

$$\gamma^k_{\alpha z} = \frac{1}{A(1 + z/R_\alpha)}\frac{\partial w^k}{\partial\alpha} + A(1 + z/R_\alpha)\frac{\partial}{\partial z}\left[\frac{u^k}{A(1 + z/R_\alpha)}\right], \tag{9.15}$$

$$\gamma_{\beta z}^k = \frac{1}{B(1+z/R_\beta)}\frac{\partial w^k}{\partial \beta} + B(1+z/R_\beta)\frac{\partial}{\partial z}\left[\frac{v^k}{B(1+z/R_\beta)}\right], \tag{9.16}$$

symbol ∂ is employed to indicate the partial derivatives. In the present chapter, we will consider only shells with constant radii of curvature (e.g., cylindrical and spherical panels) and for this reason $A = B = 1$. The above geometrical relations for shells, in the case of constant radii of curvature can be re-written as

$$\epsilon_{\alpha\alpha}^k = \frac{1}{H_\alpha}\frac{\partial u^k}{\partial \alpha} + \frac{w^k}{H_\alpha R_\alpha}, \tag{9.17}$$

$$\epsilon_{\beta\beta}^k = \frac{1}{H_\beta}\frac{\partial v^k}{\partial \beta} + \frac{w^k}{H_\beta R_\beta}, \tag{9.18}$$

$$\epsilon_{zz}^k = \frac{\partial w^k}{\partial z}, \tag{9.19}$$

$$\gamma_{\alpha\beta}^k = \frac{1}{H_\alpha}\frac{\partial v^k}{\partial \alpha} + \frac{1}{H_\beta}\frac{\partial u^k}{\partial \beta}, \tag{9.20}$$

$$\gamma_{\alpha z}^k = \frac{1}{H_\alpha}\frac{\partial w^k}{\partial \alpha} + \frac{\partial u^k}{\partial z} - \frac{u^k}{H_\alpha R_\alpha}, \tag{9.21}$$

$$\gamma_{\beta z}^k = \frac{1}{H_\beta}\frac{\partial w^k}{\partial \beta} + \frac{\partial v^k}{\partial z} - \frac{v^k}{H_\beta R_\beta}. \tag{9.22}$$

The above general geometrical relations for spherical shells degenerate into geometrical relations for cylindrical shells when R_α or R_β is infinite (that means H_α or H_β equals one), and they also degenerate into geometrical relations for plates when both R_α and R_β are infinite (that means H_α=H_β=1).

Three-dimensional linear elastic constitutive equations for orthotropic material in the structural reference system based on the orthogonal mixed curvilinear coordinates (α, β, z) are here written for a general physical k layer of the multilayered structure (see [13] for further details)

$$\sigma_{\alpha\alpha}^k = C_{11}^k\epsilon_{\alpha\alpha}^k + C_{12}^k\epsilon_{\beta\beta}^k + C_{13}^k\epsilon_{zz}^k + C_{16}^k\gamma_{\alpha\beta}^k, \tag{9.23}$$

$$\sigma_{\beta\beta}^k = C_{12}^k\epsilon_{\alpha\alpha}^k + C_{22}^k\epsilon_{\beta\beta}^k + C_{23}^k\epsilon_{zz}^k + C_{26}^k\gamma_{\alpha\beta}^k, \tag{9.24}$$

$$\sigma_{zz}^k = C_{13}^k\epsilon_{\alpha\alpha}^k + C_{23}^k\epsilon_{\beta\beta}^k + C_{33}^k\epsilon_{zz}^k + C_{36}^k\gamma_{\alpha\beta}^k, \tag{9.25}$$

$$\sigma_{\beta z}^k = C_{44}^k\gamma_{\beta z}^k + C_{45}^k\gamma_{\alpha z}^k, \tag{9.26}$$

$$\sigma_{\alpha z}^k = C_{45}^k\gamma_{\beta z}^k + C_{55}^k\gamma_{\alpha z}^k, \tag{9.27}$$

$$\sigma_{\alpha\beta}^k = C_{16}^k\epsilon_{\alpha\alpha}^k + C_{26}^k\epsilon_{\beta\beta}^k + C_{36}^k\epsilon_{zz}^k + C_{66}^k\gamma_{\alpha\beta}^k. \tag{9.28}$$

The previous geometrical relations are then included in the constitutive equations; in the obtained new form, the partial derivatives $\frac{\partial}{\partial\alpha}$, $\frac{\partial}{\partial\beta}$ and $\frac{\partial}{\partial z}$ are indicated with subscripts $_{,\alpha}$, $_{,\beta}$ and $_{,z}$, respectively. A closed form solution of 3D differential equations of equilibrium for shells can be obtained only if coefficients C_{16}^k, C_{26}^k, C_{36}^k and C_{45}^k are set to zero (this feature means orthotropic angle θ equals 0° or 90°)

$$\sigma_{\alpha\alpha}^k = \frac{C_{11}^k}{H_\alpha}u_{,\alpha}^k + \frac{C_{11}^k}{H_\alpha R_\alpha}w^k + \frac{C_{12}^k}{H_\beta}v_{,\beta}^k + \frac{C_{12}^k}{H_\beta R_\beta}w^k + C_{13}^k w_{,z}^k, \tag{9.29}$$

$$\sigma_{\beta\beta}^k = \frac{C_{12}^k}{H_\alpha}u_{,\alpha}^k + \frac{C_{12}^k}{H_\alpha R_\alpha}w^k + \frac{C_{22}^k}{H_\beta}v_{,\beta}^k + \frac{C_{22}^k}{H_\beta R_\beta}w^k + C_{23}^k w_{,z}^k, \tag{9.30}$$

$$\sigma_{zz}^k = \frac{C_{13}^k}{H_\alpha}u_{,\alpha}^k + \frac{C_{13}^k}{H_\alpha R_\alpha}w^k + \frac{C_{23}^k}{H_\beta}v_{,\beta}^k + \frac{C_{23}^k}{H_\beta R_\beta}w^k + C_{33}^k w_{,z}^k, \tag{9.31}$$

$$\sigma_{\beta z}^k = \frac{C_{44}^k}{H_\beta}w_{,\beta}^k + C_{44}^k v_{,z}^k - \frac{C_{44}^k}{H_\beta R_\beta}v^k, \tag{9.32}$$

$$\sigma_{\alpha z}^k = \frac{C_{55}^k}{H_\alpha}w_{,\alpha}^k + C_{55}^k u_{,z}^k - \frac{C_{55}^k}{H_\alpha R_\alpha}u^k, \tag{9.33}$$

$$\sigma_{\alpha\beta}^k = \frac{C_{66}^k}{H_\alpha}v_{,\alpha}^k + \frac{C_{66}^k}{H_\beta}u_{,\beta}^k. \tag{9.34}$$

9.1.3 Closed form solution for shell 3D equilibrium equations

Constitutive equations opportunely rearranged as proposed in equations (9.29)–(9.34) are substituted in 3D equilibrium equations (9.8)–(9.10) in order to obtain the displacement form of differential equations of equilibrium defined for a general k physical layer

$$\begin{aligned}
&\left(-\frac{H_\beta C_{55}^k}{H_\alpha R_\alpha^2} - \frac{C_{55}^k}{R_\alpha R_\beta}\right)u^k + \left(\frac{C_{55}^k H_\beta}{R_\alpha} + \frac{C_{55}^k H_\alpha}{R_\beta}\right)u_{,z}^k \\
&+\left(\frac{C_{11}^k H_\beta}{H_\alpha}\right)u_{,\alpha\alpha}^k + \left(\frac{C_{66}^k H_\alpha}{H_\beta}\right)u_{,\beta\beta}^k + \left(C_{55}^k H_\alpha H_\beta\right)u_{,zz}^k \\
&+ (C_{12}^k + C_{66}^k)v_{,\alpha\beta}^k + \left(\frac{C_{11}^k H_\beta}{H_\alpha R_\alpha} + \frac{C_{12}^k}{R_\beta} + \frac{C_{55}^k H_\beta}{H_\alpha R_\alpha} + \frac{C_{55}^k}{R_\beta}\right)w_{,\alpha}^k + \\
&\left(C_{13}^k H_\beta + C_{55}^k H_\beta\right)w_{,\alpha z}^k = \rho^k H_\alpha H_\beta \ddot{u}^k,
\end{aligned} \tag{9.35}$$

$$\left(-\frac{H_\alpha C_{44}^k}{H_\beta R_\beta^2}-\frac{C_{44}^k}{R_\alpha R_\beta}\right)v^k+\left(\frac{C_{44}^k H_\alpha}{R_\beta}+\frac{C_{44}^k H_\beta}{R_\alpha}\right)v_{,z}^k$$
$$+\left(\frac{C_{66}^k H_\beta}{H_\alpha}\right)v_{,\alpha\alpha}^k+\left(\frac{C_{22}^k H_\alpha}{H_\beta}\right)v_{,\beta\beta}^k+\left(C_{44}^k H_\alpha H_\beta\right)v_{,zz}^k \tag{9.36}$$
$$+\left(C_{12}^k+C_{66}^k\right)u_{,\alpha\beta}^k+\left(\frac{C_{44}^k H_\alpha}{H_\beta R_\beta}+\frac{C_{44}^k}{R_\alpha}+\frac{C_{22}^k H_\alpha}{H_\beta R_\beta}+\frac{C_{12}^k}{R_\alpha}\right)w_{,\beta}^k$$
$$+\left(C_{44}^k H_\alpha+C_{23}^k H_\alpha\right)w_{,\beta z}^k=\rho^k H_\alpha H_\beta \ddot{v}^k,$$

$$\left(\frac{C_{13}^k}{R_\alpha R_\beta}+\frac{C_{23}^k}{R_\alpha R_\beta}-\frac{C_{11}^k H_\beta}{H_\alpha R_\alpha^2}-\frac{2C_{12}^k}{R_\alpha R_\beta}-\frac{C_{22}^k H_\alpha}{H_\beta R_\beta^2}\right)w^k$$
$$+\left(-\frac{C_{55}^k H_\beta}{H_\alpha R_\alpha}+\frac{C_{13}^k}{R_\beta}-\frac{C_{11}^k H_\beta}{H_\alpha R_\alpha}-\frac{C_{12}^k}{R_\beta}\right)u_{,\alpha}^k+$$
$$\left(-\frac{C_{44}^k H_\alpha}{H_\beta R_\beta}+\frac{C_{23}^k}{R_\alpha}-\frac{C_{22}^k H_\alpha}{H_\beta R_\beta}-\frac{C_{12}^k}{R_\alpha}\right)v_{,\beta}^k \tag{9.37}$$
$$+\left(\frac{C_{33}^k H_\beta}{R_\alpha}+\frac{C_{33}^k H_\alpha}{R_\beta}\right)w_{,z}^k+\left(C_{55}^k H_\beta+C_{13}^k H_\beta\right)u_{,\alpha z}^k+\left(C_{44}^k H_\alpha+C_{23}^k H_\alpha\right)v_{,\beta z}^k$$
$$+\left(C_{55}^k\frac{H_\beta}{H_\alpha}\right)w_{,\alpha\alpha}^k+\left(C_{44}^k\frac{H_\alpha}{H_\beta}\right)w_{,\beta\beta}^k+\left(C_{33}^k H_\alpha H_\beta\right)w_{,zz}^k=\rho^k H_\alpha H_\beta \ddot{w}^k.$$

R_α and R_β are determined with respect to the reference mid-surface Ω_0 of the multilayered shell structure. H_α and H_β are defined along the thickness of the multilayered structure using equation (9.7). Equilibrium relations in equations (9.35)–(9.37) are written for spherical shell panels, they easily degenerate into equilibrium equations for cylindrical closed/open shell panels (an example is proposed in [14]) when R_α or R_β is infinite (with related H_α or H_β equal to one) and into equilibrium equations for plates (an example is shown in [15]) when both R_α and R_β are infinite (with both H_α and H_β equal to one). The proposed formulation allows a unique and general formulation for all these discussed geometries.

The closed form of equations (9.35)–(9.37) is calculated by means of the hypotheses of simply-supported shells and plates embedding isotropic or orthotropic materials with fiber orientation 0° or 90° (in both cases $C_{16}^k = C_{26}^k = C_{36}^k = C_{45}^k = 0$). The three displacement components in harmonic form that satisfy the simply-supported boundary conditions are

$$u^k(\alpha, \beta, z, t) = U^k(z)e^{i\omega t}\cos(\bar{\alpha}\alpha)\sin(\bar{\beta}\beta), \tag{9.38}$$

$$v^k(\alpha, \beta, z, t) = V^k(z)e^{i\omega t}\sin(\bar{\alpha}\alpha)\cos(\bar{\beta}\beta), \tag{9.39}$$

$$w^k(\alpha, \beta, z, t) = W^k(z)e^{i\omega t}\sin(\bar{\alpha}\alpha)\sin(\bar{\beta}\beta), \tag{9.40}$$

where $U^k(z)$, $V^k(z)$ and $W^k(z)$ are the displacement amplitudes in α, β and z directions, respectively. i is the imaginary unit, $\omega = 2\pi f$ is the circular frequency including the f frequency measured in Hz, t indicates the time. In coefficients $\bar{\alpha} = \frac{m\pi}{a}$ and $\bar{\beta} = \frac{n\pi}{b}$, m and n are the half-wave numbers and a and b are the shell dimensions in α and β directions, respectively (all these parameters are defined at the mid-surface Ω_0). Equations (9.38)–(9.40) are included in the system composed by equations (9.35)–(9.37) in order to obtain the following system of equations where the partial derivatives in α and β are correctly calculated by means of algebraic parameters $\bar{\alpha}$ and $\bar{\beta}$. Therefore, only derivatives in z remain

$$\begin{aligned}&\left(-\frac{C_{55}^kH_\beta}{H_\alpha R_\alpha^2}-\frac{C_{55}^k}{R_\alpha R_\beta}-\bar{\alpha}^2\frac{C_{11}^kH_\beta}{H_\alpha}-\bar{\beta}^2\frac{C_{66}^kH_\alpha}{H_\beta}+\rho^kH_\alpha H_\beta\omega^2\right)U^k\\&+(-\bar{\alpha}\bar{\beta}C_{12}^k-\bar{\alpha}\bar{\beta}C_{66}^k)V^k+\left(\bar{\alpha}\frac{C_{11}^kH_\beta}{H_\alpha R_\alpha}+\bar{\alpha}\frac{C_{12}^k}{R_\beta}+\bar{\alpha}\frac{C_{55}^kH_\beta}{H_\alpha R_\alpha}+\bar{\alpha}\frac{C_{55}^k}{R_\beta}\right)W^k\\&+\left(\frac{C_{55}^kH_\beta}{R_\alpha}+\frac{C_{55}^kH_\alpha}{R_\beta}\right)U_{,z}^k+\left(\bar{\alpha}C_{13}^kH_\beta+\bar{\alpha}C_{55}^kH_\beta\right)W_{,z}^k+\left(C_{55}^kH_\alpha H_\beta\right)U_{,zz}^k=0,\end{aligned}\tag{9.41}$$

$$\begin{aligned}&(-\bar{\alpha}\bar{\beta}C_{66}^k-\bar{\alpha}\bar{\beta}C_{12}^k)U^k+\left(-\frac{C_{44}^kH_\alpha}{H_\beta R_\beta^2}-\frac{C_{44}^k}{R_\alpha R_\beta}-\bar{\alpha}^2\frac{C_{66}^kH_\beta}{H_\alpha}-\bar{\beta}^2\frac{C_{22}^kH_\alpha}{H_\beta}\right.\\&\left.+\rho^kH_\alpha H_\beta\omega^2\right)V^k+\left(\bar{\beta}\frac{C_{44}^kH_\alpha}{H_\beta R_\beta}+\bar{\beta}\frac{C_{44}^k}{R_\alpha}+\bar{\beta}\frac{C_{22}^kH_\alpha}{H_\beta R_\beta}+\bar{\beta}\frac{C_{12}^k}{R_\alpha}\right)W^k\\&+\left(\frac{C_{44}^kH_\alpha}{R_\beta}+\frac{C_{44}^kH_\beta}{R_\alpha}\right)V_{,z}^k+\left(\bar{\beta}C_{44}^kH_\alpha+\bar{\beta}C_{23}^kH_\alpha\right)W_{,z}^k+\left(C_{44}^kH_\alpha H_\beta\right)V_{,zz}^k=0,\end{aligned}\tag{9.42}$$

$$\begin{aligned}&\left(\bar{\alpha}\frac{C_{55}^kH_\beta}{H_\alpha R_\alpha}-\bar{\alpha}\frac{C_{13}^k}{R_\beta}+\bar{\alpha}\frac{C_{11}^kH_\beta}{H_\alpha R_\alpha}+\bar{\alpha}\frac{C_{12}^k}{R_\beta}\right)U^k\\&+\left(\bar{\beta}\frac{C_{44}^kH_\alpha}{H_\beta R_\beta}-\bar{\beta}\frac{C_{23}^k}{R_\alpha}+\bar{\beta}\frac{C_{22}^kH_\alpha}{H_\beta R_\beta}+\bar{\beta}\frac{C_{12}^k}{R_\alpha}\right)V^k+\\&\left(\frac{C_{13}^k}{R_\alpha R_\beta}+\frac{C_{23}^k}{R_\alpha R_\beta}-\frac{C_{11}^kH_\beta}{H_\alpha R_\alpha^2}-\frac{2C_{12}^k}{R_\alpha R_\beta}-\frac{C_{22}^kH_\alpha}{H_\beta R_\beta^2}-\bar{\alpha}^2\frac{C_{55}^kH_\beta}{H_\alpha}-\bar{\beta}^2\frac{C_{44}^kH_\alpha}{H_\beta}\right.\\&\left.+\rho^kH_\alpha H_\beta\omega^2\right)W^k+\\&\left(-\bar{\alpha}C_{55}^kH_\beta-\bar{\alpha}C_{13}^kH_\beta\right)U_{,z}^k+(-\bar{\beta}C_{44}^kH_\alpha-\bar{\beta}C_{23}^kH_\alpha)V_{,z}^k+\left(\frac{C_{33}^kH_\beta}{R_\alpha}+\frac{C_{33}^kH_\alpha}{R_\beta}\right)\\&W_{,z}^k+\left(C_{33}^kH_\alpha H_\beta\right)W_{,zz}^k=0.\end{aligned}\tag{9.43}$$

The system of equations (9.41)–(9.43) can be rewritten in compact form using coefficients A_s^k for each block included in parentheses (). The index s for these blocks/coefficients goes from 1 to 19

$$A_1^k U^k + A_2^k V^k + A_3^k W^k + A_4^k U_{,z}^k + A_5^k W_{,z}^k + A_6^k U_{,zz}^k = 0, \tag{9.44}$$

$$A_7^k U^k + A_8^k V^k + A_9^k W^k + A_{10}^k V_{,z}^k + A_{11}^k W_{,z}^k + A_{12}^k V_{,zz}^k = 0, \tag{9.45}$$

$$A_{13}^k U^k + A_{14}^k V^k + A_{15}^k W^k + A_{16}^k U_{,z}^k + A_{17}^k V_{,z}^k + A_{18}^k W_{,z}^k + A_{19}^k W_{,zz}^k = 0. \tag{9.46}$$

Equations (9.44)–(9.46) are a system of three second order differential equations in z. They have been here defined for spherical shells having constant radii of curvature but they easily degenerate into equations for cylindrical shells and plates by means of simple geometrical considerations previously discussed.

9.1.4 Layer-wise solution for multilayered structures using the exponential matrix methodology

The obtained system of second order differential equations in z can be reduced to a system of first order differential equations in z by using the methodology given in details in [16] and [17]. Here, a simple example is proposed to better understand this order reduction (details are given in [16])

$$a_1 u_1'' + b_1 u_1' + c_1 u_1 = f_1, \tag{9.47}$$

$$a_2 u_2'' + b_2 u_2' + c_2 u_2 = f_2, \tag{9.48}$$

where u_r'' is employed to indicate the second order derivative in z also defined as $u_{r,zz}$ and u_r' means first order derivative in z also defined as $u_{r,z}$. f_1 and f_2 are general external loads, but in the proposed free frequency analysis they are set to zero. The following identities are used

$$a_1 u_1' = a_1 u_1', \tag{9.49}$$

$$a_2 u_2' = a_2 u_2'. \tag{9.50}$$

Therefore, equations (9.47)–(9.50) can be rearranged in the following system

$$\begin{bmatrix} a_1 & 0 & 0 & 0 \\ 0 & a_1 & 0 & 0 \\ 0 & 0 & a_2 & 0 \\ 0 & 0 & 0 & a_2 \end{bmatrix} \begin{bmatrix} u_1 \\ u_1' \\ u_2 \\ u_2' \end{bmatrix}' = \begin{bmatrix} 0 & a_1 & 0 & 0 \\ -c_1 & -b_1 & 0 & 0 \\ 0 & 0 & 0 & a_2 \\ 0 & 0 & -c_2 & -b_2 \end{bmatrix} \begin{bmatrix} u_1 \\ u_1' \\ u_2 \\ u_2' \end{bmatrix} + \begin{bmatrix} 0 \\ f_1 \\ 0 \\ f_2 \end{bmatrix}. \tag{9.51}$$

The methodology here given by means of equations (9.47)–(9.51) can also be applied to the previous system shown in equations (9.44)–(9.46)

$$\begin{bmatrix} A_6^k & 0 & 0 & 0 & 0 & 0 \\ 0 & A_{12}^k & 0 & 0 & 0 & 0 \\ 0 & 0 & A_{19}^k & 0 & 0 & 0 \\ 0 & 0 & 0 & A_6^k & 0 & 0 \\ 0 & 0 & 0 & 0 & A_{12}^k & 0 \\ 0 & 0 & 0 & 0 & 0 & A_{19}^k \end{bmatrix} \begin{bmatrix} U^k \\ V^k \\ W^k \\ U'^k \\ V'^k \\ W'^k \end{bmatrix}' = \begin{bmatrix} 0 & 0 & 0 & A_6^k & 0 & 0 \\ 0 & 0 & 0 & 0 & A_{12}^k & 0 \\ 0 & 0 & 0 & 0 & 0 & A_{19}^k \\ -A_1^k & -A_2^k & -A_3^k & -A_4^k & 0 & -A_5^k \\ -A_7^k & -A_8^k & -A_9^k & 0 & -A_{10}^k & -A_{11}^k \\ -A_{13}^k & -A_{14}^k & -A_{15}^k & -A_{16}^k & -A_{17}^k & -A_{18}^k \end{bmatrix} \begin{bmatrix} U^k \\ V^k \\ W^k \\ U'^k \\ V'^k \\ W'^k \end{bmatrix}. \quad (9.52)$$

Equation (9.52) can be rewritten in a compact form for a general k physical layer

$$\boldsymbol{D}^k \frac{\partial \boldsymbol{U}^k}{\partial \tilde{z}} = \boldsymbol{A}^k \boldsymbol{U}^k, \quad (9.53)$$

where $\frac{\partial \boldsymbol{U}^k}{\partial \tilde{z}} = \boldsymbol{U}'^k$ and $\boldsymbol{U}^k = [U^k\ V^k\ W^k\ U'^k\ V'^k\ W'^k]$. Equation (9.53) can be written as

$$\boldsymbol{D}^k \boldsymbol{U}'^k = \boldsymbol{A}^k \boldsymbol{U}^k, \quad (9.54)$$

$$\boldsymbol{U}'^k = \boldsymbol{D}^{k-1} \boldsymbol{A}^k \boldsymbol{U}^k, \quad (9.55)$$

$$\boldsymbol{U}'^k = \boldsymbol{A}^{*k} \boldsymbol{U}^k, \quad (9.56)$$

with $\boldsymbol{A}^{*k} = \boldsymbol{D}^{k-1} \boldsymbol{A}^k$.

In the case of plate configuration, coefficients A_3^k, A_4^k, A_9^k, A_{10}^k, A_{13}^k, A_{14}^k and A_{18}^k are zero because the radii of curvature R_α and R_β can be considered as infinite. The other coefficients A_1^k, A_2^k, A_5^k, A_6^k, A_7^k, A_8^k, A_{11}^k, A_{12}^k, A_{15}^k, A_{16}^k, A_{17}^k and A_{19}^k are constant inside the k layer because parametric coefficients H_α and H_β become equal to 1 and, therefore, they do not depend on the thickness coordinate $\tilde{z}$. As a consequence, matrices $\boldsymbol{D}^k$, $\boldsymbol{A}^k$ and $\boldsymbol{A}^{*k}$ are constant in each k layer of the considered multilayered plate. The solution of equation (9.56) for the plate case can be written as proposed in [17, 18]

$$\boldsymbol{U}^k(\tilde{z}^k) = \exp(\boldsymbol{A}^{*k} \tilde{z}^k) \boldsymbol{U}^k(0) \quad \text{with} \quad \tilde{z}^k \in [0, h^k], \quad (9.57)$$

where $\tilde{z}^k$ is the thickness coordinate of each layer which goes from 0 at the bottom to h^k at the top (see figure 9.2 for further details). The exponential matrix for the plate case (that means constant coefficients A_s^k) is calculated with $\tilde{z}^k = h^k$ for each k layer as

$$A^{**k} = \exp(A^{*k} h^k) = \boldsymbol{I} + A^{*k}\, h^k + \frac{A^{*k2}}{2!} h^{k2} + \frac{A^{*k3}}{3!} h^{k3} + \dots + \frac{A^{*kN}}{N!} h^{kN}, \quad (9.58)$$

where $\boldsymbol{I}$ is the 6×6 identity matrix. This expansion has a fast convergence as demonstrated in [19] and it is not time consuming from the computational point of view. An expansion order $N = 15$ can be used for one-layered cases in order to have the exact solution for each possible thickness ratio [4]. This expansion order N can be reduced in the case of multilayered structures because it is employed for each single layer. Moreover, it can be further reduced in the case of multilayered shells because several mathematical layers must be introduced in this case because of the curvature approximation. In the case of N_L physical layers for shell geometry, A^{*k} is not constant in each k physical layer because $H_\alpha(\tilde{z})$ and $H_\beta(\tilde{z})$ depend on the thickness coordinate. Therefore, $N_L - 1$ transfer matrices $T^{k-1,\, k}$ must be defined by means of the following interlaminar continuity of displacement and transverse shear/normal stress conditions for each interface

$$u_b^k = u_t^{k-1},\ v_b^k = v_t^{k-1},\ w_b^k = w_t^{k-1}, \quad (9.59)$$

$$\sigma_{zzb}^k = \sigma_{zzt}^{k-1},\ \sigma_{\alpha zb}^k = \sigma_{\alpha zt}^{k-1},\ \sigma_{\beta zb}^k = \sigma_{\beta zt}^{k-1}, \quad (9.60)$$

where each displacement and transverse stress component defined at the top (t) of the k-1 layer must be equal to the corresponding ones defined at the bottom (b) of the k layer. Equations (9.59) and (9.60) must be particularized for the shell geometry.

The interlaminar continuity of transverse shear stresses $\sigma_{\alpha z}$ and $\sigma_{\beta z}$ and transverse normal stress σ_{zz} is imposed by means of equilibrium conditions at each layer interface. The continuity of transverse shear stress $\sigma_{\alpha z}$ at each layer interface of a multilayered shell is

$$\begin{aligned} &\frac{C_{55}^{k-1}}{H_{\alpha t}^{k-1}} \bar{\alpha} W_t^{k-1} + C_{55}^{k-1} U_t'^{k-1} - \frac{C_{55}^{k-1}}{H_{\alpha t}^{k-1} R_\alpha} U_t^{k-1} \\ &= \frac{C_{55}^k}{H_{\alpha b}^k} \bar{\alpha} W_b^k + C_{55}^k U_b'^k - \frac{C_{55}^k}{H_{\alpha b}^k R_\alpha} U_b^k, \end{aligned} \quad (9.61)$$

$$\begin{aligned} U_b'^k = &\frac{1}{C_{55}^k} \left(\frac{C_{55}^{k-1}}{H_{\alpha t}^{k-1}} \bar{\alpha} - \frac{C_{55}^k}{H_{\alpha b}^k} \bar{\alpha} \right) W_t^{k-1} \\ &+ \frac{1}{C_{55}^k} \left(-\frac{C_{55}^{k-1}}{H_{\alpha t}^{k-1} R_\alpha} + \frac{C_{55}^k}{H_{\alpha b}^k R_\alpha} \right) U_t^{k-1} + \left(\frac{C_{55}^{k-1}}{C_{55}^k} \right) U_t'^{k-1}. \end{aligned} \quad (9.62)$$

The continuity of transverse shear stress $\sigma_{\beta z}$ at each layer interface of a multilayered shell is

$$\begin{aligned}&\frac{C_{44}^{k-1}}{H_{\beta t}^{k-1}}\bar{\beta}W_t^{k-1} + C_{44}^{k-1}V_t^{'k-1} - \frac{C_{44}^{k-1}}{H_{\beta t}^{k-1}R_\beta}V_t^{k-1}\\ &= \frac{C_{44}^{k}}{H_{\beta b}^{k}}\bar{\beta}W_b^{k} + C_{44}^{k}V_b^{'k} - \frac{C_{44}^{k}}{H_{\beta b}^{k}R_\beta}V_b^{k},\end{aligned} \tag{9.63}$$

$$\begin{aligned}V_b^{'k} = &\frac{1}{C_{44}^k}\left(\frac{C_{44}^{k-1}}{H_{\beta t}^{k-1}}\bar{\beta} - \frac{C_{44}^{k}}{H_{\beta b}^{k}}\bar{\beta}\right)W_t^{k-1}\\ &+ \frac{1}{C_{44}^k}\left(-\frac{C_{44}^{k-1}}{H_{\beta t}^{k-1}R_\beta} + \frac{C_{44}^{k}}{H_{\beta b}^{k}R_\beta}\right)V_t^{k-1} + \left(\frac{C_{44}^{k-1}}{C_{44}^{k}}\right)V_t^{'k-1}.\end{aligned} \tag{9.64}$$

The continuity of transverse normal stress σ_{zz} at each layer interface of a multilayered shell is

$$\begin{aligned}&-\frac{C_{13}^{k-1}}{H_{\alpha t}^{k-1}}\bar{\alpha}U_t^{k-1} + \frac{C_{13}^{k-1}}{H_{\alpha t}^{k-1}R_\alpha}W_t^{k-1} - \frac{C_{23}^{k-1}}{H_{\beta t}^{k-1}}\bar{\beta}V_t^{k-1}\\ &+ \frac{C_{23}^{k-1}}{H_{\beta t}^{k-1}R_\beta}W_t^{k-1} + C_{33}^{k-1}W_t^{'k-1} = -\frac{C_{13}^{k}}{H_{\alpha b}^{k}}\bar{\alpha}U_b^{k} + \frac{C_{13}^{k}}{H_{\alpha b}^{k}R_\alpha}W_b^{k}\\ &-\frac{C_{23}^{k}}{H_{\beta b}^{k}}\bar{\beta}V_b^{k} + \frac{C_{23}^{k}}{H_{\beta b}^{k}R_\beta}W_b^{k} + C_{33}^{k}W_b^{'k},\end{aligned} \tag{9.65}$$

$$\begin{aligned}W_b^{'k} = &\frac{1}{C_{33}^k}\left(-\frac{C_{13}^{k-1}}{H_{\alpha t}^{k-1}}\bar{\alpha} + \frac{C_{13}^{k}}{H_{\alpha b}^{k}}\bar{\alpha}\right)U_t^{k-1} + \frac{1}{C_{33}^k}\left(-\frac{C_{23}^{k-1}}{H_{\beta t}^{k-1}}\bar{\beta} + \frac{C_{23}^{k}}{H_{\beta b}^{k}}\bar{\beta}\right)V_t^{k-1} +\\ &\frac{1}{C_{33}^k}\left(\frac{C_{13}^{k-1}}{H_{\alpha t}^{k-1}R_\alpha} + \frac{C_{23}^{k-1}}{H_{\beta t}^{k-1}R_\beta} - \frac{C_{13}^{k}}{H_{\alpha b}^{k}R_\alpha} - \frac{C_{23}^{k}}{H_{\beta b}^{k}R_\beta}\right)W_t^{k-1} + \left(\frac{C_{33}^{k-1}}{C_{33}^{k}}\right)W_t^{'k-1}.\end{aligned} \tag{9.66}$$

The continuity of displacement components is imposed by means of congruence conditions at each layer interface

$$U_b^k = U_t^{k-1}, \quad V_b^k = V_t^{k-1}, \quad W_b^k = W_t^{k-1}. \tag{9.67}$$

In equations (9.61)–(9.67), t and b indicate top and bottom of $k-1$ layer and k layer, respectively. $\bar{\alpha}$, $\bar{\beta}$, R_α and R_β are referred to mid-surface Ω_0 of the whole multilayer structure. H_α and H_β are defined at each interface between the generic $k-1$ layer and the generic k layer. Equations (9.61)–(9.67) can be transformed into an opportune system

$$\begin{bmatrix} U \\ V \\ W \\ U' \\ V' \\ W' \end{bmatrix}_b^k = \begin{bmatrix} 1 & 0 & 0 & 0 & 0 & 0 \\ 0 & 1 & 0 & 0 & 0 & 0 \\ 0 & 0 & 1 & 0 & 0 & 0 \\ T_1 & 0 & T_2 & T_3 & 0 & 0 \\ 0 & T_4 & T_5 & 0 & T_6 & 0 \\ T_7 & T_8 & T_9 & 0 & 0 & T_{10} \end{bmatrix}^{k-1,\,k} \begin{bmatrix} U \\ V \\ W \\ U' \\ V' \\ W' \end{bmatrix}_t^{k-1}, \tag{9.68}$$

The compact form of equation (9.68) is

$$\boldsymbol{U}_b^k = \boldsymbol{T}^{k-1,\,k}\boldsymbol{U}_t^{k-1}. \tag{9.69}$$

The calculated $\boldsymbol{T}^{k-1,\,k}$ matrices allow one to link $\boldsymbol{U}$ at the bottom (b) of the k layer with $\boldsymbol{U}$ at the top (t) of the $k-1$ layer. Equation (9.69) can be also given in an alternative form

$$\boldsymbol{U}^k(0) = \boldsymbol{T}^{k-1,\,k}\ \boldsymbol{U}^{k-1}(h_{k-1}), \tag{9.70}$$

where $\boldsymbol{U}^k$ is defined for $\tilde{z}^k = 0$ and $\boldsymbol{U}^{k-1}$ is defined for $\tilde{z}^{k-1} = h^{k-1}$. $\boldsymbol{U}$ at the top of the k layer is linked with $\boldsymbol{U}$ at the bottom of the same k layer using the exponential matrix $\boldsymbol{A}^{**k}$, namely

$$\boldsymbol{U}^k(h^k) = \boldsymbol{A}^{**k}\ \boldsymbol{U}^k(0), \tag{9.71}$$

Equation (9.70) can be recursively introduced into equation (9.71) for the all the $N_L - 1$ interfaces in order to obtain

$$\boldsymbol{U}^{N_L}(h^{N_L}) = \boldsymbol{A}^{**N_L}\ \boldsymbol{T}^{N_L-1,\,N_L}\ \boldsymbol{A}^{**N_L-1}\ \boldsymbol{T}^{N_L-2,\,N_L-1}\ldots\ldots\boldsymbol{A}^{**2}\ \boldsymbol{T}^{1,2}\ \boldsymbol{A}^{**1}\ \boldsymbol{U}^1(0), \tag{9.72}$$

the definition of the matrix $\boldsymbol{H}_m$ for the multilayered plate allows equation (9.72) to be written as

$$\boldsymbol{U}^{N_L}(h^{N_L}) = \boldsymbol{H}_m\ \boldsymbol{U}^1(0). \tag{9.73}$$

This last equation allows one to link $\boldsymbol{U}$ defined at the top of the last N_L layer with $\boldsymbol{U}$ defined at the bottom of the first layer. In the case of multilayered plates matrices, $\boldsymbol{D}^k$, $\boldsymbol{A}^k$ and $\boldsymbol{A}^{*k}$ are constant in each k layer because R_α and R_β are infinite and, therefore, H_α and H_β assume the unit value. In the case of shell geometry, matrices $\boldsymbol{D}^k$, $\boldsymbol{A}^k$ and $\boldsymbol{A}^{*k}$ are not constant in each layer because parametric coefficients H_α and H_β depend on $\tilde{z}$ coordinate (see the geometries and definitions in figure 9.2). A first method to solve this problem could be the use of hypothesis $\frac{z}{R_\alpha} = \frac{z}{R_\beta} = 0$ (but it is valid only for very thin shells) that means $H_\alpha = H_\beta = 1$. Thanks these hypotheses, the solution is the same already seen for the plate case because matrices $\boldsymbol{D}^k$, $\boldsymbol{A}^k$ and $\boldsymbol{A}^{*k}$ are considered as constant in each k layer. However, this methodology cannot be used in this chapter because it is an approximation that is correct only for very thin shells. Therefore, it does not consider the exact geometry for the analyzed structures. The second method (used in this chapter) is based on introduction of several j fictitious or mathematical layers inside each k layer in order to exactly define H_α and H_β. Matrices $\boldsymbol{A}^{**j}$ are constant in the j layer because they can be calculated with R_α,

R_β, $\bar{\alpha}$ and $\bar{\beta}$ defined at the mid-surface Ω_0 of the whole shell structure, and H_α and H_β are defined in the middle of each j mathematical layer. Matrices $\boldsymbol{T}^{j-1,j}$ are also constant because they are evaluated with R_α, R_β, $\bar{\alpha}$ and $\bar{\beta}$ calculated in the mid-surface Ω_0 of the shell, and H_α and H_β are defined in this case at each mathematical interface. In the present chapter, each k physical layer of the multilayered shell is divided into j fictitious/mathematical layers where we can recursively apply equations (9.61)–(9.72) where a new index q is introduced in place of the physical index k. This index q considers the partition of each k physical layer in several j mathematical/fictitious layers. The thickness of each mathematical or fictitious layer is defined as h^q. Therefore, the new index q considers all the fictitious and physical layers in the structures and it goes from 1 to P. The total number of mathematical layers for a multilayered shell could vary from $P = 100$ to $P = 300$ (depending on the analyzed case in the section about results) without problems in terms of computational costs.

The analyzed structures in the next section have always simply-supported edges and they are free stresses at their top and bottom, i.e.

$$\sigma_{zz} = \sigma_{\alpha z} = \sigma_{\beta z} = 0 \quad \text{for} \quad z = -h/2, +h/2 \;\; \text{or} \;\; \tilde{z} = 0, h, \tag{9.74}$$

$$w = v = 0, \;\; \sigma_{\alpha\alpha} = 0 \quad \text{for} \quad \alpha = 0, a, \tag{9.75}$$

$$w = u = 0, \;\; \sigma_{\beta\beta} = 0 \quad \text{for} \quad \beta = 0, b, \tag{9.76}$$

Transverse shear/normal stresses defined for a generic value of $\tilde{z}$ inside the q layer are

$$\begin{aligned} \sigma^q_{zz}(\tilde{z}) &= \frac{C^q_{13}}{H_\alpha(\tilde{z})} u^q_{,\alpha} + \frac{C^q_{13}}{H_\alpha(\tilde{z}) R_\alpha} w^q + \frac{C^q_{23}}{H_\beta(\tilde{z})} v^q_{,\beta} + \frac{C^q_{23}}{H_\beta(\tilde{z}) R_\beta} w^q + C^q_{33} w^q_{,z} \\ &= -\bar{\alpha} \frac{C^q_{13}}{H_\alpha(\tilde{z})} U^q + \frac{C^q_{13}}{H_\alpha(\tilde{z}) R_\alpha} W^q - \bar{\beta} \frac{C^q_{23}}{H_\beta(\tilde{z})} V^q + \frac{C^q_{23}}{H_\beta(\tilde{z}) R_\beta} W^q + C^q_{33} W^q_{,z}, \end{aligned} \tag{9.77}$$

$$\begin{aligned} \sigma^q_{\beta z}(\tilde{z}) &= \frac{C^q_{44}}{H_\beta(\tilde{z})} w^q_{,\beta} + C^q_{44} v^q_{,z} - \frac{C^q_{44}}{H_\beta(\tilde{z}) R_\beta} v^q \\ &= \bar{\beta} \frac{C^q_{44}}{H_\beta(\tilde{z})} W^q + C^q_{44} V^q_{,z} - \frac{C^q_{44}}{H_\beta(\tilde{z}) R_\beta} V^q, \end{aligned} \tag{9.78}$$

$$\begin{aligned} \sigma^q_{\alpha z}(\tilde{z}) &= \frac{C^q_{55}}{H_\alpha(\tilde{z})} w^q_{,\alpha} + C^q_{55} u^q_{,z} - \frac{C^q_{55}}{H_\alpha(\tilde{z}) R_\alpha} u^q \\ &= \bar{\alpha} \frac{C^q_{55}}{H_\alpha(\tilde{z})} W^q + C^q_{55} U^q_{,z} - \frac{C^q_{55}}{H_\alpha(\tilde{z}) R_\alpha} U^q, \end{aligned} \tag{9.79}$$

Equation (9.74) imposed at the top (t) of the last P layer uses equations (9.77)–(9.79) with R_α, R_β, $\bar{\alpha}$ and $\bar{\beta}$ defined at the mid-surface Ω_0 of the whole multilayered shell, and with $H_{\alpha t}$ and $H_{\beta t}$ defined at top of the whole multilayered shell when $\tilde{z} = h$

$$\sigma_{zzt}^P = -\bar{\alpha}\frac{C_{13}^P}{H_{\alpha t}}U_t^P + \frac{C_{13}^P}{H_{\alpha t}R_\alpha}W_t^P - \bar{\beta}\frac{C_{23}^P}{H_{\beta t}}V_t^P + \frac{C_{23}^P}{H_{\beta t}R_\beta}W_t^P + C_{33}^P W_{t,z}^P = 0, \quad (9.80)$$

$$\sigma_{\beta zt}^P = \bar{\beta}\frac{C_{44}^P}{H_{\beta t}}W_t^P + C_{44}^P V_{t,z}^P - \frac{C_{44}^P}{H_{\beta t}R_\beta}V_t^P = 0, \quad (9.81)$$

$$\sigma_{\alpha zt}^P = \bar{\alpha}\frac{C_{55}^P}{H_{\alpha t}}W_t^P + C_{55}^P U_{t,z}^P - \frac{C_{55}^P}{H_{\alpha t}R_\alpha}U_t^P = 0, \quad (9.82)$$

Equation (9.74) imposed at the bottom (b) of the first layer ($q = 1$) uses equations (9.77)–(9.79) with R_α, R_β, $\bar{\alpha}$ and $\bar{\beta}$ defined at the mid-surface Ω_0 of the whole multilayered shell, and with $H_{\alpha b}$ and $H_{\beta b}$ defined at bottom of the whole shell for $\tilde{z} = 0$

$$\begin{aligned}\sigma_{zzb}^1 = &-\bar{\alpha}\frac{C_{13}^1}{H_{\alpha b}}U_b^1 + \frac{C_{13}^1}{H_{\alpha b}R_\alpha}W_b^1 - \bar{\beta}\frac{C_{23}^1}{H_{\beta b}}V_b^1 \\ &+ \frac{C_{23}^1}{H_{\beta b}R_\beta}W_b^1 + C_{33}^1 W_{b,z}^1 = 0,\end{aligned} \quad (9.83)$$

$$\sigma_{\beta zb}^1 = \bar{\beta}\frac{C_{44}^1}{H_{\beta b}}W_b^1 + C_{44}^1 V_{b,z}^1 - \frac{C_{44}^1}{H_{\beta b}R_\beta}V_b^1 = 0, \quad (9.84)$$

$$\sigma_{\alpha zb}^1 = \bar{\alpha}\frac{C_{55}^1}{H_{\alpha b}}W_b^1 + C_{55}^1 U_{b,z}^1 - \frac{C_{55}^1}{H_{\alpha b}R_\alpha}U_b^1 = 0. \quad (9.85)$$

Equations (9.80)–(9.82) in matrix form are (where $\boldsymbol{U}^P(h^P)$ means $\boldsymbol{U}$ defined at the top of the whole multilayered shell, last P layer with $\tilde{z}^P = h^P$)

$$\begin{bmatrix} -\bar{\alpha}\frac{C_{13}^P}{H_{\alpha t}} & -\bar{\beta}\frac{C_{23}^P}{H_{\beta t}} & \left(\frac{C_{13}^P}{H_{\alpha t}R_\alpha} + \frac{C_{23}^P}{H_{\beta t}R_\beta}\right) & 0 & 0 & C_{33}^P \\ 0 & -\frac{C_{44}^P}{H_{\beta t}R_\beta} & \bar{\beta}\frac{C_{44}^P}{H_{\beta t}} & 0 & C_{44}^P & 0 \\ -\frac{C_{55}^P}{H_{\alpha t}R_\alpha} & 0 & \bar{\alpha}\frac{C_{55}^P}{H_{\alpha t}} & C_{55}^P & 0 & 0 \end{bmatrix} \begin{bmatrix} U^P(h^P) \\ V^P(h^P) \\ W^P(h^P) \\ U'^P(h^P) \\ V'^P(h^P) \\ W'^P(h^P) \end{bmatrix} = \begin{bmatrix} 0 \\ 0 \\ 0 \end{bmatrix}. \quad (9.86)$$

Equations (9.83)–(9.85) in matrix form are (where $\boldsymbol{U}^1(0)$ means $\boldsymbol{U}$ calculated at the bottom of the whole multilayered shell, first layer 1 with $\tilde{z}^1 = 0$)

$$\begin{bmatrix} -\bar{\alpha}\dfrac{C_{13}^1}{H_{ab}} & -\bar{\beta}\dfrac{C_{23}^1}{H_{\beta b}} & \left(\dfrac{C_{13}^1}{H_{ab}R_\alpha}+\dfrac{C_{23}^1}{H_{\beta b}R_\beta}\right) & 0 & 0 & C_{33}^1 \\ 0 & -\dfrac{C_{44}^1}{H_{\beta b}R_\beta} & \bar{\beta}\dfrac{C_{44}^1}{H_{\beta b}} & 0 & C_{44}^1 & 0 \\ -\dfrac{C_{55}^1}{H_{ab}R_\alpha} & 0 & \bar{\alpha}\dfrac{C_{55}^1}{H_{ab}} & C_{55}^1 & 0 & 0 \end{bmatrix} \begin{bmatrix} U^1(0) \\ V^1(0) \\ W^1(0) \\ U'^1(0) \\ V'^1(0) \\ W'^1(0) \end{bmatrix} = \begin{bmatrix} 0 \\ 0 \\ 0 \end{bmatrix}. \quad (9.87)$$

The compact form of equations (9.86) and (9.87) in order to show the free stress state at the top and bottom of the whole multilayered shell is

$$\boldsymbol{B}^P(h^P)\,\boldsymbol{U}^P(h^P) = 0, \quad (9.88)$$

$$\boldsymbol{B}^1(0)\,\boldsymbol{U}^1(0) = 0, \quad (9.89)$$

Equation (9.73) can be substituted in equation (9.88) considering a global number of layers (mathematical and physical ones) equals P

$$\boldsymbol{B}^P(h^P)\,\boldsymbol{H}_m\,\boldsymbol{U}^1(0) = 0, \quad (9.90)$$

$$\boldsymbol{B}^1(0)\,\boldsymbol{U}^1(0) = 0, \quad (9.91)$$

therefore, equations (9.90) and (9.91) can be grouped in the following system

$$\begin{bmatrix} \boldsymbol{B}^P(h^P)\,\boldsymbol{H}_m \\ \boldsymbol{B}^1(0) \end{bmatrix} \left[\boldsymbol{U}^1(0)\right] = [0], \quad (9.92)$$

and introducing the (6×6) $\boldsymbol{E}$ matrix, the equation (9.92) can be given as

$$[\boldsymbol{E}]\left[\boldsymbol{U}^1(0)\right] = [0]. \quad (9.93)$$

equation (9.93) is also valid for plate case where the mathematical layers are not introduced and $\boldsymbol{B}^{N_L}(h^{N_L}) = \boldsymbol{B}^P(h^P)$. Matrix $\boldsymbol{E}$ has always (6×6) dimension for each possible number of global layers P, even if the method is based on an LW approach. The solution has been developed in a Matlab code where only the spherical shell method is included. In fact, such methodology easily degenerates into cylindrical open/closed shell and plate methodologies.

The free frequency analysis is based on the research of the non-trivial solution of $\boldsymbol{U}^1(0)$ in equation (9.93). This feature means the imposition of the determinant of matrix $\boldsymbol{E}$ as equal to zero

$$det[\boldsymbol{E}] = 0, \quad (9.94)$$

the meaning of equation (9.94) is to research the roots of a higher order polynomial defined in $\lambda = \omega^2$. For each imposed couple of half-wave numbers (m, n), a certain number of circular frequencies are possible. Their number depends on the order N chosen for each exponential matrix $\boldsymbol{A}^{**q}$.

When half-wave numbers m and n are imposed in the plate and shell structures, a certain number of circular frequencies ω_l can be calculated. For each frequency ω_l, it is possible to define the vibration mode through the thickness direction in terms of the three displacement components u, v and w. Then, when the frequency ω_l is substituted in (6×6) matrix $\boldsymbol{E}$, this last matrix shows six eigenvalues. We are focused on the null space of matrix $\boldsymbol{E}$, this idea means to find the (6×1) eigenvector related to the minimum value of the six calculated eigenvalues. The null space for each chosen frequency ω_l allows the calculation of the vector $\boldsymbol{U}$ defined at the bottom of the whole structure

$$\boldsymbol{U}^1_{\omega_l}(0) = \left[U^1(0) \;\; V^1(0) \;\; W^1(0) \;\; U'^1(0) \;\; V'^1(0) \;\; W'^1(0)\right]^T_{\omega_l}, \tag{9.95}$$

T indicates the transpose of the vector and the subscript ω_l means that the null space is defined for a given circular frequency ω_l.

The vector $\boldsymbol{U}^q_{\omega_l}(\tilde{z}^q)$ (with the three displacement components $U^q_{\omega_l}(\tilde{z}^q)$, $V^q_{\omega_l}(\tilde{z}^q)$ and $W^q_{\omega_l}(\tilde{z}^q)$ through the thickness) can be calculated for each q layer of the multilayered structure by employing equations (9.70)–(9.73) where the index q goes from 1 to P. The thickness coordinate $\tilde{z}$ can have all the values defined from the bottom to the top of the structure. For the plate case, the described procedure is simpler because there are not the j mathematical layers and, therefore, the index q is coincident with the index k for the physical layers (in this case, the total number of layers is defined as N_L).

9.1.5 Particular case of geometrical relations and 3D equilibrium equations for cylinders in order to analyze carbon nanotubes

The 3D shell formulation described in the present section has been developed in orthogonal mixed curvilinear coordinates (α, β, z) and it is valid for spherical shell panels. However, it automatically degenerates in formulation for cylindrical shell panels, cylinders and plates by means of simple considerations made on radii of curvature. The spherical 3D shell model has been developed in in-house academic software called 3DES (which means 3D Exact Solutions) using a Matlab environment. 3DES code is able to analyze spherical and cylindrical shell geometries, cylinders and plates.

In the case of free vibration analysis of carbon nanotubes, the presented formulation is employed and the only differences are for the geometrical relations and 3D equilibrium equations that are particularized for cylinders by considering infinite radius of curvature R_β and rectilinear coordinate β.

The strain-displacement relations of the three-dimensional theory of elasticity in orthogonal mixed curvilinear coordinates were written in equations (9.17)–(9.22) for the general physical k layer of the multilayered spherical shell, they degenerate in those for cylinders as

$$\epsilon^k_{\alpha\alpha} = \frac{1}{H_\alpha} u^k_{,\alpha} + \frac{w^k}{H_\alpha R_\alpha} \tag{9.96}$$

$$\epsilon^k_{\beta\beta} = v^k_{,\beta} \tag{9.97}$$

$$\epsilon^k_{zz} = w^k_{,z} \tag{9.98}$$

$$\gamma^k_{\beta z} = w^k_{,\beta} + v^k_{,z} \tag{9.99}$$

$$\gamma^k_{\alpha z} = \frac{1}{H_\alpha} w^k_{,\alpha} + u^k_{,z} - \frac{u^k}{H_\alpha R_\alpha} \tag{9.100}$$

$$\gamma^k_{\alpha\beta} = \frac{1}{H_\alpha} v^k_{,\alpha} + u^k_{,\beta}. \tag{9.101}$$

The radius of curvature R_β is infinite, the coordinate β is rectilinear and parametric coefficients for cylindrical shells become

$$H_\alpha = (1 + \frac{z}{R_\alpha}), \quad H_\beta = 1, \quad H_z = 1. \tag{9.102}$$

The most general form of differential equations of equilibrium for spherical shells has been given in equations (9.8)–(9.10). These equations rewritten for the case of free frequency analysis of cylinders are

$$\sigma^k_{\alpha\alpha,\,\alpha} + H_\alpha \sigma^k_{\alpha\beta,\,\beta} + H_\alpha \sigma^k_{\alpha z,\,z} + \frac{2}{R_\alpha} \sigma^k_{\alpha z} = \rho^k H_\alpha \ddot{u}^k \tag{9.103}$$

$$\sigma^k_{\alpha\beta,\,\alpha} + H_\alpha \sigma^k_{\beta\beta,\,\beta} + H_\alpha \sigma^k_{\beta z,\,z} + \frac{1}{R_\alpha} \sigma^k_{\beta z} = \rho^k H_\alpha \ddot{v}^k \tag{9.104}$$

$$\sigma^k_{\alpha z,\,\alpha} + H_\alpha \sigma^k_{\beta z,\,\beta} + H_\alpha \sigma^k_{zz,\,z} - \frac{1}{R_\alpha} \sigma^k_{\alpha\alpha} + \frac{1}{R_\alpha} \sigma^k_{zz} = \rho^k H_\alpha \ddot{w}^k \tag{9.105}$$

All the rest of the equations and the solution procedure do not change with respect to those given in previous subsections 9.1.1–9.1.4.

9.2 Results

The section about results is organized in three main parts. In the first part, free vibration analyses of SWCNTs is proposed by using the exact 3D shell model here presented. The second part shows the free vibration analyses of DWCNTs by using the described 3D LW shell model. In this case, it is remarked how short CNTs cannot be analyzed by means of beam models. In the case of DWCNT analyses, the van der Waals effects can also be investigated. The third part proposes comparisons between the proposed 3D exact shell model and several 2D classical and refined shell models developed in both LW and Equivalent Single Layer (ESL) form by using the Generalized Differential Quadrature (GDQ) model. All these models have been compared in the case of free vibration analyses of SWCNTs and DWCNTs. The results shown in this section are a summary of the results already proposed in [1]–[3], readers can refer to them for further details.

9.2.1 Free frequencies and vibration modes for SWCNTs

A preliminary assessment is proposed in the case of a simply-supported SWCNT. A beam analysis for the free frequency investigation was proposed in Simsek [20] and Aydogdu [21]. The equivalent elastic cylinder (as proposed in figure 9.3) has a very small diameter/thickness ratio. Therefore, the use of classical 2D shell models could show several difficulties in the free vibration analysis and the use of a 3D shell model could be mandatory. Moreover, in the case of frequency investigations for a large range of length/diameter ratios, beam models may become inappropriate for short SWCNTs. After the proposed preliminary assessments, the 3D shell model will be employed to find the natural frequencies of several SWCNTs having different geometries (armachair, zigzag and general chirality as shown in figure 9.4) and length/radius ratios.

The SWCNT is considered as simply-supported, the equivalent elastic cylinder proposed in figure 9.3 has equivalent elastic properties as suggested in Simsek [20]. The equivalent Young's modulus is $E = 1$ TPa and the Poisson ratio ν equals 0.3. The effective thickness combined with the proposed Young's modulus is $h = 0.35$ nm with mass density $\rho = 2300$ kg m^{-3}. The external diameter of the cylinder is $d_e = 1$ nm. Therefore, the diameter/thickness ratio is $d_e/h = 2.86$. This d_e/h ratio needs the use of beam models. In fact, several difficulties may arise when classical 2D shell models are used for the analysis of these cylinders as demonstrated in [22]. Therefore, the use of very refined 2D shell models or 3D shell models could be mandatory to overcome these limitations. The radius of curvature in α direction, with respect to the mid-surface, is $R_\alpha = d_e/2 - h/2 = 0.325$ nm. Therefore, the dimension in α direction is $a = 2\pi R_\alpha$ and the dimension b along the rectilinear coordinate β could have the following values: $L = 5$ nm, 10 nm, 20 nm, 50 nm and 100 nm that means ratios length/diameter $L/d_e = 5$, 10, 20, 50 and 100, respectively.

Table 9.1 proposes the first three dimensionless circular frequencies $\bar{\omega} = \omega L^2\sqrt{\frac{\rho A}{EI}}$ for short and long simply-supported cylinders having different L/d_e ratios. In the dimensionless form, A is the area of the ring and I is the moment of inertia of the

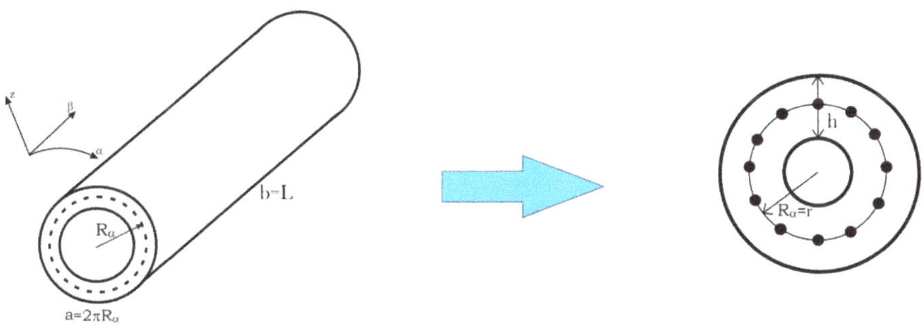

Figure 9.3. Equivalent continuum approach for a SWCNT: notation and reference system.

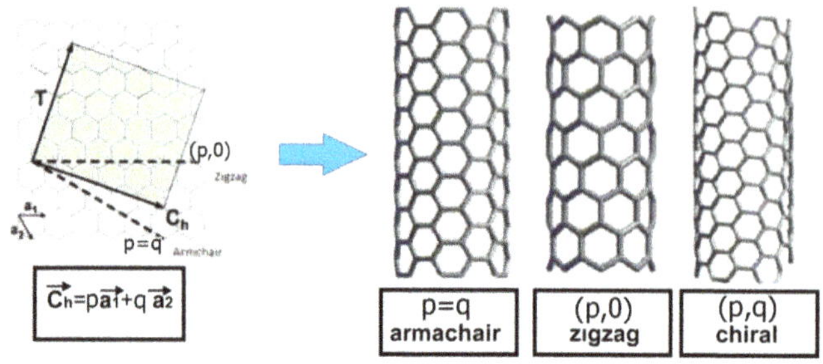

Figure 9.4. Representation of the chiral vector in the case armachair, zigzag and general chirality in SWCNTs.

Table 9.1. Preliminary assessment: comparison between the proposed 3D shell model and the EBM and the TBM as developed in [20] and [21]. The first three fundamental dimensionless circular frequencies $\bar{\omega}$ are obtained for different L/d_e ratios using appropriate values for m and n half-wave numbers.

Mode (m, n)	3D Shell	EBM [20, 21]	TBM [21]
$L/d_e = 5$			
I (2,1)	9.348 1	…	…
II (2,2)	32.917	…	…
III (2,3)	63.917	…	…
$L/d_e = 10$			
I (2,1)	9.729 5	9.869 6	9.744 3
II (2,2)	37.392	39.478	36.841
III (2,3)	79.361	88.826	57.450
$L/d_e = 20$			
I (2,1)	9.835 6	9.869 6	9.838 1
II (2,2)	38.918	39.478	38.964
III (2,3)	86.072	88.826	85.748
$L/d_e = 50$			
I (2,1)	9.863 8	9.869 6	9.864 5
II (2,2)	39.392	39.478	39.398
III (2,3)	88.375	88.826	88.415
$L/d_e = 100$			
I (2,1)	9.848 7	…	…
II (2,2)	39.488	…	…
III (2,3)	88.752	…	…

ring. The first three dimensionless circular frequencies are calculated when half-wave number $m = 2$ is imposed in α direction and half-wave numbers n in β direction vary from 1 to 3. It is clear how beam theories give correct results for long and moderately long cylinders. In contrast, shell models always give correct results for both long and short cylinders. The Euler–Bernoulli Beam Model (EBM) was given in [20] and [21] to calculate the first three frequencies when ratios L/d_e change from 10 to 50. The same analyses were also proposed in [21] by using the Timoshenko Beam Model (TBM). TBM shows more accurate results than EBM because it considers transverse shear deformation and rotary inertia effects. However, TBM could give some problems for higher order frequencies (the second and the third one) in the case of very short SWCNTs (L/d_e = 10). The proposed 3D shell model always gives satisfactory results for both long and short SWCNTs and it also gives correct free frequency response in the case of cylinders having small diameter/thickness ratios. For small diameter/thickness ratios, classical 2D shell models could show several difficulties. Table 9.1 demonstrates how the TB model obtains results very closed to those obtained via the 3D shell model. In contrast, the EB model gives larger differences. However, the TB model has several difficulties for short SWCNTs because such elements are not 1D structures. Additional results are included in table 9.1 (they were not obtained in [20] and [21] via beam models in cases of very short and very long SWCNTs). Therefore, an exhaustive overview for the SWCNT behavior is shown and these results could be used as a benchmark for the possible validation of future 1D beam and 2D shell models. The addition of new results for very short and very long SWCNTs allows researchers involved in beam and shell model analyses of SWCNTs to try to complete this table in order to validate their models. After the proposed assessment, the developed 3D shell model can be now considered as validated and it can be used with confidence for further free vibration investigations of SWCNTs.

In the previous assessment, the effective thickness h of the SWCNT was assumed to be equal to the inter-planar spacing of graphite layers (0.35 nm) and the mechanical properties were consequently defined (e.g., Young's modulus and Poisson ratio). The appropriate choice of the thickness value is fundamental in the definition of the equivalent mechanical properties. In the open literature, we can find several dissenting opinions about the choices for the equivalent mechanical properties of carbon nanotubes. The present section does not discuss these features because it is only devoted to the development of an appropriate 3D shell model able to validate future continuum approaches by other scientists and based on 1D beam and 2D shell theories. The choice of h = 0.35 nm gives thick cylinders that are very hard to be correctly analyzed by means of 1D beam and 2D shell theories. In the case of smaller thickness values and consequently higher values of Young's modulus, new benchmarks are shown and different behaviors and problems are discussed. The SWCNTs can have different geometries depending on the chiral vector defined as $\overrightarrow{C_h} = p\overrightarrow{a_1} + q\overrightarrow{a_2}$. The chiral vector is clearly explained in figure 9.4 and it proposes several radii of curvature values, as shown in table 9.2. Moreover, the change of chiral vector and radius gives different types of CNTs (see figure 9.4 and table 9.2 for further details about armachair, zigzag and general chirality CNTs). Chen and Cao [23] proposed several continuum approaches for CNT free vibration analysis by giving a completely different value for the global thickness wall h. This value was defined as h = 0.08 nm and it is related to an effective Young's modulus E = 6.85 TPa and to an effective Poisson ratio ν = 0.19. The continuum approaches proposed

Table 9.2. Geometry and properties (radius of curvature and number of atoms) for several types of SWCNTs used in the proposed analysis and those by Chen and Cao[23].

SWCNT (p,q)	(5,5)	(6,6)	(7,7)	(8,8)	(10,0)	(12,0)	(14,0)	(16,0)	(8,4)
r (nm)	0.338	0.405	0.473	0.540	0.390	0.468	0.546	0.624	0.413
n atoms	1040	1248	1456	1664	1200	1440	1680	1920	1200

in [23] used these equivalent mechanical properties for CNT analysis, and the obtained results are compared with those obtained via the Molecular Dynamic (MD) analysis showing a global difference less than 5%. For these reasons, these equivalent elastic properties have been used for the new benchmarks proposed in this section by means of the already presented 3D exact shell model. Table 9.2 gives the radius of curvature r and the number of atoms for several SWCNT configurations (armachair with $p = q$, zigzag with (p,0) and general chirality with $p \neq q$). The mass density ρ is not necessary to be defined because the circular frequencies are proposed in a dimensionless form $\omega^* = \omega L\sqrt{\frac{\rho}{E}}$ that includes ρ. Several length/radius ratios (L/r) are investigated for each SWCNT geometry ($r = R_\alpha$ is the radius of curvature in the curved α direction and it is measured with respect to the global mid-surface). Table 9.3 gives the circular frequencies for each SWCNT described in table 9.2 and in figure 9.4 by considering length/radius ratios L/r equals 20, 25, 30 and 35. The frequencies are calculated for an imposed half-wave number in α direction $m = 2$ (the cylinder is closed in this direction and only even m values can be imposed) and half-wave numbers in rectilinear β direction n from 1 to 5. For each SWCNT considered (armachair, zigzag and general chirality CNT), the dimensionless circular frequencies (in terms of the first five vibration modes) decrease when the length L of each CNT increases (that means bigger values of the L/r ratio). A longer cylinder is in general less rigid than the same cylinder having a shorter length L. The armachair SWCNTs ($p = q$) have frequencies that decrease when the value $p = q$ increases, this feature is due to the fact that armachair CNTs with higher values of $p = q$ have bigger radii of curvature (therefore, they are less rigid cylinders). This described behavior is confirmed for each investigated vibration mode (from the first mode to the fifth one) and for each L/r ratio. Zigzag SWCNTs ($p = 0$) have a radius of curvature that increases when the q value also increases (consequently the length also increases because the L/r ratio is bigger). Therefore, the behavior is the same already seen for the armachair CNTs. The general chirality (8,4) SWCNT has a behavior similar to that obtained for the (6,6) armachair SWCNT because the radii of curvature are almost the same for these two last cases. Figure 9.5 summarizes in graphical form some of the results already shown in table 9.3: in particular, the dimensionless circular frequencies for armachair and zigzag SWCNTs in relation to the L/r ratio. We can note that each dimensionless circular frequency decreases as the L/r ratio increases. These comments remain valid for both armachair (5,5) and zigzag (16,0) SWCNTs, as clearly indicated in figure 9.5.

Table 9.3. Exact 3D shell model: dimensionless circular frequencies ω^* for different SWCNTs and L/r ratios. Imposed half-wave numbers $m = 2$ in α direction and $n = 1, \ldots, 5$ in β direction.

L/r	I (2,1)	II (2,2)	III (2,3)	IV (2,4)	V (2,5)
armachair (5,5)					
20	0.340 4	1.253 5	2.526 1	3.982 5	5.513 8
25	0.275 4	1.041 3	2.163 5	3.508 8	4.978 1
30	0.230 9	0.887 0	1.880 2	3.111 4	4.496 7
35	0.198 7	0.770 9	1.656 4	2.781 5	4.077 1
armachair (6,6)					
20	0.339 7	1.250 7	2.519 9	3.971 2	5.495 3
25	0.274 8	1.039 0	2.158 5	3.499 9	4.964 1
30	0.230 4	0.885 1	1.876 0	3.104 0	4.485 1
35	0.198 2	0.769 2	1.652 8	2.775 1	4.067 1
armachair (7,7)					
20	0.339 2	1.248 9	2.516 0	3.964 3	5.483 8
25	0.274 4	1.037 6	2.155 4	3.494 5	4.955 4
30	0.230 1	0.883 9	1.873 4	3.099 4	4.478 0
35	0.198 0	0.768 2	1.650 5	2.771 1	4.061 0
armachair (8,8)					
20	0.338 9	1.247 8	2.513 5	3.959 8	5.476 4
25	0.274 2	1.036 7	2.153 4	3.490 4	4.949 8
30	0.229 9	0.883 2	1.871 7	3.096 5	4.473 4
35	0.197 8	0.767 6	1.649 1	2.768 6	4.057 1
zigzag (10,0)					
20	0.339 8	1.251 2	2.521 0	3.973 3	5.498 6
25	0.274 9	1.039 4	2.159 4	3.501 5	4.966 6
30	0.230 5	0.885 4	1.876 3	3.105 3	4.487 2
35	0.198 3	0.769 5	1.653 4	2.776 2	4.068 9
zigzag (12,0)					
20	0.339 2	1.249 0	2.516 2	3.964 7	5.484 5
25	0.274 4	1.037 7	2.155 6	3.494 8	4.955 9
30	0.230 1	0.884 0	1.873 5	3.099 7	4.478 4
35	0.198 0	0.768 3	1.650 7	2.771 4	4.061 4
zigzag (14,0)					
20	0.338 9	1.247 7	2.513 4	3.959 5	5.475 9
25	0.274 2	1.036 6	2.153 3	3.490 7	4.949 4
30	0.229 9	0.883 1	1.871 6	3.096 3	4.473 0
35	0.197 8	0.767 5	1.649 0	2.768 4	4.056 8
zigzag (16,0)					
20	0.338 7	1.246 9	2.511 5	3.956 1	5.470 3
25	0.274 0	1.035 9	2.151 8	3.488 0	4.945 2
30	0.229 8	0.882 5	1.870 3	3.094 1	4.469 6
35	0.197 7	0.767 0	1.647 9	2.776 5	4.053 8

general chirality (8,4)					
20	0.339 6	1.250 4	2.519 3	3.970 3	5.493 6
25	0.274 7	1.038 8	2.158 1	3.499 2	4.962 8
30	0.230 4	0.884 9	1.875 6	3.103 3	4.484 1
35	0.198 2	0.769 1	1.652 4	2.774 5	4.066 3

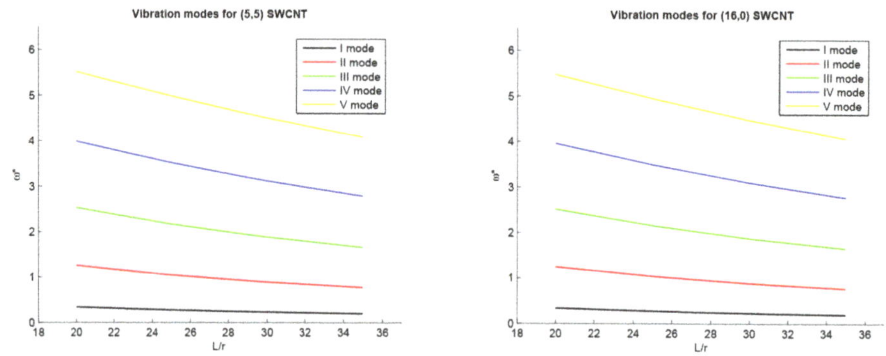

Figure 9.5. Exact 3D shell model: dimensionless circular frequencies ω^* versus L/r ratio for armachair (5,5) (on the left) and zigzag (16,0) (on the right) SWCNTs.

All the comments shown in the proposed analysis are useful for the validation of future continuum approaches developed for the free frequency analysis of SWCNTs (both cases of 1D beam and 2D shell theories). The main limitation of this proposed analysis is that the elastic properties do not change when the chirality of the SWCNT has been modified (in fact, only the geometry is considered as variable). This limitation could be overcome in the future by means of the use of an MD simulation combined with the present exact 3D shell analysis.

9.2.2 Free frequencies and vibration modes for DWCNTs

In this part of the chapter, the DWCNTs are analyzed using two different versions of the developed 3D shell model, the first version discards the effects of the van der Waals interaction while the second one includes such effects. The DWCNT analysis is based on an equivalent continuum model where the two cylinders have thickness values h_i and h_e (see figures 9.6 and 9.7 for further details). The first approach considers $P = 228$ mathematical layers to correctly approximate the curvature of the shell. The first internal cylinder having thickness h_i is divided into 114 mathematical layers (with layer index from $j = 1$ to $j = 114$) and the second external cylinder having thickness h_e is divided into the same number of mathematical layers (with layer index from $j = 115$ to $j = 228$). All the mathematical layers are linked using the interlaminar continuity conditions as already explained in the theoretical section about the 3D shell model. The same interlaminar continuity conditions are also employed to link the layer having index $j = 114$ with the layer having index $j = 115$.

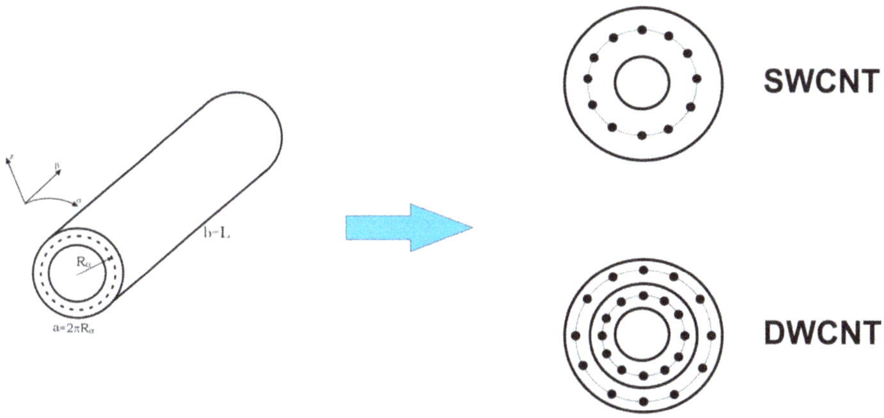

Figure 9.6. Employed notation, reference system in the continuum approach employed for the analysis of a SWCNT and a DWCNT.

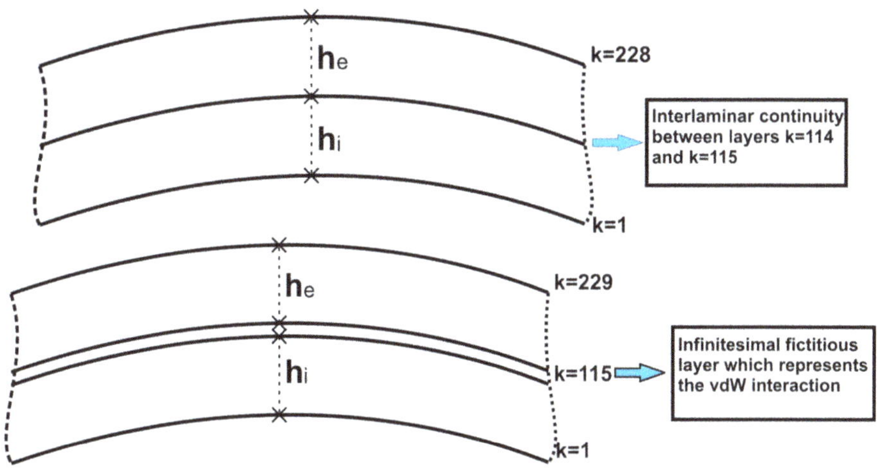

Figure 9.7. Continuum 3D shell model for DWCNT analysis. Interlaminar continuity between the two equivalent cylinders (3D model) and van der Waals interaction between the two equivalent cylinders ($3D_{vdW}$ model).

Therefore, the two cylinders are linked using classical interlaminar continuity conditions typical of the continuum approach and the van der Waals interaction effects are discarded. This first approach is indicated in the proposed results as the 3D model. The second approach here described is called in the results $3D_{vdW}$ model because it includes van der Waals interaction effects in the previously described 3D continuum shell model. The two employed continuum cylinders have thickness values h_i and h_e. The first cylinder is divided into 114 fictitious/mathematical layers (with layer index from $j = 1$ to $j = 114$) and the second cylinder is divided in the same 114 fictitious/mathematical layers (with layer index from $j = 116$ to $j = 229$ that are different from the indexes used in the first approach). Therefore, an infinitesimal layer (with index $j = 115$) must be introduced between the two equivalent continuum

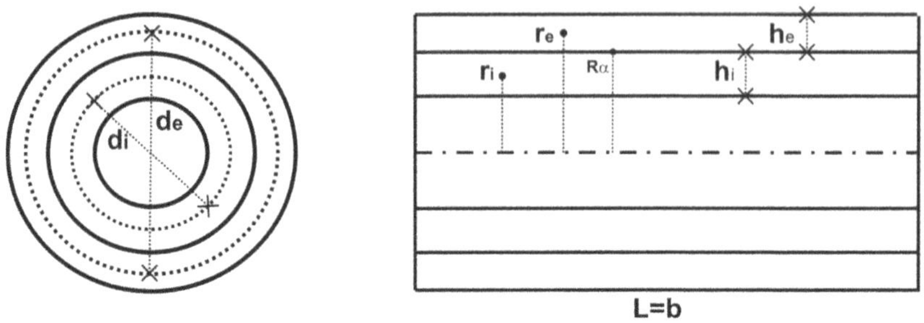

Figure 9.8. Geometrical data used in the equivalent continuum approach for the DWCNT.

cylinders in order to simulate the van der Waals interaction effects (see figures 9.6–9.8 for further details). This infinitesimal fictitious layer has a negligible thickness value calculated as $h_{115} = (h_i + h_e)/1000$ and opportune elastic properties must be chosen to opportunely simulate the van der Waals interaction. Layers $j = 114$ and $j = 115$, and layers $j = 115$ and $j = 116$ are linked by means of the interlaminar continuity conditions already explained. The fictitious layer $j = 115$ has mass density $\rho = 1.225$ kg m^{-3} (that is, the air density). The Poisson ratio is the same employed for the other layers. The van der Waals interaction coefficient c, estimated at the initial interlayer spacing, can be given as

$$c_1 = \frac{200 \text{ erg cm}^{-2}}{0.16d^2}, \tag{9.106}$$

or as

$$c_2 = \frac{320 \text{ erg cm}^{-2}}{0.16d^2}, \tag{9.107}$$

see [24] for further details about this topic. $d = 0.142$ nm is the length of C–C bond and $1 \text{ erg} = 10^{-7}$ Joule. Therefore, these coefficients could have values $c_1 = 6.199\,166\,83 \times 10^{19}$ N m^{-3} or $c_2 = 9.918\,667 \times 10^{19}$ N m^{-3}. If we consider that transverse displacements for CNT analysis have values in the order of 10^{-12} m, we can calculate the equivalent Young's modulus for the infinitesimal fictitious layer that simulates van der Waals interactions as $E_1 = 6.199\,166\,83 \times 10^7$ Pa or $E_2 = 9.918\,667 \times 10^7$ Pa, respectively. The three-dimensional approaches including van der Waals interaction are defined as $3D_{vdW}(c_1)$ and $3D_{vdW}(c_2)$.

The developed 3D shell model is firstly validated using a comparison with the Double Timoshenko Beam Model (DTBM) given in [24]. Here, the free frequency analysis of simply-supported DWCNTs was proposed. The two concentric equivalent cylinders are linked via van der Waals interaction forces. DTBM (c_1) uses the van der Waals interaction coefficient c_1 as shown in equation (9.106). DTBM (c_2) uses the van der Waals interaction coefficient c_2 as shown in equation (9.107). Therefore, the DWCNT includes two layers having the same equivalent thickness $h_i = h_e = 0.35$ nm (see figures 9.6–9.8 for further details). The elastic properties for the equivalent cylinders are Young's modulus $E = 1$ TPa and Poisson ratio $\nu = 0.25$. The mass density is supposed as $\rho = 2300$ kg m^{-3}. Four different DWCNT geometries are analyzed. The first nanotube has a reference diameter for the inner cylinder $d_i = 0.7$ nm and a reference diameter for the external cylinder $d_e = 1.4$ nm. The mean radius of curvature for the entire DWCNT is calculated as $R_\alpha = d_i/2 + h_i/2 = 0.525$ nm: this choice gives a circumferential dimension $a = 2\pi R_\alpha = 3.298\,672\,286$ nm. The second considered nanotube has a reference diameter for the inner cylinder $d_i = 2.95$ nm and a reference diameter for the external cylinder $d_e = 3.65$ nm. These values give a global mean radius of curvature for the DWCNT calculated as $R_\alpha = d_i/2 + h_i/2 = 1.65$ nm and the related circumferential dimension is $a = 2\pi R_\alpha = 10.367\,256$ nm. The third considered DWCNT geometry has a reference diameter for the inner cylinder $d_i = 4.80$ nm and a reference diameter for the external cylinder $d_e = 5.50$ nm. Using these two values, the mean radius of curvature for the DWCNT is $R_\alpha = d_i/2 + h_i/2 = 2.575$ nm, this value gives a related circumferential dimension defined as $a = 2\pi R_\alpha$=16.179 2022 nm. The fourth nanotube has $d_i = 7.0$ nm and $d_e = 7.70$ nm, they give mean radius of curvature for the DWCNT $R_\alpha = d_i/2 + h_i/2 = 3.675$ nm; this value means dimension $a = 2\pi R_\alpha = 23.090\,706$ nm. Each considered DWCNT has radius of curvature in β direction equal to infinity (rectilinear direction). The lengths $L = b$ in the rectilinear β direction are obtained from ratios $L/d_e = 5$, 10, 15, 30 and 50.

Table 9.4 shows the comparison between the proposed 3D shell model and the beam model called DTBM and developed in [24]. The fundamental frequency f is given in GHz for imposed half-wave numbers $m = 2$ and $n = 1$. The first three columns compare the DTBM and the 3D model when a van der Waals interaction coefficient $c_1 = 6.199\,166\,83 \times 10^{19}$ N m^{-3} is employed. The columns from the fourth to the sixth one compare the DTBM and the 3D model when a van der Waals interaction coefficient $c_2 = 9.918\,667 \times 10^{19}$ N m^{-3} is used. The last two columns show the 3D shell model when van der Waals interactions are discarded and the two equivalent cylinders are linked by means of the interlaminar continuity conditions typical of classical multilayered structures. The calculation of Δ_{c_1} and Δ_{c_2} values allow some important comments. The difference between the beam model and the 3D shell model decreases when L/d_e ratio increases: bigger L/d_e ratios mean longer nanotubes and, therefore, 1D conditions are better satisfied. In conclusion, the beam model is similar to the 3D shell model in the case of long nanotubes. The difference between the beam model and the 3D shell model increases when the thickness ratio R_α/h increases (where $h = h_i + h_e$). Bigger R_α/h values mean thin wall structures. As already remarked in [24], DTBM shows some difficulties due to the numerical instability for nanotube 3 having the ratio $L/d_e = 15$ (see the different values for Δ_{c_1} and Δ_{c_2}) and for nanotube 4 having ratio $L/d_e = 15$ (see the omitted value for Δ_{c_2}). Beam results for ratios L/d_e equal 30 and 50 were not calculated in [24]. Thanks to the results proposed in table 9.4, it is clear how the use of c_1 or c_2 value does not give significant differences, for this reason the proposed 3D shell model will use only the c_2 coefficient for benchmarks in next tables 9.5–9.8. Δ_{3D} values defined in table 9.4 show the van der Waals interaction effects. The frequency calculated using the 3D shell model including the interlaminar continuity conditions between the two equivalent cylinders is bigger than the frequency calculated using the 3D shell model including the van der Waals interaction between the two equivalent cylinders.

Table 9.4. Preliminary assessment. First natural frequency f [GHz] for imposed half-wave numbers $m = 2$ and $n = 1$ and different L/d_e ratios. DTBM is the Double Timoshenko Beam Model including van der Waals interaction as proposed in [24]. The beam effects are evaluated by defining $\Delta_{c_1}(\%) = \frac{\text{DTBM}(c_1) - 3\text{D}_{\text{vdW}}(c_1)}{3D_{\text{vdW}}(c_1)} \times 100$ and $\Delta_{c_2}(\%) = \frac{\text{DTBM}(c_2) - 3\text{D}_{\text{vdW}}(c_2)}{3\text{D}_{\text{vdW}}(c_2)} \times 100$. The van der Waals interaction effects are evaluated by defining $\Delta_{3D}(\%) = \frac{3\text{D} - 3\text{D}_{\text{vdW}}(c_2)}{3\text{D}_{\text{vdW}}(c_2)} \times 100$.

L/d_e	DTBM(c_1)	3D$_{\text{vdW}}$(c_1)	Δ_{c_1}(%)	DTBM(c_2)	3D$_{\text{vdW}}$(c_2)	Δ_{c_2}(%)	3D	$\Delta_{3\text{D}}$(%)
Nanotube 1 ($R_\alpha/h = 0.75$)								
5	272	270.2	0.64	273	270.8	0.81	277.4	2.44
10	72.8	72.60	0.27	72.8	72.64	0.22	73.12	0.66
15	32.8	32.75	0.15	32.8	32.76	0.12	32.85	0.27
30	\	8.266	\	\	8.267	\	8.270	0.04
50	\	2.984	\	\	2.984	\	2.984	0.00
Nanotube 2 ($R_\alpha/h = 2.31$)								
5	108	106.1	1.79	108	106.2	1.69	106.5	0.28
10	28.7	28.53	0.60	28.7	28.54	0.56	28.55	0.03
15	12.9	12.87	0.23	12.9	12.87	0.23	12.88	0.08
30	\	3.248	\	\	3.248	\	3.248	0.00
50	\	1.172	\	\	1.171	\	1.172	0.08
Nanotube 3 ($R_\alpha/h = 3.68$)								
5	73.3	71.69	2.25	73.3	71.70	2.23	71.73	0.04
10	19.5	19.33	0.88	19.4	19.33	0.36	19.34	0.05
15	8.68	8.728	−0.55	8.82	8.729	1.04	8.729	0.00
30	\	2.203	\	\	2.203	\	2.203	0.00
50	\	0.795 0	\	\	0.794 6	\	0.794 9	0.04
Nanotube 4 ($R_\alpha/h = 5.25$)								
5	53.0	51.79	2.34	53.0	51.78	2.36	51.79	0.02
10	14.3	14.00	2.14	14.3	14.00	2.14	14.00	0.00
15	6.42	6.323	1.53	\	6.323	\	6.323	0.00
30	\	1.597	\	\	1.597	\	1.597	0.00
50	\	0.576 3	\	\	0.576 2	\	0.576 1	−0.02

We can note how this difference is bigger for short cylinders (that means smaller L/d_e ratios) and it decreases for longer cylinders (that means bigger L/d_e ratios). Moreover, van der Waals effects are negligible for cylinders embedding thin walls (see results in tables proposed for nanotubes 2, 3 and 4). Tables 9.5–9.8 compare the 3D shell model developed using the interlaminar continuity conditions with the 3D shell model developed using the van der Waals interaction effects (in this last case only the c_2 coefficient case is analyzed because table 9.4 did not show significant differences for the use of c_1 and c_2 coefficients). The effect of the imposed half-wave number couples (m, n) is also analyzed in tables 9.5–9.8. Table 9.5 analyzes a nanotube with thick walls considered for the two embedded equivalent cylinders, in this case the van der Waals effect is significant and it increases when the longitudinal

Table 9.5. Benchmark: first nanotube with thickness ratio $R_\alpha/h = 0.75$. Frequency f in GHz for different half-wave numbers (m, n) and L/d_e ratios. The van der Waals effects are evaluated using $\Delta_{vdW}(\%) = \frac{3D - 3D_{vdW}(c_2)}{3D_{vdW}(c_2)} \times 100$.

(m,n)	3D	$3D_{vdW}(c_2)$	$\Delta_{vdW}(\%)$
$L/d_e = 5$			
(2,1)	277.4	270.8	2.44
(2,2)	947.3	892.3	6.16
(2,3)	1813	1628	11.4
$L/d_e = 10$			
(2,1)	73.12	72.64	0.66
(2,2)	277.4	270.8	2.44
(2,3)	579.5	554.3	4.55
$L/d_e = 15$			
(2,1)	32.85	32.76	0.27
(2,2)	128.1	126.6	1.18
(2,3)	277.4	270.8	2.44
$L/d_e = 30$			
(2,1)	8.270	8.267	0.04
(2,2)	32.85	32.76	0.27
(2,3)	73.12	72.64	0.66
$L/d_e = 50$			
(2,1)	2.984	2.984	0.00
(2,2)	11.89	11.89	0.00
(2,3)	26.65	26.60	0.19

half-wave number value n increases. This effect decreases for long nanotubes (with length/diameter ratios L/d_e from 5 to 50). These conclusions have been opportunely summarized in figure 9.9 where it is possible to see how the van der Waals effect increases when the longitudinal half-wave numbers n increase; this effect is more evident for short cylinders and it becomes less important (we can also say negligible) for long cylinders. Tables 9.6–9.8 show nanotubes with thinner layers (that mean $R_\alpha/h = 2.31$ for nanotube 2, $R_\alpha/h = 3.68$ for nanotube 3 and $R_\alpha/h = 5.25$ for nanotube 4). In these last cases, the van der Waals interaction effect is less important than the thicker layer case and, moreover, it can be considered negligible for long cylinders (that means L/d_e values usually greater than 15) even if higher longitudinal half-wave numbers n have been imposed.

9.2.3 Free frequencies and vibration modes for SWCNTs and DWCNTs: analytical versus numerical models

The added numerical models employed to perform the comparisons with the proposed exact 3D shell model are the classical and refined 2D shell models solved by Tornabene using the GDQ method. The 2D GDQ model results are added in the

Table 9.6. Benchmark: second nanotube with thickness ratio $R_\alpha/h = 2.31$. Frequency f in GHz for different half-wave numbers $(m,\ n)$ and L/d_e ratios. The van der Waals effects are evaluated using $\Delta_{\text{vdW}}(\%) = \frac{3\text{D} - 3\text{D}_{\text{vdW}(c2)}}{3\text{D}_{\text{vdW}(c2)}} \times 100$.

(m,n)	3D	$3D_{\text{vdW}}(c_2)$	$\Delta_{\text{vdW}}(\%)$
$L/d_e = 5$			
(2,1)	106.5	106.2	0.28
(2,2)	347.9	346.1	0.52
(2,3)	632.5	624.2	1.33
$L/d_e = 10$			
(2,1)	28.55	28.54	0.03
(2,2)	106.5	106.2	0.28
(2,3)	217.6	216.7	0.41
$L/d_e = 15$			
(2,1)	12.88	12.87	0.08
(2,2)	49.79	49.74	0.10
(2,3)	106.5	106.2	0.28
$L/d_e = 30$			
(2,1)	3.248	3.248	0.00
(2,2)	12.88	12.87	0.08
(2,3)	28.55	28.53	0.07
$L/d_e = 50$			
(2,1)	1.172	1.171	0.08
(2,2)	4.671	4.671	0.00
(2,3)	10.45	10.45	0.00

previous proposed tables in order to also evaluate the numerical effects in the free vibration analysis of SWCNTs and DWCNTs.

The 2D GDQ models proposed in this section consider both ESL and LW approaches implemented in the framework of the well-known unified formulation [25]. The ESL approach uses the following 2D displacement field

$$\mathbf{U} = \sum_{\tau=0}^{N_c+1} \mathbf{F}_\tau \mathbf{u}^{(\tau)} \tag{9.108}$$

where $\mathbf{U}$ is the vector for the 3D displacement components and $\mathbf{u}$ is the vector for the 2D τth generalized displacements defined at the middle surface of the shell structure [26]. $\mathbf{F}_{\tau(ij)} = \delta_{ij} F_\tau$ (for $i,\ j = 1,\ 2,\ 3$) is called as the thickness function matrix where δ is the Kronecker delta function. A first order shear deformation theory, that is based on the Reissner–Mindlin hypotheses, is here called RM-GDQ; it is obtained using $N_c = 0$. A higher order 2D model can be obtained using a fourth order expansion in z ($N_c = 4$) and it is here named with the acronym ESL-GDQ. Using the displacement

Table 9.7. Benchmark: third nanotube with thickness ratio $R_\alpha/h = 3.68$. Frequency f in GHz for different half-wave numbers (m, n) and L/d_e ratios. The van der Waals effects are evaluated using $\Delta_{\mathrm{vdW}}(\%) = \frac{3\mathrm{D} - 3\mathrm{D}_{\mathrm{vdW}(c_2)}}{3\mathrm{D}_{\mathrm{vdW}(c_2)}} \times 100$.

(m,n)	3D	$3\mathrm{D}_{\mathrm{vdW}}(c_2)$	$\Delta_{\mathrm{vdW}}(\%)$
$L/d_e = 5$			
(2,1)	71.73	71.70	0.04
(2,2)	231.6	231.5	0.04
(2,3)	416.4	415.9	0.12
$L/d_e = 10$			
(2,1)	19.34	19.33	0.05
(2,2)	71.73	71.70	0.04
(2,3)	145.8	145.6	0.14
$L/d_e = 15$			
(2,1)	8.729	8.729	0.00
(2,2)	33.67	33.66	0.03
(2,3)	71.73	71.70	0.04
$L/d_e = 30$			
(2,1)	2.203	2.203	0.00
(2,2)	8.729	8.729	0.00
(2,3)	19.34	19.33	0.05
$L/d_e = 50$			
(2,1)	0.794 9	0.794 6	0.04
(2,2)	3.168	3.168	0.00
(2,3)	7.088	7.087	0.01

field proposed in equation (9.108), the relation between generalized strains $\boldsymbol{\varepsilon}^{(\tau)}$ and displacements $\mathbf{u}^{(\tau)}$ can be obtained

$$\boldsymbol{\varepsilon}^{(\tau)} = \mathbf{D}_\Omega \mathbf{u}^{(\tau)} \quad \text{for} \quad \tau = 0, 1, 2, \ldots, N_c, N_c + 1 \tag{9.109}$$

where the explicit form of $\mathbf{D}_\Omega$ is detailed in [26]. The τth order stress resultants and the generalized strains are linked using the following relation

$$\mathbf{S}^{(\tau)} = \sum_{s=0}^{N_c+1} \mathbf{A}^{(\tau s)} \boldsymbol{\varepsilon}^{(s)} \quad \text{for} \quad \tau = 0, 1, 2, \ldots, N_c, N_c + 1 \tag{9.110}$$

where

$$\mathbf{A}^{(\tau s)} = \sum_{k=1}^{N_L} \int_{z_k}^{z_{k+1}} (\mathbf{Z}^{(\tau)})^T \bar{\mathbf{C}}^{(k)} \mathbf{Z}^{(s)} H_\alpha H_\beta dz. \tag{9.111}$$

The elastic coefficients included in the 6×6 matrix used for linear elastic constitutive equations are proposed in complete form in [26]. The governing

Table 9.8. Benchmark: fourth nanotube with thickness ratio $R_\alpha/h = 5.25$. Frequency f in GHz for different half-wave numbers (m, n) and L/d_e ratios. The van der Waals effects are evaluated using $\Delta_{\mathrm{vdW}}(\%) = \frac{3\mathrm{D} - 3\mathrm{D}_{\mathrm{vdW}(c_2)}}{3\mathrm{D}_{\mathrm{vdW}(c_2)}} \times 100$.

(m,n)	3D	$3\mathrm{D}_{\mathrm{vdW}}(c_2)$	$\Delta_{\mathrm{vdW}}(\%)$
$L/d_e = 5$			
(2,1)	51.79	51.78	0.02
(2,2)	166.2	166.2	0.00
(2,3)	296.3	296.2	0.03
$L/d_e = 10$			
(2,1)	14.00	14.00	0.00
(2,2)	51.79	51.78	0.02
(2,3)	104.9	104.9	0.00
$L/d_e = 15$			
(2,1)	6.323	6.323	0.00
(2,2)	24.35	24.35	0.00
(2,3)	51.79	51.78	0.02
$L/d_e = 30$			
(2,1)	1.597	1.597	0.00
(2,2)	6.323	6.323	0.00
(2,3)	14.00	14.00	0.00
$L/d_e = 50$			
(2,1)	0.576 1	0.576 1	0.00
(2,2)	2.296	2.296	0.00
(2,3)	5.135	5.135	0.00

equations of motion with related boundary conditions are calculated by using the Hamilton principle. In the developed general 2D models, each order τ considers a set of three motion equations: they are a function of the internal actions. These equations have the following compact matrix form

$$\mathbf{D}_\Omega^\star \mathbf{S}^{(\tau)} = \sum_{s=0}^{N_c+1} \mathbf{M}^{(\tau s)} \ddot{\mathbf{u}}^{(s)} \quad \text{for} \quad \tau = 0, 1, 2, \ldots, N_c, N_c + 1, \tag{9.112}$$

where $\mathbf{D}_\Omega^\star$ is the so-called equilibrium operator and $\mathbf{M}^{(\tau s)}$ is the well-known inertial matrix. Their explicit forms are proposed in [26]. The mass matrix $\mathbf{M}_{(ij)}^{(\tau s)} = \delta_{ij} I_0^{(\tau s)}$ includes the inertial mass terms $I_0^{(\tau s)}$ for $i, j = 1, 2, 3$

$$I_0^{(\tau s)} = \sum_{k=1}^{N_L} \int_{z_k}^{z_{k+1}} \rho^{(k)} F_\tau F_s H_\alpha H_\beta dz \quad \text{for} \quad \tau, s = 0, 1, 2, \ldots, N_c, N_c + 1, \tag{9.113}$$

where $\rho^{(k)}$ is the mass density of the material and it is defined per unit of volume of the k physical layer. Combining the kinematic equation (9.109), constitutive

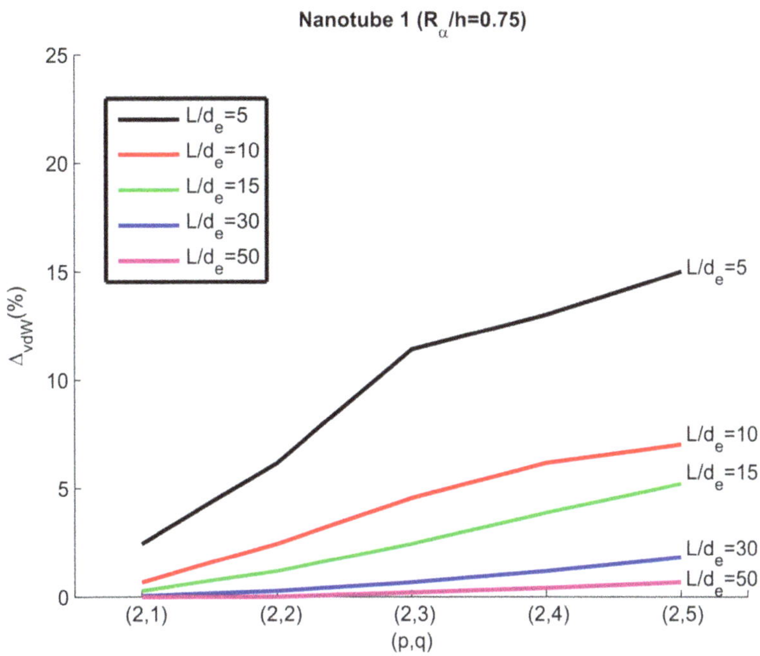

Figure 9.9. First nanotube with thickness ratio $R_\alpha/h = 0.75$: van der Waals interaction effects in relation with half-wave numbers (m, n) and nanotube lengths (L/d_e ratios).

equation (9.110) and the motion equation (9.112), the fundamental system of equations is determined using displacement parameters

$$\sum_{s=0}^{N_c+1} \mathbf{L}^{(\tau s)}\mathbf{u}^{(s)} = \sum_{s=0}^{N_c+1} \mathbf{M}^{(\tau s)}\ddot{\mathbf{u}}^{(s)} \quad \text{for} \quad \tau = 0, 1, 2, \ldots, N_c, N_c + 1, \tag{9.114}$$

where $\mathbf{L}^{(\tau s)} = \mathbf{D}_\Omega^\star \mathbf{A}^{(\tau s)} \mathbf{D}_\Omega$ is the so-called fundamental operator and it is defined in detail in [26]. Boundary conditions must be used in equation (9.114) if we want to solve the differential problem. The GDQ method is valid for any kind of boundary condition, but in the present study only simply-supported edge boundary conditions (called here S) are considered in order to perform comparisons with the 3D developed exact shell model that has been developed only for the free frequency analysis of simply-supported CNTs

$$u_\alpha^{(\tau)} = 0,\ N_\beta^{(\tau)} = 0,\ u_z^{(\tau)} = 0 \quad \text{for} \quad \tau = 0, 1, 2, \ldots, N_c, N_c + 1 \quad \text{at} \quad \beta = \beta^0 \\ \text{or} \quad \beta = \beta^1 \quad \alpha^0 \leqslant \alpha \leqslant \alpha^1. \tag{9.115}$$

The simply-supported boundary conditions are imposed on two edges, whereas compatibility equations must be written for the other two edges in order to obtain a closed geometry typical of cylinder structures

$$\begin{aligned} u_\alpha^{(\tau)}(0, \beta, t) &= u_\alpha^{(\tau)}(2\pi, \beta, t); \quad u_\beta^{(\tau)}(0, \beta, t) = u_\beta^{(\tau)}(2\pi, \beta, t); \quad u_z^{(\tau)}(0, \beta, t) \\ &= u_z^{(\tau)}(2\pi, \beta, t); \ N_{\alpha\beta}^{(\tau)}(0, \beta, t) = N_{\alpha\beta}^{(\tau)}(2\pi, \beta, t); \quad N_\alpha^{(\tau)}(0, \beta, t) \\ &= N_\alpha^{(\tau)}(2\pi, \beta, t); \quad T_\alpha^{(\tau)}(0, \beta, t) = T_\alpha^{(\tau)}(2\pi, \beta, t) \\ &\text{for} \quad \tau = 0, 1, 2, \ldots, N_c, N_c + 1 \quad \text{at} \quad \beta^0 \leqslant \beta \leqslant \beta^1. \end{aligned} \tag{9.116}$$

The higher order LWmodel is here called using the acronym LW-GDQ and it uses an expansion that is similar to the ESL approach but employing different thickness functions. The displacement vector has the following form when $N_c = 4$ is employed [27]

$$\mathbf{U}^{(k)} = \sum_{\tau=0}^{N_c+1} \mathbf{F}_\tau^{(k)} \mathbf{u}^{(k\tau)} \quad \text{for} \quad k = 1, 2, \ldots, N_L. \tag{9.117}$$

In equations (9.117) and (9.108), each LW variable is defined for each single physical layer k. In the proposed LW approach, the thickness functions are defined as a combination of Legendre polynomials that are written in a recursively way as shown in [27]. The generalized displacements $u_\alpha^{(k0)}$, $u_\beta^{(k0)}$, $u_z^{(k0)}$ for $\tau = 0$ are the displacements defined at the bottom of the kth layer ($z^{(k)} = -h_k/2$); conversely, $u_\alpha^{(k(N_c+1))}$, $u_\beta^{(k(N_c+1))}$, $u_z^{(k(N_c+1))}$ for $\tau = N_c + 1$ are the displacements defined at the top of the kth layer ($z^{(k)} = +h_k/2$). The displacement field as defined in previous equation (9.117) allows the following kinematic relations

$$\boldsymbol{\varepsilon}^{(k\tau)} = \mathbf{D}_\Omega^{(k)} \mathbf{u}^{(k\tau)} \quad \text{for} \quad \tau = 0, 1, 2, \ldots, N_c, N_c + 1, \quad k = 1, 2, \ldots, N_L. \tag{9.118}$$

CNTs are considered as continuum structures with a linear and elastic material; therefore, it is possible to use the following relations

$$\begin{aligned} \mathbf{S}^{(k\tau)} &= \sum_{s=0}^{N_c+1} \mathbf{A}^{(k\tau s)} \boldsymbol{\varepsilon}^{(ks)} \quad \text{for} \quad \tau = 0, 1, 2, \ldots, N_c, N_c + 1, \\ & k = 1, 2, \ldots, N_L \end{aligned} \tag{9.119}$$

where

$$\mathbf{A}^{(k\tau s)} = \sum_{k=1}^{N_L} \int_{-h_k/2}^{+h_k/2} (\mathbf{Z}^{(k\tau)})^T \bar{\mathbf{C}}^{(k)} \mathbf{Z}^{(ks)} H_\alpha^{(k)} H_\beta^{(k)} dz^{(k)}. \tag{9.120}$$

The τth order generalized internal action is defined as $\mathbf{S}^{(k\tau)}$ and the elastic coefficients are called as $\mathbf{A}^{(k\tau s)}$ and they are defined as demonstrated in [27].

Therefore, the proposed LW 2D shell approach is developed in a set of three equilibrium equations for each order τ

$$\begin{aligned} \mathbf{D}_\Omega^{\star(k)} \mathbf{S}^{(k\tau)} &= \sum_{s=0}^{N_c+1} \mathbf{M}^{(k\tau s)} \ddot{\mathbf{u}}^{(ks)} \\ \text{for} \quad \tau = 0, 1, 2, \ldots, N_c, N_c + 1, &\quad k = 1, 2, \ldots, N_L. \end{aligned} \tag{9.121}$$

where the equilibrium operator $\mathbf{D}_{\Omega}^{\star(k)}\mathbf{S}^{(k\tau)}$ and the inertial matrix $\mathbf{M}^{(k\tau s)}$ have already been given in explicit form in [27]. The inertial terms are put into the inertial matrix as $\mathbf{M}^{(k\tau s)} = \delta_{ij} I_0^{(k\tau s)}$ for indexes $i, j = 1, 2, 3$. Therefore, they are computed using the following relations

$$I_0^{(k\tau s)} = \int_{-h_k/2}^{+h_k/2} \rho^{(k)} F_\tau^{(k)} F_s^{(k)} H_\alpha^{(k)} H_\beta^{(k)} dz^{(k)} \quad \text{for} \quad \tau, s = 0, 1, 2, \ldots, N_c, N_c + 1, \quad k = 1, 2, \ldots, N_L, \tag{9.122}$$

where $\rho^{(k)}$ is the mass density of the material and defined per unit of volume of each kth physical layer. Therefore, the fundamental equations can be developed in terms of generalized displacements as in the following relation

$$\sum_{s=0}^{N_c+1} \mathbf{L}^{(k\tau s)} \mathbf{u}^{(ks)} = \sum_{s=0}^{N_c+1} \mathbf{M}^{(k\tau s)} \ddot{\mathbf{u}}^{(ks)} \quad \text{for} \quad \tau = 0, 1, 2, \ldots, N_c, N_c + 1, \quad k = 1, 2, \ldots, N_L, \tag{9.123}$$

where $\mathbf{L}^{(k\tau s)} = \mathbf{D}_{\Omega}^{\star(k)} \mathbf{A}^{(k\tau s)} \mathbf{D}_{\Omega}^{(k)}$ is the fundamental operator as defined in detail in [27]. The compatibility conditions must be imposed at each layer interface because the model is a layer-by-layer structural approach. Therefore, the displacements at the top of the kth physical layer must be equal to the displacements at the bottom of the $(k + 1)$ th physical layer

$$\begin{bmatrix} u_\alpha^{(k\ \text{top})} \\ u_\beta^{(k\ \text{top})} \\ u_z^{(k\ \text{top})} \end{bmatrix} = \begin{bmatrix} u_\alpha^{((k+1)\text{bottom})} \\ u_\beta^{((k+1)\text{bottom})} \\ u_z^{((k+1)\text{bottom})} \end{bmatrix} \rightarrow \begin{bmatrix} u_\alpha^{(k(N_c+1))} \\ u_\beta^{(k(N_c+1))} \\ u_z^{(k(N_c+1))} \end{bmatrix} = \begin{bmatrix} u_\alpha^{((k+1)0)} \\ u_\beta^{((k+1)0)} \\ u_z^{((k+1)0)} \end{bmatrix} \quad \text{for} \quad \begin{matrix} \tau = 0, 1, 2, \ldots, N_c, N_c + 1 \\ k = 1, 2, \ldots, N_L - 1. \end{matrix} \tag{9.124}$$

The proposed equations can be solved only if boundary conditions are correctly defined. In the present formulation, only simply-supported boundary conditions have been analyzed because the 3D exact solution is valid only for such condition types (we recall that the GDQ method based on the LW approach can work with any kind of boundary condition)

$$\begin{gathered} u_\alpha^{(k\tau)} = 0,\ N_\beta^{(k\tau)} = 0,\ u_z^{(k\tau)} = 0 \\ \text{for} \quad \begin{matrix} \tau = 0, 1, 2, \ldots, N_c, N_c + 1 \\ k = 1, 2, \ldots, N_L \end{matrix} \\ \text{at} \quad \begin{matrix} \beta = \beta^0 \quad \text{or} \quad \beta = \beta^1 \\ \alpha^0 \leqslant \alpha \leqslant \alpha^1. \end{matrix} \end{gathered} \tag{9.125}$$

CNTs are considered as closed continuum cylinders; therefore, compatibility conditions must also be correctly imposed

$$
\begin{aligned}
u_\alpha^{(k\tau)}(0, \beta, t) = u_\alpha^{(k\tau)}(2\pi, \beta, t);\quad & u_\beta^{(k\tau)}(0, \beta, t) \\
= u_\beta^{(k\tau)}(2\pi, \beta, t);\quad & u_z^{(\tau)}(0, \beta, t) = u_z^{(\tau)}(2\pi, \beta, t); \\
N_{\alpha\beta}^{(k\tau)}(0, \beta, t) = N_{\alpha\beta}^{(k\tau)}(2\pi, \beta, t);\quad & N_\alpha^{(k\tau)}(0, \beta, t) \\
= N_\alpha^{(k\tau)}(2\pi, \beta, t);\ & T_\alpha^{(k\tau)}(0, \beta, t) = T_\alpha^{(k\tau)}(2\pi, \beta, t) \\
\text{for}\quad \tau = 0, 1, 2, \ldots, N_c, N_c + 1 \quad & \text{at}\quad \beta^0 \leqslant \beta \leqslant \beta^1.
\end{aligned}
\tag{9.126}
$$

This last section of the chapter proposes results for free frequency analysis of simply-supported SWCNTs and DWCNTs where the exact 3D shell solution (previously described) is compared with the two beam theories proposed in [20] and [21] and with the 2D GDQ solutions above described (in particular, a Reissner–Mindlin model (here called RM-GDQ), a refined ESL model (here called ESL-GDQ) and a refined LW model (here called LW-GDQ)). In the preliminary assessments, the 3D exact and the 2D GDQ shell models are compared and discussed for several SWCNTs, the length L of the proposed SWCNTs is a variable parameter in order to evaluate the approximation level of the beam models for short and long CNTs. These preliminary assessments are also useful to define the correct order N for the exponential matrix expansion and the appropriate number of mathematical layers P to use in the 3D exact shell model geometry, and the number of grid points to employ in 2D GDQ shell models. In the new benchmarks for SWCNTs and DWCNTs, 2D GDQ shell models are compared with the 3D exact shell model for imposed half-wave number $m = 2$ in circular direction and longitudinal half-wave numbers n in rectilinear direction from 1 to 3 (the first mode (indicated in tables as I) is analyzed for each couple of half-wave numbers (therefore, we cannot show m,n) in all tables). The comparison between frequencies obtained via the 3D exact shell model and those obtained via the 2D GDQ shell models is possible thanks to the appropriate visualization of vibration modes: in this way, it is possible to define the opportune frequencies and half-wave numbers. The effects of the van der Waals interactions have been evaluated using the 3D exact shell model based on an LW approach.

The preliminary assessments were already proposed in previous sections and they were based on works by Simsek [20] and Aydogdu [21] that developed an EBM and a TBM. The SWCNT is considered as simply-supported, the equivalent elastic cylinder has properties as indicated in [20] and in the previous sections of the present chapter. We can recall them in a short way: equivalent Young's modulus $E = 1$ TPa, Poisson ratio $\nu = 0.3$, effective thickness combined with the proposed Young's modulus is $h = 0.35$ nm, mass density $\rho = 2300$ kg m^{-3}, external diameter $d_e = 1$ nm (that means a radius of curvature/thickness ratio $R_\alpha/h = 0.929$). If a very thick shell is analyzed, in this case the use of beam models and 3D or refined 2D shell models is mandatory. Some difficulties may arise in the case of classical 2D shell models employed for the free frequency analysis of such cylinders (see the results obtained using the RM-GDQ model). The use of very refined 2D shell models (see [22]) or 3D exact shell models (see [1, 2]) could overcome these difficulties shown by classical 2D shell models. The other geometrical data has been proposed in the previous sections but they are here summarized again for simplicity and convenience: radius of curvature in α direction and referred to the mid-surface is $R_\alpha = d_e/2 - h/2 = 0.325$ nm, the dimension in circular α direction is $a = 2\pi R_\alpha$, and the b dimension in rectilinear direction is variable as $L = 5$ nm, 10 nm, 20 nm, 50 nm and 100 nm that means ratios $L/d_e = 5$, 10, 20, 50 and 100, respectively.

Table 9.9 shows the dimensionless circular frequency given as $\bar{\omega} = \omega L^2\sqrt{\frac{\rho A}{EI}}$ (where $A = \pi(r_e^2 - r_i^2)$ and $I = \frac{\pi}{64}(d_e^4 - d_i^4)$ are the area and the moment of inertia of the ring, respectively) for short and long simply-supported SWCNTs (different L/d_e ratios). The three dimensionless circular frequencies are calculated for imposed circumferential half-wave number $m = 2$ in α circular direction and longitudinal half-wave numbers n in β rectilinear direction changing from 1 to 3 (the first mode (indicated as I in tables) is defined for each couple (m,n)). Beam models correctly work only for long and moderately long SWCNTs because they can be considered as 1D structures. In contrast, shell models give satisfactory results for both long and short SWCNTs. The added EBM was developed in [20] and [21] for ratios $L/d_e = 10$, 20, 50. The same cases were also investigated in [21] using the TBM. TBM gives more accurate results than EBM because it includes the effects of transverse shear

Table 9.9. Preliminary assessment, simply-supported SWCNTs with $R_\alpha/h = 0.929$. 3D exact and 2D GDQ shell models compared with beam models (EBM and TBM) proposed in [20, 21]). Dimensionless circular frequencies $\bar{\omega}$ for different L/d_e ratios where m and n are the imposed half-wave numbers.

Mode (m, n)	3D Exact	RM-GDQ	ESL-GDQ	EBM [20, 21]	TBM [21]
$L/d_e = 5$					
I (2,1)	9.348 1	9.367 2	9.348 0	—	—
I (2,2)	32.917	32.941	32.907	—	—
I (2,3)	63.917	63.683	63.685	—	—
$L/d_e = 10$					
I (2,1)	9.729 5	9.752 5	9.729 0	9.869 6	9.744 3
I (2,2)	37.392	37.469	37.392	39.478	36.841
I (2,3)	79.361	79.483	79.358	88.826	57.450
$L/d_e = 20$					
I (2,1)	9.835 6	9.858 4	9.833 7	9.869 6	9.838 1
I (2,2)	38.918	39.010	38.916	39.478	38.964
I (2,3)	86.072	86.264	86.070	88.826	85.748
$L/d_e = 50$					
I (2,1)	9.863 8	9.888 8	9.863 8	9.869 6	9.864 5
I (2,2)	39.392	39.485	39.386	39.478	39.398
I (2,3)	88.375	88.583	88.363	88.826	88.415
$L/d_e = 100$					
I (2,1)	9.848 7	9.893 2	9.868 2	—	—
I (2,2)	39.488	39.555	39.455	—	—
I (2,3)	88.752	88.934	88.710	—	—

deformation and rotary inertia. However, TBM shows some difficulties for second and third frequencies when short SWCNTs ($L/d_e = 10$) are analyzed. The 3D shell model gives correct results for both long and short SWCNTs and also for higher order modes in the case of SWCNTs having small radius of curvature/thickness ratios. For these smaller ratios, classical 2D shell models could exhibit some problems because the cylinders have thicker walls. Table 9.9 shows how the TBM gives similar results to the 3D shell model; in contrast, the EBM produces larger differences. However, the TBM shows some difficulties for short SWCNTs. Further results are discussed in table 9.9 for very short and very long SWCNTs using the 3D shell model and 2D GDQ models (these results were not calculated in [20, 21] via beam models). In this way, a complete overview for the SWCNT free frequency behavior is obtained and these results could be also employed as benchmarks for the validation of future 1D beam and 2D shell models developed by other scientists. 2D GDQ shell models are proposed in both classical (RM-GDQ approach) and refined (higher order ESL approach, ESL-GDQ) forms ESL-GDQ model is always coincident with the 3D exact shell solution for each proposed L/d_e ratio (length of SWCNT) and vibration mode. The RM-GDQ model shows some problems when the cylinder is very thick (see the case for $R_\alpha/h = 0.929$). The LW 2D shell model (LW-GDQ) is not employed in this preliminary assessment because SWCNTs embed only one layer. Scientists that work in the development of beam and shell continuum models for free frequency analyses of SWCNTs can try to fill this table using their results. After this proposed assessment, the 3D shell model and the 2D GDQ shell models can be considered as validated and we can use them with confidence for further analyses and new benchmarks. The 3D shell model used $P = 228$ mathematical layers for the approximation of the radius of curvature and $N = 3$ order for the expansion of the exponential matrix. 2D GDQ models used a 41×15 Chebyshev–Gauss–Lobatto grid. These verified values will be also used in the new benchmarks proposed in the following part.

The first benchmark considers a simply-supported SWCNT that has the properties as discussed in [23] and in table 9.2 and figure 9.4. The geometry of the SWCNTs depends on the chiral vector: it allows several values of the radius of curvature (see figure 9.4 and table 9.2 for further details about armachair, zigzag and general chirality CNTs here analyzed). Several continuum approaches for CNT analysis are proposed in [23] where a different value for the thickness h is given with respect to those usually proposed in the literature. This value was $h = 0.08$ nm and it means an effective Young's modulus $E = 6.85$ TPa with Poisson ratio $\nu = 0.19$. The mass density is $\rho = 9517$ kg m^{-3} (table 9.11)

The continuum models proposed in [23] were also compared with the MD models. These comparisons showed a difference which is less than 5%. Therefore, these properties can be used with confidence for the SWCNT benchmark proposed in this section where the 3D exact shell model is compared with the 2D ESL-GDQ shell models (classical and refined ones). Table 9.2 shows the radius of curvature r connected with the number of atoms for several SWCNT configurations (armchair with $p = q$, zigzag with $(p,0)$ and general chirality where $p \neq q$). Several L/r ratios are analyzed for each SWCNT where $r = R_\alpha$ is the radius of curvature in the α

Table 9.10. First benchmark, simply-supported armchair SWCNT with $R_\alpha/h = 4.225$. Frequencies f [GHz] for 2D GDQ and 3D exact shell models.

Armchair (5,5)			
Mode (m, n)	3D Exact	RM-GDQ	ESL-GDQ
$L/r = 20$			
I (2,1)	215.0	215.0	215.0
I (2,2)	791.7	791.8	791.7
I (2,3)	1595	1595	1595
$L/r = 25$			
I (2,1)	139.1	139.1	139.2
I (2,2)	526.2	526.2	526.2
I (2,3)	1093	1093	1093
$L/r = 30$			
I (2,1)	97.23	97.24	97.23
I (2,2)	373.5	373.5	373.5
I (2,3)	791.7	791.8	791.7
$L/r = 35$			
I (2,1)	71.70	71.71	71.70
I (2,2)	278.2	278.3	278.2
I (2,3)	597.9	597.9	597.9

Table 9.11. First benchmark, simply-supported zigzag SWCNT with $R_\alpha/h = 4.875$. Frequencies f [GHz] for 2D GDQ and 3D exact shell models.

Zigzag (10,0)			
Mode(m,n)	3D Exact	RM-GDQ	ESL-GDQ
$L/r = 20$			
I (2,1)	186.0	186.0	186.0
I (2,2)	684.9	684.9	684.9
I (2,3)	1380	1380	1380
$L/r = 25$			
I (2,1)	120.4	120.4	120.4
I (2,2)	455.2	455.2	455.2
I (2,3)	945.7	945.7	945.7
$L/r = 30$			
I (2,1)	84.12	84.12	84.12
I (2,2)	323.1	323.2	323.1
I (2,3)	684.9	685.0	684.9
$L/r = 35$			
I (2,1)	62.04	62.04	62.04
I (2,2)	240.7	240.7	240.7
I (2,3)	517.2	517.2	517.2

circular direction measured with respect to the mid-surface. Tables 9.10–9.12 show the frequencies f in GHz for some of SWCNTs described in table 9.2 and figure 9.4 using ratios L/r = 20, 25, 30, 35. The R_α/h ratio is bigger than the ratio proposed in the preliminary assessment and this feature means moderately thick shell analysis. Therefore, the classical 2D GDQ shell model works as well as the refined 2D GDQ and the 3D exact shell models. Only SWCNTs are investigated in this first benchmark, therefore the use of 2D GDQ LW shell models is not necessary. The 3D exact shell model is compared with the 2D GDQ shell models based on the RM-GDQ approach and on a fourth order ESL-GDQ approach. 2D GDQ shell models calculate a number of frequencies equal to the number of degrees of freedom of the employed numerical model. In tables 9.10–9.12, the comparisons of the frequencies given by the 3D exact shell model are possible when the circumferential half-wave number is m = 2 and the longitudinal half-wave number n is equal to 1, 2 and 3 (the first mode (here called I) is calculated for each couple (m,n)). This comparison is possible because the vibration modes obtained via the 2D GDQ shell models have been opportunely shown; this vibration mode visualization allows us to understand the values of m and n to be imposed in the 3D exact shell model. In tables 9.10–9.12, 3D exact frequencies are always very similar to frequencies obtained using 2D GDQ models (both classical RM-GDQ and refined ESL-GDQ ones) and this feature is confirmed for each investigated L/r ratio, for each imposed (m,n) couple and for each SWCNT structure (armachair, zigzag and general chirality). This feature is due to the fact that this SWCNT has only one isotropic layer in the equivalent cylinder.

Table 9.12. First benchmark, simply-supported SWCNT with general chirality and R_α/h = 5.162 5. Frequencies f [GHz] for 2D GDQ and 3D exact shell models.

General Chirality (8,4)			
Mode(m,n)	3D Exact	RM-GDQ	ESL-GDQ
L/r = 20			
I (2,1)	175.5	175.6	175.5
I (2,2)	646.4	646.4	646.4
I (2,3)	1302	1302	1302
L/r = 25			
I (2,1)	113.6	113.6	113.6
I (2,2)	429.6	429.6	429.6
I (2,3)	892.4	892.5	892.4
L/r = 30			
I (2,1)	79.39	79.39	79.39
I (2,2)	305.0	305.0	305.0
I (2,3)	646.4	646.4	646.4
L/r = 35			
I (2,1)	58.55	58.55	58.55
I (2,2)	227.2	227.2	227.2
I (2,3)	488.1	488.1	488.1

When smaller radius of curvature/thickness ratios will be defined (that means thicker cylinders), the use of 3D or refined 2D shell models will become mandatory in the proposed analysis.

The second benchmark considers four different simply-supported DWCNTs having four different thickness ratios R_α/h. The considered DWCNTs have two layers with the same equivalent thickness $h_i = h_e = 0.35$ nm (see figures 9.6, 9.7 and 9.8 for all the necessary data). The equivalent elastic properties are the same as already given in the previous sections: $E = 1$ TPa, $\nu = 0.25$ and $\rho = 2300$ kg m^{-3}. Four different DWCNT geometries are investigated (see figures 9.6 and 9.8 for reference notations). We can recall all the data already seen in the previous section for 3D exact shell model results. The first nanotube has a reference diameter for the inner cylinder $d_i = 0.7$ nm and a reference diameter for the external cylinder $d_e = 1.4$ nm. Therefore, the mean radius of curvature is $R_\alpha = d_i/2 + h_i/2 = 0.525$ nm that means $a = 2\pi R_\alpha = 3.298\,672\,286$ nm. The second nanotube has $d_i = 2.95$ nm and $d_e = 3.65$ nm, a mean radius of curvature $R_\alpha = d_i/2 + h_i/2 = 1.65$ nm and $a = 2\pi R_\alpha = 10.367\,256$ nm. The third nanotube has $d_i = 4.80$ nm $d_e = 5.50$ nm; therefore, $R_\alpha = d_i/2 + h_i/2 = 2.575$ nm and $a = 2\pi R_\alpha = 16.179\,2022$ nm. The fourth nanotube has the following geometrical data: $d_i = 7.0$ nm, $d_e = 7.70$ nm, $R_\alpha = d_i/2 + h_i/2 = 3.675$ nm and $a = 2\pi R_\alpha = 23.090\,706$ nm. The four described DWCNTs have an infinity radius of curvature in β direction that is rectilinear. Therefore, the lengths $L = b$ in β direction are defined using the following values for the ratios L/d_e: 5, 10, 15, 30 and 50. The DWCNT is analyzed via an equivalent continuum approach where two cylinders having thickness values h_i and h_e are included (this idea is summarized in figures 9.6, 9.7 and 9.8). As in the previous sections, the 3D exact shell model uses $P = 228$ mathematical layers to correctly approximate the curvature of the shell. The internal cylinder (thickness h_i) includes 114 mathematical layers (from $j = 1$ to $j = 114$) and the external cylinder (thickness h_e) includes a further 114 mathematical layers (from $j = 115$ to $j = 228$). In this first case, the two cylinders are linked using the interlaminar continuity conditions between layer $j = 114$ and layer $j = 115$. The related model is called 3D model in the proposed results. ESL-GDQ and LW-GDQ shell models consider the DWCNT composed by two layers with thickness $h_i = h_e = 0.35$ nm: ESL-GDQ model is based on an ESL approach and LW-GDQ model is based on an LW approach where the two cylinders are linked by means of the interlaminar continuity conditions. The second 3D exact shell model here proposed is called the $3D_{vdW}$ model because van der Waals interaction effects are included in the proposed approach, as already described in the previous sections: the two cylinders having thickness values h_i and h_e are divided by means of 114 fictitious layers (from $j = 1$ to $j = 114$) in the case of thickness h_i and a further 114 fictitious layers (from $j = 116$ to $j = 229$) in the case of thickness h_e. Moreover, an infinitesimal layer (with index $j = 115$) is introduced between the two cylinders in order to simulate the van der Waals interaction. This infinitesimal fictitious layer that simulates the van der Waals interaction effects has already been described in the previous section when DWCNTs have been analyzed using only the 3D exact shell model. An alternative method could be the inclusion of van der Waals forces directly

Table 9.13. Second benchmark, simply-supported DWCNTs (first geometry with ratio $R_a/h = 0.75$). Frequencies f [GHz] for 2D GDQ and 3D exact shell models.

Nanotube 1				
Mode (m, n)	3D exact	$3D_{vdW}$ exact	ESL-GDQ	LW-GDQ
$L/d_e = 5$				
I (2,1)	277.4	270.8	277.4	277.4
I (2,2)	946.2	892.3	946.2	946.2
I (2,3)	1785	1628	1785	1785
$L/d_e = 10$				
I (2,1)	73.12	72.64	73.12	73.12
I (2,2)	277.4	270.8	277.4	277.4
I (2,3)	579.4	554.3	579.4	579.4
$L/d_e = 15$				
I (2,1)	32.85	32.76	32.85	32.85
I (2,2)	128.1	126.6	128.1	128.1
I (2,3)	277.4	270.8	277.4	277.4
$L/d_e = 30$				
I (2,1)	8.266	8.266	8.266	8.266
I (2,2)	32.85	32.76	32.85	32.85
I (2,3)	73.12	72.64	73.12	73.12
$L/d_e = 50$				
I (2,1)	2.980	2.980	2.980	2.980
I (2,2)	11.89	11.89	11.89	11.89
I (2,3)	26.65	26.60	26.65	26.65

in governing equations for the 3D exact shell model. The method of the fictitious layer has been used in the present analysis because the direct inclusion of van der Waals forces gave numerical problems in the solution of the 3D exact equations. In contrast, the method including the van der Waals forces and their coefficients c has been used by the first author in [22] in the case of refined 2D models. The method using the fictitious layer, that represents the van der Waals forces, has been validated and tested in [2] by the first author for the 3D exact model. Tables 9.13–9.16 show results for the four different DWCNT geometries previously described. Comparisons are performed between the 3D exact model and the refined 2D GDQ models for several length-to-diameter ratios (from very short DWCNTs to very long DWCNTs). 2D GDQ models are used to obtain the 3D exact frequencies (only the first mode I) when $m = 2$ and $n = 1,2,3$. 2D GDQ models give several frequencies: in order to understand which are the frequencies to be compared, a vibration mode analysis must be performed as shown in the examples proposed in figure 9.10, it allows one to understand the values of half-wave numbers (m,n) to be imposed in the 3D exact shell model. In tables 9.13–9.16, 3D exact frequency values are always very similar to frequencies calculated by means of 2D GDQ models. RM-

Table 9.14. Second benchmark, simply-supported DWCNTs (second geometry with ratio $R_\alpha/h = 2.36$). Frequencies f [GHz] for 2D GDQ and 3D exact shell models.

Nanotube 2				
Mode (m, n)	3D exact	$3D_{vdW}$ exact	ESL-GDQ	LW-GDQ
$L/d_e = 5$				
I (2,1)	106.5	106.2	106.5	106.5
I (2,2)	347.7	346.1	347.7	347.7
I (2,3)	629.9	624.2	629.9	629.9
$L/d_e = 10$				
I (2,1)	28.55	28.54	28.55	28.55
I (2,2)	106.5	106.2	106.5	106.5
I (2,3)	217.6	216.1	217.6	217.6
$L/d_e = 15$				
I (2,1)	12.88	12.87	12.88	12.88
I (2,2)	49.79	49.74	49.79	49.79
I (2,3)	106.5	106.2	106.5	106.5
$L/d_e = 30$				
I (2,1)	3.248	3.248	3.248	3.248
I (2,2)	12.88	12.87	12.88	12.88
I (2,3)	28.55	28.53	28.55	28.55
$L/d_e = 50$				
I (2,1)	1.172	1.171	1.172	1.172
I (2,2)	4.671	4.671	4.671	4.671
I (2,3)	10.45	10.45	10.45	10.45

GDQ model is not here given because it shows big errors for cylinders with small R_α/h values and it is coincident with the refined 2D GDQ models in the case of bigger R_α/h values. Both refined ESL and LW 2D GDQ models are coincident in terms of frequency results, even if a DWCNT is considered, because the two cylinders are isotropic and made of the same material, therefore, non zigzag affects are shown. Refined 2D GDQ models always give very accurate results for each DWCNT geometry, length/diameter ratio, radius of curvature/thickness ratio and vibration mode. In tables 9.13–9.16, a $3D_{vdW}$ exact shell model is also added in order to evaluate the van der Waals interaction effects. Therefore, the obtained frequencies are compared with those calculated via the 3D exact model where the interlaminar continuity conditions between the two cylinders have been imposed in place of the van der Waals interaction effects. From tables 9.13–9.16, it is clear how the van der Waals effects are usually very small and they are often negligible. However, these effects are zero for long DWCNTs and/or low frequencies (see for example results for $m = 2$ and $n = 1$); however, they become more important for short cylinders and/or high frequencies (see for example cases for $m = 2$ and $n = 2$ or 3). In conclusion, van der Waals interaction effects are bigger in tables 9.13 and 9.14

Table 9.15. Second benchmark, simply-supported DWCNTs (third geometry with ratio $R_\alpha/h = 3.68$). Frequencies f [GHz] for 2D GDQ and 3D exact shell models.

Nanotube 3				
Mode (m, n)	3D exact	$3D_{vdW}$ exact	ESL-GDQ	LW-GDQ
$L/d_e = 5$				
I (2,1)	71.74	71.70	71.74	71.74
I (2,2)	231.5	231.3	231.5	231.5
I (2,3)	414.6	414.4	414.6	414.6
$L/d_e = 10$				
I (2,1)	19.34	19.33	19.34	19.34
I (2,2)	71.74	71.70	71.74	71.74
I (2,3)	145.8	145.6	145.8	145.8
$L/d_e = 15$				
I (2,1)	8.729	8.729	8.729	8.729
I (2,2)	33.67	33.66	33.67	33.67
I (2,3)	71.74	71.70	71.74	71.74
$L/d_e = 30$				
I (2,1)	2.203	2.203	2.203	2.203
I (2,2)	8.729	8.729	8.729	8.729
I (2,3)	19.34	19.33	19.34	19.34
$L/d_e = 50$				
I (2,1)	0.794 9	0.794 6	0.794 9	0.794 9
I (2,2)	3.168	3.168	3.168	3.168
I (2,3)	7.088	7.087	7.088	7.088

where cylinders have a smaller R_α/h ratio. In contrast, in tables 9.15 and 9.16 (that are for bigger R_α/h ratios), van der Waals effects are always negligible for each investigated length (L/d_e ratio) and vibration mode.

9.3 Main conclusions

The present chapter has proposed a 3D exact shell model and several 2D refined and classical GDQ shell models for free frequency analyses of simply-supported SWCNTs and DWCNTs. The continuum approach for SWCNTs is possible using an equivalent elastic isotropic cylinder in place of the typical discrete molecular structure of SWCNTs. The 3D exact shell solution is possible only for simply-supported CNTs. In contrast, 2D GDQ shell models allow several boundary conditions investigations. Different types of SWCNTs were analyzed: armchair, zigzag and general chirality geometries. Different radius/thickness and length/radius ratios were considered in the free frequency analyses in order to show the limitations of 1D and 2D classical models. In fact, 1D beam models correctly work only for long or moderately long cylinders (this feature means higher values of length/radius ratio). In contrast, for short cylinders (that means smaller values of length/radius

Table 9.16. Second benchmark, simply-supported DWCNT (fourth geometry with ratio $R_\alpha/h = 5.25$). Frequencies f [GHz] for 2D GDQ and 3D exact shell models.

Nanotube 4				
Mode (m, n)	3D exact	$3D_{vdW}$ exact	ESL-GDQ	LW-GDQ
L/d_e=5				
I (2,1)	51.79	51.78	51.79	51.79
I (2,2)	166.1	166.1	166.1	166.1
I (2,3)	295.6	295.6	295.6	295.6
L/d_e=10				
I (2,1)	14.00	14.00	14.00	14.00
I (2,2)	51.79	51.78	51.79	51.79
I (2,3)	104.9	104.9	104.9	104.9
L/d_e=15				
I (2,1)	6.323	6.323	6.323	6.323
I (2,2)	24.35	24.35	24.35	24.35
I (2,3)	51.79	51.78	51.79	51.79
L/d_e=30				
I (2,1)	1.597	1.597	1.597	1.597
I (2,2)	6.323	6.323	6.323	6.323
I (2,3)	14.00	14.00	14.00	14.00
L/d_e=50				
I (2,1)	0.576 1	0.576 1	0.576 1	0.576 1
I (2,2)	2.296	2.296	2.296	2.296
I (2,3)	5.135	5.135	5.135	5.135

ratio), the use of refined 2D or 3D shell models is mandatory. Classical 2D shell models can exhibit some problems for shells having small values of radius/thickness ratio; in these cases, the use of refined 2D shell models or 3D elasticity models is necessary. The proposed 3D exact elastic shell model could be used as a reference solution for the validation of future 1D beam and 2D shell theories developed by other scientists and applied to the free frequency analysis of SWCNTs. The present authors are aware of the limitations of continuum mechanics models applied to these types of analyses; in fact, our future works will be devoted to the integration of an MD analysis with the proposed continuum 3D exact and 2D numerical shell models.

In the second part of the proposed chapter, the free frequency analysis of simply-supported DWCNTs was proposed using the same continuum exact 3D shell model. In this case, the LW approach of the proposed model allowed interlaminar continuity or van der Waals interaction conditions to be imposed between the two concentric cylinders embedded in the analyzed DWCNTs. In this way, the van der Waals interaction effect can be evaluated in free frequency analysis of DWCNTs. We can notice how this effect is important for higher half-wave number values, thick and/or short DWCNTs and in all these cases it cannot be discarded. In contrast, for

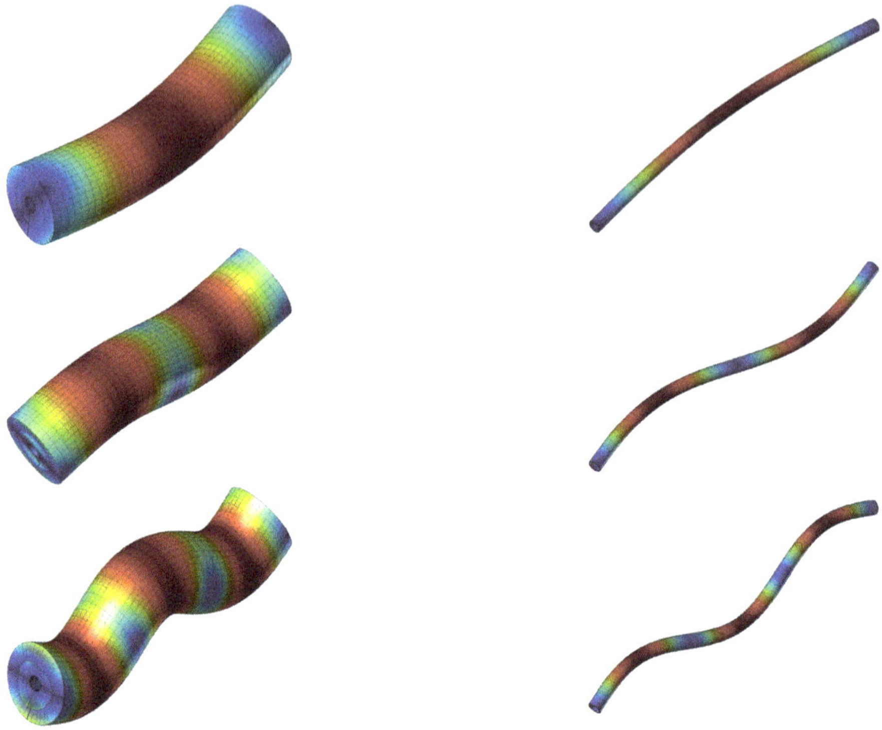

Figure 9.10. DWCNT benchmark: first geometry for $L/d_e = 5$ on the left side and fourth geometry for $L/d_e = 30$ on the right side. Vibration modes defined via the LW-GDQ shell model for imposed half-wave numbers $m = 2$ and $n = 1,2,3$.

lower half-wave number values, thin and/or long DWCNTs, van der Waals interaction effect can be considered as negligible. Moreover, the proposed 3D exact shell model was compared with a TBM proposed in the literature: we can conclude that the beam model is valid for long DWCNTs and/or thick layers; in contrast, for short DWCNTs and/or thin layers for the cylinders, the use of the 3D exact shell model is mandatory.

The third part of the chapter proposed the free frequency analysis of simply-supported SWCNTs and DWCNTs using the 3D exact shell model and several 2D GDQ models. The employed 2D numerical models are the classical RM-GDQ approach and two refined fourth order models (as ESL-GDQ approach and LW-GDQ approach). The first proposed 3D exact shell model is an LW approach that uses the interlaminar continuity conditions between the two cylinders employed to simulate DWCNTs. The second proposed 3D exact shell model is called $3D_{vdW}$ and in it the two isotropic cylinders, which represent the DWCNT, are linked by using van der Waals forces: these forces are simulated using an opportune fictitious layer between the two cylinders. Comparisons in terms of frequencies between the 3D exact and the $3D_{vdW}$ exact shell models allow one to understand the van der Waals effects. These effects are very small and, in general, they can be neglected. However, they become important for short DWCNTs, small radius of curvature/thickness ratios and/or higher frequencies. ESL-GDQ and LW-GDQ shell models work very well and they always give the same results as the 3D exact shell model in terms of frequency for each investigated DWCNT. This feature happens because these structures are made of two isotropic cylinders without any transverse anisotropy. Therefore, an ESL approach is coincident with an LW approach when the same order of expansion is used for the displacement components. It was also confirmed that for SWCNT analysis, the use of beam models is inappropriate for short CNTs. In this last case, refined 2D or 3D shell models are mandatory, in particular, when small radius of curvature/thickness ratios were investigated.

References

[1] Brischetto S 2014 A continuum elastic three-dimensional model for natural frequencies of single-walled carbon nanotubes *Composites* **B61** 222–8

[2] Brischetto S 2015 A continuum shell model including van der Waals interaction for free vibrations of double-walled carbon nanotubes *Comput. Model. Eng. Sci.* **104** 305–27

[3] Brischetto S, Tornabene F, Fantuzzi N and Bacciocchi M 2015 Refined 2D and exact 3D shell models for the free vibration analysis of single- and double-walled carbon nanotubes *Technologies* **3** 259–84

[4] Brischetto S 2014 Three-dimensional exact free vibration analysis of spherical, cylindrical and flat one-layered panels *Shock Vib.* **2014** 1–29

[5] Brischetto S 2013 Exact elasticity solution for natural frequencies of functionally graded simply-supported structures *Comput. Model. Eng. Sci.* **95** 391–430

[6] Brischetto S 2014 An exact 3D solution for free vibrations of multilayered cross-ply composite and sandwich plates and shells *Int. J. Appl. Mech.* **6** 1–42

[7] Brischetto S and Torre R 2014 Exact 3D solutions and finite element 2D models for free vibration analysis of plates and cylinders *Curved Layer. Struct.* **1** 59–92

[8] Leissa A W 1969 *Vibration of Plates SP-160* NACA, Washington

[9] Leissa A W 1973 *Vibration of Shells SP-160* NACA, Washington

[10] Hildebrand F B, Reissner E and Thomas G B 1949 *Notes on the Foundations of the Theory of Small Displacements of Orthotropic Shells Technical Note No. 1833* NACA, Washington

[11] Tornabene F 2012 *Meccanica delle Strutture a Guscio in Materiale Composito* (Bologna: Societá Editrice Esculapio)

[12] Soedel W 2004 *Vibration of Shells and Plates* (New York: Marcel Dekker)

[13] Reddy J N 2004 *Mechanics of Laminated Composite Plates and Shells. Theory and Analysis* 2nd edn (New York: CRC Press)

[14] Soldatos K P and Ye J 1995 Axisymmetric static and dynamic analysis of laminated hollow cylinders composed of monoclinic elastic layers *J. Sound Vibr.* **184** 245–59

[15] Messina A 2009 Three dimensional free vibration analysis of cross-ply laminated plates through 2D and exact models *3rd Int. Conf. on Integrity, Reliability and Failure (Porto, 20–24 July)*

[16] Open document *Systems of Differential equations* free available on http://www.math.utah.edu/gustafso/ (accessed 23rd May 2013)

[17] Boyce W E and DiPrima R C 2001 *Elementary Differential equations and Boundary Value Problems* (New York: Wiley)

[18] Zwillinger D 1997 *Handbook of Differential equations* (New York: Academic)

[19] Molery C and Van Loan C 2003 Nineteen dubious ways to compute the exponential of a matrix, twenty-five years later *SIAM Rev.* **45** 1–46

[20] Simsek M 2010 Vibration analysis of a single-walled carbon nanotube under action of a moving harmonic load based on nonlocal elasticity theory *Physica* **E43** 182–91

[21] Aydogdu M 2009 A general non local beam theory: its application to nanobeam bending, buckling and vibration *Physica* **E41** 1651–5

[22] Cinefra M, Carrera E and Brischetto S 2011 Refined shell models for the vibration analysis of multiwalled carbon nanotubes *Mech. Adv. Mater. Struct.* **18** 476–83

[23] Chen X and Cao G 2006 A structural mechanics study of single-walled carbon nanotubes generalized from atomistic simulation *Nanotechnology* **17** 1004–15

[24] de Borbón F and Ambrosini D 2012 On the influence of van der Waals coefficient on the transverse vibration of double walled carbon nanotubes *Comput. Mater. Sci.* **65** 504–8

[25] Carrera E, Brischetto S and Nali P 2011 *Plates and Shells for Smart Structures: Classical and Advanced Theories for Modeling and Analysis* (New Delhi: Wiley)

[26] Tornabene F, Viola E and Fantuzzi N 2013 General higher-order equivalent single layer theory for free vibrations of doubly-curved laminated composite shells and panels *Compos. Struct.* **104** 94–117

[27] Tornabene F 2016 General higher order layer-wise theory for free vibrations of doubly-curved laminated composite shells and panels *Mech. Adv. Mater. Struct.* **23** 1046–67

IOP Publishing

Modeling and Computation in Vibration Problems, Volume 1
Numerical and semi-analytical methods
S Chakraverty, F Tornabene and J N Reddy

Chapter 10

Crack and interface interaction under quasi-static and dynamic loading

Dhaladhuli Pranavi, K S S Reddy, Amirtham Rajagopal and J N Reddy

A thermodynamically consistent phase field formulation for modeling the interactions between interfacial damage and bulk fracture in heterogeneous materials having matrix and inclusion phases with a matrix-inclusion interface is presented. A regularization scheme is considered for both the interface and the crack phase field. A coupled exponential cohesive zone law is adopted to model the interface which has the contributions of both normal and tangential displacement jump components. A novel nonlocal approach is devised to evaluate the smoothened values of jump at the regularized interface using element specific geometric information. The effects of stiff and soft interface on the mechanical response and the crack propagation are studied. An isotropic model is studied under dynamic loading. The performance of the proposed formulation is illustrated through representative numerical examples.

Symbols

Ω_m	Matrix
Ω_{inc}	Inclusion
ϕ	Crack phase field
α	Interface phase field
Γ_c	Sharp crack
Γ_I	Sharp interface
Γ_ϕ	Diffused crack
Γ_α	Diffused interface
t^n	Normal traction
j^n	Normal separation
t_u^n	Maximum normal traction
δ^n	Characteristic normal opening length
t^t	Tangential traction

 © IOP Publishing Ltd 2021

j^t	Tangential separation
t_u^t	Maximum tangential traction
δ^t	Characteristic tangential opening lengths
E_m	Modulus of elasticity of matrix
E_{inc}	Modulus of elasticity of inclusion
ν	Poisson's ratio
$\mathcal{G}_c^m$	Fracture toughness of matrix
$\mathcal{G}_c^{\text{inc}}$	Fracture toughness of inclusion
$\mathcal{G}_c^I$	Fracture toughness interface
ρ	Mass density of the material
v_0	Applied velocity

10.1 Introduction

The computational modeling of crack nucleation, growth, and propagation has become significant in understanding the failure mechanisms in engineering materials. Heterogeneous materials made up to two or more materials have better mechanical and fracture properties compared to homogeneous materials and can be designed according to the specified purpose. An interface exists between the individual materials which plays a key role in load transfer between the materials and the response of the whole system.

Griffith [1] has laid the foundation for studying the fracture mechanics in brittle materials. His approach is based on the minimization of global energy in cracked solids. The notable works of Irwin [2] and Rice [3] have contributed to the development of brittle fracture models. The fracture mechanism at bimaterial interfaces was studied by Evans *et al* [4] and interface fracture energy is evaluated [5]. Cohesive zone models defined by Dugdale [6] and Barenblatt [7] have been adopted to model the interface using cohesive zone laws relating traction and separation across the interface [8]. A surface damage model is developed to model delamination initiation and growth in a laminated composite (see [9] and [10]). Other numerical techniques, namely, XFEM (see [11] and [12]) and Thick Level-Set (TLS) (see [13] and [14]) methods have been developed and implemented to model embedded cohesive interfaces and crack propagation in composites.

The phase-field modeling is a diffuse approach [15–18] based on variational formulation by energy minimization. This approach is more attractive as it can simulate complicated fracture processes, such as crack initiation, propagation and branching. Propagation of cracks is automatically tracked by the evolution of the crack field which leads to a significant advantage over the discrete fracture methods, whose numerical implementation requires the handling of discontinuities. The derivation of the phase-field evolution equation and constitutive equation are based on the concepts of variational formulation by Francfort and Marigo [18], which extends the Griffiths theory of fracture and is regularized by Bourdin *et al* [19].

Quasi-static phase-field formulation has been implemented for modeling brittle fracture (see [20] and [21]), ductile fracture (see [22] and [23]), dynamic fracture (see [24] and [25]), and transient dynamic problems (see [26] and [27]). Further, PFM (Phase-field Model) combined with cohesive zone approach is adopted to model cohesive fracture (see [28, 29], and [30]). The displacement field, crack phase field and an auxiliary field representing displacement jump is considered for modeling cohesive fracture. This model requires an additional constraint on the auxiliary field that needs to be satisfied [28]. Both the interface and crack are considered to be diffused in [29], where the displacement jump across the interface is evaluated using level-set method. There exists an interaction between the length scales corresponding to crack and interface (see [17] and [31]). A single energy considering both bulk energy and cohesive surface energy is adopted in [30], where the displacement jump depends on the crack phase field.

Cohesive zone models are developed by treating the fracture phenomenon as gradual where the separation takes place near the extended crack tip. A cohesive law is a traction–separation relation that represents the stress transmitted between the crack faces in a cohesive zone. The relation is such that, across the interface, with increase in the separation, the traction increases, reaches a peak value, decreases and then eventually ceases. There are various forms of cohesive zone laws such as polynomial, linear, exponential among others. In some cases, the interaction/coupling between normal and tangential components of traction and or separation is not considered (see [32] and [33]). But to model the realistic interface behavior, a coupling between the modes has to be considered [34]. The most meaningful form of coupled interface behavior is obtained from thermodynamic considerations (see [8] and [35]). In the present work, an exponential cohesive zone law is considered (see figure 10.1)

The present chapter is organized as follows. Section 10.2 explains briefly the methodology of the proposed model and, the governing equations to be solved to obtain the displacement and crack phase field. In section 10.3, the coupling between the equilibrium and evolution equations through the degradation function is shown. Numerical studies on dynamic fracture and crack–interface interaction are conducted in section 10.4. The concluding remarks are discussed in section 10.5.

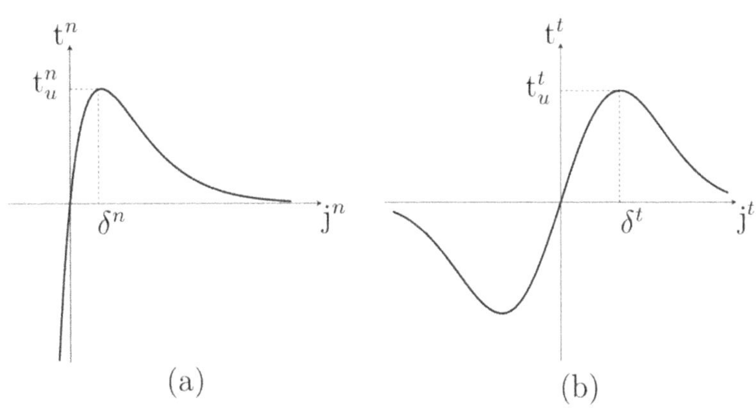

Figure 10.1. Exponential cohesive zone law depicting (a) normal traction separation and (b) tangential traction separation.

10.2 Methodology

Consider a two-phase solid body $\Omega \in \mathbb{R}^d$ $(d \in 1, 2, 3)$ composed of matrix Ω_{m} and inclusion Ω_{inc} with an interface between them Γ_{I} (see figure 10.2). The crack surface in the body is denoted as Γ_{c}. $\mathbf{F}_{\mathrm{u}}$ are the body forces and $\mathbf{F}_{\mathrm{t}}$ is the prescribed traction over $\partial\Omega_{\mathrm{F}}$. The displacement and force boundary conditions are prescribed as:

$$\mathbf{u} = \bar{\mathbf{u}} \quad \text{on} \quad \partial\Omega_{\mathrm{u}} \quad \text{and} \quad \mathbf{F}_{\mathrm{t}} = \boldsymbol{\sigma} \cdot \mathbf{n} \quad \text{on} \quad \partial\Omega_{\mathrm{F}} \tag{10.1}$$

where $\mathbf{n}$ denotes the outward normal vector to the body Ω and $\boldsymbol{\sigma}$ is the Cauchy stress tensor.

The free energy functional is then defined as

$$\begin{aligned} \mathsf{E} = & \int_{\Omega} \psi^e(\boldsymbol{\varepsilon}(\mathbf{u}))d\Omega + \int_{\Gamma_{\mathrm{c}}} \mathcal{G}_c d\Gamma_{\mathrm{c}} \\ & + \int_{\Gamma_{\mathrm{I}}} \psi^I([\![\mathbf{u}]\!], \boldsymbol{\kappa})\, d\Gamma_{\mathrm{I}} - \int_{\Omega} \mathbf{F}_{\mathrm{u}} \cdot \mathbf{u} d\Omega - \int_{\partial\Omega_F} \mathbf{F}_{\mathrm{t}} \cdot \mathbf{u} dA \end{aligned} \tag{10.2}$$

In the above equation, $\mathcal{G}_c$ is the fracture toughness, ψ^I is a strain density function associated with the interface Γ_{I} and $\boldsymbol{\kappa}$ is a history parameter. $[\![\mathbf{u}]\!]$ is the displacement jump which is denoted as $\mathbf{j}(\mathbf{x})$. The infinitesimal strain tensor can be additively split into a bulk local strain and other a nonlocal regularized strain at the interfaces and is given by $\boldsymbol{\varepsilon} = \boldsymbol{\varepsilon}^e + \boldsymbol{\varepsilon}^{\mathrm{j}}$. Equation (10.2) gets modified as

$$\begin{aligned} \mathsf{E} = & \int_{\Omega} \psi^e(\boldsymbol{\varepsilon}^e(\mathbf{u}, \alpha), \phi)d\Omega + \int_{\Omega} [1 - \alpha(\mathbf{x})]\mathcal{G}_c\gamma_\phi(\phi)d\Omega \\ & + \int_{\Omega} \psi^I(\mathbf{j}, \boldsymbol{\kappa})\gamma_\alpha(\alpha)d\Omega - \int_{\Omega} \mathbf{F}_{\mathrm{u}} \cdot \mathbf{u} d\Omega - \int_{\partial\Omega_F} \mathbf{F}_{\mathrm{t}} \cdot \mathbf{u} dA \end{aligned} \tag{10.3}$$

where $\gamma_\phi(\phi)$ and $\gamma_\alpha(\alpha)$ are crack density and interface density functions per unit volume [16]. For a dynamic problem, the kinetic energy of the body is given by

$$T(\dot{\mathbf{u}}) = \int_{\Omega} \frac{1}{2}\rho\dot{\mathbf{u}} \cdot \dot{\boldsymbol{u}}\, dV \tag{10.4}$$

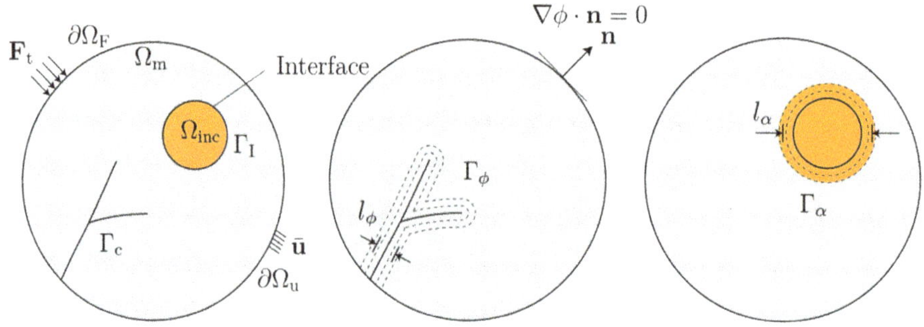

Figure 10.2. (a) Schematic representation of the domain Ω having sharp crack and interface; (b) regularized crack; (c) regularized interface.

where ρ is the mass density of the material and $\dot{u}$ denotes the local velocity The Lagrangian (L) is expressed as the difference between kinetic energy (T) and potential energy (E) as, (body forces and traction are neglected) :

$$\begin{aligned} L = T - \mathsf{E} = \int_{\Omega} \frac{1}{2}\rho\dot{\mathbf{u}} \cdot \dot{\mathbf{u}}\, dV - \int_{\Omega} \psi^e(\boldsymbol{\varepsilon}^e(\mathbf{u}, \alpha), \phi) d\Omega \\ - \int_{\Omega} [1 - \alpha(\mathbf{x})]\mathcal{G}_c\gamma_\phi(\phi) d\Omega - \int_{\Omega} \psi^I(\mathbf{j}, \boldsymbol{\kappa})\gamma_\alpha(\alpha) d\Omega \end{aligned} \tag{10.5}$$

The equilibrium equation is obtained by varying E with respect to $\mathbf{u}(\mathbf{x})$.

$$\int_{\Omega} \frac{\partial \psi^e}{\partial \boldsymbol{\varepsilon}^e} : \boldsymbol{\varepsilon}^e(\delta\mathbf{u}) d\Omega + \int_{\Omega} \frac{\partial \psi^I(\mathbf{j}, \boldsymbol{\kappa})}{\partial \mathbf{j}} . \, \delta\mathbf{j}\gamma_\alpha(\alpha) d\Omega = \int_{\Omega} \rho\ddot{\mathbf{u}}(\delta\mathbf{u}) d\Omega \tag{10.6}$$

The evolution equation is obtained by varying E with respect to $\phi(\mathbf{x})$.

$$\int_{\Omega} \frac{\partial \psi^e}{\partial \phi} d\Omega + \int_{\Omega} [1 - \alpha]\mathcal{G}_c\delta\gamma_\phi(\phi) d\Omega = 0 \tag{10.7}$$

10.3 FEM formulation

The energy storage function is defined as

$$\psi^e(\boldsymbol{\varepsilon}^e, \phi) = [g(\phi) + k_o]\,\psi^{e+}(\boldsymbol{\varepsilon}^e) + \psi^{e-}(\boldsymbol{\varepsilon}^e) \tag{10.8}$$

Considering the fracture process is occurring in tension regions, the degradation function $g(\phi)$ is considered to act only on the tensile (positive) part of the stored energy. The function has the form $g(\phi) = (1 - \phi)^2$ that follows the basic properties $g(0) = 1$, $g(1) = 0$, $g'(1) = 0$.

The evolution equation from equation (10.7) can be written as

$$-2(1 - \phi)\psi^{e+} + [1 - \alpha]\mathcal{G}_c \frac{1}{l_\phi}[\phi - l_\phi^2 \Delta\phi] = 0 \tag{10.9}$$

10.3.1 Exponential cohesive zone law

The interface is characterized by existence of a potential function ψ^I which describes the coupled traction-separation completely at all stages of loading and is defined as

$$\mathbf{t} = -\frac{\partial \psi^I}{\partial \mathbf{j}}, \quad \mathbf{j} = [\mathrm{j}^n, \mathrm{j}^t]^T \tag{10.10}$$

The potential function ψ^I for exponential cohesive zone law is defined as [36]

$$\psi^I(\mathrm{j}^n, \mathrm{j}^t) = \psi_n^I \left[1 - \left(1 + \frac{\mathrm{j}^n}{\delta^n}\right) \exp\left(-\frac{\mathrm{j}^n}{\delta^n}\right) \exp\left(-\left(\frac{\mathrm{j}^t}{\delta^t}\right)^2\right)\right] \tag{10.11}$$

10.3.2 Displacement jump

The displacement jump vector can be evaluated by multiplying the nodal displacement vector $\mathbf{u}_I$ with a difference array $\mathbf{L}$, which gives the difference between the displacements of upper nodes of matrix element and lower nodes of inclusion element across the interface, and shape functions array $\mathbf{M}$ [37]. Accordingly, within the finite element discretization we can write

$$\mathbf{j} = \mathbf{M}\mathbf{L}\mathbf{u}_I \tag{10.12}$$

10.4 Numerical examples

In this section, the performance of the proposed formulation is demonstrated by (a) crack–interface interaction studies for stiff-soft interface cases in a composite system, (b) dynamic fracture studies in concrete.

10.4.1 Stiff–soft interface in a micro structure of a composite system

In this example, a representative volume element of a cracked heterogeneous microstructure that can be considered as a multiple inclusion composite system with stiff or soft interface and crack is considered. The geometry, boundary conditions and discretization are shown in figure 10.3. Four-noded quadrilateral elements and plane strain conditions are assumed for the analysis.

The purpose of this example is to understand the evolution of crack phase field and its interaction with interfaces and inclusions, in terms of the dominant failure mode of constituents of the composite system when the interface is either stiff or soft. The length scale corresponding to both crack and interface is taken as 0.02 mm. The modulus of elasticity of matrix and inclusion are taken as $E_m = 21.23$ GPa and $E_{\text{inc}} = 62.26$ GPa, respectively. The fracture energy values for matrix and inclusion are taken as $\mathcal{G}_c^m = 0.000\,3$ GPa, and $\mathcal{G}_c^{\text{inc}} = 0.008$ GPa. Poisson's ratio $\nu = 0.27$ and fracture strength for cohesive interface $t_u^n = t_u^t = 0.01$ GPa are considered for the

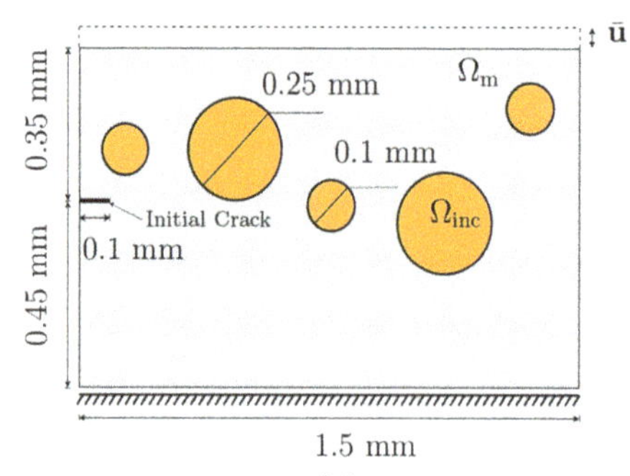

Figure 10.3. Multiple inclusion composite system with stiff or soft interface. (a) Geometry and boundary conditions. (b) Discretization used in the analysis.

analysis. Two cases are considered for the analysis: (1) when the interface is stiff with $\mathcal{G}_c^I = 0.003$ GPa; and (2) when the interface is soft with $\mathcal{G}_c^I = 0.000\,075$ GPa.

The evolution of the crack phase field is plotted in figure 10.4. In the first case when the interface is stiff, the crack phase field is deflected into the matrix phase and the predominant mode is by cracking of matrix, as shown in figure 10.4(a). This is characterized by a higher value of the peak load deflection curve, as it augments the overall stiffness of the material. The post peak behavior is characterized by a near sudden drop, as shown in figure 10.5. In the second case, when the interface is soft, the crack is initiated and evolves along the interface and between the inclusions, resulting in a predominantly interface debonding and matrix cracking between inclusions, as shown in figure 10.4(b). The load deflection diagram is characterized by a lower peak load and a slightly gradual post peak behavior, as shown in figure 10.5. Similar observations have also been obtained in experimental studies available in literature (see [38]). From figure 10.4, it is evident that for a stiff interface, matrix cracking will be predominant and for a soft interface, interfacial

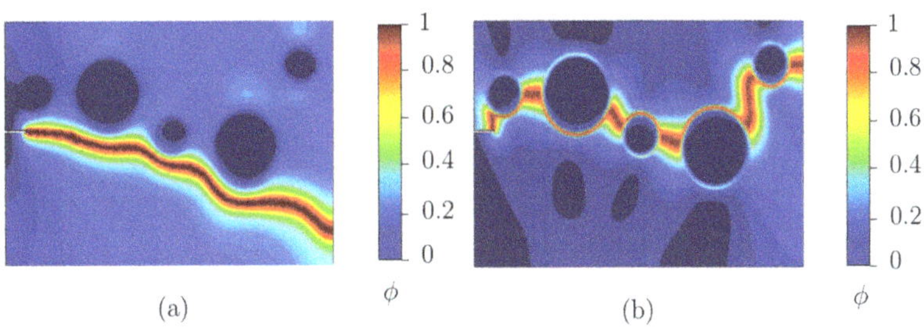

Figure 10.4. Evolution of crack phase field $\phi(\mathbf{x})$ for (a) stiff interface and (b) soft interface.

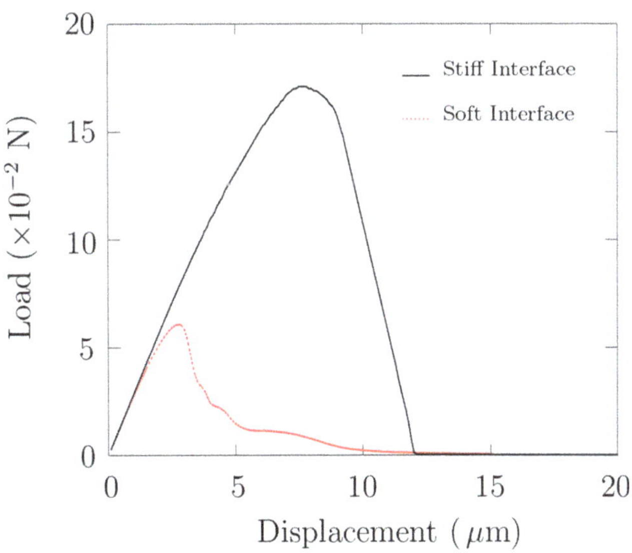

Figure 10.5. Plot of the load versus the displacement for crack in a multiple inclusion composite system.

cracking dominates. In the case of soft interface, crack is initiated and propagated along the interface and the failure load is less (see figure 10.5), whereas when the interface is stiff the crack propagates in the matrix phase and prevents interfacial debonding. The failure load is comparatively very high in this case.

10.4.2 Compact tension tests in concrete

To test the phase field model for dynamic fracture, compact tension tests are performed on concrete [39]. The geometry and the boundary conditions are shown in figure 10.6. The velocity v_0 is applied on the right edge of the notch while the left edge is fixed.

The material properties are ρ = 2400 kg m^{-3}, E = 36 GPa, ν = 0.18 and G_c = 65 J m^{-2}. The length scale parameter is set to 1 mm, element size is taken as 0.5 mm. The simulations are done for three different velocities 0.304 m s^{-1}, 1.375 m s^{-1} and 3.3 m s^{-1}. The velocity condition given in [39] is imposed while applying the velocity boundary condition (the velocity is increased from zero to v_0 until t_0 = 100 μs and then kept constant till the end of stimulation). Figure 10.7

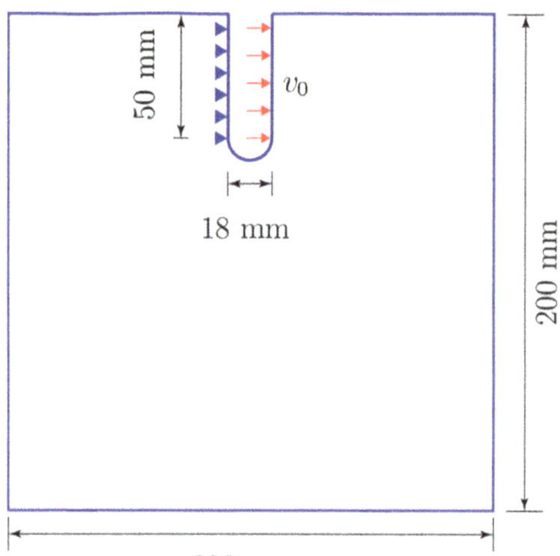

Figure 10.6. Geometry and boundary conditions for compact tension test.

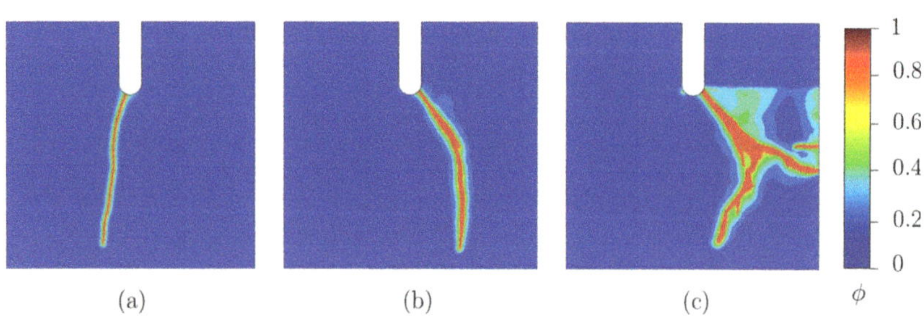

Figure 10.7. Damage profile for compact tension tests in concrete (a) v_0 = 0.304 m s^{-1}, (b) v_0 = 1.375 m s^{-1} and (c) v_0 = 3.3 m s^{-1}.

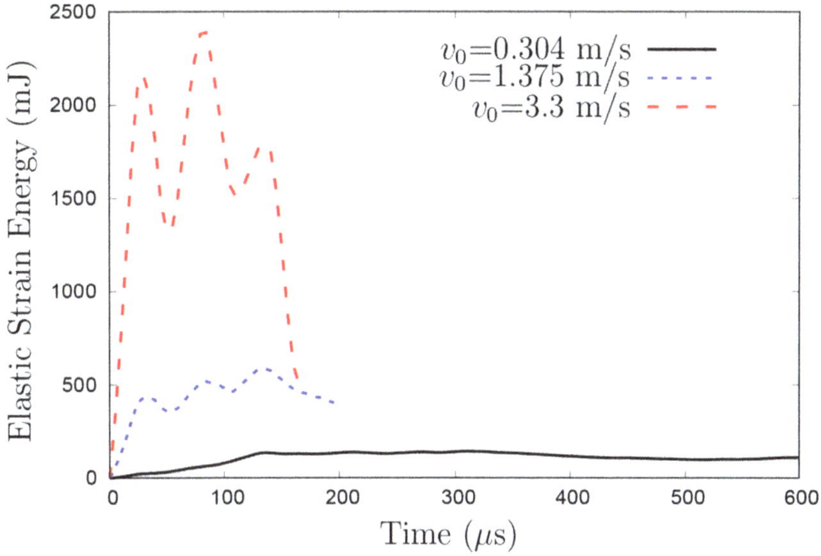

Figure 10.8. Strain energy curves for compact tension tests in concrete.

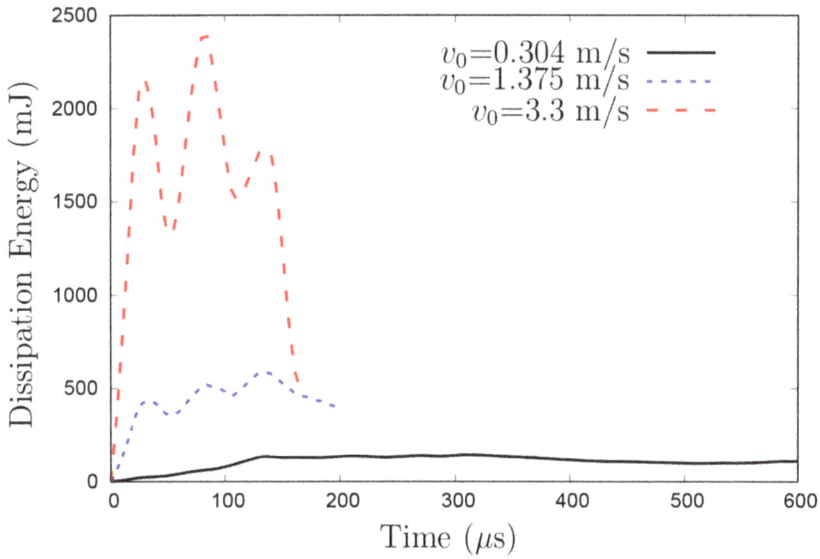

Figure 10.9. Dissipation energy curves for compact tension tests in concrete.

shows crack patterns for three different velocities. Figure 10.8 shows the elastic strain energy curves for three velocities. The maximum energy storage capacity is higher for higher rate of loading and it decreases with the decrease in loading rate. Figure 10.9 presents the dissipation energy curves. The dissipation energy increases as the time increases for all the velocities considered. The energy dissipated for higher rate of loading is more than the lower rates of loading. From the energy curves it can be inferred that the crack surface area formed for higher rate of loading is more than the lower rates of loading.

10.5 Conclusions

In the first simulation, the interaction of crack with stiff–soft interfaces are analyzed. From the evolution of crack phase field, it can be observed that the predominant failure is matrix failure, for the case of stiff interface, and interface failure, for the case of soft interface. From the graph, it can be seen that a higher peak load is observed for stiff interface but the failure is gradual for soft interface. From the second simulation, it can be observed that: (a) as the rate of loading decreases, the maximum energy storage capacity also decreases; (b) the crack surface area formed increases with increase in rate of loading.

References

[1] Griffith A A 1921 The phenomena of rupture and flow in solids *Phil. Trans. R. Soc.* **A221** 582–93

[2] Irwin G R 1957 Analysis of stresses and strains near the end of a crack traversing a plate *J. Appl. Mech.* **24** 351–69

[3] Rice J R 1968 A path independent integral and the approximate analysis of strain concentrations of notches and cracks *J. Appl. Mech.* **35** 379–86

[4] Evans A G, Dalgleish B J, He M and Hutchinson J W 1989 On crack path selection and the interface fracture energy in bimaterial systems *Acta Metall.* **37** 3249–54

[5] Evans A G, Rühle M, Dalgleish B and Charalambides B G 1990 The fracture energy of bimaterial interfaces *Metall. Trans.* **A21** 2419–29

[6] Dugdale D S 1960 Yielding of steel sheets containing slits *J. Mech. Phys. Solids* **8** 100–4

[7] Barenblatt G I 1959 The formation of equilibrium cracks during brittle fracture. General ideas and hypotheses *J. Appl. Math. Mech.* **23** 434–44

[8] Needleman A 1990 An analysis of decohesion along an imperfect interface *Int. J. Fract.* **42** 21–40

[9] Allix O and Ladevèze P 1992 Interlaminar interface modelling for the prediction of delamination *Compos. Struct.* **22** 235–42

[10] Allix O and Corigliano A 1996 Modeling and simulation of crack propagation in mixed-modes interlaminar fracture specimens *Int. J. Fract.* **77** 111–40

[11] Belytschko T and Gracie R 2007 On XFEM applications to dislocations and interfaces *Int. J. Plast.* **23** 1721–38

[12] Sukumar N, Möes N, Moran B and Belytschko T 2000 Extended finite element method for three-dimensional crack modelling *Int. J. Numer. Methods Eng.* **48** 1549–70

[13] Cazes F and Möes N 2015 Comparison of a phase-field model and of a thick level set model for brittle and quasi-brittle fracture *Int. J. Numer. Methods Eng.* **103** 114–43

[14] Latifi M, van der Meer F P and Sluys L J 2017 An interface thick level set model for simulating delamination in composites *Int. J. Numer. Methods Eng.* **111** 301–400

[15] Bourdin B, Francfort G A and Marigo J J 2008 *The Variational Approach to Fracture* (Berlin: Springer)

[16] Miehe C, Welschinger F and Hofacker M 2010 Thermodynamically consistent phase-field models of fracture: variational principles and multi-field FE implementations *Int. J. Numer. Methods Eng.* **83** 1273–11
[17] Mumford D B and Shah J 1989 Optimal approximations by piecewise smooth functions and associated variational problems *Commun. Pure Appl. Math.* **42** 577–685
[18] Francfort G A and Marigo J J 1998 Revisiting brittle fracture as an energy minimization problem *J. Mech. Phys. Solids* **46** 1319–42
[19] Bourdin B, Francfort G and Marigo J J 2000 Numerical experiments in revisited brittle fracture *J. Mech. Phys. Solids* **48** 797–826
[20] Kasirajan P, Bhattacharya S, Rajagopal A and Reddy J N 2020 Phase field modeling of fracture in quasi-brittle materials using natural neighbor Galerkin method *Comput. Methods Appl. Mech. Eng.* **366** 113019
[21] Rajagopal A, Fischer P, Kuhl E and Steinmann P 2010 Natural element analysis of the Cahn–Hilliard phase-field model *Comput. Mech.* **46** 471–93
[22] Ambati M, Gerasimov T and De Lorenzis L 2015 Phase-field modeling of ductile fracture *Comput. Mech.* **55** 1017–40
[23] Borden M J, Hughes T J R, Landis C M, Anvari A and Lee I J 2016 A phase-field formulation for fracture in ductile materials: finite deformation balance law derivation, plastic degradation, and stress triaxiality effects *Comput. Methods Appl. Mech. Eng.* **312** 130–66
[24] Borden M J, Verhoosel C V, Scott M A, Hughes T J R and Landis C M 2012 A phase-field description of dynamic brittle fracture *Comput. Methods Appl. Mech. Eng.* **217-20** 77–95
[25] Schlüter A, Willenbücher A, Kuhn C and Müller R 2014 Phase field approximation of dynamic brittle fracture *Comput. Mech.* **54** 1141–61
[26] Zhou S, Rabczuk T and Zhuang X 2018 Phase field modeling of quasi-static and dynamic crack propagation: COMSOL implementation and case studies *Adv. Eng. Softw.* **122** 31–49
[27] Raghu P, Rajagopal A and Reddy J N 2020 Nonlocal transient dynamic analysis of laminated composite plates *Mech. Adv. Mater. Struct.* **27** 1076–84
[28] Verhoosel C V and Borst R de 2013 A phase-field model for cohesive fracture *Int. J. Numer. Methods Eng.* **96** 43–62
[29] Nguyen T T, Yvonnet J, Zhu Q Z, Bornert M and Chateau C 2016 A phase-field method for computational modeling of interfacial damage interacting with crack propagation in realistic microstructures obtained by microtomography *Comput. Methods Appl. Mech. Eng.* **312** 567–95
[30] Paggi M and Reinoso J 2017 Revisiting the problem of a crack impinging on an interface: a modeling framework for the interaction between the phase field approach for brittle fracture and the interface cohesive zone model *Comput. Methods Appl. Mech. Eng.* **321** 145–72
[31] Ambrosio L and Tortorelli V M 1990 Approximation of functional depending on jumps by elliptic functional via t-convergence *Commun. Pure Appl. Math.* **43** 999–1036
[32] Tvergaard V and Hutchinson J W 1993 The influence of plasticity on mixed mode interface toughness *J. Mech. Phys. Solids* **41** 1119–35
[33] Mohammed I and Liechti K M 2000 Cohesive zone modeling of crack nucleation at bimaterial corners *J. Mech. Phys. Solids* **48** 735–64
[34] Cui W C, Wisnom M R and Jones M 1992 A comparison of failure criteria to predict delamination of unidirectional glass/epoxy specimens waisted through the thickness *Composites* **23** 158–66

[35] Xu X P and Needleman A 1993 Void nucleation by inclusion debonding in a crystal matrix *Model. Simul. Mater. Sci. Eng.* **1** 111–32

[36] Van den Bosch M J, Schreurs P J G and Geers M G D 2006 An improved description of the exponential Xu and Needleman cohesive zone law for mixed-mode decohesion *Eng. Fract. Mech.* **73** 1220–34

[37] Reinoso J and Paggi M 2014 A consistent interface element formulation for geometrical and material nonlinearities *Comput. Mech.* **54** 1569–81

[38] Krishnan A and Xu L R 2013 Experimental studies on the interaction among cracks, notches and interfaces of bonded polymers *Int. J. Solids Struct.* **50** 1583–96

[39] Li J, Hai L and Wu J Y 2020 A phase-field damage model with micro inertia effect for the dynamic fracture of quasi-brittle solids *Eng. Fract. Mech.* **225** 106821

IOP Publishing

Modeling and Computation in Vibration Problems, Volume 1
Numerical and semi-analytical methods
S Chakraverty, F Tornabene and J N Reddy

Chapter 11

Vibration of compliant robotic grippers and wrists

Debanik Roy

Parts of this chapter have been adapted from [1] with permission from Springer.

11.1 Introduction

Industrial robotic systems have two essential functional sub-systems, namely gripper and wrist, that are prone to vibration-induced characteristics. This vibration is randomized in real-time and can be *in situ* and/or external impulse-based. Compliant Robotic Gripper (CRG) belongs to a selected niche of the first sub-system and these grippers are modular, semi-flexible and often small-enveloped with multi-task enabled ability. The vibration in CRG is completely built-in type, design invariant and self-propagating, which does not follow analytical modeling and rule-base in all applications. Several designs of CRG have been attempted by researchers in the past decade in order to alleviate this vibration but most of those trials have been unsuccessful. In-line with CRG, Compliant Robotic Wrist (CRW) is another niche for the study of *in situ* vibration. The only difference between these two sets, i.e. vibration signature for CRG vis-à-vis CRW, is related to the aspect of self-balancing of the end-effector assembly of the robot. In other words, inherent compliance of the CRW helps reducing the external vibration at times, much like the principle of Remote Centre Compliance. Nonetheless, compliance of these two sub-systems often complicates the vibration attenuation problem from the aspect of control system design.

The ensemble domain of CRG caters for real-time aspects like rheology (stress-strain paradigms), *in situ* vibration, sensor fusion and non-linear coupled dynamics for control. As some of these features are inherent in CRG, prototyping a multiple degrees-of-freedom customized compliant gripper is highly challenging. To add to this, micro-CRGs are prone to *in situ* deflection that cannot be eliminated from the control loop. Irrespective of all hardware-types, this inherent vibration of CRGs gets

doi:10.1088/978-0-7503-3483-9ch11

© IOP Publishing Ltd 2021

realized in the form of mild to severe trembling of the slender links and/or shaking or twisting of the link-joint interface and inter-spaced joints. Slenderness of the gripper-links plays a very important role in self-generation of this trembling. We will discuss modeling of this *in situ* vibration as well as solving design issues using a customized topology optimization model for the firmware of customized CRGs in miniaturized form. Study of the turning phenomena of the CRG, due to the fall-out effect of system tremble/jerk/vibration will also be delineated analytically. Paradigms will be nearly similar for CRW, barring structural design and deflection characteristics.

The vibration synthesis (VS) aspect of CRG and CRW will encompass three modules in real-time, viz. (i) vibration signature; (ii) dynamic model and (iii) turning model. The VS for CRG dwells primarily with grasp prehension that occurs through four cases of realization of the external geometry of the jaw surface and object surface. We will detail an analytical model of VS for two types of gripper jaws: (a) flat jaw; and (b) curvilinear jaw. Likewise, two envelopes of the object surfaces will be considered in the modeling. For flat-surfaced objects, we need to consider their 'circular envelope', while 'curved approximation' of flat jaw will be adhered to for a flat-jaw gripper. The strategy for modeling is based on the analytical treatment of both jaws as well as object surfaces as curves. The advantage of this sort of modeling is the subtleness of the 'point of contact' between the mating surfaces of the gripper jaw and object grasped. We will delineate three types of grasp-induced vibration models: (a) asperity contact-based; (b) contact mechanics vector-based and (c) adhesive mechanics-based. As CRW does not involve mating parts, we will highlight four different design metrics of it so far as its VS is concerned. We have made use of two types of deflection model for CRG and CRW, namely, using (i) only spring members and (ii) spring-dashpot-damper members.

The 'vibration signature' of CRG and CRW gets manifested in two ways: modal frequency and Eigen value. It has been observed that vibration in CRG & CRW is not time-dependent and the duration and periodicity of it cannot be correlated with the task-space of the robotic system. Moreover, for CRG, this vibration gets induced in the successive member(s) till the jaw-plate of the gripper. The problem gets more complicated when we attempt for multi-link design of the CRG, wherein various kinds of coupled effect and non-linearity come in. For CRW, the vibration percolates from the upper plate to the lower one via inter-spaced links. The tuple of stiffness coefficient and viscous damping of the spring-dashpot-damper model have been expounded in Taylor's series for functional evaluation of the natural frequency of vibration of the CRG and CRW and subsequently, modal frequencies. The vibration signature in CRG and CRW gets assessed in real-time through multiple force sensors, spread over links, body and joints.

The ensemble paradigm of vibration signature in CRGs or CRWs and its control in real-time is rooted in model-driven research and validation. One of the effective tools towards characterization of this in-built vibration of the CRGs used hitherto is Finite Element Analysis (FEA). The other crucial tool for this study is somewhat system-level realization of this vibration, when the CRG and/or CRW gets interfaced with the robotic manipulator and performs the intended task of grasping in real-time. However, the vibration signature of CRG will be different depending upon the type of robotic manipulator wherein the CRG-assembly gets interfaced. The effect of vibration will be much dominant if the said interfacing is done with a Flexible Robotic System (FRS) in comparison to a Semi-Flexible Robotic Manipulator (SFRM) or Multi-Gripper Assistive Robot (MGAR). We will bring out the jist of some of the important past research literature in the domain of vibration: starting from its identification and going deep into its manifestation in real-life experimentation via suitable modeling and simulation interfaces.

Qassab and Ali Sultan [2] analyzed the FEA results on stress and deflection for a two-fingered robotic hand and compared those with the results obtained from theoretical model-based calculations as well as experimentation. An interesting FEM-based modal analysis outcome of the robotic arm reported that a circular-shaped cross-section of the arm can sustain higher vibration than a square-shaped robot arm structure, irrespective of the type of cross-section (solid or hollow) [3]. The principle of FEA was extended to fine-tune the design of a three-link slender flexible robotic manipulator, especially for the links and revolute joints [4]. Details of vibration signature and associated deflection of the links of the said indigenously-designed flexible manipulator have been studied by the authors. FEM-based simulation thereof helped in pin-pointing the support locations and related kinetics between the base-assembly and the first link of the said flexible manipulator. Various attributes of parametric design and dynamic simulation of the indigenously-developed controller of a novel MGAR were reported along with results of test-runs [5]. The ensemble programming logic for the robot was developed towards controlling in-built vibration of the robot in real-time. The hardware of the robotic manipulator was accomplished in a way so as to minimize the inherent shaking of the manipulator arms. Roy [6] discussed the differences in vibration characteristics of multi-link flexible robot from that of a single-link system which is due to the coupling effects of joints and flexible shafts. The proposed methodology builds up an optimal foundation for analyzing inherent vibration of flexible robots using strain gauge-based measurement as well as stochastic model-based fusion of sensory data. A methodology for reducing tare-weight and minimizing structural deformation in 3D space of a 6-axis articulated robot (ARISTO®) was addressed, based on the calculation of the loading forces applied in a static study [7]. An interesting treatise on the study of the natural frequency of a robotic arm in free vibration case was reported, wherein maiden analysis was made using the classical analytical method of well- known Euler–Bernoulli theory [8]. Chouhan and Kanwal [9] presented the procedure of design and fabrication of a stepper motor controlled robotic gripper to be used in industry for handling small objects. A multi-objective evolutionary algorithm was used to solve a modified dual-objective problem and to optimally find the dimensions of links and joint angle of a robot gripper [10]. The study was extended to find suitable relationships between the decision variables therein and the objective functions. A force-voltage relationship could be obtained from each of the non-dominated solutions which helped to determine the voltage to be applied for the actuation of the gripper based on the application. An optimal design procedure was used to synthesize an adaptive monolithic compliant two-finger robotic gripper with high mechanical advantage for grasping irregular objects [11]. The study addressed novel numerical methods to synthesize compliant mechanisms (made up of silicon rubber) with higher output force. A novel process for modeling robot grippers and optimizing their structure was reported, wherein an equivalent Jacobian matrix was derived to find the kinematic model while the vibration-laced dynamic model was obtained using Lagrange formulation [12]. Based on these models, a structural multi-objective optimization (MOO) problem was formalized in the static configuration of the gripper. We have found that the real-life design problem of two-finger grippers has been approached and formulated as an optimization problem by using the basic characteristics of grasping mechanisms in presence of *in situ* trembling of the gripper system [13]. A specific case of study has been reported in this treatise, by using revolute-prismatic paired linkages as a proposed grasping mechanism. The intricacy of vibration-induced design synthesis of robotic gripper system goes deeper as and when we come across grippers with complex kinematics under full payload semantics, such as: grippers with large re-orientations (more than ð/2 radian) [14]; industrially-used grippers with wide-spanned payloads in production-line [15]; mechanics of grasping and finger-object contact interactions pertinent to grasp analysis, simulation and synthesis; [16] and FE-software–based static and dynamic analysis of the robotic gripper [17]. A detailed analysis on the origin, source apportionment, real-time evaluation and control of inherent vibration in a multi-degrees-of-freedom flexible manipulator can be referred for assessment as well as estimation of the system-level vibration of a robotic gripper system [18]. The maiden design and firmware of an extremely light-weight small-enveloped robotic jaw gripper alongwith its integration with an assistive-type multi-degrees-of-freedom planar flexible robot was delineated by Roy [19].

Vibration signature and control issues of FRS have gained research attention over the last few decades, which deal with novel techniques of control of system dynamics in real-time [20], inclusive of harnessing self-propagating vibration in the system [21]. While perturbation method was tried for fine-tuning the FRS-controller [22], direct real-time feedback from strain gauges was experimented with too [23]. It is true that a robust dynamic model becomes very effective in understanding the behavior of FRS in real-time and the same becomes crucial for a multi-link FRS [24], [25]. Feliu *et al* attempted the control issue of a three degrees-of-freedom FRS using the methodology of inverse dynamics in contrast to strain gauge-based control [26, 27]. Specific metrics related to reduction of system vibration of a robotic gadget were attributed by Singer and Seering [28]. Various techniques for vibration attenuation and control in FRS have been reported hitherto, such as sliding mode theory [29], adaptive resonant control [30], online frequency and damping estimation [31] and integral resonant control [32]. New paradigms on guidance and control of flexible robotic systems have been experimented with, using linear-quadratic-Gaussian design methods [33] and closed-form solutions for feedback control [34].

One of the fundamental paradigms of the run-time vibration of the CRG as well as CRW is the dynamics of rotation of the constituent members/links. Two complementary aspects for the realization of this dynamics of rotation for jointed semi-flexible structure are: (a) stabilization of run-time constraints for the motion; and (b) evaluation of integrands for the motion after normalization of the participating parameters. Baumgarte [35] addressed the issue of attaining stability of the constraints of a real-time rigid-body dynamic system *ab initio* and formulated the grassroot-level analytical models for the rotational motion of such systems. In line with this fundamental research, Argyris [36] addressed various aspects of multi-body rotations from basic formulations, pertaining to their extent and variability. On the other hand, new propositions and analytical models for finite rotation in computational solid mechanics are reported as part of follow-up research [37].

The off-shoot of the research on modeling of dynamics of rotation for rigid bodies has culminated in better insight towards ensemble dynamics of flexible systems, inclusive of deflection and rotation. In fact, dynamics of flexible systems and to that effect, dynamics of complaint bodies go hand-in-hand in various metrics. The inter-coupling of the deflection and rotation of flexible systems, which is the nucleus of *in situ* vibration in such 'elastic' mechanisms, remains a challenging domain for the modeling and real-time control [38]. While rotational aspects of this real-time dynamics of the compliant systems have been tackled via beam theory-based proposition under small strain [39], the translational aspect of the non-linear dynamics is addressed through displacement along the curved flexible axis of the respective members [40]. It may be mentioned here that, by and large, numerical integration remains the well-accepted technique for evaluation of the non-linear dynamics of elastic multi-body systems [41], including modeling of the dynamic friction in such compliant and elastic systems [42].

Indigenous design, system modeling, FEA and worthy hardware development of two 'test-beds', respectively, for three-degrees-of-freedom planar and articulated-type FRS, fitted with mini-grippers, have been reported by our group in [4-6, 43]. In both of these 'test-beds', we explored the subtleness of the design for the ultimate objective of grasping of payload by the end-of-arm tooling of the FRS, namely the miniaturized jaw-type gripper. However, vibration modeling of the multi-link CRG or CRW using compliant sub-assemblies like spring-dashpot-damper remains an open research domain to date.

The chapter will address modeling of vibration of CRG and CRW *ab initio*, for both *in situ* as well as external excitation. In line with the principal theme of the topic, the chapter has been composed in six sections. An overview on various designs of the CRGs is presented in the next section. Section 11.3 details the fundamental design schemes for the indigenous hardware of the CRWs that are compatible with different types of compliant grippers upon interfacing. Section 11.4 addresses different types of grasp-induced vibration models of the CRGs. The parlances of vibration signature for the CRG and CRW will be delineated in section 11.5. The secondary-stage vibration at the CRG due to payload and force closure phenomenon will also be addressed in this section. Section 11.6 will highlight the development of turning model for the VS in real-time that will be applicable to CRG and CRW in unison. We will also discuss the local effects of vibration in this section, which are in the form of vibration accumulation at CRW and the consequences of external forcing on the robotic manipulator. Case-studies will be reported in section 11.7, supported by test results for a table-top small-sized semi-flexible robotic system, augmented with a tailor-made CRG. Finally, section 11.8 will conclude the chapter.

11.2 Overview on various indigenous designs of the compliant robotic grippers

11.2.1 Metrics of the indigenous design

We will begin our journey towards characterization and modeling of the *in situ* vibration of the compliant robotic grippers through an overview of the designs of those, all accomplished indigenously. The fundamental motto for conceptualizing these designs was to investigate various possible pathways of vibration that may creep in due to the payload of the gripper. In that sense, the designs of these CRGs are made with a consideration of intrinsic real-time vibration only, devoid of any external forcing function. However, we have considered these designs with great variability in terms of 'type of payload' as well as 'form closure of the grasp'. Ten different designs of the CRGs will be delineated in this section, with reference to the relative complexity of the form closure of the grasp in serial order. All of the designs were conceptualized from fundamentals, be it multi-body dynamics or grasp prehension or augmentation of non-linearity. It may be stated that all of these ten designs have been realized in physical hardware, developed indigenously. The uniqueness of these designs with respect to characterization of *in situ* vibration will be described through 10 different design metrics for the indigenously developed CRGs. These metrics are: (i) payload: tare weight, type and ensemble shape; (ii) kinematic layout of the linkage; (iii) material for fabrication: links and jaws; (iv) inclusion of compliance via spring or pin-joints; (v) sensory instrumentation; (vi) adapter plate: as connection to CRW; (vii) base plate: for mounting the drive-motor; (viii) jaws; (ix) jaw holding mechanism and (x) jaw accessories: rubber padding. It may be noted that the first five of these metrics are related to functional attributes of the CRGs involving overall drive-train, fabrication and payload. In contrast, the remaining five metrics are related to operational/run-time attributes of the CRGs involving interfacing with the CRW and drive-motor and form closure of the grasp at the jaws. We will elucidate on these two groups of design metrics in detail now.

The first and foremost design principle that has been invoked in all of these designs is the range of payload that the CRG will handle. The synthesis of design as well as source apportionment of inherent vibration was carried out based on the type and nature of the payload. A wide spectrum of payloads was tested with these CRGs in order to ascertain the subtleness of those grasps with relation to the output vibration of the CRG. It was observed quite vividly that besides tare-weight of the payload, its ensemble shape (outer periphery) was equally contributive towards vibration of the jaw-holding mechanism of the CRG. And, of course, the type of payload did matter a lot in ascertaining the vibration signature, the effect of which was found to be decoupled from the tare-weight or shape of the payload. We carried out a wide spectrum of experiments with the developed CRGs using different varieties of payloads, e.g. paper, clip, pin, needle, cellotape, writing with ball pen and marker pen, computer mouse, mechanical bush, bearing, metallic coupler, medical syringe, cotton, plastic ball, medicine strip, plastic cup, bottle etc. While some of these payloads are metallic, others are made of various non-metallic materials.

The next important metric of the design is the kinematic layout of the linkages that will be responsible for generating the mechanical motion of the CRG. We have used several different kinematic chains for the creation of the motion, after the drive begins off motor-shaft of the CRG-motor. Few designs are based on modified a four-bar type mechanism while some have a layout with gear-trains. The kinematic chain with gears is again branched in various types, such as full-round spur gear pair, sector gear pair and rack and pinion pair. We have also used kinematic drive-chain with spring-supported linkage in one of the designs. All of these kinematic layouts have due importance towards generation of vibration in the CRG-system based on the nature of the transfer-motion.

The third crucial factor of the design metric is the choice of material for fabrication of the links and jaws of the CRGs. We have used both metal as well as non-metal as the choice of material for the mechanical hardware of the CRGs. Naturally, we have observed change in vibration pattern between the CRGs made up with metallic materials vis-à-vis CRGs fabricated out of non-metallic materials. We even had designs with a mix of metallic and non-metallic materials. While aluminum and mild steel were used under metallic materials in some of the designs due to lower density and better mass compliance, the rest of the designs are built largely with non-metallic materials, such as nylon, teflon, acrylonitrile butadiene styrene (ABS), polypropelene, neoprene rubber and plastic. Mass compliance was the major design yardstick in the selection of material for fabrication, based on the ensemble vibration characteristics of the CRGs.

The next salient design metric under functional attribute is inclusion of compliance in the overall design of the CRG. We have customized the designs with two types of members in order to imbibe sufficient compliance in the ensemble system during form closure of the grasp. Although invoking higher compliance may be a deterrent to vibration harnessing, but, we need to use some compliant-members in the system to ensure near-perfect grasp. As a matter of fact, this sort of grasp is required to ensure good form closure, especially for light-weight and fragile objects. The first group of member(s) for imbibing compliance is spring sub-systems that are used as intermediate members in-between links of the CRGs. The other group of member(s) is pin-joints. These pin-joints are made adhesive-type so as to have good mating contact as well as planar rotations between the links of the CRGs.

The fifth design metric under functional attribute is the sensory instrumentation of the CRG. Placement and integration of various tiny sensor-elements play a crucial role in the design conceptualization of the CRGs. The physical size and disposition of these sensor-elements need to be accounted for and fine-tuning of the design with respect to drive and transmission system of the CRGs will follow thenafter. The sensor-ensemble of the CRGs will be responsible for three types of tasks: (a) sensing of the gripping force, either in totality or partially; (b) indication of the incipient slippage of the payload; and (c) identification of the payload at the grasp-zone. The primary sensor-element that is responsible for sensing the gripping force, viz, miniaturized load cell, will be fitted at the jaws of the CRGs. Ideally, we will use a pair of load cells, fitted at the left-hand-side and right-hand-side jaws of the CRGs in order to maintain symmetry in the sensor data/output (in milli-volt or micro-volt, depending on the tare-weight of the payload). It is to be noted that the load cells fitted at the jaws will record the griping force in totality, baring loss due to instrumentation and/or white noise. However, in order to have better clarity over the grasping phenomena of the CRGs in the midst of *in situ* vibration we have integrated force-sensing sensor-elements (not load cell) at other locations of the CRGs, namely at the jaw-holder and jaw-connecting links. The sensor-cells at these locations will give data supporting gripping force in partial form. Nonetheless, these data are also important for overall understanding of the 'force closure' of the grasp phenomena, in case fitment of load cell is not feasible in the jaw-zone due to difficulty in manufacturing and assembly. Besides, sensor-elements at other locations will help redundancy, in case of malfunctioning of the load cells. Besides gripping force, the other crucial aspect of grasp synthesis is the ideation on the incipient slippage of the payload. The incipient slippage is quantitatively manifested through two parameters, viz, (i) slipped distance and (ii) slip velocity. The sensor-elements for arresting the slippage-related information will be mounted at the distal links of the CRG or at the base-plate, depending upon the design. Slippage is a real-time phenomenon that involves multiple members of the CRG and thus, it is difficult to pinpoint one particular sensor-element for its detection. The identification of the presence of payload at the grasp-zone is made through a pair of infra-red sensors (emitter and detector), to be fitted at the bottom-most part of the jaws. Since infra-red sensors, in the form of light emitting diodes (LEDs) are very tiny, the fitment is crucial and it varies between the CRGs, depending upon the exact size and shape of the jaws.

The operational attributes of the design of CRGs mainly concentrate on the critical elements of the CRG, which are responsible for the motion of the jaws. The prime-most attribute is the design of the adapter plate that is to be used as the direct connection with the CRW and thereby with the robotic manipulator. The design of the adapter plate is critical in terms of vibration accumulation as well as being the junction-point between the robotic manipulator and the gripper. The next crucial attribute is the design of the base plate that is to be used for mounting of the drive-motor. This component/sub-assembly of the CRG is instrumental in the assessment of the overall tare-weight of the CRG as well as incipient vibration. Good design practice for CRG will always be to reduce the mass of the base plate sub-assembly and make it as minimal as possible just to contain the drive-motor and mounting of the sensory instrumentation for checking the grasp force and/or incipient slippage. In fact, unlike the adapter plate, design of the base plate needs several iterations to optimize the overall ensemble. Since quantum of natural vibration is also linked with these design iterations of the base plate, topology optimization procedure is being followed specially for fine-tuning the base plate design. The next three functional attributes of the design of CRGs are related to the crucial-most part of CRG, namely, jaws. The prime-most significant attribute of the jaw design is the design of the jaw-plates, both left-hand-side as well as right-hand-side. We have incorporated two types of jaw-plates in our indigenous designs, viz. (a) straight parallel-type jaw plates and (b) curvilinear non-parallel-type jaw plates. Irrespective of the types, the primary objective of the design of jaw-plates is mass compliance. It is important here to note that while straight parallel jaw-plates will always have parallel motion of the jaws during grasp operation, the curvilinear-jaw type CRGs can have either parallel motion or non-parallel motion of the jaws during a stable grasp. In either of the jaw-types, the nature of final motion at the jaw-plate is governed by a different functional attribute, namely jaw holding mechanism. We have incorporated six different jaw holding mechanisms in ten designs of the CRGs, which can be described as: (i) serial-linkage; (ii) parallel-bar linkage; (iii) crossed parallel-bar linkage; (iv) gear-connection linkage: type I; (v) gear-connection linkage: type II and (vi) spring-actuated linkage. All of these six variants of jaw holding mechanisms possess characteristic nature of motion transmission that makes the jaws actuate in a specific path-way. Besides generation of motion for the jaws in real-time, these linkage mechanisms are also instrumental in creating *in situ* vibration in the CRGs in multiple manifestations. For example, vibration signatures in the CRGs that involve gear-driven transmission chains are found to be distinctly different from those in mechanical linkage-driven transmission chains. As a matter of fact, the assessment of vibration in the CRGs is made from a practical standpoint and with real-life experience that was earned from our indigenous firmware. The last functional attribute of the design is related to jaw accessories that are fabricated almost as an 'integral' part of the jaw-plates. The salient jaw accessory that has been used in the majority of our CRGs is rubber padding over the jaw-plates. The rubber padding is self-compliant by nature and it provides not only better stability during force closure of the grasp but also helps harnessing *in situ* vibration, produced on or just before the grasp. The other accessory that is instrumental in abetting *in situ* localized vibration at the jaws is a spring system, connecting the jaw-plate and the linkage mechanism. This spring-based accessory has been used in one of the designs of our CRGs.

11.2.2 Classification of the indigenous designs

Indigenous firmware of ten CRGs will be discussed in three characteristic groups, based on the jaw-types. The reason for putting weightage on the type of jaw is predominantly from the viewpoint of end-use. We will now go through these groupings systematically, as: (a) flat jaw CRGs; (b) curvilinear jaw CRGs and (c) contoured jaw CRGs. Out of ten indigenously-developed CRGs, four are of flat jaw type with varying geometries, sizes and articulations. Three CRGs have been manufactured with curvilinear jaws, comprising metallic as well as non-metallic types. The remaining three CRGs have been fabricated in customized fashions so as to generate various contours, i.e. some sort of mix of flat and curved surfaces, either at jaw surfaces and/or final motion of the gripper. These CRGs require specialties in the jaw-profile to suit the grasp of selected payloads. Interestingly, these payloads are of wide range: starting from thin paper-sheet to welding electrode.

The flat-jaw CRGs have been designed with various motion-transferring mechanisms so as to impart characteristic motion in a straight line during the grasping operation. Out of the four CRGs that were fabricated with flat jaws, two grippers were manufactured in miniaturized form using metallic components. In fact, these two indigenously-designed CRGs are technologically novel because of the size limitation and tare-weight. Although similar such CRGs could have been manufactured from non-metallic materials to reduce the tare-weight but strength and resilience would have been compromised. These two metallic flat jaw miniature CRGs are used in flexible manipulators for the best assessment of run-time performance amidst *in situ* vibration (both at gripper level as well as robotic system level). The other two flat jaw CRGs are manufactured using non-metallic materials having a small envelope. These two grippers use novel linkage mechanisms for the transmission of motion to the jaws.

The curvilinear jaw CRGs were designed and manufactured with geared mechanisms for the motion transfer. As the jaws are curved, linkage type mechanisms were found to be somewhat obstructive in making the final assembly of the gripper system. We have used both metal as well as non-metal for the fabrication of these three CRGs (metallic type: (1) non-metallic type: (2)). The manufacturing of the curvilinear components do require specialized machine tool support with tight control on the engineering dimensions (tolerance and surface finish to be particular). The form closure of the grasp gets more defined and robust by virtue of the motion of the rightly-manufactured curvilinear jaws of these CRGs.

The contoured jaw CRGs are very typical in design, both at the jaw-plates as well as motion-transferring mechanism. Both metallic and non-metallic materials were used for the fabrication of three CRGs of this type. The prime design concept behind these CRGs is related to the end-use. The characteristic features of the grasping action, as required by the application, govern the design here, notwithstanding the traditional methodology for designing drive-train mechanism of the said CRG. The contoured jaw CRGs were designed for three different applications: (i) grasp of ultra-thin objects; (ii) grasp of tiny objects inside a closed conduit; and (iii) grasp of variable-length welding electrode. It may be mentioned here that all of these three applications are unique in their respective work-envelope, control semantics and grasp-programming. The design of the jaw-plates of these three CRGs was made in such a way as to conform to the external shape of the graspable object and the incipient motion of the jaw sub-assembly post-grasp. The crucial-most aspect of these CRGs is the mathematical analysis of the output-motion of the jaws, which is normally non-parallel motion. The vibrational paradigm of these CRGs is dependent on the output-motion as per the nature of the grasp.

11.2.3 Firmware of the flat-jaw type CRGs

We will describe the firmware of four flat-jaw type indigenously designed CRGs in chronological fashion so as to bring forward the design intricacies that were tackled as well as knowledge quotient that was added to the ensemble design process, specifically with respect to vibration control. A technical epilogue of these CRGs will also open up the design novelties that were tried in combatting *in situ* vibration in research mode of development. For obvious reasons, our journey began with two variants of these grippers with small external envelope. The experience gained from the hardware manifestation of these two grippers helped us to proceed further to miniaturization of the mechanical design as well as sensory instrumentation. Naturally, the task of harnessing *in situ* vibration became tougher with miniaturization of the external envelope of these CRGs. We will delineate the firmware of these four CRGs in the order of research attainment and vibration abetment.

11.2.3.1 Link-type non-metallic jaw robotic gripper—type I (LiNJRoG-I)

Our very first attempt of designing a moderately small-volume jaw-type robotic gripper took off on the journey with non-metallic materials for fabrication, in order to keep the tare-weight to a minimum. The crux of the design was rooted with the linkage system of assembling various components that took the onus of transmission of requisite motion to the jaw-plates, originated from the d.c. servomotor system. The fabrication paradigm of the sub-assemblies, namely, (a) linkagae; (b) jaws and (c) fixation (for motor and sensory elements) was largely attained through rapid prototyping and wire-cut electro-discharge machining processes. Figure 11.1 illustrates the photographic view of this gripper namely, Link-type Non-metallic Jaw Robotic Gripper-Type I (LiNJRoG-I) post-fabrication.

Now, as per the design and fabrication of LiNJRoG-I, prime source-elements for *in situ* vibration are at 'C' and 'D', respectively. Several design-factors are instrumental in creating this vibration: (i) lengths of the linkage-segments; (ii) fixation-elements (pin-type non-rotary attachments); (iii) span of the linkage assemblies with fully-open jaws; and (iv) method of fixation with the central anchor. Since cross-section of these linkage-elements is very small (rectangular slender section), its effect on vibration estimation was neglected for analytical modeling.

11.2.3.2 Link-type non-metallic jaw robotic gripper—type II (LiNJRoG-II)

The next hardware of link-type non-metallic jaw robotic gripper (LiNJRoG-II) that was attempted with indigenous design had more intricate semantics of the linkages. This version of the hardware was made with similar pre-requisites of the earlier one with major design alterations in the linkage system. The rest of the sub-assemblies/ components were kept unaltered. The design philosophy of this variant was made

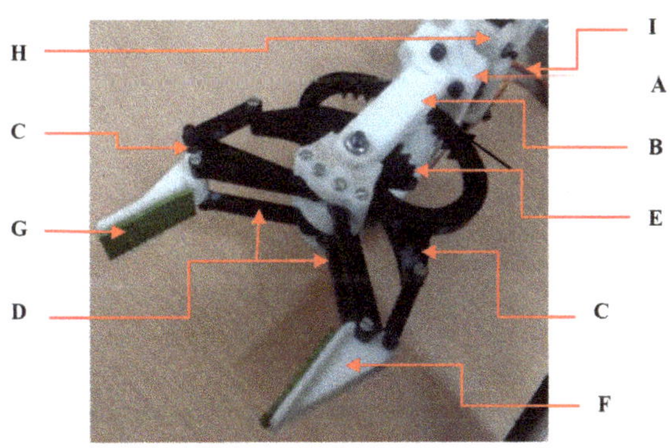

Index: **A:** Adapter Plate; **B:** Central Anchor; **C:** Upper-row Linkage; **D:** Lower-row Linkage; **E:** Gear Assembly; **F:** Jaws; **G:** Rubber Pad; **H:** Servomotor Assembly; **I:** Motor Connection Wire

Figure 11.1. Photographic view of the link-type non-metallic jaw robotic gripper-type I.

with more intricate and slender structure for the upper-row and lower-row linkages. The transmission of motion from d.c. servomotor system was channelled via these linkages till the jaw-plate. Like LiNJRoG-I, here too, the fabrication process of the sub-assemblies was realized through rapid prototyping and wire-cut electro-discharge machining processes. Figure 11.2 illustrates a post-fabrication photographic view of the gripper, namely, LiNJRoG-II.

The main difference in design between LiNJRoG-I and LiNJRoG-II is the element, ''G', i.e. linkage anchor. This part was fabricated as an additional item in order to provide requisite strength to the lower-row linkage pair. As illustrated in figure 11.2, design of both upper-row linkage and lower-row linkage were intricate with more slenderness added. Although this slender structure will be effective for reduction of tare-weight of the system, it will create additional metrics of *in situ* vibration. Hence so far as harnessing of vibration is concerned, LiNJRoG-II is more critical than its predeccessor, LiNJRoG-I.

11.2.3.3 Patient assistant miniaturized robotic gripper (PAMRoG)

The experience gathered through indigenous design and hardware development of the linkage-type robotic grippers was instrumental in taking up the design challenge of fabricating a truly miniaturized ultra-low payload two-jaw type robotic gripper. The principal design objective of this gripper was rooted at low-payload grasp with adequate form and force closure paradigms. The prototype gripper hardware was manufactured mainly to take care of various medical necessities for patient(s) at hospitals/health care centres. This gripper, christened as Patient Assistant Miniaturized Robotic Gripper (PAMRoG) works in a tiny volume, approximately 26.5 cm^3, with reference to its external dimensions (length: 50 mm; width: 35 mm and breadth: 15 mm). Figure 11.3 shows the photographic view of the developed prototype of PAMRoG, wherein the novelty of miniaturized manufacturing can be ascertained. The payload capacity of this miniature gripper is nearly 800 gm, which

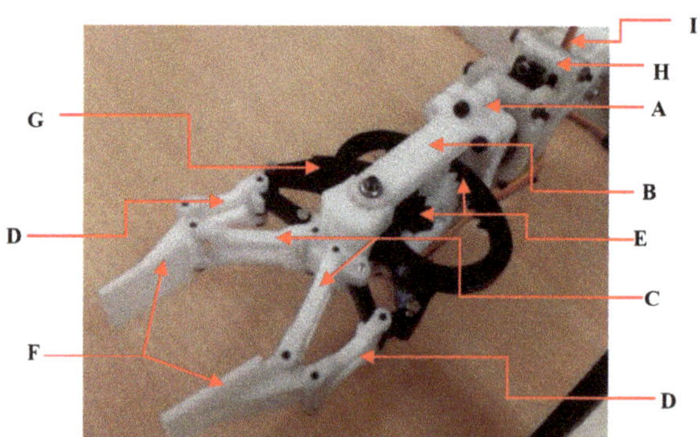

Index: **A:** Adapter Plate; **B:** Central Anchor; **C:** Upper-row Linkage; **D:** Lower-row Linkage; **E:** Gear Assembly; **F:** Jaws; **G:** Linkage Anchor; **H:** Servomotor Assembly; **I:** Motor Connection Wire

Figure 11.2. Photographic view of the link-type non-metallic jaw robotic gripper-type II.

is much consistent in comparison to its tare-weight. In fact, payload-to-tare-weight ratio is quite high for the prototype PAMRoG.

The prototype PAMRoG has various characteristic features commensurate to its tiny size and actuation of the jaws. It is needless to state that because of its size miniaturization, the ensemble design is prone to a higher level of vibration, both in standalone condition of the jaws as well as during grasp. Unlike the previous two versions of flat-jaw gripper, this gripper has a backbone, called, 'base plate'. As illustrated in figure 11.3, the design of the base plate was customized keeping in mind the overall size minimization and also the optimum strategy for fitment with the robot wrist. In that sense, the function of adapter plate (as designed in the earlier two types of grippers) gets subsumed in the improvised design of the base plate of the PAMRoG. This integration is very helpful not only from the angle of size minimization but also from the standpoint of compactness and self-weight. However, with size minimization, PAMRoG is prone to acute vibration that needs to be controlled through the programming for gripper actuation. The open-loop control algorithm of PAMRoG was customized to get an assessment of the real-time vibration, generated due to the grasp-load.

11.2.3.4 Sensor instrumented miniaturized robotic gripper (SIMRoG)

The prime design semantics of PAMRoG were utilized to design a more sophisticated intelligent gripper system with adequate sensory instrumentation and minor size miniaturization. This version of our gripper, namely, Sensor Instrumented Miniaturized Robotic Gripper (SIMRoG) was also aimed at low-payload grasp with adequate form and force closure. Like PAMRoG, the prototype hardware of SIMRoG was manufactured mainly to take care of various medical necessities for patient(s) as well as tiny objects. The prototype SIMRoG works in a tiny volume, approximately 26.5 cm^3, with reference to its external dimensions (length: 50 mm; width: 35 mm and breadth: 15 mm). Figure 11.4(a) shows the Computer Aided Design (CAD) model view of the developed prototype of SIMRoG, wherein the novelty of sensory instrumentation of heretogenous type may be appreciated besides

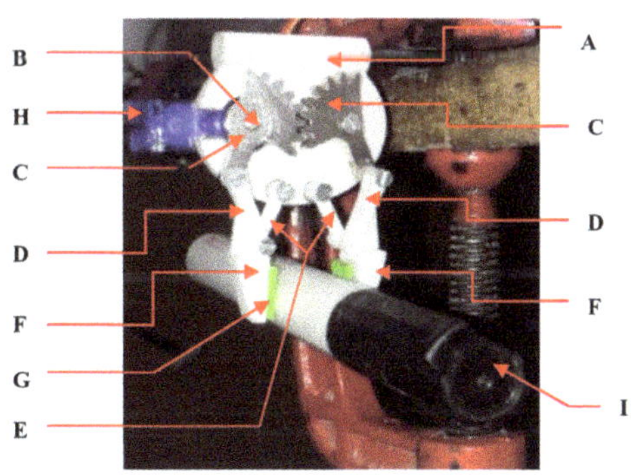

Index: **A:** Base Plate (with easy attachment with robot wrist / external anchor); **B:** Fixation for Servomotor; **C:** Sector Gears; **D:** Intermediate Linkage; **E:** Jaw Plate Connecting Links; **F:** Jaws; **G:** Rubber Pad; **H:** Servomotor Assembly; **I:** Grasped Object

Figure 11.3. Photographic view of the patient assistant miniaturized robotic gripper.

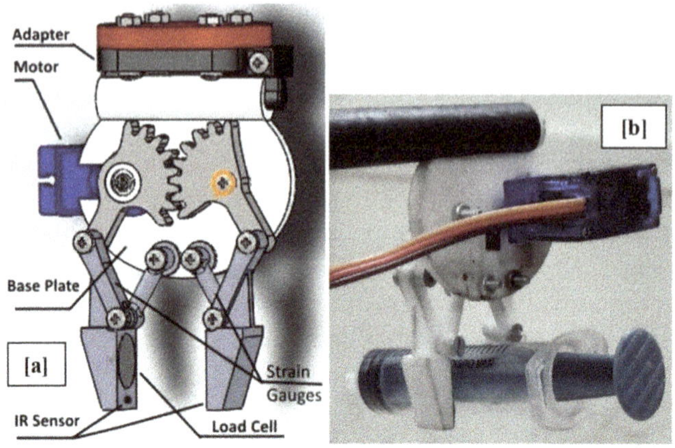

Figure 11.4. Views of the sensor instrumented miniaturized robotic gripper: (a) CAD model; (b) back-side of prototype.

intricacies of miniaturized manufacturing. The back-side view of the developed prototype with a payload (syringe) is illustrated in figure 11.4(b).The payload capacity of this miniature gripper is nearly 1 kg, which is an improvement due to better torque rating of the d.c, servomotor system used. As before, payload-to-tare-weight ratio of the prototype SIMRoG is quite high. The embedded sensor-cells, namely load cells, flexi-force sensors, infra-red sensors and strain gauges, will be helpful is real-time assessment of the vibration of SIMRoG.

The prototype SIMRoG has challenges from the viewpoint of sensory instrumentation and data fusion. The design is affected by *in situ* vibration, both in standalone mode as well as during grasp. As illustrated in figure 11.4(a), we have an additional member in SIMRoG, viz. adapter sub-assembly. The design of this adapter sub-assembly (component and fixture) was customized keeping in mind the overall size minimization and also the optimum strategy for fitment with the base plate at one end and the robot wrist on the other. Nonetheless, SIMRoG is prone to higher-level of vibration because of sensory instrumentation/wiring that needs to be abeted with the help of programming. The closed-loop sensor-infused control algorithm of SIMRoG was customized to get an assessment of the real-time vibration, generated due to the grasp-load.

11.2.4 Firmware of the curvilinear-jaw type CRGs

We will now report the firmware of three indigenously designed curved-jaw type CRGs in chronological fashion in order to unearth the design intricacies with respect to control of vibration in real-time. All of these grippers have specialties in jaw design for obvious reasons. We started the fabrication of two non-metallic type curved-jaw grippers first so as to gain confidence in tackling form closure under the purview of a small external envelope. The experience gained from the hardware manifestation of these two non-metallic grippers drove us to take up the challenge of making the metallic jaw version of the gripper for enhanced robustness of the grasp.

Nonetheless, all these three varieties of CRGs were made in accordance with the sensory instrumentation, as described in earlier sub-sections. Harnessing of *in situ* vibration is comparatively tricky with curvilinear shape of the jaw and also overall miniaturization of the external envelope. These curvilinear-jaw grippers have been designed with gear-train type mechanism with direct transmission of motion to the jaws. Accordingly, the first two non-metallic type curvilinear-jaw grippers are labeled as: (i) Gear-driven Non-metallic Curved-jaw Robotic Gripper—Type I (GeNCRoG-I) and (ii) Gear-driven Non-metallic Curved-jaw Robotic Gripper—Type II (GeNCRoG-II). Figure 11.5 illustrates the photographic view of GeNCRoG-I and its assembled view with a small-size table-top robot. Likewise, a CAD model as well as a photographic view of the GeNCRoG-II in standalone form are shown in figure 11.6. In comparison to these two non-metallic type curved-jaw grippers, the metallic variety is more robust and slightly larger in ensemble mechanical design. Figure 11.7 illustrates the photographic view of the Gear-driven Metallic Curved-jaw Robotic Gripper, GeMCRoG along with its CAD model. The nature of vibration abetment is more or less the same as pointed out for flat-jaw type CRGs.

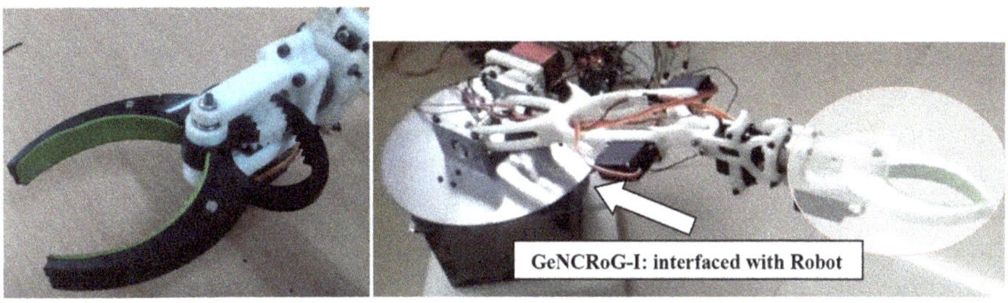

Figure 11.5. Photographic view of the gear-driven non-metallic curved-jaw robotic gripper—type I [1], copyright (2020), with permission of Springer.

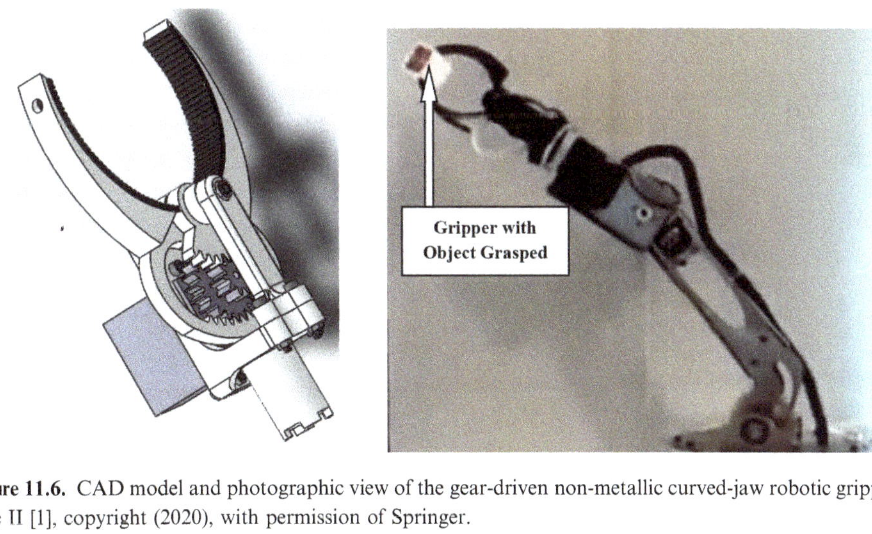

Figure 11.6. CAD model and photographic view of the gear-driven non-metallic curved-jaw robotic gripper—type II [1], copyright (2020), with permission of Springer.

11.2.5 Firmware of the contoured-jaw type CRGs

We will now describe the highlights of design and firmware of three unique variants of our CRGs having jaws with contours. The designs of these CRGs were fine-tuned with an aim towards the specific end-applications, viz. internal inspection of cylindrical pipe, flexible robotic systems and robotic welding. All of these CRGs are susceptible to run-time vibration with varying magnitude depending on the end application. The details of the hardware of these three CRGs will be described in the following sub-sections.

11.2.5.1 Rack and pinion-driven miniaturized robotic gripper (RPMRoG)

The first hardware of this kind, namely, Rack and Pinion-driven Miniaturized Robotic Gripper (RPMRoG) is meant for interfacing with a compliant universal Robotic Leech (CURL). The ensemble motion of the gripper will be serpentine, because of its assemblage with the hardware of cylindrical-shaped CURL. However, the fundamental local-frame motion of the jaws of this gripper will be linear due to the incorporation of the rack and pinion mechanism. Figure 11.8 illustrates the photographic view of the final prototype of RPMRoG and various CAD models of it.

Out of the four CAD models, model ‘A’ highlights the overall disposition of the gripper with base plate, adapter plate and jaw plate. Model ‘B’ is more focused on the design of the jaws with the external ensemble and its final fitment with the servomotor. The location of servomotor and its assemblage with the mechanical system of the gripper is crucial due to cylindrical shape of RPMRoG in order to maintain compatibility with the CURL. In fact, the compact assemblage of the servomotor may be observed in CAD models ‘A’ and ‘B’ as well as in the photographic view at the leftmost part of figure 11.8. The CAD model ‘C’ illustrates the indigenous design of the rack assembly of the jaws with its size compactness. This model is significant because the fittment of servomotor plays an important role in controlling the linear motion of the jaws. The last CAD model, viz, ‘D’ shows the relative positioning of the pinion inside the rack sub-assembly of RPMRoG. As can be idealized from the views of figure 11.8, the sources of *in situ* vibration are primarily at the adapter plate and base plate. These two components are thus crucial in vibration harnessing of the prototype gripper, RPMRoG.

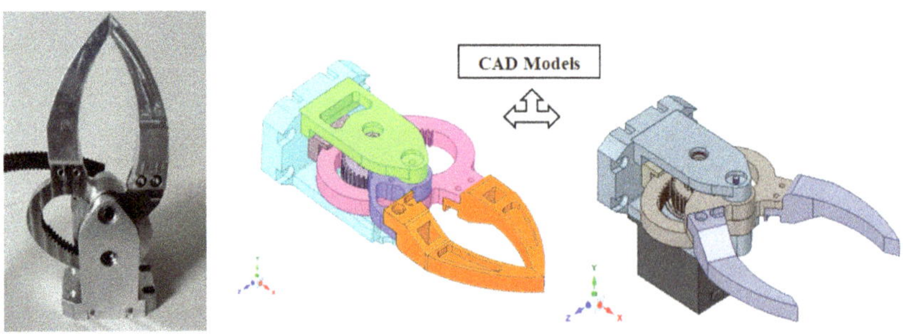

Figure 11.7. Photographic view of the gear-driven metallic curved-jaw robotic gripper and its CAD models.

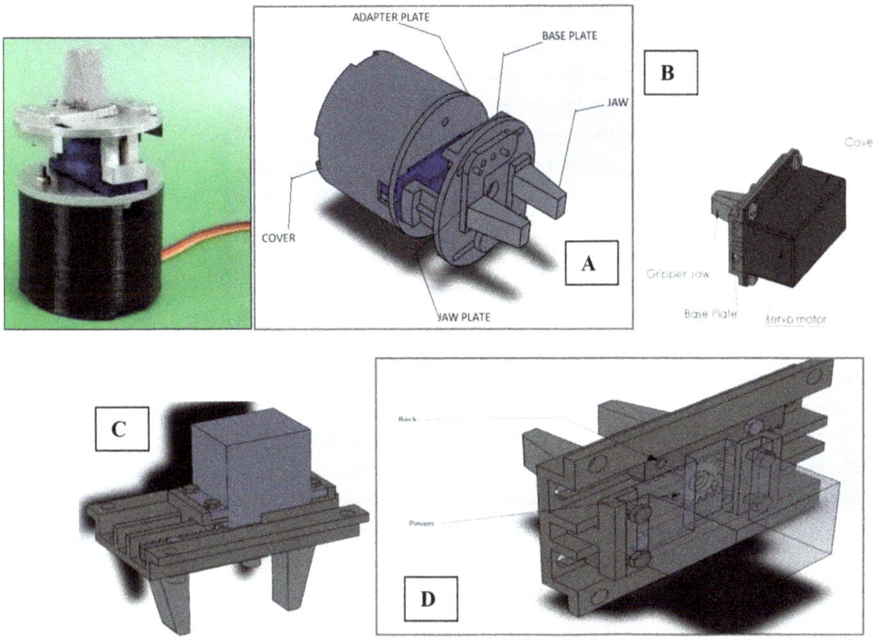

Figure 11.8. Photographic view and various CAD models of the rack and pinion-driven miniaturized robotic gripper.

11.2.5.2 Sensor instrumented miniaturized suspended jaw-plate robotic gripper (SIMSJRoG)

The next hardware under contoured-jaw CRG is complex design-wise, which is a very tiny lightweight system meant for assemblage with planar serial-chain Flexible Robotic System (FRS). Figure 11.9 illustrates the photographic views of the hardware of Sensor Instrumented Miniaturized Suspended Jaw-plate Robotic Gripper (SIMSJRoG), snapped from two different view-angles.

The prototype hardware of SIMSJRoG reveals various interesting features, pertaining to indigenously-designed jaw-plate and jaw-link sub-assembly (refer to 'E' & 'I' of figure 11.9). The jaw-plates were contoured and these were fabricated through human skill out of specialized pre-shaped material. The crux of the design is a suspended type jaw-plate, which has immense impact on the overall vibration of the gripper system. The sensory instrumentation is another feature of this gripper, wherein 2 nos. flexi-force sensors and 2 nos. mini-load cells are used, as shown in figure 11.9 (refer to 'E' & 'G'). The tare-weight of SIMSJRoG is nearly 100 gm and it can grasp a payload of 10–20 gm.

11.2.5.3 Welding instrumented sensor and spring-supported robotic gripper (WISSRoG)

Our third indigenous prototype under contoured-jaw CRG is deployed for robotic welding. This prototype is very interesting, not only because of its innovative architecture but also for its heterogenous design-modules. The prototype, namely,

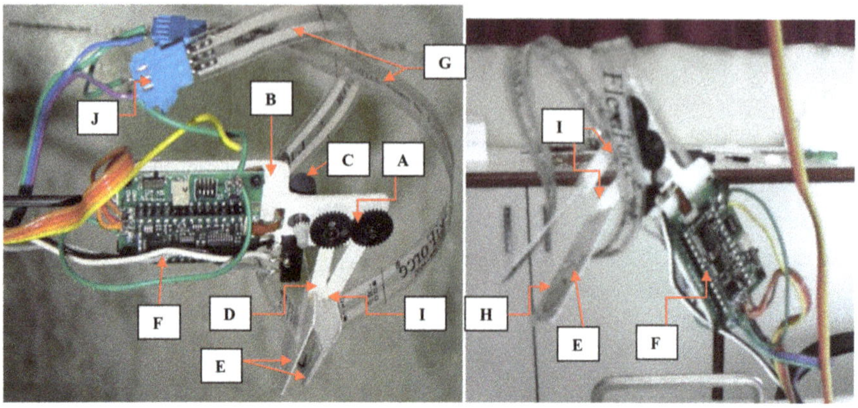

Index: **A:** Spur Gear pair; **B:** Base Plate; **C:** Servomotor; **D:** Jaw-holding Linkages; **E:** Plate-type Fingers; **F:** Gripper Controller; **G:** Flexi-force Sensors; **H:** Mini-Load Cells; **I:** Finger-Linkage assembly; **J:** Cable Connector

Figure 11.9. Photographic views of the sensor instrumented miniaturized suspended jaw-plate robotic gripper.

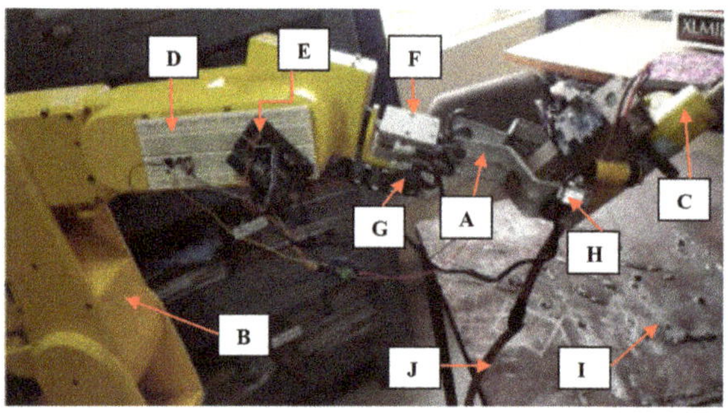

Index: **A:** Monolithic Gripper Body & Jaws; **B:** Industrial Robot; **C:** Welding Gun; **D:** Gripper Controller Board; **E:** Junction-board for Sensors; **F:** Base assembly of Gripper; **G:** Spring assembly; **H:** Flexi-force Sensor & Vibration Sensor assembly; **I:** Weld Beads & Seam; **J:** Cable Connector for Arc Welding Machine

Figure 11.10. Photographic view of the welding instrumented sensor and spring-supported robotic gripper. Reprinted from [44], copyright (2020), with permission from Elsevier.

Welding Instrumented Sensor & Spring-supported Robotic Gripper (WISSRoG) has three design features that are characteristic to its destined task/operation in real-time. These design features are: (i) welding instrumentation; (ii) sensory instrumentation; and (iii) spring-supported actuation of jaws. Figure 11.10 shows the photographic view of the prototype WISSRoG in action (robotized arc welding). Besides the standard apparatus for arc welding and allied welding instrumentation, WISSRoG is merged with an additional sensor system for weld/grasp synthesis. This sensor system is developed with the help of 2 nos. flexi-force sensors and 2 nos. piezo-ceramic vibration sensors (refer to 'H' of figure 11.10), embedded over the monolithic gripper-body and jaws.

The performance of WISSRoG has significant contribution from control inputs, both welding system and sensory instrumentation. Piezo-ceramic sensors are used for the measurement of run-time vibration. These sensors are very suitable especially

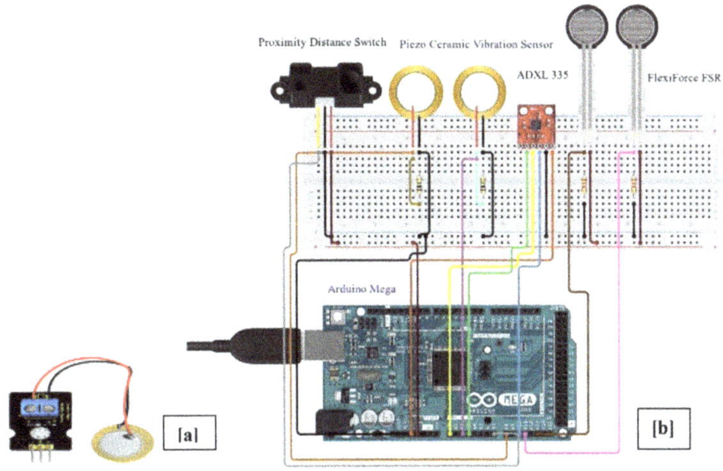

Figure 11.11. Photographic view of: (a) piezo-ceramic vibration sensor and (b) junction board for the sensors for WISSRoG.

for low-amplitude vibration as they generate a low voltage (micro-volt level) upon external excitation. Figure 11.11(a) shows the photographic view of the piezo-ceramic vibration sensor that has been used in the prototype WISSRoG. The hardware details of the junction board of sensory instrumentation for WISSRoG are illustrated schematically in figure 11.11(b). The ensemble run-time vibration of the gripper is inevitable but can be controlled to a large extent through the spring-supported sub-assembly, designed indigenously after a series of trials.

11.2.6 Miniaturized CRGs: a wider horizon

The hardware development of Sensor-Instrumented Miniaturized Robotic Gripper (SIMRoG), described under sub-section 11.2.3.4, can be entrusted as the 'foundation' of the complete indigenous R&D on the *Minaturized Complaint Robotic Gripper* technology. The 'wider horizon' of the ensemble development has a total of six variants of sensor-instrumented robotic grippers, with different external dimensions and payload capacities. These six variants of SIMRoG have different challenges with respect to identification of source, estimation, measurement and harnessing of *in situ* vibration. However, the commonality between these six variants pertains to uniformity in sensory instrumentation for the pick-up and measurement of run-time vibration. Apart from other parameters, external dimensions (length, breadth and width) and tare-weight play great role in the assessment of payload capacity and also run-time vibration of these miniaturized CRGs. The listing of this indigenous firmware of miniaturized CRGs is provided in the table 11.1.

The tuple of load cell, flexi-force sensor, strain gauge and infra-red sensors has dual responsibility of detection and measurement of grip force and *in situ* vibration of the miniaturized CRGs. The payload requirements of these grippers are commensurate to the ensemble structural layout and respective tare-weight. The extent of miniaturization, aided by advanced manufacturing processes and

Table 11.1. Wider horizon of miniaturized compliant robotic grippers.

Sl. No.	Description of the miniaturized CRG	Overall external dimensions (Approx.)	Sensory instrumentation	Tare-weight (with motor) and payload
1	Advanced sensor instrumented miniaturized robotic gripper **(ASIMRoG)**	Length: 45 mm; breadth: 30 mm; width: 15 mm (excluding motor)	Load cell (at jaw), flexi-force sensor (at base plate); strain gauge (at links); infra-red sensors (at jaw-tip)	Tare-weight: ~ 1 kg; payload: ~ 600 gm
2	Sensor instrumented curvilinear-jaw robotic gripper **(SICRoG)**	Length: 80 mm; breadth: 40 mm; width: 15 mm (excluding motor)	Load cell (at jaw), flexi-force sensor (at base plate); strain gauge (at base plate and links); infra-red sensors (at jaw-tip)	Tare-weight: ~ 2.5 kg; payload: ~ 1 kg
3	Sensor instrumented parallel-bar robotic gripper: type-I **(SIPRoG-I)**	Length: 80 mm; breadth: 50 mm; width: 15 mm (excluding motor)	Load cell (at jaw), flexi-force sensor (at base plate); strain gauge (at base plate and links); infra-red sensors (at jaw-tip)	Tare-weight: ~ 2.5 kg; payload: ~ 1 kg
4	Sensor instrumented parallel-bar robotic gripper: type-II **(SIPRoG-II)**	Length: 90 mm; breadth: 50 mm; width: 15 mm (excluding motor)	Load cell (at jaw), flexi-force sensor (at base plate & parallel-bar linkage); strain gauge (at base plate and links); infra-red sensors (at jaw-tip)	Tare-weight: ~ 2.5 kg; payload: ~ 1 kg
5	Sensor instrumented finger-type robotic gripper **(SIFRoG)**	Length: 120 mm; breadth: 40 mm; width: 25 mm (excluding motor)	Load cell (at jaw), flexi-force sensor (at base plate & finger-linkage); strain gauge (at base plate, links and fingers); infra-red sensors (at jaw-tips)	Tare-weight: ~ 3 kg; payload: ~ 1.5 kg
6	Sensor instrumented spherical-body robotic gripper **(SISRoG)**	Length: 50 mm; breadth: 50 mm; width: 25 mm (excluding motor)	Load cell (at jaw), flexi-force sensor (at spherical body surface); strain gauge (at jaws); infra-red sensors (at jaw-tips)	Tare-weight: ~ 3 kg; payload: ~ 2 kg

technologies, will play another important role in the estimation of run-time vibration of these CRGs. It is also interesting to note here that a wide variety of the miniaturized CRGs are compatible for interfacing with CRW. In other words, a cluster of mini-CRGs and CRWs is the most desirable choice for vibration attenuation and equalization of modal frequencies. We will address various design templates of CRWs now that are adaptable to design of CRGs, described hitherto.

11.3 Indigenous design of the compliant robotic wrists

11.3.1 Fundamental facets of the indigenous design

As delineated earlier, the CRW is another important element is designing intelligent grippers. A successful prototype of CRW will aid the process of vibration attenuation and smoothening of the overall control system architecture of the CRG and the industrial robotic system to a great extent. The self-vibration of CRW will be in tune with the natural frequency of vibration of the CRG, which will help harnessing the external vibration to the ensemble system. In other words, the conjugate tuple of CRW and CRG will be the most ideal and optimal combination towards dampening the run-time vibration of the robotic system. It may be mentioned here that unlike CRGs, the fundamental facets of design of CRWs are identical. The common thread of these facets is generation of enough 'compliance' inside CRW-structure, namely the housing. Hence, we have considered various indigenous designs of CRWs with sufficient variability in terms of 'type of internal members' as well as 'actuating mechanism'. Four different designs of the CRWs will be reported in this section, with reference to the relative complexity of the actuating mechanism(s) to generate compliance in serial order.

The designs of our CRWs were conceptualized from fundamentals, such as kinematics of the actuating mechanism, multi-body dynamics for compliance or infusion of non-linearity. These four designs of CRWs have been modeled and animated successfully and indigenously developed beta version prototypes have also been realized. The uniqueness of these designs with respect to harnessing of *in situ counter-vibration* will be described through seven different design metrics for the indigenously developed CRWs. These metrics are: (i) ensemble shape and tare-weight; (ii) kinematic layout of the internal actuating mechanism; (iii) material for fabrication: housing and compliant members; (iv) inclusion of compliance via spring or pin-joints; (v) housing (comprising of top plate and bottom plate) and arrangement for mounting the drive-motor, if any; (vi) sensory instrumentation; and (viii) adapter plate: as connection to the distal arm of the robot/robot-wrist. It may be noted that the first six of these metrics are related to functional attributes of the CRWs involving overall actuation and drive mechanism, housing and manufacturing. And, the last metric pertains to operational/run-time attribute of the CRW involving interfacing with the robot-wrist.

We will concentrate on the fundamental metrics of the indigenous designs of CRWs towards characterization and modeling of the *in situ* vibration. In all practical purposes, our designs of CRWs are accomplished with a cylindrical structure in compact volume, which has design similarity from the exterior with commercially available Remote Centre Compliance (RCC) devices. The fundamental objective towards conceptualizing these designs was to investigate various possible pathways of nascent compliance that will creep in to combat the vibration of the combined architecture of CRW and CRG in real-time. In fact, the designs of these CRWs are made with a consideration of intrinsic real-time compliance/ counter-vibration, devoid of any external compressive forcing function. One important aspect of this *in situ* compliance of CRWs is its manifestation. Our novel designs of CRWs are made in a way so that a requisite and proportionate amount of compliance evolves in both axial as well as radial directions.

The first and foremost design principle that has been invoked in all of these designs of CRWs is the ensemble shape and tare-weight. The synthesis of design as well as source apportionment of inherent compliance and counter-vibration was carried out based on the ensemble cylindrical shape and housing enclosure of the CRW. Fundamentally, the cylindrical-shaped housing of our CRWs has been conceptualized with two circular plates, namely, top plate and bottom plate. Figure 11.12 illustrates the schematic disposition of the basic design metric and design-volume of the CRWs. Fundamentally, the CRW will have two Newtonian co-ordinate systems, viz. Wrist Co-ordinate System : [X_w, Y_w, Z_w] and Gripper Co-ordinate System: [X_g, Y_g, Z_g]. Our design of CRW is conceived with the association of these two co-ordinate systems with bottom plate and top plate, respectively, as depicted in figure 11.12. The actuating mechanism of the CRW, responsible for creating compliance, will be positioned within the fixed volume between these two plates under mechanical shielding.

It may be noted here that the disposition of CRW as per figure 11.12(a) is the ideal layout wherein the dimensions of the top and bottom plates are the same. However, we have adopted a slightly modified version of this layout in our beta version prototyping of the CRWs, as shown in figure 11.12(b). Herein, the dimensions of the plates are not identical and in-line with our customized design, the size of the top plate is significantly larger than that of the bottom plate. However, the cylindrical shape and dimension of the housing will remain the same in both layouts.

A wide spectrum of static dead-load was tested with these CRWs in order to ascertain the subtleness of *in situ* compliance with relation to the output vibration of the mating CRGs. It was found during such testings that besides tare-weight of the CRW, its ensemble shape (outer periphery) was equally contributive towards

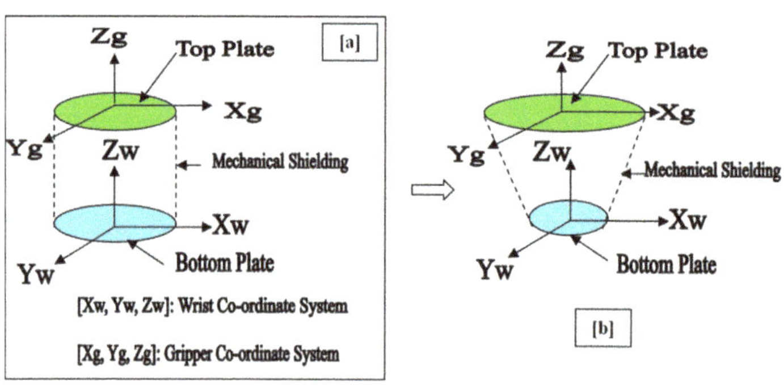

Figure 11.12. Conceptual schematic of the ensemble disposition of the customized CRW: (a) ideal layout; (b) beta version.

generating compliance to its own structure as well as counter-vibration to CRG. It is true that the type of external impulse/compressive force does matter significantly in ascertaining the compliance signature, the effect of which is found to be decoupled from the tare-weight or shape of the CRW. We carried out a range of experiments with the developed *beta versions* of the CRWs using different external compressive forcing functions through scientific weights and qualitative impulse by hand-forcing.

The next important metric of the design is the kinematic layout of the internal actuating mechanism that will be responsible for generating the mechanical motion of the CRW and also for the transmission of the *in situ* compliance. We have used four different kinematic chains of spring-elements for the creation of *in situ* compliance, once the external excitation and/or compressive forcing get imbibed through the robot-wrist. The third crucial factor of the design metric is the choice of material for fabrication of the housing and compliant members of the CRWs. We have used both metal as well as non-metal as the choice of material for the mechanical hardware of the CRWs (beta version). We did observe change in compliance/counter-vibration pattern between the CRWs made up with metallic materials vis-à-vis CRWs fabricated out of non-metallic materials (except housing). While aluminum and mild steel were used for the housing by and large, low carbon steel was used under metallic materials in some of the designs of internal components due to lower density and better mass compliance. Nonetheless, some of the designs are fabricated largely with non-metallic materials, such as nylon, teflon, ABS, polypropelene, neoprene rubber and plastic. Mass compliance was the major design yardstick in the selection of material for fabrication, based on the ensemble compliance as well as vibration characteristics of the CRWs.

The next crucial design metric under functional attribute is inclusion of compliance in the overall design of the CRW. We have customized the designs with two types of members in order to imbibe sufficient compliance in the ensemble system, namely: springs and pin-joints. Although infusing large compliance can be a dampener for vibration harnessing, we need to use a sufficient number of compliant-members in the system to ensure near-perfect counter-vibration of the CRG. The first group of member(s) for imbibing compliance is spring sub-systems that are used as intermediate members between the plates of the CRWs (refer to figure 11.12). The other group of member(s) is pin-joints that are made adhesive-type so as to have good mating contact as well as planar rotations between the axial and radial components of the CRWs.

The fifth design metric under functional attribute is fabrication of the housing of the CRW and arrangement for fixing the drive-motor, if any. The metallic housing of the CRW will comprise circular-shaped top plate and bottom plate and a mechanical shield. Although miniaturization of the housing is an issue that needs to be accomplished in order to make it compatible with the CRG, manufacturing of the housing is very straightforward. On the flip side, manufacturing of the mechanical shield, comprising a variety of spring members, will be challenging. Some of our CRWs need a drive-motor for finer actuation sequences and thus, placement of the motor inside the housing will become a big task due to the availability of relatively merged volume.

The sixth design metric under functional attribute is the sensory instrumentation of the CRW. Fitting and integration of various tiny sensor-elements play a crucial role in the design conceptualization of the CRWs. The physical size and disposition of these sensor-elements need to be accounted for and fine-tuning of the design with respect to drive and transmission system of the members inside the 'mechanical shield' of the CRWs will follow thenafter. The sensor-ensemble of the CRWs will be responsible for three types of tasks: (a) sensing of the external compressive force, either in totality or partially; (b) indication of the incipient compliance of the system; and (c) identification of the counter-vibration at various contact-zones (between CRW and CRG). The primary sensor-element that is responsible for sensing the compressive force, generated in real-time through external forcing function is semiconductor strain gauges. Ideally, we will use these strain gauges in a pair, fitted at symmetrical disposition along the radial and axial members of the CRWs in order to maintain symmetry in the sensor data/output (in milli-volt or micro-volt, depending on the magnitude of the external forcing and its point of application over the top plate. It is to be noted that the strain gauges fitted over the members will record the deflection/deformation of the CRW due to the compressive force in totality, baring loss due to instrumentation and/or white noise. However, in order to have better clarity over the compliance and counter-vibration phenomena of the CRWs in the midst of *in situ* vibration of CRGs, we have integrated flexi-force sensors at other locations of the CRWs as well. The flexi-force sensors at those locations will give relevant data supporting the counter-vibration forcing force in partial form. Nonetheless, these data are also important for overall understanding of the 'force closure' of the grasp phenomena by the CRGs.

The run-time attribute of the design of CRWs is the design of the adapter plate that is to be used as the direct connection with the robotic manipulator. The design of the adapter plate is critical in terms of vibration accumulation. This component is instrumental in the assessment of the overall tare-weight of the CRW as well as incipient compliance. Like CRGs, the good design practice for CRW will always be to reduce the mass of the mechanical shield sub-assembly and make it as minimal as possible just to contain the drive-motor, if any and mounting of the sensory instrumentation.

11.3.2 An overview on the varieties of indigenous designs

Various designs of CRW, the miniaturized compliant wrist, having requisite compliance in axial and radial directions, have been developed by the author's group. Out of several feasible layouts, indigenous design of four representative CRWs will be reported here. These designs and subsequently beta version prototypes can be classified on the basis of the layout of multiple number *spring elements* in axial and radial directions. Accordingly, our indigenously designed and customized CRWs are grouped on the basis of the disposition of spring-elements, viz. (a) compression spring-elements; (b) passive spring-elements; (c) pneumatic spring-elements and (d) sensor-based spring-elements. It is to be noted that in all of the four schemes, the compliant members (i.e. the springs or the spring-like elements) need to be placed between the two plates inside the housing, namely top plate and bottom plate (refer to figure 11.12(a)). However, the dimensions of these elements will be governed by the size of the bottom plate (refer to figure 11.12(b)). As illustrated in figure 11.12, an outer membrane will house the plates and the spring-systems inside and will act as a mechanical shielding. Pathways for sensory instrumentation for the pick-up of compressive force and deformation of CRW will also be accommodated inside the said housing. Strain gauge-based instrumentation has been invoked for the quantitative analysis of the deformation rheology as well as estimation on the resonant frequency (as a measure of counter-vibration for the CRG).

The nucleus of our indigenous design of CRWs is assimilation of sensor-instrumented 'spring-elements' in radial and axial directions. This sort of symmetric disposition of the spring-elements ensures the spread of the compressive force evenly inside the mechanical shielding. Fulfilment of this criterion also signifies somewhat even distribution of compressive force inside the housing and deflection in the members thereafter. Spring-element-based sensor-instrumented members will be evenly disposed in axial and radial directions; which will be four nos. each in three of our designs. This number can be altered depending upon the ensemble size of the CRW and its physical attachment with the CRG. We have used three types of 'spring-like elements' in our designs: (a) passive spring; (b) pneumatic spring; and (c) sensor-based spring. The passive spring-based system invokes miniature pneumatic cylinders while pneumatic spring-based system is essentially a self-instrumented fluidic muscle. The sensor-based spring system is conceptualized with various elemental parts and sensors in assemblage. By virtue of the design, the characteristic motions of these spring-based members will be non-linear and coupled. Sizes of the spring-based members are indeed crucial, and have been finalized through topology optimization. The backbone structures of the spring-elements in all of these designs are made with non-metallic materials for weight reduction. It may be mentioned here that all of these four designs of CRWs are unique in their respective work-envelope, control semantics and counter-vibration protocol (aided by control programming for the CRGs). The layout of the spring-elements in axial and radial directions was made in such a way as to conform to the external shape of the CRW as well as its incipient undulation and/or elastic deformation. The vibrational paradigm of the CRW is dependent on the input-motion, as receivable from the external excitation in real-time.

11.3.3 Details of the firmware

We will elucidate on the fundamentals of the firmware of the four variants of our CRWs in this section. Since these customized CRWs are more sensitive with respect to internal mechanisms, we will focus the detailing on those mechanisms while dealing with the harnessing of vibration of CRGs.

11.3.3.1 Compression spring-element-based robotic wrist (CSRoW)

The compression spring-element-based CRW has been designed with a total of eight 'compression spring-element members', spread out equally along axial and radial directions with uniform angular spacing. All of these members are of identical dimensions as well as content. Figure 11.13 shows the schematic view of CSRoW, with major external dimensions as per the beta version firmware.

The inset of figure 11.13 illustrates the actual disposition of four axial springs (A1 to A4) and four radial springs (R1 to R4) inside the mechanical shielding of CSRoW. The relative positioning of the axial and radial springs is important from the point of view of assembling the spring-elements at an equal angular spacing of 45°. The height of the mechanical shield of CSRoW is also crucial in determining the transmission route for the external compressive force. The assembly of one representative compressive spring-element spring will consist of one compression spring and two strain gauges at both sides of the spring (disposed symmetrically). The assembly will be attached with the top and bottom plates as rigid 'pin' type attachments. Figure 11.14 illustrates the schematic view of a representative spring assembly of CSRoW.

11.3.3.2 Passive spring-element-based robotic wrist (PSRoW)

The passive spring-element-based CRW has been designed with a total of eight 'passive spring-element members', spread out equally along axial and radial directions with uniform angular spacing. All of these members are of identical dimensions as well as content. However, hardware manifestation of 'passive spring' is skill-based and tricky from the point of view of manufacturability. Figure 11.15 shows the schematic view of the Passive Spring-element-based Robotic Wrist, PSRoW, with major external dimensions as per the beta version firmware.

Unlike the previous design of CSRoW, the assembly of 'Passive Spring' calls for engineering skill on hardware realization. The passive spring is constituted from: (i) hollow tube (in the form of a hollow cylinder); (ii) spring; and (iii) sensing element, i.e. strain gauge. As illustrated in figure 11.15, the final assembly of the passive spring will be attached with the top plate and bottom plate through 'pin'-type attachment. The hollow tube, actuated through pneumatics, will aid the spring in subsuming externally-applied compressive force. Two strain gauges will be mounted on the connecting cable of the passive spring assembly, equally disposed on either side of the hollow tube. The strain gauges will sense the deflection of the

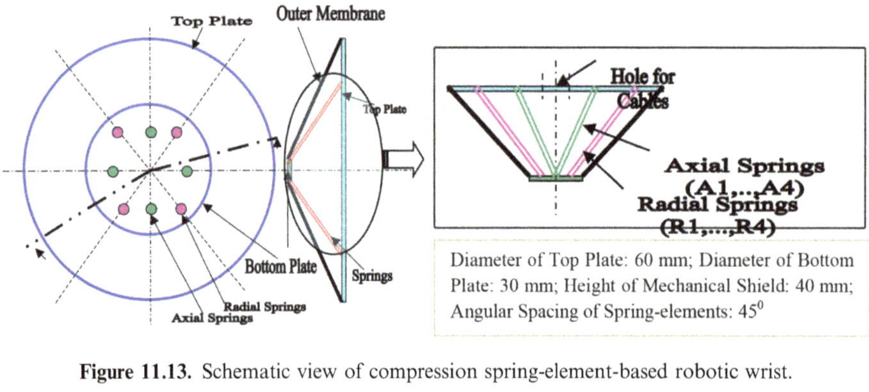

Figure 11.13. Schematic view of compression spring-element-based robotic wrist.

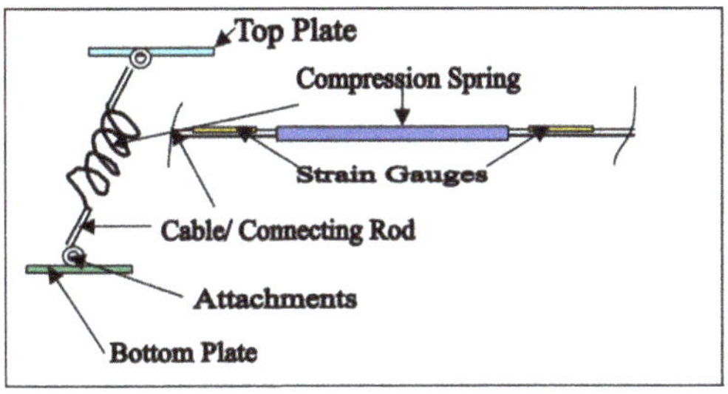

Length of Spring: 24 mm; Length of Connecting Rod: 5 mm. (on either side); Diameter of Connecting Rod: 2 mm; 'Pin' type Attachment: 3 mm. (on either side); Length of Strain Gauges: 2mm. (max.).

Figure 11.14. Schematic view of a representative spring assembly of CSRoW with hardware specifications.

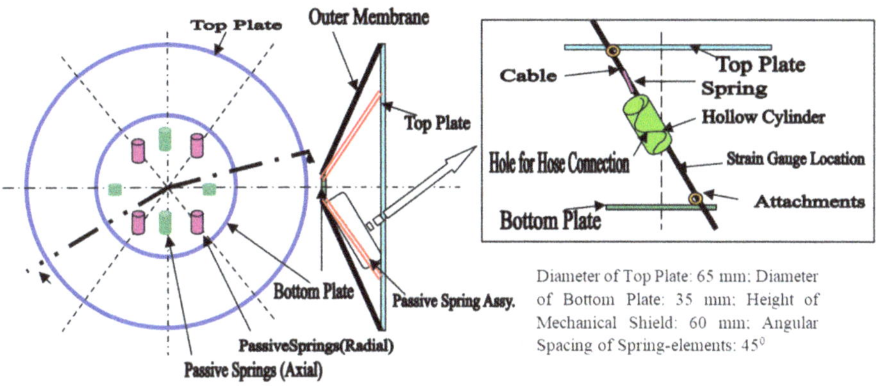

Diameter of Top Plate: 65 mm; Diameter of Bottom Plate: 35 mm; Height of Mechanical Shield: 60 mm; Angular Spacing of Spring-elements: 45^0

Length of 'Passive Spring': 54 mm; Length of Hollow Cylinder: 40 mm; Diameter of Hollow Cylinder: 15 mm; 'Pin' type Attachment: 3 mm. (on either side); Length of Cable: 7 mm. (on either side); Length of Spring: 5 mm; Length of Strain Gauges: 2 mm. (max.).

Figure 11.15. Schematic view of passive spring-element-based robotic wrist with hardware specifications.

respective passive spring system(s) and will show cumulatively the overall undulation of the top plate of the CRW.

The hollow tubes of PSRoW will be actuated by means of small-sized moderate stroke-length double acting pneumatic cylinders of miniaturized size. Thus, the ensemble pneumatic circuitry of PSRoW will comprise of eight double-acting cylinders, a solenoid-based actuation system, pressure-regulating valves, pressure relief values and allied instrumentation. Now, depending upon the fluid pressure inside the hollow cylinders and the corresponding flotation of those, the spring will be either compressed or elongated. This will doubly ensure the compliance in the system and the corresponding deflection (of the passive spring) will be sensed by the strain gauges. The hollow cylinders will have inlet hole, suitable for fixing the hose and the other end of the (flexible) hose will be attached to the double-acting cylinder.

11.3.3.3 Pneumatic spring-element-based robotic wrist (PuSRoW)

The third innovative design, namely the pneumatic spring-element-based CRW, has been designed with a total of six 'pneumatic spring-element members', arranged in a symmetric matrix of Stewart Platform layout along axial and radial directions with uniform angular spacing. In other words, the six pneumatic spring members will be placed at equal angular intervals in a regular hexagon. All of these members are of identical dimensions as well as content. However, hardware manifestation of the 'pneumatic spring' is also a skill-based activity like the 'passive spring' in PSRoW. As a matter of fact, manufacturing of these pneumatic springs is even harder than in the case of passive springs. Figure 11.16 shows the schematic view of the Pneumatic Spring-element-based Robotic Wrist, PuSRoW, with major external dimensions as per the beta version firmware.

Since the pneumatic springs are essentially *fluidic muscles*, these members do not need any other sensory system. Thus, there is no need of mounting a separate sensing element in PuSRoW, as pneumatic springs are capable of sensing the fluid pressure applied on the respective element. The system compliance of PuSRoW will be manifested by the extension/compression of the pneumatic spring-elements. In summary, these pneumatic springs will function like single-acting actuators. The inset of figure 11.16 schematically illustrates the positioning and fixation of the pneumatic spring with the top and bottom plates of PuSRoW.

11.3.3.4 Sensor-based spring-element-driven robotic wrist (SeSRoW)

The fourth novel design, namely, sensor-based spring-element-driven CRW has been designed with a total of eight 'customized spring-element members', arranged in a radially-symmetric layout along axial and radial directions with uniform angular spacing. The 'customized spring' elements have been designed and fabricated from a very thin plate with sensing element mounted on it. The optimal design of sensor-based spring-element-driven robotic wrist, SeSRoW, will incorporate four customized springs, along radial and axial directions. All of these members are of identical dimensions as well as content. However, hardware manifestation of a 'customized spring' is also a skill-based activity like the 'passive spring' in PSRoW or 'pneumatic spring' in PuSRoW. Figure 11.17 shows the schematic view of the SeSRoW, with major external dimensions as per the beta version firmware.

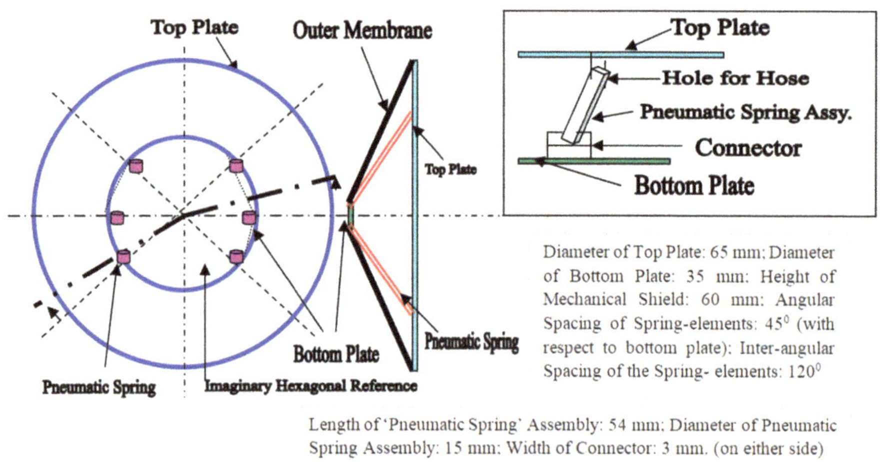

Figure 11.16. Schematic view of pneumatic spring-element-based robotic wrist with hardware specifications.

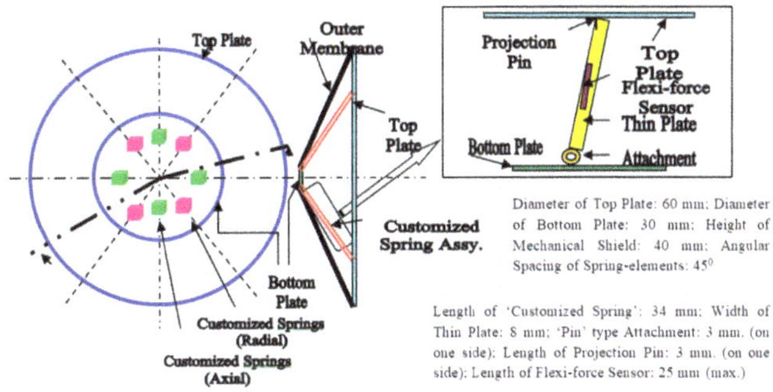

Figure 11.17. Schematic view of sensor-based spring-element-driven robotic wrist with hardware specifications.

It may be noted that the change in output response (in milli-volts) of the flexi-force sensors (placed inside these sensor-based spring members) will represent the overall undulation of the top plate, i.e. the wrist assembly in general. Unlike other schemes of CRW, the sensor-integrated spring members of SeSRoW will be 'free' at one end, namely, in the direction of the top plate (refer to the inset of figure 11.17). However, the other end of the member, viz. along the bottom plate, will be anchored via a 'pin'-type attachment. One interesting design-metric of SeSRoW is the augmentation of 'Projection Pins' underneath the top plate, which will induce vibration/excitation to the sensor-integrated spring members. The overall disposition of the customized spring assembly has been made commensurate with the height of the mechanical shield of SeSRoW.

11.4 Grasp-induced vibration models of the compliant robotic grippers

11.4.1 An overview of vibration synthesis

Real-time functionalization of a CRG is, no doubt, highly challenging from the purview of control system architecture. In fact, control of CRG is one of the niche ensembles of present-day robotics research that deals with various modes of vibrations, inherent in the system. The vibration, so referred, is completely built-in type and thus it is design invariant. By nature, this very vibration in CRG is self-propagating and does not follow analytical modeling and rule-base in practically all end applications. As delineated in section 11.2, our novel CRGs do have an advantage of very low tare-weight that is quite befitting for a large number of applications. But, on the flip side, the major bottleneck is the inherent vibration of the units. And, due to this inherent aspect of CRG, prototyping multiple degrees-of-freedom customized compliant gripper is highly challenging.

Several designs of CRG have been attempted by the researchers in the past decade in order to alleviate this vibration but most of those trials have been unsuccessful. In contrast to previous vibration-combatting modeling attempts, our endeavour for vibration synthesis of CRG will be in a holistic fashion that caters for real-time aspects like rheology (stress-strain paradigms), *in situ* vibration, sensor fusion and non-linear coupled dynamics for control.

The domain of VS is branched out into two interlinked domains: (i) Grasp Synthesis (GS) and (ii) Real-Time Dynamics (RTD). The present treatise reports modeling of this *in situ* vibration as well as solving design issues for the firmware of customized CRGs. It is needless to state that this novel modeling for the run-time synthesis of customized miniature CRGs will be carried out with an aim towards analytical modeling for GS and RTD. Fundamentally, an ensemble of GS will be governed by the appropriate 'grasp model', involving the principles of form closure and force closure. With this backdrop, the domain of GS comprises three fundamental paradigms, viz. (a) object geometry; (b) contact mechanics-based model; and (c) gripper kinematics. Thus, modeling for GS will revolve around the tuple of object and gripper, involving geometrical and kinematic features. Semantics of *contact mechanics* is the key for the formulation of GS in run-time condition of the CRG. The facets of RTD are complementary to that of GS and these are more focused towards the vibration signature aspect during the robotic grasp. The domain of RTD is developed with the help of three metrics: (a) vibration signature (natural frequency and Eigen vectors); (b) dynamic model of the gripper; and (c) turning model of the gripper. We will now go through the details of GS and RTD in subsequent sections.

11.4.2 Paradigms of grasp synthesis

GS for compliant robotic grippers dwells primarily with 'grasp prehension', which occurs through four cases of realization of the external geometry of the jaw surface and object surface. A grasp-induced vibration model will depend on the specific type of this prehension in real-time. Hence, geometrical features of the external surfaces of the CRG-jaws as well as the object being gripped are very crucial in building up this model. Of course, the process of this modeling will involve smooth transformation of surface geometry from a pure geometrical standpoint to analytical perspectives.

We will detail the analytical model for two types of jaws of our CRGs, as described in section 11.2, namely (a) flat jaw and (b) curvilinear jaw. Likewise, two envelopes of the object surfaces will be considered in the modeling. As per the basic proposition of our model, we need to consider the 'circular envelope' for flat-surfaced objects. Similarly, 'curved approximation' of flat jaw will be adhered to for flat-jaw gripper. However, we do not need any curvilinear approximation for curved-jaw CRGs, as the geometries are already defined. And, analytical treatment for objects having curved surface will be straightforward also. Hence, the strategy for modeling is based on the analytical treatment of both jaws as well as object surfaces as *curves*.

The advantage of this sort of modeling is the subtlety of the *point of contact* between the mating surfaces of the gripper jaw and object grasped. Our grasp-induced model will consider a 2-D planar model for the mating between CRG-jaw and the object, and thus we will require co-ordinate location of the said 'point of contact' for easy identification of the centroid for the form closure of the grasp. Accordingly, four cases of 'grasp prehension' can occur as per our modeling semantics, viz. (a) *Case I:* Curvilinear Jaw surface and Curved Object (CJCO);

(b) *Case II*: Curvilinear Jaw surface and Flat-surfaced Object (CJFO); (c) *Case III*: Flat Jaw-surface and Curved Object (FJCO); and (d) *Case IV*: Flat Jaw-surface and Flat-surfaced Object (FJFO). It may be noted that while Case I and Case II are valid for Curvilinear Jaw Gripper, Case III and Case IV are valid for Flat Jaw Gripper.

We will delineate three types of grasp models: (A) Asperity Contact-based (applicable for FJCO and FJFO types of grasp); (B) Contact Mechanics Vector-based (applicable for all four types of grasp); and (C) Adhesive Mechanics-based (applicable for CJCO and CJFO types of grasp). The details on the formulation of these three grasp-induced vibration models are presented below.

11.4.3 Development of the grasp models

11.4.3.1 Asperity contact-based grasp model (ACGM)

The Asperity Contact-based Grasp Model (ACGM) is applicable essentially for flat-jaw CRGs towards comprehending FJCO and FJFO types of grasp. The asperity contact analysis of the flat-jaw compliant robotic gripper and its postulation thereof is realized through polynomial syntax of the surfaces of the object and gripper. Prior to dealing with the subtlety of the model, we will first look at the contact condition between the gripper-jaw and the object from an analytical viewpoint. Figure 11.18 explains the scenario of the contact condition, prevailing at the gripper–object interface, at the instant of grasping.

It may be noted here that while the final asperity-level contact demarcates the location of the contact point, CP, its position gets defined by the line of contact, pq, It may be noted here that while the final asperity-level contact demarcates the location of the contact point, CP, its position gets defined by the line of contact, pq↔, which is the common tangent between CRGS and OS. On the other hand, both C_1 and C_2 are equally likely locations on the Jaw surface of CRG, from the stability aspect. In general, CRGS and OS can be expressed mathematically in $y = f(x)$ form using polynomials in a 2-D Cartesian co-ordinate system. The respective first derivatives of the said polynomials for CRGS and OS must be identical for a

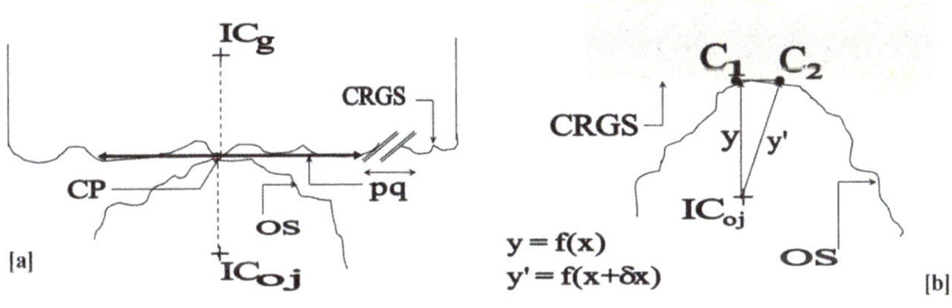

Legends:

CRGS: Compliant Robotic Gripper Surface; **OS**: Object Surface; **CP**: Contact Point; **pq**: Line of Contact; **IC_g**: Instantaneous Centre of Gripper; **IC_{oj}**: Instantaneous Centre of Object; **C_1 & C_2**: Consecutive Stable Locations on the Gripper Jaw Surface; **x**: Length of Asperity on CRGS & OS; **y & y'**: Mathematical Expressions for the Locations on the Object Surface at time 't' & '(t+δt)' respectively.

Figure 11.18: Schematics of the contact facets of ACGM, vide (a) asperity contact and (b) stability during contact.

perfect asperity contact at the line of contact of object(s) to be grasped. The following two lemmas can be proposed in extension to the concept of ACGM:

LEMMA 1:
While first derivatives of the said polynomials ensure a *stable grasp*, the second derivatives signify *slip-free grasp*.

LEMMA 2:
The higher-order derivatives are therefore, non-entity, as both 'force closure' and 'form closure' of robotic grasp are ensured.

Analytically 'y_1' and 'y_2' represent the CRGS and OS, respectively, composed by sets of polynomial coefficients, vide $\{a_0, a_1, \ldots, a_n\}$ and $\{b_0, b_1, \ldots, b_n\}$ that can be expressed in polynomial series form as shown below:

$$y_1 = f_1(x) = a_0 + a_1x + a_2x^2 + a_3x^3 + \ldots \ldots + a_nx^n \tag{11.1a}$$

$$y_2 = f_2(x) = b_0 + b_1x + b_2x^2 + b_3x^3 + \ldots \ldots + b_nx^n \tag{11.1b}$$

and,

$$\frac{dy_1}{dx} = \frac{dy_2}{dx} \tag{11.1c}$$

Now, the relative displacement vector between the gripper and the object, $[\delta_{goj}]$ can be expounded with respect to displacement vectors pertaining to gripper and object separately as,

$$[\delta_{goj}] = [\delta_g] + [\delta_{oj}] \tag{11.2}$$

Now, for flat jaw surface of CRG, $[\delta_g]$ will be zero and in those cases (vide Case III and Case IV of section 11.4.2: FJCO and FJFO), $[\delta_{oj}]$ is evaluated as,

$$[\delta_{oj}] = f_2(x + \delta x) - f_2(x) = \delta x . \frac{dy_2}{dx} + (\delta x)^2 . \frac{d^2y_2}{dx^2} + \ldots \ldots \tag{11.3}$$

The final computing expression for $[\delta_{goj}]$ after simplification will be:

$$\begin{aligned}[\delta_{goj}] = (\delta x)[b_1 + 2b_2x + 3b_3x^2 + \ldots + nb_nx^{n-1}] \\ + \frac{(\delta x)^2}{2}[2b_2 + 6b_3x + \ldots + n(n-1)b_nx^{n-2}]\end{aligned} \tag{11.4}$$

It is important to realize the physical significance of $[\delta_{goj}]$ in light of the posterior calculation of the grasping force per unit length of the CRG-jaw under run-time vibration scenario. The uniqueness of our CRGs is the uniform distribution of force concentration matrix over the jaw surfaces so as to ensure a perfect force closure and also, form closure. The computed value of the grasping force per unit length of the jaw surface or the ensemble grasping force in case we can determine the total surface area of the CRG-jaws will be the deciding factor for estimating the natural frequency of vibration of the CRG.

The force concentration matrix will be modeled as a concentrated force vector in 2-D, wherein the resultant force vector passes through the centroid of the force closure. We will use Newtonian frame of reference for the modeling of this force concentration vector, $\{B_g\}$ and accordingly, the distribution of $\{B_g\}$ will be, $\{B_g\} = [B_{g\_x}{}^2 + B_{g\_y}{}^2]^{1/2}$, in a simplified form, having its dimension as ML^{-1} T^{-2} (=> Force/Area : MLT^{-2}/L^2).

Thus, we get the final expression for the grasping force per unit length under ACGM as,

$$[F_g] = \{B_g\}[\delta_{goj}] = [B_{gx}{}^2 + B_{gy}{}^2]^{\frac{1}{2}}\{(\delta x)(b_1 + 2b_2x + 3b_3\,x^2) + \frac{(\delta x)^2}{2}(2b_2 + 6b_3x)\} \quad (11.5)$$

It may be noted that we have excluded the higher-order terms in equation (11.5) in order to avoid computational complexity. The formulation of equation (11.5) is dependent on the value of 'x' and the numerical quantification of the Cartesian components of $\{B_g\}$. In fact, both B_{g_x} and B_{g_y} can be evaluated through experimentation on grasp performance of the CRGs, through the readings of load cells and/or strain gauges.

11.4.3.2 Contact mechanics vector-based grasp model (CMVGM)

The Contact Mechanics Vector-based Grasp Model (CMVGM) is applicable for all four types of grasp prehensions, viz. CJCO, CJFO, FJCO & FJFO. The point-contact phenomena between object and gripper jaw surfaces and postulation thereof will be evaluated in this model through two distinct methodologies: (a) Contact Vector Approach and (b) Talyor series-based Approach. CMVGM is primarily devised on the aspects of 'point contact' pheonemoa, which is a deviation from the 'asperity contact' scenario. Hence, we will delineate the postulation of CMVGM under two approaches on the basis of Point Contact-based Contact Mechanics model for the graspable object. From the proposition of asperity-based contact model in sub-section 11.4.3.1, let us take a detour to look at the scenario of contact mechanics at a slightly larger level. The present model, CMVGM, is effective in analyzing the grasp phenomena in real-time, which is based on the metric of contact vectors between the gripper surface and object surface. Figure 11.19 schematically illustrates the postulation of the model, highlighting the nature of contact (mating) prevailing in the CRG and the object(s) to be gripped.

The following nomenclatures are used in the formulation of the model (CMVGM):

$[X_g, Y_g, Z_g]^T$: Newtonian/Cartesian Co-ordinate system of the gripper;
$[X_{gi}, Y_{gi}, Z_{gi}]^T$: as above, at the ith time-instant;
O_g, O_{gi}: respective origins of the above systems;
[r_g]: contact radius vector of the gripper;
[r_{gi}]: as above, at the ith. time-instant;
$\delta\theta_{gi}$: angular shift in position at the ith. instant;
$[B_g]$: force concentration matrix of the gripper;
$[X_{oj}, Y_{oj}, Z_{oj}]^T$: Newtonian/Cartesian co-ordinate system of the object;
$[X_{oji}, Y_{oji}, Z_{oji}]^T$: as above, at the ith. time-instant;
O_{oj}, O_{oji}: respective origins of the above systems;
$[r_{oj}]$: contact radius vector of the object;

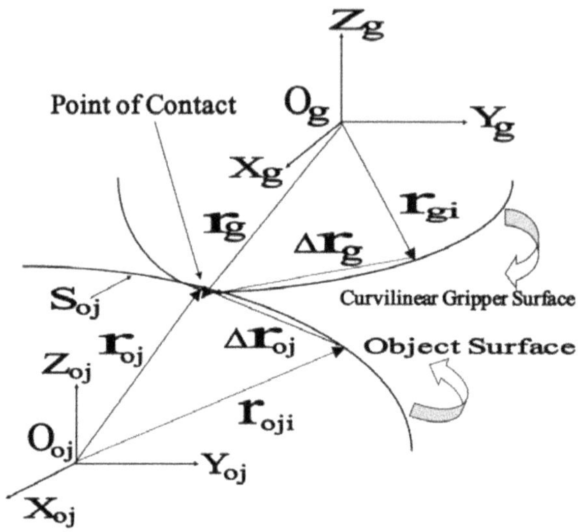

Figure 11.19. Schematics of the contact mechanics vector-based grasp model of CRGs.

$[r_{oji}]$: contact radius vector at the *i*th time-instant;
$\delta\theta_{oji}$: change in angular position of the object;
$[r(s)]$: generalized radius vector at the gripper-object contact surface;
S_{oj}: curvilinear segment at the contact-zone of the object and gripper.

It is to be noted that details of $[X_{gi}, Y_{gi}, Z_{gi}]^T$, $[X_{oji}, Y_{oji}, Z_{oji}]^T$, O_{gi}, O_{oji}, $[r(s)]$, $\delta\theta_{gi}$ and $\delta\theta_{oji}$ are not shown in figure 11.19 explicitly in order to avoid cluttering. The significance of these legends will be amplified in the modeling. Also, origins of the systems (for object and gripper) are meant for calculating the contact radius vectors only; those are not the actual centres of the respective objects.

With this general schema in the background, now we will adopt two approaches for building up this contact mechanics-based model for grasp synthesis, as stated earlier. The first approach is based on *vector signature* of various domain-indices and grasp-related parameters, such as r_{oj}, r_{oji}, r_g, r_{gi} etc, wherein we will develop expressions and equations in co-planar vectors. The second approach is based on *Taylor series expansion* of the domain-indices and grasp-specific parameters, as stated before.

11.4.3.2.1 Contact mechanics-based model: vector signature-based approach

In this method, we will develop vector expression for the grasp-related parameters by considering two-dimensional simplification of the above paradigm, as depicted in figure 11.19. Thus, we have the following two vector representations (using standard *direction cosines*) for the object surface:

$$r_{oj} = (r_{oj} \text{ Cos } \alpha_{oj})\,\mathbf{i} + (r_{oj} \text{ Sin } \alpha_{oj})\,\mathbf{j} \tag{11.6a}$$

$$r_{oji} = (r_{oj} \text{ Cos } \alpha_{oji})\,\mathbf{i} + (r_{oj} \text{ Sin } \alpha_{oji})\,\mathbf{j} \tag{11.6b}$$

where,

$$\alpha_{oji} \sim \alpha_{oj} = \delta\theta_{oji} \tag{11.6c}$$

{**i**, **j**} are the respective unit vectors along X and Y axes.

Likewise, the vector representation for the gripper surface will be:

$$r_g = (r_g \operatorname{Cos} \alpha_g)\,\mathbf{i} + (r_g \operatorname{Sin} \alpha_g)\,\mathbf{j} \tag{11.7a}$$

$$r_{gi} = (r_g \operatorname{Cos} \alpha_{gi})\,\mathbf{i} + (r_g \operatorname{Sin} \alpha_{gi})\,\mathbf{j} \tag{11.7b}$$

where,

$$\alpha_{\mathrm{gi}} \sim \alpha_{\mathrm{g}} = \delta\theta_{\mathrm{gi}} \tag{11.7c}$$

As before, {**i**, **j**} are the respective unit vectors along X and Y axes.

Hence, we can compute the differential radii vectors, Δr_{oji} and Δr_{gi} for small angular deviations as follows:

$$\begin{aligned}\Delta\, r_{oji} &= r_{oji} - r_{oj} = [r_{oj} \operatorname{Cos}(\delta\theta_{oji})]\,\mathbf{i} + [r_{oj} \operatorname{Sin}(\delta\theta_{oji})]\,\mathbf{j}\\ &\approx [r_{oj}\,(1 - \delta\theta^2_{oji})]\,\mathbf{i} + [r_{oj}\,(\delta\theta_{oji})]\,\mathbf{j}\end{aligned} \tag{11.8}$$

In-line with the formulation of equation (11.8), we can deduce the following for the gripper surface:

$$\begin{aligned}\Delta\, r_{gi} &= r_{gi} - r_g = [r_g \operatorname{Cos}(\delta\theta_{gi})]\,\mathbf{i} + [r_g \operatorname{Sin}(\delta\theta_{gi})]\,\mathbf{j}\\ &\approx [r_g(1 - \delta\theta^2_{gi})]\,\mathbf{i} + [r_g\,(\delta\theta_{gi})]\,\mathbf{j}\end{aligned} \tag{11.9}$$

Therefore, we get the algebraic expression for ($\Delta r_{oji} - \Delta r_{gi}$) as:

$$\Delta r_{oji} - \Delta r_{gi} = [(r_{oj} - r_g) - r_{oj}.\ \delta\theta^2_{oji} + r_g.\ \delta\theta^2_{gi}]\,\mathbf{i} + [r_{oj}(\delta\theta_{oji}) - r_g(\delta\theta_{gi})]\,\mathbf{j} \tag{11.10}$$

Hence, relative displacement vector between the gripper and the object can be simplified as,

$$\begin{aligned}d_{oj\_gi} &= \Delta\, r_{oji} - \Delta\, r_{gi}\\ &\approx [r_g\, \delta\theta^2_{gi} - r_{oj}\, \delta\theta^2_{oji}]\,\mathbf{i} + [r_{oj}\,(\delta\theta_{oji}) - r_g(\delta\theta_{gi})]\,\mathbf{j}\end{aligned} \tag{11.11}$$

Considering the distribution of force concentration vector of the gripper $[B_g]$ in 2-D planar matrix as, $[B_g] = B_{g\_x}\ \mathbf{i} + B_{g\_y}\ \mathbf{j}$, in a simplified form, having its dimension as $ML^{-1}\ T^{-2}$ (=> Force/Area: MLT^{-2}/L^2), we get the final expression for the grasping force per unit length (of the gripper) as,

$$[F_G] = [B_g]\,[d_{ojgi}] = [\{r_g\, \delta\theta^2_{gi} - r_{oj}\, \delta\theta^2_{oji}\}\, B_{gx}]\,\mathbf{i} + [\{r_{oj}.\ (\delta\theta_{oji}) - r_g(\delta\theta_{gi})\}\, B_{gy}]\,\mathbf{j} \tag{11.12}$$

The final formulation of grasping force of the CRG per unit length of the jaw surface, as per equation (11.12) clearly shows the parlances of point contact-based grasp with in-built force concentration vector.

11.4.3.2.2 Contact mechanics-based model: Taylor series-based approach

Let us consider the geometry of figure 11.19 again and have a close look over 'S_{oj}'. Now, from the geometry we can expound Δr_{oji} in the form of Taylor Series as:

$$\begin{aligned}\Delta r_{oji} &= r_{oji}(s + \delta s) - r_{oj}(s) = \left[r_{oj}(s)\right]_i - r_{oj}(s) \equiv \left[r_{oj}(s)\right]_i - r(s)\\ &= [r_{oj}(s) + \delta s.\, dr_{oj}/ds + (\delta s)^2/2!\, d^2r_{oj}/ds^2 + \dots\dots\dots\dots]_i - r_{oj}(s)\end{aligned} \tag{11.13}$$

On simplification, the equation yields to:

$$\Delta r_{oji} = \delta s. dr_{oj}/ds + d^2r_{oj}/ds^2(\delta s)^2/2 \tag{11.14}$$

By taking into account all four cases of grasp, namely, CJCO, CJFO, FJCO & FJFO, we can arrive at an approximation towards parametric modeling for the curvilinear pattern of the grasp (pick) at the jaw surface of the compliant gripper after a successful pick-up yields the following equation:

$$S_{oj} = k1 \text{ Sin } \phi + k2 \text{ Cos } \phi \tag{11.15}$$

where, ϕ denotes the angle of shift, measured between the tangents at the inside and outside boundary curves, as conceptualized in figures 11.19 and $k1$, $k2$ are numeric parameters, vide $\{k1, k2\} \in (0,1)$.

Thus, on simplification of equation (11.15), δS_{oj} (equals δs) can be approximated as,

$$\delta s = [k1(1 - \phi^2/2! + \phi^4/4!) - k2(\phi - \phi^3/3!)]\, \delta\phi \tag{11.16}$$

Using the same analogy of equation (11.11), as stated earlier in sub-section 11.4.3.2.1, we get the relative translation of the gripper with respect to the object as,

$$d_{ojgi} = \Delta r_{oji} - \Delta r_{gi} = \delta s\, [dr_{oj}/ds - dr_g/ds] + (\delta s)^2/2 \left[(d^2r_{oj}/ds^2 - d^2r_g/ds^2)\right] \tag{11.17}$$

It may be noted here that the contact radius vector of the object, $r_{oj}(s)$ and its complimentary on the gripper surface, i.e. $r_g(s)$ are identical. Both $r_{oj}(s)$ and $r_g(s)$ are equal to and the same as $r(s)$, which is the contact radius vector between the gripper and object surfaces.

Considering the distribution of force concentration vector of the gripper [Bg] in 2-D planar matrix as, $[Bg] = [B_{g\_x}{}^2 + B_{g\_y}{}^2]^{1/2}$, in a simplified form, having its dimension as $ML^{-1}T^{-2}$ (=> Force/Area : MLT^{-2}/L^2), we get the final expression for the grasping force per unit length as,

$$\begin{aligned}[F_G] &= [B_g]\,[d_{ojgi}] = [B_{gx}{}^2 + B_{gy}{}^2]^{1/2}\\ &\quad .\ [\delta s\,(dr_{oj}/ds - dr_g/ds) + (\delta s)^2/2\,(d^2r_{oj}/ds^2 - d^2r_g/ds^2)]\end{aligned} \tag{11.18}$$

Here too, the final formulation of grasping force of the CRG per unit length of the jaw surface, as per equation (11.18) shows the paradigms of point contact-based grasp with in-built force concentration vector without ambiguity.

11.4.3.3 Adhesive mechanics-based grasp model (AMGM)

The Adhesive Mechanics-based Grasp Model (AMGM) is applicable for all curvilinear jaw-type grippers with two types of grasp prehensions, viz. CJCO and CJFO. The point-contact phenomena and postulation thereof, as expounded in sub-section 11.4.3.2, can be extended to another important dimension wherein we need to provide a technological solution for gripping of: (a) very lightweight and tiny objects; (b) cluster of tiny objects in soft encapsulation and (c) consortia of micro-substances in permeable covering with the help of a curvilinear jaw compliant robotic gripper. The subtle point-contact phenomena between object and gripper jaw surfaces in such small lightweight objects or soft encapsulation of multiple tiny objects will be evaluated in this model through two distinct methodologies: (a) Contact Vector Approach and (b) Talyor series –based Approach.

At times, these tiny objects are typical in shape and made of customized design. Likewise, in cases where specific tiny/micro-objects are put under suitable soft encapsulations, e.g. polythene packets, gel-like covering, soft epoxy coating etc, we need to use a specific gripper-jaw as well as grasp model based on adhesion-type mating. Hence, we need to study the design features of such objects/encapsulated object-mass in detail, before going through the postulation of the grasp model. In general, for curvilinear objects, we will consider inner and outer radii for the modeling. The quantum of force concentration vector of the gripper and its dissipation over the jaw surface will depend on the shape and contour of the curvilinear jaw gripper. The ensemble of force concentration and dissipation will be guided by the adhesive mechanics-based model of the grasp action.

The parametric analysis has been made having inputs on the contact mechanics of the mating parts and the kinetics of the gripping action. Thus, the model postulates the formulae for the quantitative estimation of the gripping force as a function of various geometric parameters of the tiny object(s) and the gripper. The 'object' is assumed to have curved surface with marginal thickness, approximated through mathematical function. However, the model is equally applicable for objects with flat surfaces. Since the nature of grasp is analyzed using the ideation of adhesive-contact (between gripper surface and object surface), we will adhere to the systematic procedure of using location vectors for the respective surfaces. The force of gripping gets initiated from the in-built stiffness of the contact-zone. We will model this stiffness in the pathway of spring stiffness, using vector form. Thus, the present model, AMGM is essentially a point contact-based model with in-built force concentration vector plus stiffness vector ('adhesion'). Figure 11.20 illustrates the geometrical paradigms of the adhesive (contact) mechanics-based grasp of the CRG.

We have used the following nomenclatures in developing the AMGM:

$[X_{gr}, Y_{gr}, Z_{gr}]^T$: Newtonian/Cartesian Co-ordinate system of the gripper;
$[X_{gri}, Y_{gri}, Z_{gri}]^T$: as above, at the ith time-instant;
O_{gr}, O_{gri}: respective origins of the above systems;
$[r_{gr}]$: contact radius vector of the gripper;
$[r_{gri}]$: as above, at the ith time-instant;
$\delta\theta_{gri}$: angular shift in position at the ith instant;
$[B_{gr}]$: force concentration vector of the gripper;

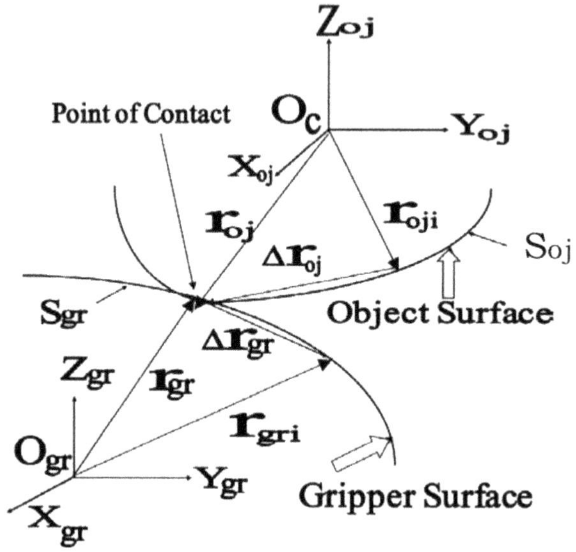

Figure 11.20. Schematics of the adhesive mechanics-based grasp model of CRGs.

$[K_{gr}]$: stiffness matrix of the gripper;
$[B\text{-}K]_{gr}$: resultant force vector at the gripper;
$[X_{oj}, Y_{oj}, Z_{oj}]^T$: Newtonian/Cartesian co-ordinate system of the object ;
$[X_{oj}, Y_{oji}, Z_{oji}]^T$: as above, at the ith time-instant;
O_{oj}, O_{oji}: respective origins of the above systems;
$[r_{oji}]$: contact radius vector at the ith time-instant;
$\delta\theta_{oji}$: change in angular position of the object ;
$[r^*(s)]$: generalized radius vector at the gripper-object contact surface;
R_{gi}: inner radius of the gripper 'adhesion-zone';
R_{go}: outer radius of the gripper 'adhesion-zone';
R_{gm}: mean radius of the gripper 'adhesion-zone';
S_{gr}: curvilinear segment of the gripper surface at the contact-zone;
S_{oj}: curvilinear segment of the object surface at the contact-zone.

It is to be noted that details of $[X_{gri}, Y_{gri}, Z_{gri}]^T$, $[X_{oji}, Y_{oji}, Z_{oji}]^T$, O_{gri}, O_{oji}, $[r^*(s)]$, R_{gm}, R_{gi}, R_{go}, $\delta\theta_{gri}$ and $\delta\theta_{oji}$ are not shown in figure 11.20 explicitly in order to avoid cluttering. Like the previous case of Contact Mechanics-based model, the significance of these legends will be amplified in the modeling. It may also be noted that the origins of the systems (for the object and gripper) are meant for calculating the contact radius vectors only; those are not the actual centres of the respective entities. We will proceed for developing the Adhesive Mechanics-based model for grasp synthesis using two approaches, namely, (a) Vector Signature-based approach and (b) Taylor series-based approach. The expressions and equations will be developed in co-planar vectors under Vector Signature-based approach. In contrast, the second approach is based on Taylor series expansion of the domain-indices and grasp-specific parameters. In both the approaches, we will investigate the final expression of the gripping force in X, Y and Z directions through 'Part-Mating' Model, using 'Adhesive Mechanics'. We will postulate the specific modeling thematics under these two approaches in the next sub-sections.

11.4.3.3.1 Adhesive mechanics-based model: vector signature-based approach

As explained earlier, Vector Signature-based approach relies on vector signature of various domain-indices and grasp-related parameters as per the indices of figure 11.20, such as r_{oj}, r_{oji}, r_{gr}, r_{gri} etc. Considering two-dimensional simplification of the grasp paradigm, we can arrive at the following equations for the geometry of the gripper (using standard direction cosines),

$$r_{gr} = (r_{gr}\cos\alpha_{gr})\mathbf{i} + (r_{gr}\sin\alpha_{gr})\mathbf{j} \tag{11.19a}$$

$$r_{gri} = (r_{gr}\,\mathrm{Cos}\,\alpha_{gri})\mathbf{i} + (r_{gr}\,\mathrm{Sin}\,\alpha_{gri})\mathbf{j} \tag{11.19b}$$

where,

$$\alpha_{gri} \sim \alpha_{gr} = \delta\theta_{gri} \tag{11.19c}$$

Likewise, the vector representation for the object surface will be:

$$r_{oj} = (r_{oj}\,\mathrm{Cos}\,\alpha_{oj})\,\mathbf{i} + (r_{oj}\,\mathrm{Sin}\,\alpha_{oj})\,\mathbf{j} \tag{11.20a}$$

$$r_{oji} = (r_{oj}\,Cos\,\alpha_{oji})\,\mathbf{i} + (r_{oj}\,\mathrm{Sin}\,\alpha_{oji})\,\mathbf{j} \tag{11.20b}$$

where,

$$\alpha o_{ji} \sim \alpha_{oj} = \delta\theta_{oji} \tag{11.20c}$$

In both of the above sets of formulation, $\{\mathbf{i}, \mathbf{j}\}$ are the respective unit vectors along X and Y axes.

Hence, we can compute the differential radii vectors, Δr_{oji} and Δr_{gri} for small angular deviations as follows:

$$\begin{aligned}\Delta r_{oji} = r_{oji} - r_{oj} &= [r_{oj}\,\mathrm{Cos}\,(\delta\theta_{oji})]\,\mathbf{i} + [r_{oj}\,\mathrm{Sin}\,(\delta\theta o_{ji})]\,\mathbf{j}\\ &\approx [r_{oj}\,(1 - \delta\theta o_{ji}^{2})]\,\mathbf{i} + [r_{oj}\,(\delta\theta o_{ji})]\,\mathbf{j}\end{aligned} \tag{11.21}$$

$$\begin{aligned}&\Delta r_{gri} = r_{gri} - r_{gr} = [r_{gr}\,\mathrm{Cos}\,(\delta\theta_{gri})]\,\mathbf{i} + [r_{gr}\,\mathrm{Sin}\,(\delta\theta_{gri})]\,\mathbf{j}\\ &\approx [r_{gr}\,(1 - \delta\theta_{gri}^{2})]\,\mathbf{i} + [r_{gr}\,(\delta\theta_{gri})]\,\mathbf{j}\end{aligned} \tag{11.22}$$

Therefore, we get the algebraic expression for $(\Delta r_{oji} - \Delta r_{gri})$ as:

$$\Delta r_{oji} - \Delta r_{gri} = [(r_{oj} - r_{gr}) - r_{oj}.\;\delta\theta_{oji}^{2} + r_{gr}.\;\delta\theta_{gri}^{2}]\,\mathbf{i} + [r_{oj}\,(\delta\theta_{oji}) - r_{gr}\,(\delta\theta_{gri})]\,\mathbf{j} \tag{11.23}$$

Hence, relative displacement vector between the gripper and the object can be simplified as,

$$d_{oj\_gri} = \Delta r_{oji} - \Delta r_{gri} \approx [r_{gr}\,\delta\theta_{gri}^{2} - r_{oj}\delta\theta_{oji}^{2}]\,\mathbf{i} + [r_{oj}\,(\delta\theta_{oji}) - r_{gr}\,(\delta\theta_{gri})]\,\mathbf{j} \tag{11.24}$$

In this model, apart from force concentration vector, the adherence of the objects inside the gripper adhesive-zone is primarily due to the contact stiffness of the gripper surface. In that respect, we will encounter a coupled effect of 'Force Concentration Vector' and 'Spring-like Contact Force', applicable at the 'adhesive zone' of the gripper. We will consider the resultant of these two vectors for the final modeling of the grasping force per unit length of the gripper surface.

We will consider the distribution of force concentration vector $[B_{gr}]$ (dimension: $ML^{-1}\ T^{-2}$) and spring stiffness vector $[K_{gr}]$ (dimension: $MLT^{-2}/L \rightarrow MT^{-2}$) as,

$$[B_{gr}] = B_{gr\_x}\ \mathbf{i} + B_{gr\_y}\ \mathbf{j} \tag{11.25}$$

$$[K_{gr}] = K_{gr\_x}\ \mathbf{i} + K_{gr\_y}\ \mathbf{j} \tag{11.26}$$

The resultant force per unit area of the gripper, $[B\text{-}K]_{gr}$, having its dimension as $ML^{-1}\ T^{-2}$ (=> Force/Area: MLT^{-2}/L^2), will be computed approximately as:

$$[B - K]_{gr} = \left[B_{grx}{}^2 + \left(\frac{K_{grx}}{2\pi R_{go}} \right)^2 \right]^{1/2} \hat{i} + \left[B_{gry}{}^2 + \left(\frac{K_{gry}}{2\pi R_{go}} \right)^2 \right]^{1/2} \hat{j} \tag{11.27}$$

The above expression for $[B\text{-}K]_{gr}$ is slightly non-optimal in the sense that we have approximated the circumference of the gripper 'adhesive-zone' as a smooth circular zone of radius R_{go}. But since this 'adhesive-zone' is an exact replica of the object being grasped, we need to fine-tune this expression in the final calculation of the grasping force per unit length of the gripper. The number of 'adhesive-zones' in the gripper-object contact plane will be instrumental in this recast calculation. The 'circumference' of the ensemble 'adhesive-zone' will not be smooth circular, but, a cumulative factor.

We will now go through the engineering metrics of the 'adhesive-zones' in order to proceed further with the grasp model, AMGM. Fundamentally, the concept of 'adhesive-zones' will occur in cases of grasping amorphous and/or soft tiny objects, such as cotton-ball, wool, small garments etc. Besides, we will come across the concept of 'adhesive-zones' in situations of grasping multiply tiny/micro-objects inside soft encapsulation, such as safety-pins, clips, nails, buds, threads, needles in isolation/clustered form inside poythene/plastic wrappers. The tuple of 'adhesive-zones' will be viable also for micro-objects encapsulated in gel-like covering/ wrapper. Additionally, the principle of 'adhesive-zone' will be effective for tiny substances covered by soft epoxy amorphous outer casing. As a matter of fact, today's manufacturing arena is booming with efficient and accurate handing of such critical components under soft outer coverings. These activities in industry come under the purview of robotized gripping, by and large, governed by a fine-tuned grasp modeling and control algorithm/gripper program. Our conceptualization of *Adhesive Mechanics* will be instrumental in analyzing grasp phenomena of such components/cluster of objects/consortia of micro-substances. Figure 11.21(a) schematically illustrates the disposition of multiple adhesive-zones that will be formed over the surface of a soft tiny object/amorphous substance/cluster of micro-objects inside soft encapsulation/consortia of micro-substances in a permeable capsule.

Some representative photographic views of these objects are shown in figure 11.21(b) for ready reference.

It is interesting to note the varieties of graspable objects that follow the principle of 'adhesive-zone' during grasp, as illustrated through 11 representative views under three groups, viz. (a) amorphous and soft objects; (b) cluster of tiny objects in soft encapsulation; and (c) consortia of micro-substances in permeable capsule. Multiple adhesive-zones will be formed during grasp of 'amorphous and soft objects (# 1 to 4 of figure 11.21(b)) irrespective of the form of gripping. In contrast, defined number of adhesive-zones will be created during grasp of soft-encapsulated objects (# 5 to 9 of figure 11.21(b)) and consortium of objects under permeable covering (# 10 and 11 of figure 11.21(b)). However, exact number of adhesive-zones will be dependent on the type, dimension and natural coherence of the objects. At times, it will also vary depending on the nature of encapsulation.

Considering number of 'adhesive-zones' in the gripper as N_{gr}, the analytical expression for the circumference of the 'gripper-object contact plane' C_{gr} will be (refer to the geometry of figure 11.21(a)):

$$[C]_{gr} \approx 2N_{gr}. \sqrt{\left[R_{go} - R_{gi}\cos\left(\frac{\pi}{N_{gr}}\right)\right]^2 + \left[R_{gi}\sin\left(\frac{\pi}{N_{gr}}\right)\right]^2} \tag{11.28}$$

Accordingly, we get the final expression for the grasping force of the gripper per unit length as,

$$\begin{aligned}[F_{gr}] = [B - K]_{gr}.\ [d_{oj\_gri}] = &\left[(r_{gr}\delta\theta_{gri}^2 - r_c\delta\theta_{ci}^2)\left\{\left(B_{gr\_x}^2 + \frac{K_{gr\_x}^2}{C_{gr}^2}\right)^{1/2}\right\}\right] \\ &\hat{i} + \left[(r_{oj}\delta\theta_{oji} - r_{gr}\delta\theta_{gri})\left\{\left(B_{gr\_y}^2 + \frac{K_{gr\_y}^2}{C_{gr}^2}\right)^{1/2}\right\}\right]\hat{j}\end{aligned} \tag{11.29}$$

Hence, the ensemble grasping force for the entire gripper will be:

$$[F_{gr}]^{\text{Total}} = [F_{gr}].\ 2\,\pi.\ R_{gm} \tag{11.30}$$

The mean radius of the 'adhesive-zone' of the gripper is the measured for the developed circumference of the 'adhesive-zone' and accordingly, the ensemble grasping force can be computed quickly using R_{gm} as input.

It is to be noted here that being very small in size, the evaluation of F_{gr} needs to be further substantiated by the 'part-mating model', as applicable for the 'adhesive-zone'. Accordingly, we will be able to find analytical expression for the X, Y and Z components of F_{gr}, namely, $[F_{gr}]^X$, $[F_{gr}]^Y$ and $[F_{gr}]^Z$ using the parlances of the said 'part-mating model'.

11.4.3.3.2 Adhesive mechanics-based model: Taylor series-based approach

Let us consider the geometry of the 'adhesive-zone' again and have a close look over 'S_{oj}' and 'S_{gr}' (refer to figures 11.20 and 11.21(a)). Now, from the geometry we can expand Δr_{oji} in the form of Taylor series as:

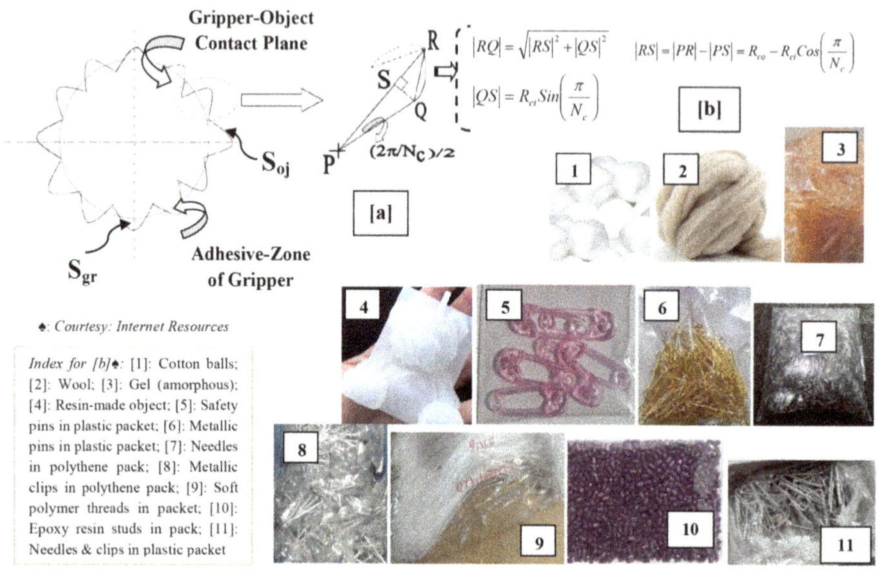

Figure 11.21. Disposition of multiple adhesive-zones during grasp: (a) schematic; (b) real-life view of the graspable objects.

$$\begin{aligned}\Delta\, r_{oji} &= r_{oji}\,(s + \delta s) - r_{oj}\,(s) = [r_{oj}(s)]_i - roj\,(\text{s}) \equiv [r_{oj} * (s)]_i - r * (\text{s}) \\ &= [r_{oj}(s) + \delta s.\; dr_{oj}/ds + (\delta s)^2/2!d^2r_{oj}/ds^2 + \ldots\ldots\ldots\ldots.]_i - r_{oj}\,(s)\end{aligned} \tag{11.31}$$

On simplification, the above equation yields,

$$\Delta r_{oji} = \delta s\; .dr_{oj}/ds + d^2r_{oj}/ds^2(\delta s)^2/2 \tag{11.32}$$

Now, from the geometry of the ‘adhesive-zone’, we have,

$$\begin{aligned}\Delta r_{gri} &= r_{gri}\,(\text{s} + \delta s) - r_{gr}\,(s) = [r_{gr}(s)]_i - r_{gr}\,(s) \equiv [r_{gr} * (s)]_i - r * \;(s) \\ &= [r_{gr}(s) + \delta s.\; dr_{gr}/ds + (\delta s)^2/2!d^2r_{gr}/ds^2 + \ldots\ldots\ldots\ldots.]_i - r_{gr}\,(s)\end{aligned} \tag{11.33}$$

On simplification, the above equation yields,

$$\Delta r_{gri} = \delta s\; .dr_{gr}/ds + d^2r_{gr}/ds^2(\delta s)^2/2 \tag{11.34}$$

An approximation for parametric modeling for the curve pattern of the ‘adhesive-zone’ yields following equations:

$$S_{oj} = k_{oj} \sin \upsilon + l_{oj} \cos \upsilon \tag{11.35a}$$

$$\text{S}_{\text{gr}} = \text{k}_{\text{g}}\text{r Sin}\;\upsilon + \text{l}_{\text{gr}}\;\text{Cos}\;\upsilon \tag{11.35b}$$

where, 'υ' denotes the angle of twist, measured between the tangents of the inner and outer curves and $\{k_{oj}, l_{oj}, k_{gr}, l_{gr}\}$ are numeric parameters ranging between 0 and 1, i.e. $\{k_{oj}, l_{oj}, k_{gr}, l_{gr}\} \in (0,1)$.

It is to be noted here that the geometric equation for the boundary curve will be identical for both object and gripper ('adhesive-zone'). As we are modeling the contact mating phenomenon, previous equations will hold true for either of the mating members.

Differentiating equations (11.35a) and (11.35b) we get,

$$dS_{oj} / d\,\upsilon = k_{oj} \cos \upsilon - l_{oj} \sin \upsilon \quad (11.36a)$$

$$dS_{gr} / d\,\upsilon = k_{gr} \cos \upsilon - l_{gr} \sin \upsilon \quad (11.36b)$$

Thus, on simplification, the algebraic expression for 'ds' under the respective cases of object and gripper ('adhesive-zone') can be approximated as,

$$ds_{oj} = [k_{oj}(1 - \upsilon^2/2! + \upsilon^4/4!) - l_{oj}\,(\upsilon - \upsilon^3/3!)]\, d\upsilon \quad (11.37)$$

$$ds_{gr} = [k_{gr}\,(1 - \upsilon^2/2! + \upsilon^4/4!) - l_{gr}\,(\upsilon - \upsilon^3/3!)]\, d\upsilon \quad (11.38)$$

Thus, we get the final expressions for differential translation of the object and gripper adhesive-zone as,

$$\Delta r_{oji} = \delta s_{oj}.\, dr_{oj}/ds + d^2 r_{oj}/ds^2 (\delta s_{oj})^2/2 \quad (11.39)$$

$$\Delta r_{gri} = \delta s_{gr}\,.dr_{gr}/ds + d^2 r_{gr}/ds^2 (\delta s_{gr})^2/2 \quad (11.40)$$

And, we get the relative translation of the gripper adhesive-zone with respect to the object as,

$$\begin{aligned} d_{ojgi} &= \Delta r_{oji} - \Delta r_{gri} = \\ &[\delta s_{oj} dr_{oj}/ds - ds_{gr} dr_{gr}/ds] + \left[(\delta s_{oj})^2/2 d^2 r_{oj}/ds^2 - (\delta s_{gr})^2/2.\ d^2 r_{gr}/ds^2\right] \end{aligned} \quad (11.41)$$

Considering the modulus values of the force concentration vector of the gripper $[B_{gr}]$ in 2-D planar matrix as, $[B_{gr}] = [B_{gr\_x}{}^2 + B_{gr\_y}{}^2]^{1/2}$ and that of the stiffness vector $[K_{gr}]$ in 2-D planar matrix as, $[K_{gr}] = [\{K_{gr\_x}{}^2/C_{gr}{}^2\} + \{K_{gr\_y}{}^2/C_{gr}{}^2\}]^{1/2}$, we get the final expression for the grasping force per unit length as,

$$\begin{aligned} &[F_{gr}] = [\mathrm{B} - \mathrm{K}]_{gr}.\ [\mathrm{d}_{ojgri}] = \\ &\left[\left(B_{grx} + \frac{K_{grx}}{C_{gr}}\right)^2 + \left(B_{gry} + \frac{K_{gry}}{C_{gr}}\right)^2\right]^{1/2} \times \\ &\left[\delta S_{oj}.\ \frac{dr_{oj}}{ds} - \delta S_{gr}.\ \frac{dr_{gr}}{ds}\right] + \left[\frac{(\delta S_{oj})^2}{2}.\ \frac{d^2 r_{oj}}{ds^2} - \frac{(\delta S_{gr})^2}{2}.\ \frac{d^2 r_{gr}}{ds^2}\right] \end{aligned} \quad (11.42)$$

Hence, the ensemble grasping force for the entire 'adhesive-zone' of the gripper will be:

$$[F_{gr}]^{\mathrm{Total}} = \left[\left(B_{gr\_x} + \frac{K_{gr\_x}}{C_{gr}}\right)^2 + \left(B_{gr\_y} + \frac{K_{gr\_y}}{C_{gr}}\right)^2\right]^{1/2} \times$$
$$\left\{\int_{s=0}^{s=2\pi R_o}\left[\delta S_{oj}.\ \frac{dr_{oj}}{ds} + \frac{(\delta S_{oj})^2}{2}.\ \frac{d^2 r_{oj}}{ds^2}\right]ds\right. \tag{11.43}$$
$$\left. - \int_{s=0}^{s=2\pi R_m}\left[\delta S_{gr}.\ \frac{dr_{gr}}{ds} + \frac{(\delta S_{gr})^2}{2}.\ \frac{d^2 r_{gr}}{ds^2}\right]ds\right\}$$

Equation (11.43) uses two different limits of integration, while estimating the ensemble grasping force of the gripper adhesive-zone. Unlike the earlier case of estimating the ensemble grasping force using Vector Signature-based approach, in this case we need to integrate the infinitesimal curvilinear contact zones of the object and adhesive-zone of the gripper in a meaningful manner. Accordingly mathematical entities have been arranged in two groups: one for the object and the other for gripper adhesive-zone. Although object and 'adhesive-zone' are of identical design, the limits of integration are not the same. While the differential segment, '*ds*' varies from zero to the outer circumference of the object ($2\pi R_o$), the limits for the differential segment for gripper adhesive-zone varies from zero to the 'mean' circumference of the object ($2\pi R_m$), in order to portray minor variations in actual dimensions due to fabrication.

11.4.4 Facets on real-time dynamics of grasp model

We will focus now on another pertinent aspect of grasp model of CRG, viz. the real-time dynamics. The phenomena of Real-Time Dynamics (RTD) are governed by two paradigms of CRGs: (i) *in situ* vibration and (ii) turning. Ideologically, these two paradigms are independent of each other and both are self-sustaining by nature. Nonetheless, we do observe some level of coupled behaviour between these two aspects at times during real-time operation of the CRG. The coupled interaction between *in situ* vibration and turning of the CRG can be studied through FEA. The common thread of analysis of these two facets of RTD is the study on the '*spring members*' of the CRG-system. As a matter of fact, these imaginary spring members are the nucleus of estimating the deflection hue of the CRG-assembly. The parenthesis of spring-structural model of the CRG is thus crucial for the RTD and overall vibration-free grasp of the CRG. The details of this spring-induced modeling will be delineated in the next section.

Fundamentally, we need to perform the following three metrics in order to have a successful RTD-based grasp model of the CRG, namely: (A) study of *in situ* real-time deflection of the CRG and associated robot link(s); (B) study of turning phenomena of the CRG, due to the fall-out effect of system tremble/jerk/ vibration; and (C) sensitivity analysis of certain key parameters pertaining to the CRG. It may be mentioned here that a significant domain of CRG, especially micro-grippers, are prone to *in situ* deflection that cannot be eliminated from the control loop. We will initiate our discussion with an aim towards understanding the significance of the governing design variables that are instrumental for the *in situ* deflection of the constituent members of the CRG and finally the ensemble, as it is the common thread to all three facets stated above. The *in situ* vibration of the CRG-assembly is truly wholesome with each of the constituent members participating in the process. The final modeling for RTD will ensure accounting of component-level deflection, based on the type, size, cross-section and material of the said component of the CRG-assembly. The most useful physically realizable parameter for deflection hue is the angle of deflection with respect to the base-axis and/or central axis of the component. In fact, the angle of deflection will become an important element of the analytical model of RTD of CRGs. The other salient metric for the evaluation of deflection hue of CRG is the change of cross-section of the constituent member and of course, the length of the member. It is true that by and large, links of the CRG are the members that will undergo deflection and hence it is customary to articulate the design of CRG-links in an optimal manner. FEA is the pathway for such optimization of the CRG-links based on the basic design. Mass parameters and Moment of Inertia of CRG-members/links are two important attributes in the development of an RTD-based grasp model of a prototype CRG. The payload is also crucial in evaluating the overall deflection of the CRG-assembly and in the majority of the cases, we consider the wide range of payloads as 'lumped end-mass' in order to ease out the complexity of the RTD-model. With this backdrop, figure 11.22 schematically illustrates the deflection hue of five representative 'compliant' members of the CRG-assembly, having varying cross-sections and design elements (may not occur simultaneously in a single hardware).

It may be stated here that alongwith payload, damping of the ensemble CRG-system also comes into play and the same becomes instrumental in deciding the attenuation of vibration within a finite time-range. Besides, spring stiffness factor influences the ensemble dynamic model of the CRG-system, which, in turn, is dependent upon the deflection hue of the CRG-members/links.

With this preamble, we will postulate the fundamental equation of control dynamics of the CRG, using measurable parameters. As illustrated in figure 11.22, we will treat the CRG-assembly as an *open-end cantilever structure*, thereby imbibing spring-effects at its links and pin-joints. The ensemble deformation at the CRG-joints will be accounted for as friction torque while stiffness will be estimated through viscous damping of the joints.

If we take a closer look over the traditional dynamic equation of a rotational body involving inertia, damping, stiffness effect and finally torque, then the same can be expanded for CRG too as shown below, which is the fundamental equation of real-time control of any robotic system per se:

$$M.\ \frac{d^2\varphi}{dt^2} + B.\ \frac{d\varphi}{dt} + K.\ (\Delta\varphi) = \Delta\tau \qquad (11.44a)$$

Equation (11.44a) is self-explanatory wherein the parameters, 'M', 'B', 'K', 'ϕ' and 'τ' have their natural nomenclatures. For instance, 'M' is the mass moment/inertia of the system; 'B' is the viscous damping; 'K' is the spring constant of the naturally vibrating system; 'ϕ' is the joint angle movement and 'τ' is the torque function. Now, let us take a deeper understanding of this equation and we will observe that it is essentially composed of two broad groups, viz. (a) material-specific

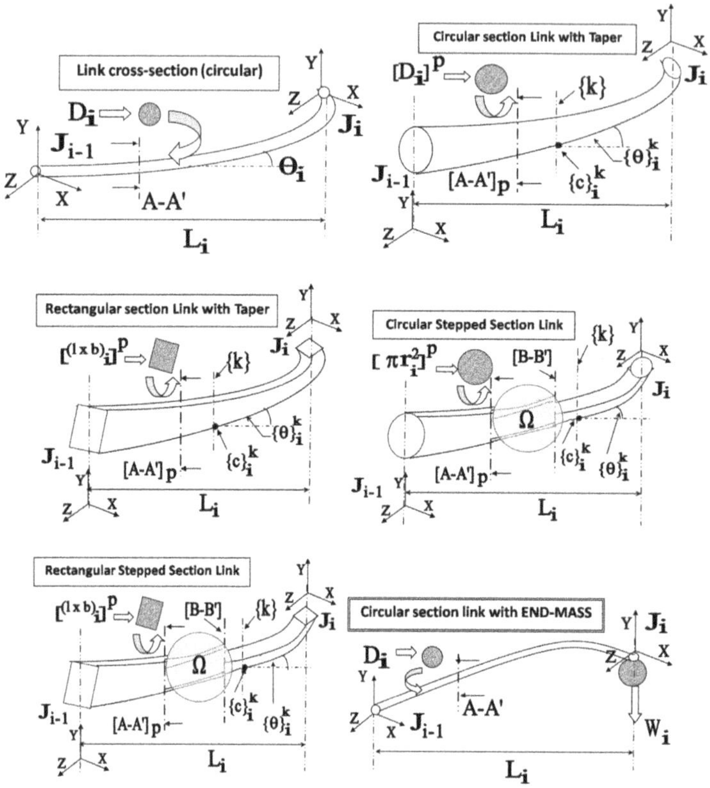

Index: $\mathbf{L_i}$: Length of the ith CRG-member / link; $\mathbf{D_i}$: Diameter of the ith CRG-member / link having circular cross-section; $\{\mathbf{J_{i-1}}, \mathbf{J_i}\}$: End-joints of the ith CRG-member / link; $\boldsymbol{\theta_i}$: Angle of deflection of the ith CRG-member / link; **[A-A']**: Intermediate section-plane for CRG-links with uniform cross-section; $[\mathbf{D_i}]^p$: Diameter of the ith CRG-member / link having circular cross-section with taper at pth location; **{k}**: Imaginary plane of reference for the evalution of angle of deflection ;$\{\boldsymbol{\theta_i}\}^k$: Angle of deflection of the ith CRG-member / link with varying cross-section at kth location; $\{\mathbf{c_i}\}^k$: Deflected point of reference of the ith CRG-member / link with varying cross-section at kth location; $[\mathbf{A\text{-}A'}]^p$: Intermediate section-plane at pth location of the CRG-link; **[B-B']**: Intermediate section-plane for CRG-links with stepped cross-section; $[(\mathbf{l} \times \mathbf{b})_i]^p$: Dimensions of the rectangular cross-section of the ith CRG-member / link with taper / stepped section at pth location; $[\pi r_i^2]^p$: Dimensions of the circular cross-section of the ith CRG-member / link with stepped section at pth location; $\mathbf{W_i}$: Payload at the ith CRG-member / link; **(X,Y.Z}**: Cartesian co-ordinate system reference frame; $\boldsymbol{\Omega}$: Enlarged view of the stepped cross-section zone of the CRG-member / link.

Figure 11.22. Schematics of the deflection hue of five representative compliant members of the CRG.

equation and (b) torque-specific equation. We can appreciate that equation (11.44a) is in the format of a generalized dynamic equation of motion, namely,

$$M.\,\ddot{q} + B.\,\dot{q} + K.\,q = \tau \tag{11.44b}$$

where, '*q*' signifies the joint variable in generalized co-ordinates. While the parameters pertaining to the material-specific equation move around the matrices of '*M*', '*B*' and '*K*', the torque-specific equation gives weightage to assimilation of various kinds of torque components appearing in the CRG. We will take an in-depth look into the expanded form of control dynamics equation of the CRG, by amalgamating equations (11.44a) and (11.44b) and inducing a few measurable parameters in subsequent sections.

11.5 Vibration signature of the compliant robotic grippers and compliant wrists

11.5.1 Paradigms of vibration signature

The inherent vibration of CRG and counter-vibration/compression of the CRW are totally built-in type, structure independent and get manifested in two ways, namely: Modal Frequency and Eigen Value. It has been also observed that vibration in CRG is not time-dependent and the duration and periodicity of it cannot be correlated with the task-space of the robotic system. Moreover, this vibration is self-propagating by nature and it gets induced to the successive member of the CRG till the jaw-plate of the gripper. The problem gets more complicated when we attempt for multi-link design of the CRG, wherein various kinds of coupled effect and non-linearity creep in. The tuple of stiffness coefficient and viscous damping of the spring-dashpot-damper model have been expounded in Taylor's series for functional evaluation of the natural frequency of vibration of the CRG and subsequently, modal frequencies. Irrespective of all hardware-types, this inherent vibration of CRGs gets realized in the form of mild to severe trembling of the slender links and/or shaking or twisting of the link-joint interface and inter-spaced joints. Slenderness of the gripper-links plays a very important role in self-generation of this trembling.

Due to all these characteristics, it is very difficult to obtain a generic analytical model for the vibration in CRG. Usual rule-bases for adopting control algorithm for the end applications are also unviable. The asynchronous data fusion, emanating out of CRG is a challenging research paradigm to date, primarily due to the inherent characteristics in quantifying the output response of the system. Real-time assessment of vibration signature in CRG is a pre-requisite for establishing a reliable control system for any real-life application. The vibration signature in CRG gets assessed in real-time through multiple force sensors, spread over its links, body and joints. The formulation, modeling and instrumentation for such deflection measurement are pre-characterized and the same has been achieved with the help of miniature strain gauges and flexi-force sensors, through excitation data of these sensors in real-time.

It is true that under natural condition of joint rotations of the ensemble robotic system, consisting of CRW and CRG, there will be built-in oscillations in the arms of the manipulator as well as end-unit, viz. CRG plus CRW. These oscillations and ensemble trembling of the manipulator arms are not coupled all the time and due to this, we get different deflection hues for the manipulator. There are various pathways to study the patterns of deflection hues and subsequent oscillations of the CRG and CRW, which get translated to robust analytical models for evaluating vibration-related parameters under real-time dynamics. Nonetheless, in its simplest form, we will consider here undamped free vibration of the CRG, without any external forcing. This free vibration model has been selected in order to study the natural pattern of deflection/trembling of the CRG due to its own inherent dynamics, namely, rotation of its links via the servomotor system. This natural oscillation paradigm of CRG is also supplemented with two facets, viz. (a) no external forcing on any of the CRG-member and/or CRW in any form (i.e. condition of 'forced vibration' is excluded) and (b) no viscous damping from exterior of the CRG and/or CRW (i.e. condition of 'viscous damping' is excluded). Our study, experimental investigation and finally, numerical assessment of natural frequency of vibration of the CRG–CRW tuple is based on this basic paradigm of undamped free natural vibration.

11.5.2 Development of spring-induced geometric models

It is customary to adhere to a traditional 'spring model' to designate free vibration of the CRG-members and/or CRW-system. However, criticalities in vibration signature will eventually emerge due to several natural causes, thereby producing varying amplitude-frequency of the induced vibration in different members of the CRG. This is a very obvious process and it is system-generated. However, the problem of induced vibration is comparatively less in CRW, because of its end-connections. Nonetheless, the challenge lies with accurate and effective modeling of this varying vibration using the spring model. We will discuss here the intricacies of the vibration model using the novel pathway of designing various innovative layouts for these spring elements. It may be noted that the said layout schemes for the spring elements are segregated on the basis of modeling for the CRG or CRW, namely: (i) individual component-level; (ii) sub-assembly level: concept of equivalent spring; and (iii) system assembly-level. These three ordered levels of modeling will be detailed in the following three sub-sections, respectively.

11.5.2.1 Layout schemes for the spring elements

The physical parameter of a spring that goes instrumental in characterizing its performance, i.e. its elongation and subsequently natural pathway of oscillation in vertical plane, is nothing but its stiffness matrix. In uni-axial deformation model of the spring, this stiffness matrix is analogous to spring constant '*K*', having dimension: MT^{-2} (force per unit length). As all other parameters of the CRG-members and that of CRW-system are essentially material-specific or design-specific, it is the value of '*K*' that controls the numerical value of the natural frequency of vibration. However, the value of '*K*' is the effective value, which is to be obtained through various combinations of placement of the spring(s) in the model. Further, the value of '*K*' may not be equal in all such combinations; and, we may use non-identical springs with different values of the spring constant. In a nutshell, the design of the spring elements is a vital aspect of the design ensemble of the CRG as well as CRW. In fact, there can be various design-architectures of the spring elements/members that are feasible from the point-of-view of manufacturing. These are essentially out-of-the-box solutions for the spring layout of CRG and CRW, which will be mounted directly over the link/member of the CRG and/or CRW in the model. Figure 11.23 schematically demonstrates ten novel layouts for the spring elements that can be used as feasible and realizable options to model *in situ* vibration in the specific members of CRG or CRW. The physical disposition of the representative member of CRG or CRW may be noted along with the interfacing of pair(s) of strain gauges, as part of sensory instrumentation.

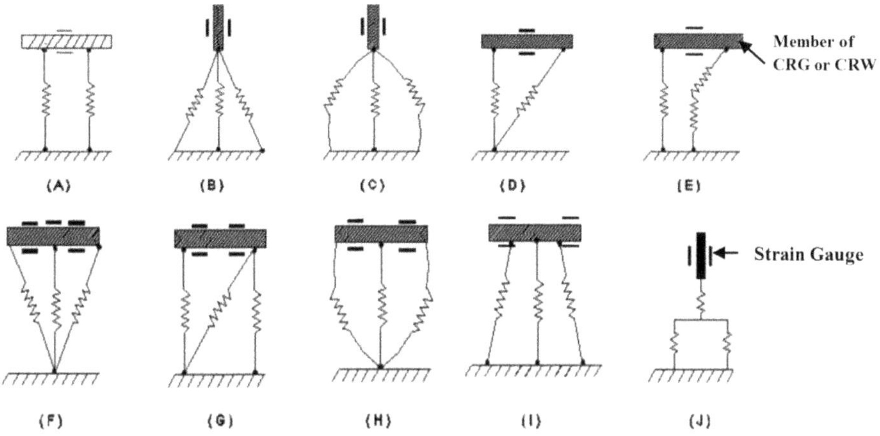

Figure 11.23. Schematics of feasible layouts of spring elements symbolizing *in situ* vibration of CRG-member/CRW.

Two types of design layouts of springs are emerging so far as the original disposition of the member of CRG or CRW is concerned, as per the schematic of figure 11.23 above. These are: (a) *Type-I:* horizontal member: spring-mounted (design schemes: 'A', 'D', 'E', 'F', 'G', 'H', 'I') and (b) *Type-II:* vertical member: spring-mounted (design schemes: 'B', 'C', 'J'). It is to be noted that irrespective of the type of layout, spring-members are acting in one direction only in order to characterize the vibration pathway. In other words, spring rheology (tension and/or compression) is based on uni-directional arrangement. It is to be noted here that design options for vertical/column members are limited, as the layout of the springs will not undergo major variations. However, horizontal members of CRG or CRW will have a number of variations possible, owing to the layout of the spring member (s). Another interesting feature of this spring mounted modeling structure is augmentation of a couple of 'branches' (just like 'tree branches') of the different spring elements and/or spring- dashpot system, in extended layout. The fundamental aspect of the disposition of the spring metric for horizontal members is essentially multi-spring type, which is the crux of the spanning layouts (refer to 'A', 'D', 'E', 'F', 'G', 'H' & 'I' of figure 11.23). As a matter of fact, multiple springs give rise to options for enhanced compliance in the overall structural geometry of CRG or CRW. Moreover, with various permutations of spring-layout, flexibility of modeling of the overall CRG/CRW-structure gets boosted. Nonetheless, individuality of the spring-elements, irrespective of the members (horizontal or vertical) thereof is always maintained. It may be noted here that all spring-elements in the layouts of figure 11.23 are straight and not bonded with any fellow spring. In contrast to this, we will now deal with another layout, wherein disposition of the spring-elements can be functionally bonded and/or crisscrossed with fellow spring-element(s). Figure 11.24 schematically illustrates these layouts.

It may be noted that crisscrossed layout of the spring elements is a unique call of design, wherein two spring-members are 'crossing' each other to form an ensemble.

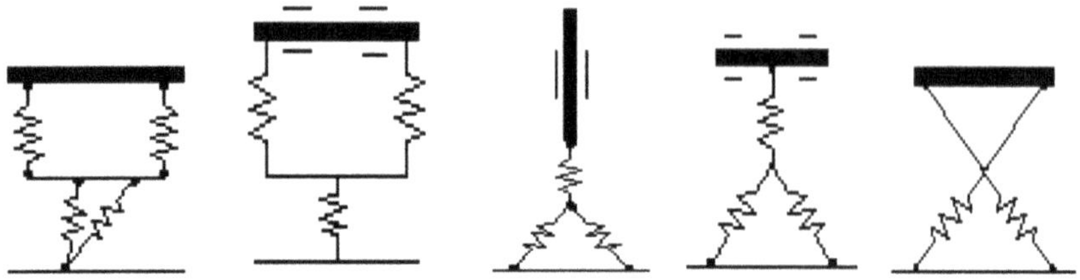

Figure 11.24. Schematics of functionally bonded spring elements under *in situ* vibration of CRG-member/CRW.

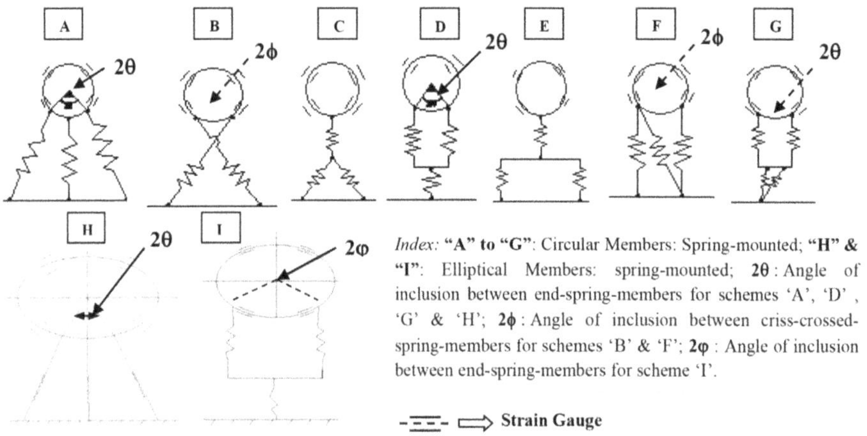

Figure 11.25. Disposition schematics of circular and elliptical spring elements under *in situ* vibration of CRG-member/CRW.

These layouts are entrusted to provide more subtle input to the characterization of real-time damping of the CRG-member as well as CRW. Let us now investigate two other types of design layouts under *in situ* vibration-based layout, viz. 'circular member: spring-mounted' and 'elliptical member: spring mounted', as illustrated schematically in figure 11.25.

The layouts of the spring elements, shown in figures 11.23 to 11.25, are the essential pre-requisites for the subsequent modeling. Mathematical paradigms of the layouts of these spring elements need to be ascertained *a priori* to capture the characteristics of the free undamped vibration of the CRG-members and/or CRW-system.

11.5.2.2 Schematic dispositions of equivalent spring element

The spring-based geometric layouts of an individual member of CRG and/or CRW with strain gauge instrumentation, as explained in the previous sub-section need system-level culmination so as to subsume in the vibration model for the robotic grasp. Accordingly, ensemble vibration (for CRG) and damping/counter-vibration (for CRW) have to be modulated through the concept of *Equivalent Spring Element* (ESE). The characteristics of ESE will be wholesome and intrinsic so as to represent the ensemble CRG or CRW effectively under run-time condition. It is thus noteworthy to highlight various possible combinations of ESE, which can be conceptualized as extended manifestation of series and parallel layout. Figure 11.26 illustrates schematic disposition of the undamped free natural vibration scheme of CRG or CRW in sub-assembly mode, along with ten possible layouts of ESE thereof (refer to # I to X). It may be stated that standard formulae of the series-parallel combinations will be used to compute ensemble spring constant, considering equal values of '*K*' in all these 10 layouts. Numerical treatment will remain the same in cases where multiple different springs with unequal values of '*K*', say K_1, K_2, K_3, …etc are used in the model.

We know that numerically, natural frequency of vibration is a function of '*K*' and 'mass' of the vibrating member, as per the postulation of natural vibration. As there is no alteration in the 'mass' ('*m*') of the vibrating member, sole dependency of natural frequency of vibration (f_n) will be on the 'equivalent spring constant' (K_{eq}) of the CRG-member or CRW-link system. This proposition justifies the lemma of figure 11.26 with its 10 different models of spring layout. This direct proportional relationship between 'f_n' and 'K_{eq}' is the crux of the engineering aspect of the study of the undamped free vibration of CRW-link(s) and CRG-members. Thus, it essentially boils down to the near-optimal numerical evaluation of 'K_{eq}' using the known parameters of CRG or CRW.

It is interesting to observe the dispositions of the spring-elements in the layout of figure 11.26 and to be specific, the gradual variation of complexity of the layouts. By default, the first layout is the simplest with only one spring-element, having spring-constant '*K*'. The next two schemata, viz. layouts #2 and 3 represent series and parallel combination of spring-elements in simplest form. Layouts #4 and 5 involve

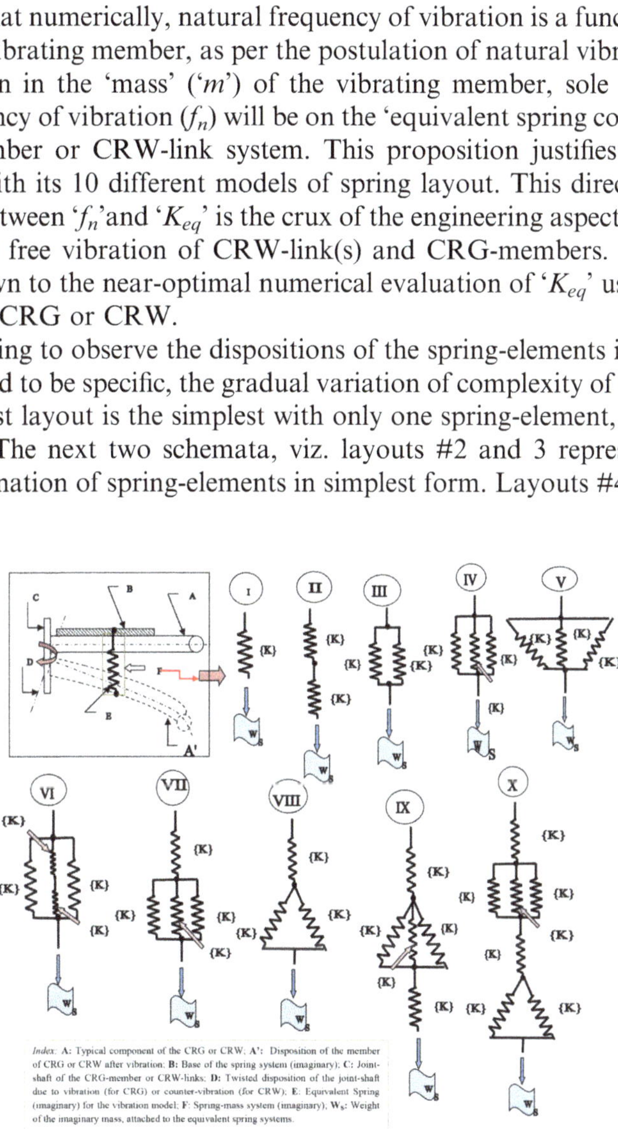

Figure 11.26. Schematic disposition of undamped free natural vibration schemes of equivalent spring element.

three spring-elements each, disposed in two geometrical variants. Two extended versions of layout #4 are depicted in the next two schemata, namely, layouts #6 and 7. The complexity of the computation of 'K_{eq}' begins from layout #6 and continues thereafter. We introduce new sub-schemata of '*star-delta connection*' of spring-elements in the next three layouts, viz. layouts #8, 9 and 10. While layout #8 shows the basic star-delta type spring-elements, complexity gets introduced in layout #9 with additional spring-elements, adjoining the basic star-delta. The last layout is a Boolean combination of layouts #7 and 8 and it is, by virtue, the representative of all such possible Boolean combinations of layouts of the spring-elements in CRW-link or CRG-members.

11.5.2.3 Development of the spring—structural model

After comprehending the component-level spring layouts followed by layouts for the equivalent spring element, we will move on to detailing of the system assembly-level model of CRW or CRG-system. This modeling is essentially customized type, entitled, 'Spring Structural Model' (SSM). The root of the customized SSM is related to the development of a 'spring structure' for the ensemble, i.e. CRG or CRW, divided into specific numbers of functional sub-assemblies, decided *a priori*. These functional sub-assemblies are created as virtual entities with ESE, with a goal towards efficient run-time modeling of the control system of CRG or CRW. It is important to have geometric symmetry of the design of the CRG in order to construct the SSM. By virtue of the customized design of two-fingered type CRGs, all prototypes of our CRGs are symmetrical and thus it is relatively easier to construct an SSM for those designs. Nonetheless, the technique can be adopted for asymmetric designs of CRGs as well with proper selection of ESEs. In either of the cases, dynamics-based governing control system equation of the CRG will subsume the effect of each ESE separately with the input of torque-components.

We will illustrate the customized SSM for two representative CRGs, vide PAMRoG (discussed in sub-section 11.2.3.3) and SIMSJRoG (discussed in sub-section 11.2.5.2), each with three functional sub-assemblies with ESEs. Figure 11.27 shows the disposition of SSM for PAMRoG and the same for SIMSJRoG is

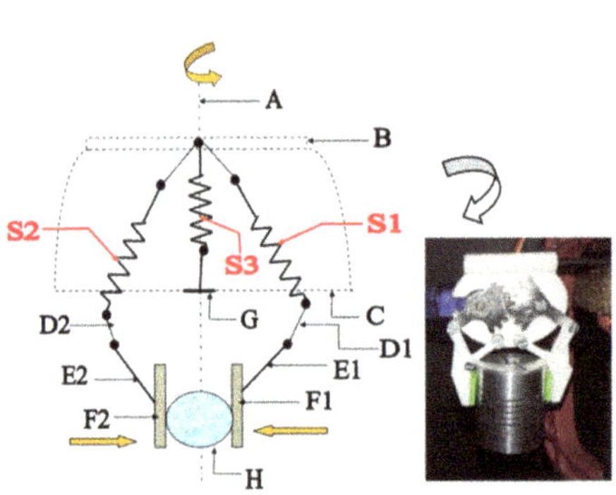

Index: **A:** Central axis of the CRG (placement of servo-motor); **B:** Adapter plate; **C:** Base plate assembly.; **D1:** Upper link (right-hand-side) assembly.; **D2:** Upper link (left-hand-side) assembly.; **E1:** Lower link (right-hand-side) assembly.; **E2:** Lower link (left-hand-side) assembly.; **F1:** Jaw plate (right-hand-side); **F2:** Jaw plate (left-hand-side); **G:** Imaginary grouting (of spring); **H:** Payload; **{S1, S2, S3}:** Imaginary spring systems -ESEs

Figure 11.27. Spring-structural model of the representative compliant robotic gripper assembly (PAMRoG).

presented in figure 11.28. The prototypes of the respective CRGs are also included in the figures for easy resemblance of the model with the actual.

As evident through the illustrated view of SSM in figure 11.27, the backbone of the SSM consists of three ESEs, located symmetrically within the mechanical ensemble of PAMRoG. The form and force closures are ensured through the actuation of the ESEs, as shown in the photographic view of the grasp of a solid cylindrical object.

Although the backbone of SSM for SIMSJRoG consists of three ESEs, as shown in figure 11.28, the link sub-system is different from that of PAMRoG. The ESEs are located symmetrically within the mechanical ensemble of SIMSJRoG. In contrast to the SSM of PAMRoG, we will encounter a special swing motion ('J') of the jaw plates due to the customized design of the jaws. This motion is dependent on the modeling of two ESEs, viz. 'S1' and 'S2'. The link sub-system ('D') has greater significance here as direct transmission of motion in being made as an outcome of the excitations of the ESEs.

It may be stated here that the complementary aspect of the SSM is sensory instrumentation that works as a measured for real-life experimentation towards evaluation of the *in situ* vibration of the CRG-system. The real-life design model for the *in situ* vibration of the CRG involves requisite sensory sub-systems for the measurement of strain in the CRG-members. Hence, direct ramification of vibration model of CRG is the sensor-based SSM. Figure 11.29 schematically illustrates the design of a sensor-based SSM for the enhanced design model for CRG and/or CRW.

The basic architecture of the sensor-instrumented SSM of CRG or CRW-system comprises four distinct members, viz. 'A', 'B', C' and 'D', as illustrated in figure 11.29(a). We have shown two successive 'strain' members in figure 11.29(a) so as to highlight the coupling effect between constituent members of the CRG or

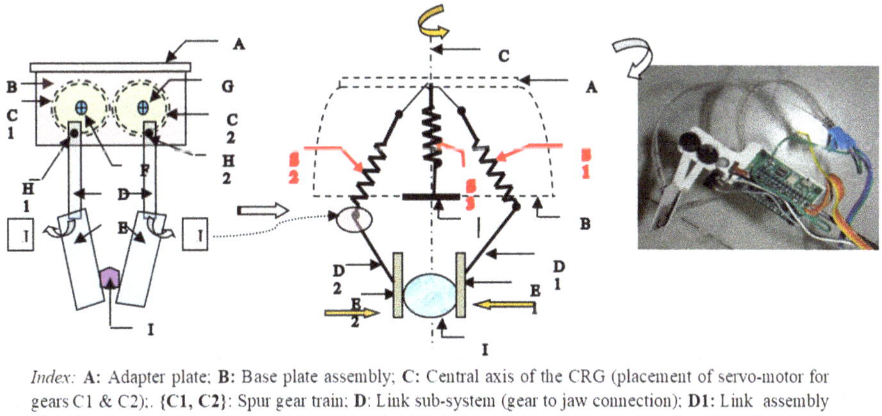

Index: **A:** Adapter plate; **B:** Base plate assembly; **C:** Central axis of the CRG (placement of servo-motor for gears C1 & C2);. **{C1, C2}**: Spur gear train; **D**: Link sub-system (gear to jaw connection); **D1:** Link assembly (right-hand-side); **D2:** Link assembly (left-hand-side); **E**: Jaw plates sub-system; **E1:** Jaw plate (right-hand-side); **E2:** Jaw plate (left-hand-side); **F**: Fixation of left-hand-side driving gear and Imaginary grouting (of spring); **G:** Fixation of driven gear; **H1**: Gear-link connecting pin (left-hand-side); **H2**: Gear-link connecting pin (left-hand-side); **I:** Payload; **J**: Swing motion of the jaw plates; **{S1, S2, S3}:** Imaginary spring systems -ESEs

Figure 11.28. Spring-structural model of the representative compliant robotic gripper assembly (SIMSJRoG).

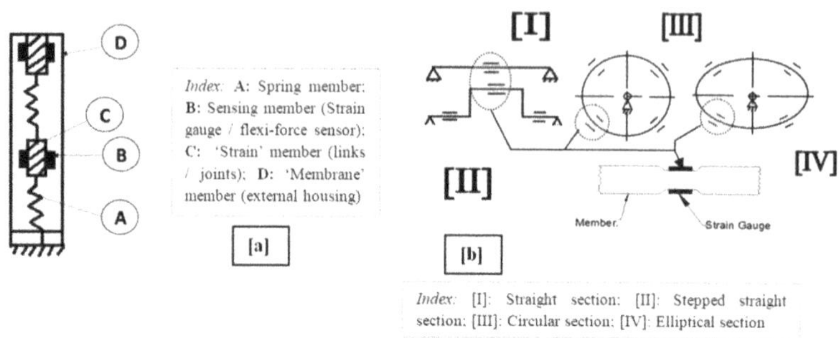

Figure 11.29. Sensor-instrumented spring-structural model of CRG or CRW: (a) basic architecture; (b) strain gauge layout.

CRW-system in actual hardware. It is needless to state that 'sensing' member is the most crucial element of the system and its modeling is dependent on the layout of the sensor-cells (e.g. strain gauges), as shown in figure 11.29(b). The paradigm of *in situ* vibration of members/links of CRG or CRW-system can be symbolized optimally as the real-time deflection of a hollow beam having four representative cross-sections with mounting of strain gauges, as shown in figure 11.29(b). The locations of placement of strain gauges have been marked by '=' symbol in four possible design variations (straight, stepped, circular, elliptical). The fulcrum location of the beam cross-section is denoted by 'Δ'. As illustrated in figure 11.29(b), the CRG/ CRW-member will have *micro-beam* and the force sensing mechanism will be based on beam deflection principle. A particular member may have multiple beams embedded in it, each having its own characterization. The placement of those beams inside the CRG/CRW-member is also another technological challenge. Thus, these as-modeled spring-mounted members will act as *in situ* vibration source for the CRG/CRW-system, which will be helpful in assessing the overall deflection of the system.

11.5.3 Development of spring-supported vibration model

We will now formulate a customized vibration model for the CRG and/or CRW-system using the concept of ESE. The prime-most attribute of this spring-supported vibration model is natural frequency of vibration. Numerical evaluation of natural frequency of vibration is traditional theory-based and for a specific material and known mass of CRG-member or CRW-link, it is directly proportional to the spring constant of ESE, i.e. 'K_{eq}', depending upon the layout schemata. Interestingly, 'K_{eq}' is also dependent on geometry of the CRG-member/CRW-link as well as its material. In a way we get a tuple to evaluate natural frequency of vibration, f_n, of the CRG/CRW-link system, as shown below:

$$f_n = \frac{1}{2\pi}\sqrt{\frac{K_{eq}}{m}} \tag{11.45}$$

$$K_{eq.} = \frac{Y.A}{L} = \frac{F}{\delta} \tag{11.46}$$

where, m: mass of the CRG-member or CRW-link; F: in-built force acting on the CRG-member to cause vibration; A: cross-sectional area of the CRG-member or CRW-link; L: length of the CRG-member or CRW-link; Y: Young's modulus of the material of the CRG-member or CRW-link; δ: deflection of the CRG-member or CRW-link (in the direction of in-built forcing function).

Thus, as per theory, the working formula for the numerical evaluation of 'f_n' becomes (using equations (11.45) and (11.46)),

$$\begin{aligned} f_n &= \frac{1}{2\pi}\sqrt{\frac{Y.A/L}{m}} = \frac{1}{2\pi}\sqrt{\left(\frac{Y.A}{L}\right)\left(\frac{4}{\rho.L.\pi.D^2}\right)} \\ &= \frac{1}{\pi^{3/2}L.D}\sqrt{\frac{Y.A}{\rho}} = \frac{1}{2\pi L}\sqrt{\frac{Y}{\rho}} \end{aligned} \tag{11.47}$$

where, D: diameter of the CRW-link or average diameter of the CRG-members and ρ: density of the material of the CRW-link or CRG-member. Other symbols have been nomenclated earlier in equations (11.45) and (11.46). As an off-shoot to the evaluation of f_n we can deduce the working formula for the 'force of vibration ('F'), which is essentially a theoretical and imaginary attribute. Since we are not considering the situation of forced vibration here we will only look at the expression for 'F' as a direct function of the average strain obtained (ε_{av}). in the CRW-link(s) or CRG-member(s), as shown below:

$$F = \frac{Y.A.\delta}{L} = Y.\,A.\,\{\varepsilon_{av}\} = \frac{\pi.D^2 Y\{\varepsilon_{av}\}}{4} \tag{11.48}$$

Experimental determination of the natural frequency of vibration of the CRW-link(s) and/or CRG-member(s) is crucial. We need to evaluate it with the help of mere strain gauge data. There is no other way out and besides strain gauge data (of dynamic strain), we cannot even arrange any other instrumentation in real-time for such assessment. This is the greatest irony of a CRG or CRW and we have to deal with it so far as estimation of *in situ* vibration/natural frequency is concerned. The experimental runs are essentially made with the help of activation of the joint servomotors of the CRG in a selected pattern. In that sense, the experimental runs are representative of pesudo-forced vibration of the CRG and counter-vibration for CRW. Although there is no external forcing function acting on the CRG-members, the vibration is being forced into the system by means of rotational motion of the joint motor of CRG and also the joint-motors of the robotic manipulator interfaced with.

To begin with, we will investigate the phenomena of pesudo-forced vibration of the CRG-member(s) under the excitation created by only one servomotor of the manipulator-joint. Here we assume that all joints of the robotic manipulator are identical in all respects hardware-wise and accordingly, all joint-servomotor systems are equally likely. We also have the pre-condition that during the activation of the specific joint-servomotor, the rest of the joint-motors are not actuated and thus, those remain in the then state of angular postures. With this proposition, the computational formula for natural frequency of vibration of the CRG or

CRW-system due to the excitation/rotation of the *k*th joint-motor of an 'N'-link serial-chain robotic manipulator under this state of pseudo-forced vibration can be deduced from equations (11.45) and (11.46), as described below:

$$f_n^{(k)} = \frac{1}{2\pi}\sqrt{\frac{K_{eq}}{M}} = \frac{1}{2\pi}\sqrt{\frac{\frac{F^{v}_{\text{link\_N}}}{\delta_k}}{M}}$$
$$= \frac{1}{2\pi}\sqrt{\frac{\frac{F^{v}_{\text{link\_N}}}{\varepsilon^{(k)}_{Av} \cdot L_{\text{CRG/CRW}}}}{\rho \cdot A_{\text{CRG/CRW}} \cdot L_{\text{CRG/CRW}}}} = \frac{1}{2\pi L}\sqrt{\frac{F_{\text{link\_N}}{}^{v}}{\varepsilon_{Av}{}^{(k)} . \rho . A}} \quad (11.49)$$
$$= \frac{1}{2\pi L}\sqrt{\frac{\frac{\tau^{(k)}_{\text{motor}}}{\sum_{i=k}^{i=N} L_i}}{\varepsilon_{Av}{}^{(k)} . \rho . A}}$$

where, $f_n^{(k)}$: natural frequency of vibration of the CRG or CRW-system due to excitation of *k*th joint-motor of the N-link serial-chain robotic manipulator; K_{eq}: equivalent spring constant for the CRG or CRW-system; M: mass of the CRG or CRW-system; $F_{\text{link\_N}}{}^{v}$: Pesudo force of vibration at the CRG or CRW-system due to excitation of *k*th joint-motor of the N-link serial-chain robotic manipulator; δ_k: deflection occurring at the CRG or CRW-system due to excitation of *k*th joint-motor of the N-link serial-chain robotic manipulator; $\varepsilon_{Av}{}^{(k)}$.: average strain obtained at the CRG or CRW-system due to excitation of *k*th joint-motor of the N-link serial-chain robotic manipulator; $L_{\text{CRG/CRW}}$: length of the CRG or CRW-system; ρ: density of the material of fabrication of the CRG or CRW-system; $A_{\text{CRG/CRW}}$: Cross-sectional area of the CRG or CRW-system; $\tau_{\text{motor}}{}^{(k)}$: run-time torque of the *k*th servo-motor of the robotic manipulator ($\forall k = 1,2,\ldots N$).

It may be observed that equation (11.49) is the generic format of evaluation of the run-time natural frequency of vibration of the CRG or CRW-system when fitted with a multi-jointed robotic manipulator. Here we are considering the situation wherein only one servomotor of the robotic manipulator is being activated to create the pesudo forced vibration scenario. Hence, source of the vibration, so generated, is the excitation of the system due to the activation of any one of the revolute joints of the robotic manipulator. The very servomotor is responsible for the actuation of the corresponding revolute joint of the manipulator. However, choice of a particular servomotor that is to be activated alongwith its extent of activation (joint movement in degrees) is user-selectable. During this programmed activation, the rest of the servomotors are kept under non-activation mode. It is to be noted here that equation (11.49) is valid if we do not consider any retarding torque due to the joint rotation of the preceding link(s) of the manipulator. The formulation will alter under a real-life situation wherein the pseudo-forced vibration takes into account impedance of the joint that is being actuated by the servomotor. This joint impedance results in a retardation torque and its numerical extent gets defined by a factor, named as 'joint retardation coefficient'. The effect of the joint retardation will be on the preceding link(s) of the robotic manipulator, with respect to the joint-servomotor that is being activated. The revised formula for the evaluation of natural frequency of vibration of the CRG/CRW-system under this real-life situation of joint-retardation is stated below:

$$f_{n*}^{(k)} = \frac{1}{2\pi L}\sqrt{\frac{\left\{\frac{\tau_{\mathrm{motor}}^{(k)}}{\sum_{i=k}^{i=N} L_i - \lambda . \sum_{i=1}^{i=k-1} L_i}\right\}}{\varepsilon_{Av}{}^{(k)} . \rho . A}} \tag{11.50}$$

where, $f_{n*}{}^{(k)}$: real-life natural frequency of vibration of the CRG or CRW-system due to excitation of kth joint-motor of the N-link serial-chain robotic manipulator under joint-retardation effect; $\varepsilon_{A_v}{}^{(k)}$: average strain obtained at the CRG or CRW-system due to excitation of kth joint-motor of the N-link serial-chain robotic manipulator; L : length of CRG or CRW-system; ρ : density of the material of fabrication of the CRG or CRW-system; A: cross-sectional area of CRG or CRW-system; $\tau_{\mathrm{motor}}{}^{(k)}$: Run-time torque of the kth servo-motor of the robotic manipulator ($\forall k$ = 1,2,…N); λ: Joint Retardation Coefficient (JRC), 0< λ⩽0□1.

The syntax of the natural frequency of vibration under a pesudo-forced vibration situation gets computationally complex if two or more servomotors of the robotic manipulator are in simultaneous activation. In such cases, the effective force of vibration on the CRG or CRW-system will be multi-sourced and evaluation has to be made accordingly, keeping the basic format of equation (11.50) intact.

11.5.4 Modeling paradigms and control dynamics for secondary-stage vibration

11.5.4.1 An insight into secondary-stage vibration

The nucleus of run-time control dynamics of CRG as well as CRW-system is the built-in vibration and its incarnations. However, the ground-level vibration gets created inside the envelope of the CRG or CRW-system by virtue of the slenderness of the design. This vibration is primarily deducible through the spring-structural model and formulation, as described in the previous section. But the run-time nature of this nascent vibration gets multiplied as soon as the CRG-plus-CRW-system gets integrated with the robotic manipulator. The synergistic effect of this interfacing results in a higher level of vibration along with its modal frequencies. This vibration is termed as 'secondary-stage vibration' and it has multiple pathways of realization. In fact, the secondary-stage inherent vibration of the robotic manipulator, integrated with CRW and CRG-systems, is directly proportional to the number of degrees-of-freedom of the manipulator. In all practical situations of serial-chain robotic manipulator with CRG, this secondary-stage vibration is evaluated in terms of natural frequency of vibration (ω_n) and the higher-order modal frequencies. On the other hand, the dynamic strain vector (ε_i, $\forall i$ = 1,2,3,…,k: where 'k' is the link-number of the manipulator) is evaluated from structural analysis of the robotic manipulator. Thus, the paradigm of control dynamics of an integrated CRG plus CRW-system in real-time essentially involves investigation of strain vis-à-vis vibration tuple. In the case of 'direct-to-joint-drive' serial-chain robotic manipulator, the instantaneous real-time displacement will be prudent in the links in the form of 'deflection'.

Hence we can say that the material-specific deformation of any constituent member of the robotic manipulator at any time-instant is a combined effect of natural frequency of vibration and total strain produced inside the said member. The principal axis of such deformation and/or deflection is curvilinear. It may be noted that modal frequency distribution pattern in the robotic manipulator and/or CRG-system can be obtained through FE simulation as well as experimentation, up to tenth mode separately for each member, with a distinct flavor of 'deflection hue'.

The multi-modal nature of the secondary-stage vibration of the robotic system (manipulator plus CRG) needs subtle mathematical modeling, over and above sensor-instrumented SSM. The spring-damper-dashpot design scheme for the link and joint of CRG or CRW-system is the most optimal tool for the said secondary-level vibration analysis as well as real-time vibration control issues. The spring-damper system is self-compensated and the resultant vibration, generated thereof, will be the optimal amount required for stimulating the manipulator-ensemble. The design model of this enhanced sensor-instrumented SSM of the CRG or CRW-system is schematically shown in figure 11.30.

As shown in figure 11.30(a), CRG or CRW-body will have member(s) mounted on mechanical springs, in order to realize spring-dashpot-based modeling. The design architecture of the real-life sensor-augmented SSM of CRG or CRW-system towards enhancement of pick-up of secondary-stage vibration consists of four distinct members, viz. 'A', 'B', C' and 'D', as illustrated in figure 11.30(a). Like the earlier disposition (refer to figure 11.29(a), two successive 'strain' members are shown in figure 11.30(a) as well so as to highlight the coupling effect between constituent members of the CRG or CRW-system in actual hardware. In this model too, 'sensing' member is the most crucial element of the system and its modeling is dependent on the layout of the sensor-cells (e.g. strain gauges), as shown in figure 11.30(b). The paradigm of the *in situ* secondary-stage vibration of members/links of CRG or CRW-system can be symbolized optimally as the real-time

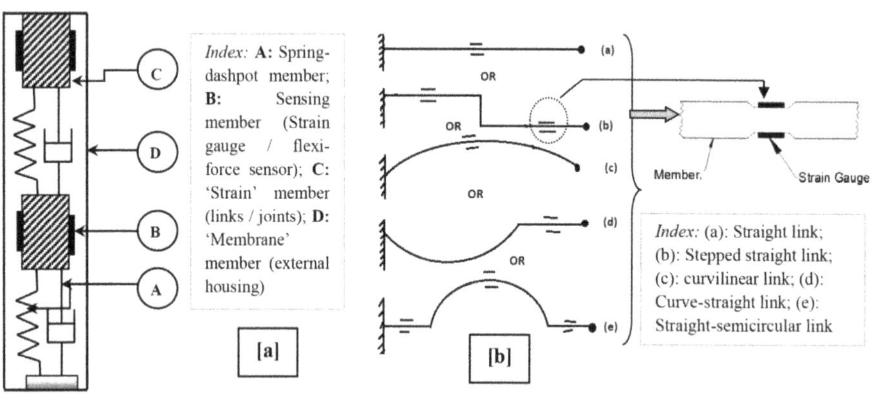

Figure 11.30. Real-life sensor-augmented vibration model of CRG or CRW: (a) design architecture; (b) strain gauge layout.

deflection of a cantilever beam having five representative cross-sections with mounting of strain gauges, as shown in figure 11.30(b). Cantilever-type CRG/ CRW-members do possess easy potential for affixing sensing element(s) as well as better relief at the non-fixed end. The locations of placement of strain gauges have been marked by an '=' symbol in the figure, in five possible design variations, viz. straight, stepped straight, curvilinear, curve-straight and straight-semicircular.

As evident from figure 11.30(a), 'A' and 'D' have mutual sharing of kinetics, while the effect of that will be arrested by 'B' and 'C'. This conjugate spring-damper-dashpot model is the key to the assessment of the modal frequencies (1st to *N*th mode). It is important to note that an efficient dynamic model of CRG or CRW-system will be able to characterize *deformation* and *deflection* of its members and/or sub-assemblies in real-time. While links of CRG and/or CRW are more prone to deflection with certain frequency of vibration, the joints are subjected to mild deformation. At times these two entities are quite unseparable and we need to rely on FEA simulation and trial run of the hardware only. In most of the designs of CRGs that we have explored as well as manufactured, the instantaneous real-time displacement, in the form of 'deflection', was observed to be prudent in the CRG-links. Nonetheless, we can arrive at the following summary on the effect of secondary-stage vibration of CRG/CRW-system, as per table 11.2 below.

Before detailing the modeling scheme, we need to take a closer look at the characteristics of a CRG/ CRW-member when subjected to *in situ* vibration (both inherent as well as external). This phenomenon has been simulated through the actuation of various spring-elements (for *in situ* vibration) and spring-dashpot elements (for secondary-stage/external vibration) in unison, which are attached to the CRG/ CRW-member.

11.5.5 Modeling force-displacement tuple

The multi-dimensionality of the secondary-stage *in situ* vibration of the CRG/CRW-system and/or the ensemble robotic manipulator system needs subtle mathematical modeling. As stated earlier, the spring-damper-dashpot design scheme is the most optimal tool for the vibration analysis of CRG/CRW-system as it has inherent uncertainties and real time vibration control issues. Since this design is related to 'forcing function' of spring system or spring-mass-damper system, we have enough

Table 11.2. Nature of modal frequencies of secondary-stage vibration in CRG/CRW-system.

Type	Nomenclature of vibration mode	Observation and remarks
I	1st mode	Basic natural frequency of vibration (ω_n)
II	2nd and 3rd mode	Gradual increase of ω_n
III	4th and 5th mode	Progressive gradual increase of ω_n
IV	6th, 7th and 8th mode	Threshold of failure → higher stress → deformation pattern is like 'deflection hue' in FEA simulation
V	9th and 10th mode	Sharp increase of ω_n → failure/undesirable condition

room to imbibe the concept of *interval mathematics* to model and solve the force displacement tuple. As a matter of fact, this *in situ* vibration-based method has become prudent in design synthesis for CRG/CRW due to its in-built nature of the range of dimension of the parameters. We will explore the said postulation of vibration-based design method, with reference to the schematics of figure 11.30(a).

We have for each design variable, 'X' in '$\mathfrak{R}$' ('$\mathfrak{R}$': 3D space in real-time), a close association of parametric range (end-values) such that **{X}:** → $[\mathbf{X_{min}}, \mathbf{X_{max}}]$ in '$\mathfrak{R}$'. These two range-values of 'X' will be experimentally determined and/or simulated *a priori*. Thus, we will have a combination of four design variables, pertaining to 'A', 'B', 'C' and 'D' (as per the Real-time Spring-Damper-Dashpot Design Model for *in situ* Vibration of CRG/ CRW-system).

'A':→ {Spring-Constant} →**{K}:** → $[\mathbf{K_{min}}, \mathbf{K^{max}}] \in \mathfrak{R}$

'B'→ {Strain Gauge Resistance & Gauge Length} → $\{{R_{SG} \atop L_G}\} \rightarrow [\begin{smallmatrix} \underline{R} & \bar{R} \\ \underline{L} & \bar{R} \end{smallmatrix}] \rightarrow [\begin{smallmatrix} R_{SG\,\min} & R_{SG\,\max} \\ L_{G\,\min} & L_{G\,\max} \end{smallmatrix}] \in \mathfrak{R}$

'C'→ {Member Length and Member Width } → $\{{l \atop w}\}: \rightarrow [\begin{smallmatrix} \underline{l} & \bar{l} \\ \underline{w} & \bar{w} \end{smallmatrix}] \rightarrow [\begin{smallmatrix} l_{\min} & l_{\max} \\ w_{\min} & w_{\max} \end{smallmatrix}] \in \mathfrak{R}$

'D' → {Joint Strength: Membrane} → **{J$_\mathbf{M}$}:** → $\left[\underline{J}_M\ ,\ \bar{J}_M\right]\left[J_{M_{\min}}\ ,\ J_{M_{\max}}\right] \in \mathfrak{R}$

The formulation and computation of these interval matrices are crucial for the calibration of the sensing elements of the CRG or CRW-system. Now, while dealing with a group of design variables like in this case of A', 'B', 'C' and 'D' (i.e. four variables) or in expanded fashion six variables $[K, l, w, R_{SG}, L_G, J_M]$ we can define '*Constraint Satisfaction Problem*' (CSP). By definition, CSP is formulated by a set of variables, $\{V\} = \{x_1, x_2, \ldots. x_n\}$ along with a set of 'Constraints' or equations, $\{E\} = \{c_1, c_2, \ldots\ldots c_n\}$ over interval domains, $\{[x_1]\,[x_2]\,\ldots..\,[x_n]\}$. The formulation of CSP is essentially design-specific, i.e. for a particular design of the CRG/CRW-member we will have one CSP formulated. Hence, each CSP can tackle a group of new variables, pertaining to the tuple of $\{V\}$ and $\{E\}$. A damping model can be established mathematically using this lemma.

11.5.6 Modeling of real-time control dynamics

With this preamble, we will postulate the fundamental equation of control dynamics of the CRG/ CRW-system, using measurable parameters. We will treat the CRG/ CRW-system as an open-end cantilever structure, thereby imbibing spring-effects at the links and joints. The ensemble deformation at the CRG/CRW-joints will be accounted for as friction torque while stiffness will be estimated through viscous damping of the joints. The root of this model is with the SSM (refer to sub-section 11.5.2.3) of the representative CRGs with three functional sub-assemblies, as illustrated in figure 11.31 below.

We will analyze the schematic disposition of figure 11.31 from the viewpoint of control system layout of CRG/CRW-system. To begin with, let us take an in-depth

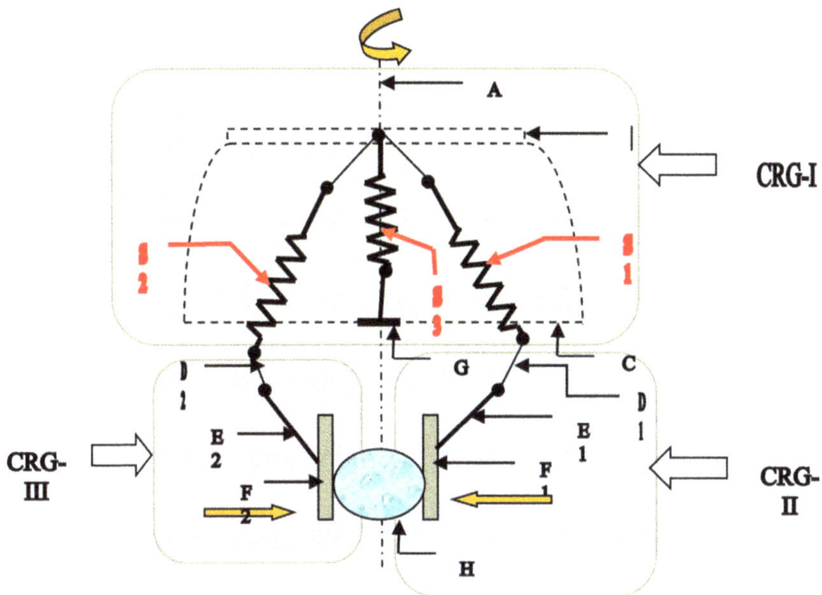

Figure 11.31. Spring—structural model of the representative CRG with functional sub-assemblies.

look into the expanded form of the control dynamics equation of the CRG/ CRW-system, viz. equation (11.44), by amalgamating both parts of the equation and inducing a few measurable parameters. The following generic equation is arrived at in support of the real-time dynamics of the CRG/CRW-system under the influence of secondary-stage vibration (based on the SSM):

$$[M_i]\frac{d^2\theta}{dt^2} + \{B_i\}\frac{d\theta}{dt} + [K_p + K_i](\theta - \delta)_i + (\tau_f)_i = (\tau_L)_i \;\; \forall\, i \in N,\, p \subset (J_{i-1}, J_i) \tag{11.51}$$

where, i: CRG sub-assembly number; p: joint index (CRG-II and CRG-III in this case); $\{J_i, J_{i+1}, J_{i+2}\}$: consecutive 'pin' joints of the CRG-I,II,III (in this case); θ: rotation of the link/spring of the CRG-sub-assembly; M: desired moment of inertia matrix of the CRG-sub-assembly; B: desired viscous damping (friction) coefficient of the CRG-sub-assembly; K_p: desired stiffness of the 'pin'-joint; K_i: Desired stiffness of the CRG-sub-assembly; δ: initial angular position of the CRG-sub-assembly and/or 'pin'-joint; τ_f: frictional torque; τ_L: load torque; N: real numbers ('I', 'II', 'III').

As a finer assessment of the control dynamics of the CRG/CRW-system, viscous damping will also be modeled as two metrics, located at CRG-sub-assembly and at 'pin'-joints. This conjugate approach of modeling the spring constant and viscous damping has been used in order to ascertain the intricacies of the real-life actuation of the CRG. The ensemble deformation at the CRG-joints will be accounted for as friction torque while stiffness will be estimated through viscous damping of the joints as per the equation below in support of the real-time dynamics of the CRG:

$$[M_i]\frac{d^2\theta}{dt^2} + \{B_i + B_p\}\frac{d\theta}{dt} + [K_p + K_i + K_{ip}](\theta - \delta)_i + (\tau_f)_i = (\tau_L)_i$$
$$\forall\ i \in N, p \subset (J_i, J_{i+1}, J_{i+2}) \tag{11.52}$$

where, i: CRG-sub-assembly number; p: joint index (CRG-II & CRG-III in this case); $\{J_{i,}\ J_{i+1},\ J_{i+2}\}$: consecutive 'pin' joints of CRG-I,II,III (in this case) ; θ: rotation of the link/spring of the CRG-sub-assembly; M: desired moment of inertia matrix of the CRG-sub-assembly; B_i: desired viscous damping (friction)coefficient of the ith CRG-sub-assembly; B_p: desired viscous damping coefficient of the pth 'pin' joint of the CRG-sub-assembly; K_p: desired stiffness of the pth 'pin' joint; K_i: desired stiffness of the ith CRG-sub-assembly; K_{ip}: Desired stiffness of the CRG-sub-assembly-joint interface (ith sub-assembly and pth joint in mating condition); δ: initial angular position of the CRG-sub-assembly and/or 'pin' joint; $\tau_{f:}$ frictional torque; τ_L: load torque; N: real numbers ('I', 'II', 'III').

It may be observed that the control dynamic model of the CRG, as stipulated above, is essentially a torque-induced material equation, wherein selection of appropriate material for the CRG-sub-assemblies (especially the links as well as the joints) is very crucial. The terms of the dynamic model equation includes vectors of angular velocity and angular acceleration that need to be determined through real-time experiments. The so-called tiny *flexible* link of the CRG and/or CRW can be conceptualized as a 'free elastic body' in an ideal scenario, but in all practical purposes we can treat it as a 'cantilevered elastic body' (because of the motor overhung), which acts as the prime source of *in situ* vibration of the CRG and/or CRW. The secondary-stage vibration of CRG/CRW-system is evaluated in terms of natural frequency of vibration (w_n) and the higher-order modal frequencies.

We may note that *in situ* secondary-stage vibration of CRG or CRW or the entire robotic manipulator is directly proportional to its number of degrees-of-freedom. The study on the movement/instantaneous motion of CRG-links should ideally be restricted to low joint speeds, thereby neglecting the second-order effects of elastic deformation. In fact, in all such cases, second order and higher terms of control equation need to be avoided for successful control of the CRG. On the other hand, the dynamic strain vector (e_i, $\forall i =$ I,II,III,…,k: where 'k' is the sub-assembly-number of the CRG) is evaluated from structural analysis of the CRG. Thus, the paradigm of control dynamics of CRG in real-time essentially involves investigation of strain vis-à-vis vibration tuple.

We have bestowed importance to the characteristics of viscous damping and spring constant by augmenting influence of both link and joint and finally, the link–joint interface of the CRG/CRW-system. Accordingly, we can model viscous damping coefficient of joint and stiffness of link-joint interfaces of the CRG/CRW-system as:

$$B_p = \xi(B_i) \text{ and } K_{ip} = \gamma(K_p.\ K_i) \tag{11.53}$$

In contrast to earlier control dynamics equation, we may deduce the generalized expression for the torque-specific equation for the CRG/ CRW-system as:

$$F(\theta, t)_i + \left\{\sum \tau_f\right\}_i + \left\{\sum \tau_f\right\}_p = \tau_L \ \forall\, i \in N, p \subset (J_i, J_{i+1}, J_{i+2}) \tag{11.54}$$

where 'F' denotes the real-time function of 'q' and time-instant 't', which is obtainable from the generalized terms of the control dynamics equations. Thus we can express $F(q,t)$ for any CRG/CRW-sub-assembly as per the following expression, utilizing material-specific nomenclature of the CRG (viz. 'α', 'β', 'λ'):

$$F(\theta, t)_i = \left[\alpha.\ \frac{d^2\theta}{dt^2} + \beta.\ \frac{d\theta}{dt} + \lambda.\ \theta\right]_i \tag{11.55}$$

The spring-damper (dashpot)-based control dynamics model (refer to equations (11.51) and (11.52)) will be self-compensated in nature and the resultant vibration, generated thereof, will be the optimal amount required for stimulating the CRG/ CRW-ensemble. In fact, disposition of a specific CRG/ CRW-member under *in situ* as well as secondary-stage vibration can have several incarnations so far as the damping model for the control system is concerned.

It may be concluded that the overall design architecture of CRG/CRW-system will have a decent deflection scheme: the first is the inherent deflection/vibration of itself and the second one is the deflection of the links and joints of the associated robotic manipulator. This conjugate deflection paradigm is the crux of the controller design of the CRG or CRW-system in real-time.

11.6 Development of turning model for vibration synthesis of compliant robotic gripper and wrist system

11.6.1 Facets of vibration synthesis

At this point it is worth mentioning that a complete dynamic model of the serial-chain robotic manipulator, integrated with CRW plus CRG-system must incorporate both link and joint rotation as well as joint twist. While the paradigm of real-time rotation of link and/or joint of CRG/CRW-system will be tackled by the control dynamics model delineated hitherto, vide equations (11.52)–(11.55), modeling of joint twist will call for in-depth analysis of the twisting phenomena, comprising twist of the joint and twist at the link–joint interface. Twist of the links and joints of CRG/ CRW-system is complex and it will involve modeling of the stiffness in quadrature plane(s) at a particular time-instant. It may be noted that the root–control system model, vide equations (11.54) and (11.55) do not involve time function and thus the model is perfectly suited for joint flexure only.

The real-time dynamics of the joints of the CRG will comprise three independent phenomena, namely: (a) twist; (b) turn; and (c) angular displacement. Out of these three facets, 'turning of joints' is the most significant in the case of CRG. Likewise, turning phenomenon is significant for the CRW too. It may be noted that the turning of CRG and/or CRW-system is manifested essentially at the local frame of references and these phenomena differ from the turning/twisting of joints of the serial-chain robotic manipulator, integrated with a combined CRG system. It is to be noted that 'joint twist' phenomenon is not significant in the case of CRG as the majority of the twisting torque will be absorbed by the wrist of the CRG, i.e. essentially the CRW-system. On the other hand, 'joint turn' will be somewhat significant; and, it will be realized through 'turning of the jaws' of CRG or links associated with the jaws. We will now go through the analytical modeling of the turning phenomena of the CRG and/or CRW-system, as obtainable in real-time conditions.

11.6.2 Formulation of the turning model

We may note that joint dynamics of the CRG-system and/or CRW play an important role in defining the '*impedance*' at the link–joint interfaces. These impedances are intra-joint type and get activated as and when the joint of the CRG is in motion. This joint impedance of turning phenomenon of CRG can be modeled as:

$$I_{tu\text{CRG}} = \frac{\lambda_{\text{CRG}} \cdot \tau_{tu\text{CRG}}}{\rho^{av.}_{\text{CRG}} \cdot V_{\text{CRG}}} \tag{11.56}$$

where, I_{tu_CRG}: impedance of turning of the CRG-joint; V_{CRG}: volume of the 'turned' segment(s) of the CRG; ρ_{CRG}: average value of the density of the CRG-material(s); τ_{tu_CRG}: turning torque of the CRG; λ_{CRG}: dissipation factor. It may be noted that although equation (11.56) is indexed for CRG, it is valid for CRW as well, as the formulation has been made using basic principles of enegy and mass balance.

However, since the CRG is fabricated from different components and sub-assemblies having various kinds of materials, all segments of the CRG are not identical. Hence, volume of the 'turned' segment(s) of the CRG can be estimated analytically by the cumulative sum of '*j*' number of segments ($j = 1,..,p$):

$$V_{\text{CRG}} = \sum_{j=1}^{j=p} (V_{\text{CRG}})_j \tag{11.57}$$

Now, theoretically, each of these '*j*' number of 'turned' segments inside the physical enclosure of the CRG and/or CRW-system will undergo small-scale turning and the torque, responsible for those small-scale turning can also be estimated. The ensemble 'turning torque' of the CRG can be evaluated as:

$$\tau_{tu\text{CRG}} = \sum_{k=1}^{k=p} (F_{tu\text{CRG}})_k \cdot (r_{\text{CRG}})_k \tag{11.58}$$

where, $(F_{tu\_\text{CRG}})_k$: force responsible for the turning phenomenon of the *k*th segment of the CRG and $(r_{\text{CRG}})_k$: radius of turning (imaginary) of the *k*th segment of the CRG. We assume all of these elemental forces for turning, i.e. F_{tu_CRG} are co-planar and hence the 'moments', so generated at each of the *k*-segments ($k = 1,...p$) can be added in order to evaluate the ensemble turning torque. We can obtain angular torque of turning as well, by adhering to a different procedure. As angular torque will be intermixing in nature and it will span evenly between the segments, we will use infinitesimal angular torques for the evalaution of the ensemble. In other words, we won't go by the segmental distribution of the masses of the CRG, unlike the case of the estimation of the linear turning torque. The ensemble angular torque of turning will be estimated within the thresholds of the angular velocity of the CRG-system. The formulation is shown below:

$$\tau_{tu\text{CRG}|ang\rangle} = \int_{\omega_{\text{CRG}}=0}^{\omega_{\text{CRG}}=\omega^*} (I_{\text{CRG}}.\ \omega_{\text{CRG}}) d\omega_{\text{CRG}} \tag{11.59}$$

where, $\tau_{tu\_\text{CRG}|ang\rangle}$: angular torque of turning of the CRG; I_{CRG}: moment of inertia of the CRG; ω_{CRG}: angular velocity of the CRG-system; $d\omega_{\text{CRG}}$: infinitesimal angular velocity of the CRG.

The theoretical formulation of the turning torque of the CRG, vide equations (11.58) and (11.59), will be just sufficient to have an overall estimation of the power requirement of the servomotor of the CRG or prime-movers for the CRW-system. However, the actual evlaution of the turning toque needs to be made with the help of experimentation in real-time grasp of payload. This real-life evalation of turning torque will be dependent on the design features of the CRG as well as the disposition of the SSM. In other words, such real-life evaluation will be made with respect to strain gague-based sensory instrumentation of the CRG/CRW-system. The final form of the said indigenously-developed *impedance model* of the CRG, based on the SSM of figure 11.31 as well as experimental evaluation of turning torque will be:

$$\begin{aligned}\tau_{tu\text{CRG}}(t) = K_{bp\text{CRG}}(t).\ \{\theta_{bp\text{CRG}} - \theta_0\} + K_{ll\text{CRG}}(t).\ \{\theta_{ll\text{CRG}} - \theta_0\} \\ + K_{rl\text{CRG}}(t).\ \{\theta_{rl\text{CRG}} - \theta_0\}\end{aligned} \tag{11.60}$$

where, $\tau_{tu\_\text{CRG}}(t)$: total torque of turning of the CRG at tth time-instant; $K_{bp\_\text{CRG}}(t)$: effective spring-constant of the base plate sub-assembly of the CRG at tth time-instant as per spring-structural model; $K_{ll\_\text{CRG}}(t)$: effective spring-constant of the left link sub-assembly of the CRG at tth time-instant as per spring-structural model; $K_{rl\_\text{CRG}}(t)$: effective spring-constant of the right link sub-assembly of the CRG at tth time-instant as per spring-structural model; θ_0: initial angular position of the CRG system; θ_{bp_CRG}: intermediate angular position of the base plate sub-assembly of the CRG at tth time-instant; θ_{ll_CRG}: intermediate angular position of the left link sub-assembly of the CRG at tth time-instant; θ_{rl_CRG}: intermediate angular position of the right link sub-assembly of the CRG at tth time-instant. It is crucial to have proper experimental validation of the turning torque of the CRG/CRW-system as per the model of equation (11.60) in a repeated manner.

11.6.3 Accumulation of vibration in robotic wrist

It is important to note here that a sizeable portion of the ensemble turning torque of the CRG gets accumulated inside the housing of CRW during grasp operation so as to give enough resilience to the CRG for beter form and force closure. In fact, this accumulation of run-time vibration will go beyond the physical structure of CRW and eventually, it will reach the robotic manipulator that is being interfaced with the CRG-plus-CRW-system. This accumulation process will lead to generation of counter-vibration in CRW and it will help harnessing the trembling of the CRG during the grasp. The overall forcing function of this couner-vibration will be compresive type, by and large. Interestingly, this compresive type forcing function, responsible for generating counter-vibration will be highy time-dependent and it will act only for a specific duration of the grasp in order to ensure slip-free force closure. Thus, activation protocol of the counter-vibration of CRW is analogous to spring-dashpot-damper system of the spring-structural model (of the CRG). In other words, occurrence of counter-vibration in CRW will be realized via a suitable deflection model, aided by: (i) spring member (ESE) and (ii) spring-dashpot-damper member. With reference to the layout of figure 11.31, we can extend the paradigm of compressive forcing function of CRW analytically as shown below:

$$M.\,\ddot{x} + B.\,\dot{x} + K.\,x = F_C(t) \Rightarrow M.\,\ddot{x} + \{B_0\,\mathrm{Sin}(t)\}.\,\dot{x} + \{K_0\,\mathrm{Cos}(t)\}.\,x = F_C(t) \tag{11.61}$$

where, x: displacement of the CRW-members under the influence of compressive forcing function; M: mass of the CRW-system; B: viscous damping coefficient of CRW-system as cyclic function of time-instant, 't'; K: eqivalent spring constant of CRW-system as cyclic function of time-instant; $F_C(t)$: Compressive force function at tth time-instant. It may be noted that we have modeled both 'B' and 'K' as cyclic functions that are differentiable on a continuous time-spectrum. Further, by principle, both 'B' and 'K' should be function of 't' according to the energy disipation theory of viscoelastic dampers. This is a deviation from our earlier postulations, namely, equations (11.51), (11.52), (11.54), (11.55) and (11.60), wherein dynamic models of CRG/CRW-system have been developed by considering angluar shift of the gripper assembly. The aspect of differentiability on a continuous time-scale engulfing the entire grasping process by the CRG has propelled us to think of suitable continuous functions for 'B' and 'K' from an analytical angle. Accordingly, we have zeroed on sine and cosine functions for 'B' and 'K', respectively, which culminates in the following governing equation for the vibration accumulation at CRW, viz.:

$$M.\,\ddot{x} + B_0\left[t - \frac{t^3}{3!} + \frac{t^5}{5!} - \frac{t^7}{7!} + \ldots.\right].\,\dot{x} + K_0\left[1 - \frac{t^2}{2!} + \frac{t^4}{4!} - \frac{t^6}{6!} + \ldots..\right].\,x = F_C(t) \tag{11.62}$$

We may note that sensory instrumentation plays a big role in the assessment of the counter-vibration of CRW-system in real-time. It is also important to study the change of value of $F_C(t)$, numerically during the entire time-stamping of the grasp. Hence, a re-arranged form of equation (11.62) will be the useful disposition of the evaluation the ensemble compressive force of the CRW-system, as detailed below:

$$(M.\,\ddot{x} + K_0.\,x)t^0 + (B_0.\,\dot{x})t^1 - \left(\frac{K_0}{2}x\right)t^2 - \left(\frac{B_0}{6}\dot{x}\right)t^3 + \left(\frac{K_0}{24}x\right)t^4 - \ldots\ \ldots.. = F_C(t) \tag{11.63}$$

Thus, equation (11.63) is expounded in functional form of 't', wherein positive-signed terms and negative-signed terms will appear for any two consecutive terms together. Naturally, equation (11.63) is the practical form of the control dynamics of CRW-system, prior to fitting experimental data.

11.7 Case-studies and experimental results

11.7.1 Robotic system used for the case-study

In order to maintain good parity in size as well as external disposition, we have used a small-sized table-top robotic manipulator as the case-study. This indigenously developed robotic system is semi-flexible type, having low tare-weight. The beauty of this semi-flexible table-top robotic system lies in its quick adaptability to a variety of CRG, especially, our indigenous designs. The curvilinear-jaw type CRGs are ideally suited for grasping cylindrical or spherical objects, such as bottle, cup, tumbler, ball etc. Accordingly, provision has been made to interface any one of those CRGs (refer to sub-section 2.4) to this Test Robotic System. Figure 11.32 shows the photographic view of the indigenously-developed robotic system, fitted with a curvilinear-jaw CRG.

The Test Robotic System (TRS) has a characteristic design feature that is commensurate to its customized controller. The hardware of the robotic manipulator was accomplished in a way to minimize the inherent shaking of the manipulator arms. Besides, all three variants of the curvilinear-jaw CRGs, augmented with this test robotic system, have been activated with the aim of synchronization of dynamics of the manipulator body. The performance of TRS was ensured after repeated trials in order to control the trembling of links to the best possible extent.

11.7.2 Experimental synopsis

Two posture-driven strategies for control command syntax for the TRS have been formulated as part of the present research, pertaining to two different phases of the slip-free grasp of an object: (a) in-plane grasping and (b) off-plane lifting. As part of in-plane grasping, the second link and the gripper have been kept horizontal and parallel to the ground during the entire course of gripping of the object. In contrast,

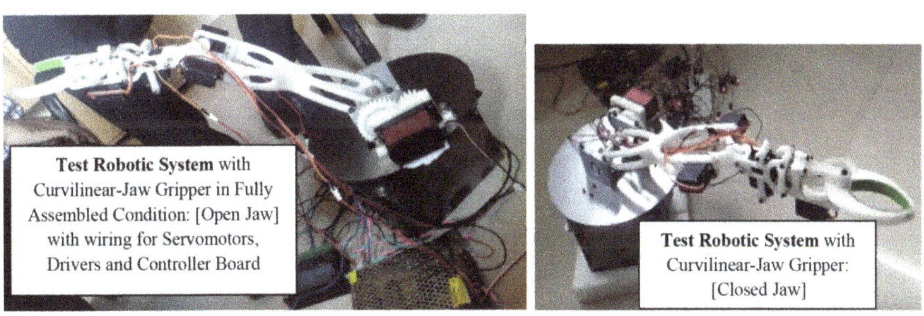

Figure 11.32. Photographic view of the test robotic system fitted with curvilinear-jaw gripper [1], copyright (2020) with permission of Springer.

the second link and first link of the TRS must move through the same angle but in opposite directions for proper off-plane lifting. Figure 11.33(a) illustrates the real-life experimental view of the TRS undergoing in-plane grasping of a cylindrical object, while the real-life experimental view of the off-plane lifting is shown in figure 11.33(b).

The paradigms of in-plane grasping of 'nested' task, e.g. shake-free gripping of a glass filled with water, have been achieved by the TRS through dedicated control semantics. Although gripping of glass can be accomplished by a regular control scheme, advanced programming is needed to ensure shake-free gripping so that dripping of water from the glass can be avoided. We have successfully carried out this sort of nested task with the TRS wherein the coupled state vector of the control system has been solved. Similarly, for off-plane lifting, it is being ensured through the customized control semantics that the robotic manipulator is rotated till destination co-ordinates are reached and the robotic system is able to place the object without jerking. The other classical problem of off-plane lifting that has been experimented with is serving/pouring liquid inside a tumbler. In this case, the angle of the robot-wrist gets finalized in order to serve the content without spilling. Figure 11.34 shows the nested experiment of off-plane lifting and sideways-pouring, as performed by the TRS, along with exterior lengths of the links and gripper.

The selection of rotational speed of the motor of the TRS is equally crucial in all these experiments so as to launch the control system appropriately. In fact, object manipulation has been carried out with a slow speed at the robot-wrist, which is

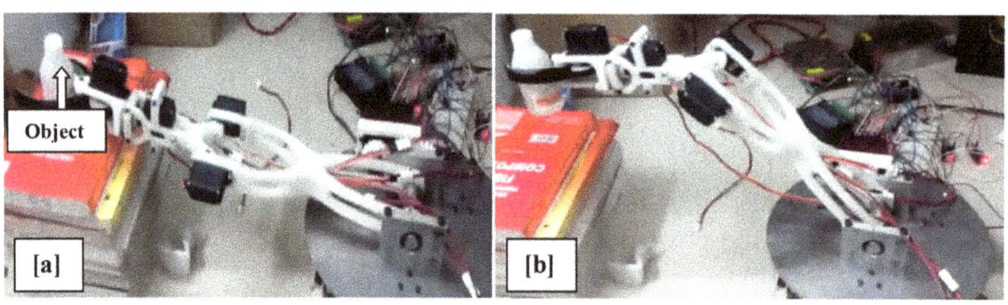

Figure 11.33. Photographic view of: (a) in-plane grasping and (b) off-plane lifting of object by the test robotic system [1], copyright (2020) with permission of Springer.

Figure 11.34. Real-time testing of nested work-module by the test robotic system [1], copyright (2020) with permission of Springer.

approximately 25% of the base rotation speed (~3–4 rpm). A time delay is imbibed between every incremental angle of rotation so as to ensure jerk-free movement of the wrist and gripper assembly during task execution. The customized program module has been made interactive in order to systemize augmentation of multiple grippers in quick succession. The real-life performance of the TRS with slip-free robust grasp by the CRG can be visualized through the following video-clips at: https://www.youtube.com/watch?v=lpkQ8JM1-rY; https://www.youtube.com/watch?v=E5G_veak8tU; https://www.youtube.com/watch?v=K6OZYSr-8pk.

11.7.3 Sensory instrumentation and test results

We have used strain gauge-based instrumentation all along the TRS, both at manipulator-links as well as at the CRG. The backbone of the instrumentation is realized through *National Instruments*™-made products, e.g. system chassis, data acquisition card and signal conditioner. This instrumentation has easy handling with the sensor-cells at one end and computer system at the other side for seamless real-time processing of sensory data. The hardware instrumentation has compatibility with a matching software routine, provided by *LabView*®. Figure 11.35 schematically illustrates the hardware-level layout of the sensory instrumentation for the TRS, alongwith a representative program-module in *LabView*®.

The dynamic strain values were computed based on the readings of the strain gauges. A total of six strain gauges were mounted on the TRS: four on the first link and the remaining two on the second link. Likewise, two strain gauges were affixed on the CRG. The strain values, so obtained, were instrumental in computing the natural frequency of vibration of the CRG and also the same for the TRS-link(s). Table 11.3 presents the ensemble dynamic strain values of 6 nos. strain gauges, as obtained during various time-intervals under different trial-runs of the TRS. The

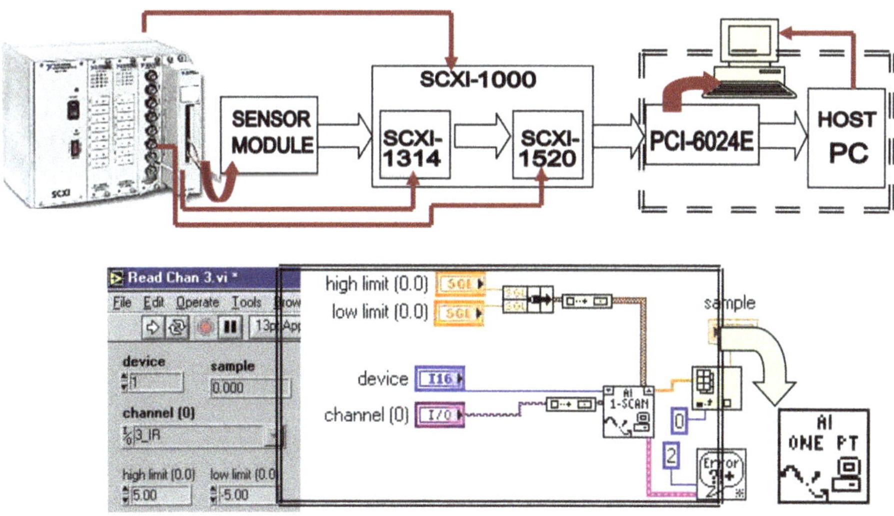

Figure 11.35. Schematic of the sensory instrumentation of the test robotic system for real-time grasp experimentation.

Table 11.3. Test results on the computation of dynamic strain for the TRS-links. [♠: actual experimental data = tabulated data × 10^{-5}]

	Strain values for the 1st link of the TRS ♠					Strain values for the 2nd link of the TRS ♠		
Trial No.	S1	S2	S3	S4	$[Av.\varepsilon]_1$	S5	S6	$[Av.\varepsilon]_2$
1	0.0532	0.0575	0.0548	0.0586	0.0560	0.0134	0.0148	0.0141
2	0.0546	0.0567	0.0543	0.0578	0.0561	0.0145	0.0145	0.0145
3	0.0537	0.0534	0.0548	0.0589	0.0552	0.0138	0.0148	0.0143
4	0.0542	0.0587	0.0539	0.0584	0.0563	0.0137	0.0152	0.0144
5	0.0563	0.0544	0.0537	0.0583	0.0556	0.0136	0.0146	0.0141
6	0.0537	0.0564	0.0542	0.0574	0.0554	0.0138	0.0152	0.0145
7	0.0485	0.0559	0.0545	0.0579	0.0542	0.0165	0.0147	0.0156
8	0.0539	0.0575	0.0558	0.0576	0.0562	0.0153	0.0139	0.0146
9	0.0526	0.0548	0.0546	0.0569	0.0547	0.0155	0.0142	0.0148
10	0.0528	0.0557	0.0534	0.0573	0.0548	0.0157	0.0145	0.0151

Table 11.4. Test results on the computation of dynamic strain for the TRS-gripper.

	Trial Number [♠:Actual experimental data = tabulated data × 10^{-5}]									
Strain values ♠	1	2	3	4	5	6	7	8	9	10
S7	0.0232	0.0234	0.0189	0.0236	0.0245	0.0243	0.0179	0.0157	0.0178	0.0233
S8	0.0212	0.0256	0.0167	0.0154	0.0237	0.0214	0.0165	0.0146	0.0168	0.0246
$\{Av.\varepsilon\}_{CRG}$	0.0222	0.0245	0.0178	0.0195	0.0241	0.0228	0.0172	0.0151	0.0173	0.0239

dynamic strain values, generated from the strain gauge readings are indexed as 'S1', 'S2','S3' and 'S4' for the first link and 'S5' and 'S6' for the second link of the TRS. The average of the strain values in these two links are labeled as $[Av.\varepsilon\ ]_1$ and $[Av.\ \varepsilon]_2$ respectively. Likewise, the ensemble dynamic strain values of 2 nos. strain gauges ('S7' and 'S8'), fitted on the CRG-body against various trial-runs of the TRS are computed, alongwith average value of the strain, $[Av.\varepsilon\ ]_{CRG}$ thereof. Table 11.4 presents these strain values for the CRG-unit of the TRS.

The average strain values, so obtained, have an important role to play so far as the evaluation of natural frequency of vibration of the CRG-system and/or TRS are concerned. Fundamentally, these strain values (refer to data of tables 11.3 and 11.4) are used to compute natural frequency of *in situ* primary vibration of the CRG, as per equations (11.45) and (11.47). However, in all practical purposes we will come across some sort of 'pesudo-forced vibration' scenario, because our system under case-study is a conjugate form of robotic system. The excitation of the TRS-ensemble will be effected by the real-time activation of either of its joint-servomotors, namely, motors for rotation of joint 1 or joint 2. The natural frequency of this primary *in situ* type pesudo-forced vibration can be evaluated through the formulation of equation (11.49), wherein input joint-torque will be in Nm and output natural frequency will be expressed in Hz. Accordingly, experimentation was made to observe the pesudo-forced vibration of the CRG under the excitation of the first joint-servomotor of the TRS. Table 11.5 below reports the evaluated values of the natural frequency of primary *in situ* pesudo-forced vibration of the CRG, '$f_n^{(1)}$', against the run-time torque of the first joint-motor of the TRS, '$\tau 1$'.

We will now proceed towards evaluation of natural frequency of vibration at the secondary stage, i.e. effect of pesudo-forced vibration at the TRS-links. We will calculate the said natural frequencies using the formulation of equation (11.50). Table 11.6 presents the evaluated values of the natural frequency of secondary pesudo-forced vibration of the CRG, '$f_{n*}^{(1)}$' in Hz, against the run-time torque of the first joint-motor of the TRS, '$\tau 1$', in Nm.

It may be noted here that the issue of joint retardation will not be effective in this case, as the excitation is made through joint1 servomotor. The said effect of joint retardation will be more prominent in cases of flexible robotic systems having multiple slender links and joints. In the present case-study of the TRS, estimated values of natural frequencies will not alter much even if we consider excitation via the second joint-servomotor.

Table 11.5. Evaluation of natural frequency of primary in-situ pesudo-forced vibration for the TRS-gripper **(excitation via first joint–servomotor).**

	Trial number [♠:Actual experimental data = tabulated data × 10^{-5}]									
Variables	1	2	3	4	5	6	7	8	9	10
τ1	1.504	1.476	1.348	1.421	1.498	1.465	1.496	1.503	1.387	1.436
{Av.ε}$_{CRG}$♠	0.0222	0.0245	0.0178	0.0195	0.0241	0.0228	0.0172	0.0151	0.0173	0.0239
$f_n^{(1)}$♣	4004.6	3776.4	4233.9	4153.3	3835.8	3900.0	4537.5	4854.1	4356.4	3771.3

[♣: length of TRS-gripper: 120 mm; area of cross-section of the TRS-gripper: 50 mm × 15 mm = 75 × 10^{-5} m^2; density of material of TRS-gripper: 2.2 gm/c.c.; overall length of the TRS (refer to figure 11.34): (250 + 80 + 120) mm = 0.45 m]

Table 11.6. Evaluation of natural frequency of secondary pesudo-forced vibration for the TRS-2nd link **(Excitation via first joint–servomotor).**

	Trial number [♠:Actual experimental data = tabulated data × 10^{-5}]									
Variables	1	2	3	4	5	6	7	8	9	10
τ1	1.504	1.476	1.348	1.421	1.498	1.465	1.496	1.503	1.387	1.436
{Av.ε}$_2$♠	0.0141	0.0145	0.0143	0.0144	0.0141	0.0145	0.0156	0.0146	0.0148	0.0151
$\mathbf{f_{n*}^{(1)}}$♣	9731.7	9505.8	9147.6	9539.4	9711.3	9470.3	9226.4	9559.5	9120.9	9187.9

[♣: length of TRS-2nd link: 80 mm; area of cross-section of the TRS-2nd link: 30 mm × 15 mm = 45 × 10^{-5} m^2; density of material of TRS-2nd link: 2.2 gm/c.c.; overall length of the TRS (refer to figure 11.34): (250 + 80 + 120) mm = 0.45 m]

11.8 Conclusions

In-depth design, modelling with analysis and hardware development of ten customized compliant robotic grippers and a novel lightweight table-top robotic system with adaptability to multiple grippers have been delineated in the chapter. We have carried out the present research with the prime objective of demonstration of the capabilities of the CRGs and also the TRS in real-time, alleviating inherent system vibration. Links as well as joints of the CRG/CRW-system have been designed and manufactured in such a way so as to control the *in situ* trembling of the robotic unit. Improvisation of the design with miniaturization as well as wider selection of materials for fabrication of the CRGs will be taken up as future milestones of the research. The run-time performance of the slip-free robust grasp of some of the developed CRGs can be observed through the following video-clips: (i) https://youtu.be/Nwbt94E8Tds; (ii) https://youtu.be/29gz3_Im6lg; (iii) https://www.youtube.com/watch?v=048N9Xqsseg and (iv) https://www.facebook.com/SVRINFOTECH.NET/videos/346400913019942.

Acknowledgments

The author expresses sincere thanks to Mr Pratik Chothe, Mr Aniket Bhelsaikar and Mr Khushaldas Badhan of M/s SVR Infotech, Pune, India for design synthesis, FEA and co-ordination on hardware development of the flat-jaw and curvilinear-jaw CRGs and RPMRoG. The technical contribution of Mr Anas Islam, Mr Aman Sharma and Mr Rohit Sharma of GLA University, Mathura, India in developing the welding-gripper, WISSRoG is highly acknowledged. Sincere appreciation is due to Mr Pankaj Pandit, Mr Prathamesh Warude and Mr Shreyash Gajrekar of Rajarambapu Institute of Technology, Maharashtra, India for effective contribution towards firmware of the TRS and PAMRoG. The contribution of Mr Uday Phalnikar, Ex-Bhabha Atomic Research Centre, Mumbai, India and Ms Vinaya Patil of Savitribai Phule Pune University, India towards fabricating SIMSJRoG and RPMRoG, respectively, is acknowledged. Financial support from Department of Atomic Energy, Govt. of India, Professor Vicente Feliu Battle of Universidad Castilla La Mancha, Ciudad Real, Spain, Mr Viinod Atpadkar of M/s SVR Infotech, Pune and Professor Kamal Shrma of GLA University, India is greatly acknowledged. Besides, the contribution of various intern student groups and research scholars of the author in context of the experimentation with the CRGs is deeply acknowledged.

References

[1] Chakraverty S and Biswas P 2020 *Recent Trends in Wave Mechanics and Vibrations Lecture Notes in Mechanical Engineering* (Cham: Springer)

[2] Qassab S Y and Ali Sultan H 2013 Modal analysis of pneumatic two finger robotic hand by finite element analysis and experimental testing *Int. J. Eng. Technol. IJET-IJENS* **13** 1–9

[3] Jeevan A N R 2015 Modeling and analysis of robot arm using ANSYS *Int. J. Sci. Eng. Technol. Res.* **4** 6692–97

[4] Warude P, Patel M, Pandit P, Patil V, Pawar H, Nate C, Gajlekar S, Atpadkar V and Roy D 2019 On the design and vibration analysis of a three-link flexible robot interfaced with a

mini-gripper *Recent Trends in Wave Mechanics and VibrationsLecture Notes in Mechanical Engineering* (Singapore: Springer) pp 29–45

[5] Pawar H, Nate C, Patel M, Pandit P, Patil V, Warude P, Wankhede N, Chothe P, Atpadkar V and Roy D 2019 Design, dynamic simulation and test-run of the indigenous controller of a multi-gripper revolute robot by minimizing system trembling *Recent Trends in Wave Mechanics and VibrationsLecture Notes in Mechanical Engineering* (Singapore: Springer) pp 47–60

[6] Roy D 2019 Towards the control of inherent vibration of flexible robotic systems and associated dynamics: new proposition and model *Int. J. Robot. Res. Appl. Autom.* **1** 6–17

[7] Chitte P G and Bansode S S 2016 Structural and vibrational analysis of six axis ARISTO robot using ANSYS *Int. J. Innov. Res. Technol.* **3** 291–96

[8] Shareef I R and Zayer I A 2018 Natural frequency estimation of a free vibrated robotic arm using artificial intelligence *Int. J. Appl. Eng. Res.* **13** 5393–99

[9] Chouhan R and Kanwal F 2014 Design and development of a prototype robotic gripper *Proc. of the 2014 Int. Conf. on Robotics and Emerging Allied Technologies in Engineering (iCREATE) (Islamabad, April 22–24)* pp 317–20

[10] Datta R and Pradhan S 2015 Analysis and design optimization of a robotic gripper using multiobjective genetic algorithm *IEEE Trans. Syst. Man Cybernet.: Syst.* **46** 1–11

[11] Liu C-H 2019 Topology and size–shape optimization of an adaptive compliant gripper with high mechanical advantage for grasping irregular objects *Robotica* **37** 1383–400

[12] Abomoharam A H M 2017 Modeling and design optimization of a robot gripper mechanism *Rob. Comput. Integr. Manuf.* **46** 94–103

[13] Lanni C 2019 An optimum design algorithm for mechanisms in two-finger grippers *Proc. of the 13th WSEAS Int. Conf. on SYSTEMS* pp 63–70

[14] Bircher W G 2017 A two-fingered robot gripper with large object reorientation range Proc. of the IEEE Int. Conf. on Robotics and Automation (ICRA) *(Singapore, May 29–June 3)* pp 3453–460

[15] Sahu S and Choudhury B B 2017 Static analysis of a 6-axis industrial robot using finite element analysis *Int. J. Mech. Eng. Technol. (IJMET)* **8** 49–55

[16] Bicchi A and Kumar V 2000 Robotic grasping and contact: a review *Proc. of the IEEE Int. Conf. on Robotics and Automation* pp 348–53

[17] Badadhe S 2018 FEM based dynamic analysis of robot end gripper mechanism *Int. J. Eng. Sci. Invent. (IJESI)* **7** 50–67

[18] Roy D 2019 Control of inherent vibration of flexible robotic systems and associated dynamics *Recent Trends in Wave Mechanics and VibrationsLecture Notes in Mechanical Engineering* (Singapore: Springer)

[19] Roy D 2020 Design, modeling and indigenous firmware of a patient assistance flexible robotic system-type I: beta version *Adv. Robotics Mech. Eng.* **2** 148–59

[20] Benosman M and Vey G 2004 Control of flexible manipulators: a survey *Robotica* **22** 533–45

[21] Singer N C and Seering W C 1990 Preshaping command inputs to reduce system vibration *J. Dyn. Syst. Meas. Control-Trans. ASME* **112** 76–82

[22] Fraser A R and Daniel R W 1991 *Perturbation Techniques for Flexible Manipulators* (Norwell, MA: Kluwer)

[23] Luo Z H 1993 Direct strain feedback control of a flexible robot arm: new theoretical & experimental results *IEEE Trans. Autom. Control* **38** 1610–22

[24] Chen W 2001 Dynamic modelling of multi-link flexible robotic manipulators *Comput. Struct.* **79** 183–95

[25] Subudhi B and Morris A S 2002 Dynamic modeling, simulation and control of a manipulator with flexible links and joints *Rob. Autom. Syst.* **41** 257–70
[26] Feliu V, Somolinos J A and Garcia A 2003 Inverse dynamics based control system for a three degrees-of-freedom flexible arms *IEEE Trans. Robotics Automat.* **19** 1007–14
[27] Feliu V and Ramos F 2005 Strain gauge based control of single-link flexible very light weight robots robust to payload changes *Mechatronics* **15** 547–71
[28] Singer N C and Seering W C 1990 Preshaping command inputs to reduce system vibration *J. Dyn. Syst. Meas. Control-Trans. ASME* **112** 76–82
[29] Chen Y P and Hsu H T 2001 Regulation and vibration control of an FEM-based single-link flexible arm using sliding-mode theory *J. Vibrat. Control* **7** 741–52
[30] Tjahyadi H and Sammut K 2006 Multi-mode vibration control of a flexible cantilever beam using adaptive resonant control *Smart Mater. Struct.* **15** 270–78
[31] Trapero-Arenas J R, Mboup M, Pereira-Gonalez E and Feliu V 208 Online frequency and damping estimation in a single-link flexible manipulator based on algebraic identification *Proc. of the 16th. Mediterranean Conf. on Control & Automation (IEEE)* pp 338–43
[32] Pereirea E, Aphale S S, Feliu V and Moheimani S O R 2011 Integral resonant control for vibration damping and precise tip-positioning of a single-link flexible manipulator *IEEE ASME Trans. Mech.* **16** 232–40
[33] Gupta N K 1980 Frequency-shaped cost functionals: extension of linear-quadratic-gaussian design methods *J. Guid. Control* **3** 529–35
[34] Juang J-N, Turner J D and Chun H M 1985 Closed-form solutions for feedback control with terminal constraints *J. Guid. Control* **8** 39–43
[35] Baumgarte J 1972 Stabilization of constraints and integrals of motion in dynamic systems *Comput. Method Appl. Mech. Eng.* **1** 1–16
[36] Argyris J 1982 An excursion into large rotations *Comp. Method Appl. Mech. Eng.* **32** 85–155
[37] Atluri S and Cazzani A 1995 Rotations in computational solid mechanics *Arch. Comput. Meth. Eng.* **2** 49–138
[38] Fraeijs de Veubeke B 1976 The dynamics of flexible bodies *Int. J. Eng. Sci.* **14** 895–913
[39] Danielson D and Hodges D 1988 A beam theory for large global rotation, moderate local rotation and small strain *ASME J. Appl. Mech.* **55** 179–84
[40] Li D and Likins P 1987 Dynamics of a multi-body system with relative translation on curved flexible tracks *J. Guid.* **10** 299–306
[41] Bauchau O, Damilano G and Theron N 1995 Numerical integration of non-linear elastic multi-body systems *Int. J. Numer. Methods Eng.* **38** 2727–51
[42] Oden J and Martins J 1985 Models and computational methods for dynamic friction phenomena *Comput. Method Appl. Mech. Eng.* **52** 527–634
[43] Rauniyar A, Roy D, Pandit P and Atpadkar V 2018 Design model for the drive and actuator of the test set-up of a novel flexible robotic system *Proc. of the 2018 IEEE Int. Conf. on Computational Intelligence and Computing Research (IEEE-ICCIC-2018) (Madurai, Tamil Nadu, India, December 13–15)* [Awarded Best Paper Prize]
[44] Islam A, Sharma S, Sharma K, Sharma R, Sharma A and Roy D 2020 Real-time data monitoring through sensors in robotized shielded metal arc welding *Mater. Today: Proc.* **26** 2368–73

Chapter 12

A study on mode shape-based approaches for health monitoring of a reinforced concrete beam under transverse loading

S K Panigrahi and Ajay Chaurasia

Reinforced concrete (RC) beams in buildings are considered one of the most critical members of building structures. The existence of primary cracks and occurrence of distress for any reason results in the degradation of the stiffness of the structure which reduces the strength. The mode shape-based approaches are very efficient techniques in damage identification in structural elements. This present study provides a detailed review on the various types of damage prediction approaches based on mode shape approaches. The current state of the art along with different damage identification models for RC structures proposed in the past are also presented. In this chapter, firstly an analytical approach has been developed for the identification of damage on a beam model considering damage on the beam. The mode shapes and modal curvatures have been computed both for undamaged and damaged beam structures for damage identification. For validation of the approach, simply-supported RC beams have been considered and were subjected to static two-point loading in steps till the occurrence of ultimate failure. The sensors for vibration measurement are mounted at the critical locations and readings were taken after each step of loading. The plots for mode shapes and modal curvature have been drawn to find out the location of developed crack pattern on the beam.

12.1 Preamble

Damage is the permanent change in the physical properties of a structure and it affects the performance of the structure. Ageing, harsh conditions, fatigue, overloading, corrosion of reinforcing bars in concrete, excessive vibration etc are the causes for change in physical properties of the structure. To avoid a disaster due to this, it is essential to locate and to know the progression of damage in a structure.

doi:10.1088/978-0-7503-3483-9ch12 © IOP Publishing Ltd 2021

This will also give an idea for remaining useful life of the structure. This has been attempted by acquiring vibration data and system load testing.

Damage identification process consists of four levels, viz:

Level 1: Damage detection: determination of presence of damage.

Level 2: Damage location: determination of geometric location of damage.

Level 3: Damage typification: determination of the type of damage.

Level 4: Damage extent: prediction of severity of damage.

Damage in a structure is a local phenomenon which changes the physical properties such as stiffness, mass and damping. The change in physical properties, changes the vibration behavior of the structure. Vibration based structural health monitoring techniques are based on the change in vibration characteristics. However, extensive research is going on to assess the damage of a structure in global scenario (Lee *et al* 1987 and Farrar and Jauregui 1996).

Stubbs and Osegueda (1990) studied the global damage detection approaches and their experimental verification. Chatterjee *et al* (1994) had studied the vibration characteristics of a bridge structure, subjected to live loads. Some related research works on vibration based damage identification are by Santos *et al* (2000), Dulieu-Barton *et al* (2003) and Darpe *et al* (2004). Bandyopadhyay and Bhattacharyya (2007) presented a statistical system identification technique for health monitoring of structure from noisy dynamic response. Subsequently, the authors extended the methodology for identification of damage with limited noisy data (Bandyopadhyay and Bhattacharyya 2008). For dealing with noise in experimental data, Panigrahi *et al* (2009, 2013) optimized the residual force vector using genetic algorithm and extended the work for a shear structure with both noisy and sparse modal information. Panigrahi *et al* (2015) presented a new methodology for damage identification in shear structures with noisy modal information. Doebling *et al* (1998) and Carden and Fanning (2004) presented detailed reviews of vibration-based damage detection methods. Humar *et al* (2006) and Fan and Qiao (2011) reviewed the vibration based approaches for identification of damage for structural systems. The review suggests that the damage identification in structures such as bridges and offshore platforms using only change in frequencies may not be feasible. To address this issue, Rizos *et al* (1990) and Fox (1992) focused on the use of mode shape measures to evaluate damage. Based on changes in frequencies and mode shapes of vibration, a damage identification technique was proposed by Ren and Roeck (2002) for predicting damage location and severity. Gandomi *et al* (2008) reviewed the use of mode shape method for damage identification.

Pandey *et al* (1991) presented and demonstrated a new approach based on modal curvature of damage identification. Further, Wahab (1999), Ratcliffe (2000) and Ko *et al* (2002) presented the use of modal curvature for damage identification. Saleh *et al* (2004) used the change in curvature mode shapes to locate damage of non-prismatic beams. The change in curvature was obtained from the displacement mode shape (displacement mode curve) of a finite element (FE) model of the beam and cracks were detected by surface observation. Unger *et al* (2005) developed a finite element (FE) model updating technique and implemented for damage detection of a gradually damaged pre-stressed concrete beam. Modal curvature with eigen frequencies and mode shapes were used in the model to determine the damage. To demonstrate the efficacy of modal curvature method, Foti and Campian (2007) compared 'modal curvature method' with the 'change in mode shape method'. The comparison was established with reference to a simply supported bridge model using (FE) technique in which damage was simulated by reducing the flexural stiffness *EI*. Cao and Qiao (2009) proposed a Laplacian scheme from which an improved damage identification algorithm was developed and was tested using the data generated by an analytical crack beam model. The applicability of the algorithm was validated with an experimental program of a delaminated composite beam using scanning laser vibrometer (SLV) to acquire mode shapes.

Ciambella *et al* (2011) proposed a technique for damage assessment based on measurements of eigen frequencies and eigen vector curvatures. A cantilever Euler beam mounted with piezoelectric transducers set has been referred to for illustration of the proposed approach. It is observed that the modal curvature difference approach is very effective for damage detection for a one-dimensional beam. The proposed approach was further improved by Zhang (2011) by introducing coordinate transformation technique to overcome the calculation difficulty of modal curvature approach. Damage detection of a plane grid structure similar to 2D plate bending has been presented by Yu et al (2012) by calculating modal curvature changes along the bi-directions of the damaged structure. Ciambella and Vestroni (2015) proposed a filtering procedure to be used in modal curvature difference approach for an accurate estimate of single and multiple damage shapes by using only the minimum number of information conveyed in the first few eigen modes.

A lot of work by various researchers can be seen in the literature for damage detection in reinforced concrete structures (Unger 2005, Srinivasa *et al* 2013) using experimental/ simulated modal information. The authors have previously identified the damage on RC beams using modal curvature approach (Chaurasia *et al* 2018). In this chapter, the detailed analytical approach and various mode shape approaches including modal curvature have been studied on identification of damage. In this study, first of all an analytical approach has been developed for identification of damage on a beam model considering damages on the beam. The mode shapes and curvature damage factors (CDFs) from modal curvatures have been computed both for undamaged and damaged beam structure for damage identification. For validation of the approach, simple-supported RC beams were subjected to two-point static loading in steps till the occurrence of ultimate failure. Vibration data have been extracted in each load step from vibration sensors mounted at critical locations. After each load step, modal tests on the beam were conducted to determine the modal parameters, namely frequencies and mode shapes. For the validation of the mode shape approaches, CDF values from the modal curvatures have been computed to compare with the obtained crack patterns for each case on the beam. The current study presents an overview of the behavior of damaged RC beams with different loaded situations using various mode shape approaches.

12.2 Mathematical model

It is found by Pandey *et al* (1991) that absolute value of the changes in modal curvature is very much sensitive to damage, and so can be used as a damage indicator. Using central difference approximation, the curvature (k) is generally calculated from the measured eigen vectors/mode shapes i.e.

$$k = \frac{u_{i+1} - 2u_i + u_{i-1}}{l^2} \tag{12.1}$$

Where l is the distance between the two measured locations, i represents the node number and u_i is the deflection of the beam at the nodal point. So, curvature mode shape can be commutated from eigen vector/ mode shape and so, the curvature of a particular mode shape may be derived from the vertical displacement of the same mode shape.

CDF as introduced by Wahab and Roeck (1999) is the average of the difference in modal curvature of all modes that are considered. It is most suitable for multi-damage scenarios and gives a clear indication of damaged locations. The curvature damage factor is expressed as,

$$\boldsymbol{CDF} = \frac{\mathbf{1}}{\boldsymbol{N}}\sum_{i=1}^{N}\left|\boldsymbol{u}_{0i}'' - \boldsymbol{u}_{di}''\right| \tag{12.2}$$

where N is the total number of modes considered, u_0'' is the curvature mode shape of the undamaged structure and u_d'' is that of the damaged structure.

The method is explained by considering different depths of damage, thereby establishing the efficacy of various mode shape approaches for damage identification.

12.3 Illustrative example

A simply-supported RC beam with uniform area of cross section (figure 12.1) is considered for the damage detection using modal curvature approach. The rough estimation of the extent of damage is made considering the variations in CDF values from the graph. First, the beam is simulated numerically with a finite element (FE) model taking twenty elements. Each element has both translation and rotational degrees of freedom at each nodal point with a total of forty two nodal points. As the supported point degree of freedom (DOF) has zero translation movements, this results in forty DOFs. The properties of the beam are: modulus of elasticity $E = 31.622$ GPa; cross-sectional area $A = 0.015\ \mathrm{m}^2$; moment of inertia $I = 28\,125 \times 10^{-9}\ \mathrm{m}^4$; mass density $\rho = 2500\ \mathrm{kg\ m}^{-3}$; total length of the beam $l = 1.65$ m

Figure 12.2 shows the second element of the beam. Q3 and Q5 are the transverse displacement and Q4 and Q6 are the rotational vectors, having four DOF. For the element, $le = 0.825$ m and other properties are the same as that of the entire beam.

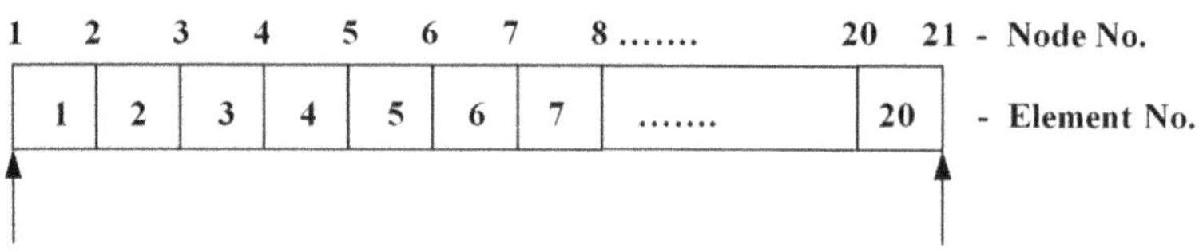

Figure 12.1. Simply-supported RC beam.

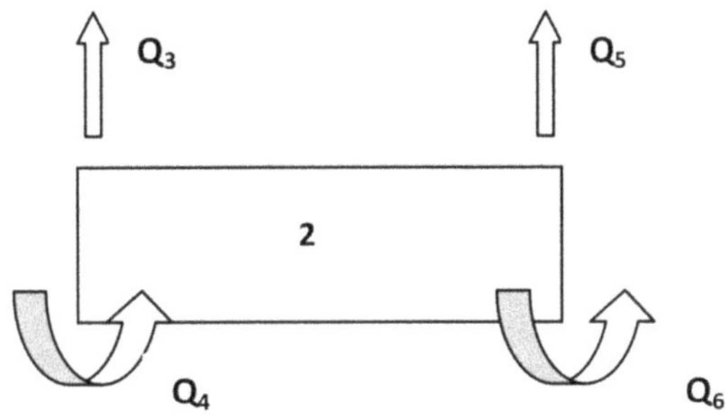

Figure 12.2. State of beam element.

The *j*th eigen value equation for ambient vibration is given by

$$[K]\{\varphi_j\} - \lambda_j[M]\{\varphi_j\} = 0 \tag{12.3}$$

where $[M]$ and $[K]$ are $(m \times m)$ system mass and stiffness matrices respectively, m is the total DOF, λj and ϕj are the *j*th eigen value and corresponding eigen vector. In the finite element model of the structure, the global stiffness and mass matrices can be represented as

$$[K] = \sum_{i=1}^{n}[k^e]_i \text{ and } [M] = \sum_{i=1}^{n}[m^e]_i \tag{12.4}$$

Where n is the total number of elements and $[k^e]$iand $[m^e]$ may be expressed as

$$k^e = \frac{EI}{l_e^3}\begin{bmatrix} 12 & 6l_e & -12 & 6l_e \\ & 4l_e^2 & -6l_e & 2l_e^2 \\ & & 12 & -6l_e \\ \text{sym} & & & 4l_e^2 \end{bmatrix} \text{and } m^e = \frac{\rho A_e l_e}{420}\begin{bmatrix} 156 & 22l_e & 54 & -13l_e \\ & 4l_e^2 & 13l_e & -3l_e^2 \\ & & 156 & -22l_e \\ \text{sym} & & & 4l_e^2 \end{bmatrix} \tag{12.5}$$

$$[K_d] = \sum_{i=1}^{n}\beta_i[k]_i \tag{12.6}$$

As a result of damage in structure, stiffness is reduced, with no change in mass. Thus, the stiffness matrix for damaged structure may be given by

The value of the damage factors lies in the range of 0 to 1 where $\beta_i = 1$ indicates an undamaged element and $\beta_i < 1$ denotes a partially damaged element. The measurement noise is ignored first and then for simulating noise tainted experimental measurement; the vibration characteristics are introduced using random noise of 2% on frequency value and a random noise of 3% on each modal component of partial mode shape. Four damage scenarios were considered by introducing damage to validate the CDF approach:

1. Element 4 damaged partially to an extent of 30% and element no. 17 by 30%.
2. Element 3 damaged partially to an extent of 20% and element no. 8 by 50%.
3. Element 3 damaged partially to an extent of 15% and element no. 14 by 30%.
4. Element 3 damaged partially to an extent of 30% and element no. 14 by 15%.

FE analysis was performed to simulate damage location and compare the results with CDF approach. The CDF values versus element location plots for different damage scenarios are shown in figure 12.3.

Table 12.1 shows the ratio of CDF values at damage locations are closely in agreement with the ratio of percentage (%) of the extent of damage at corresponding

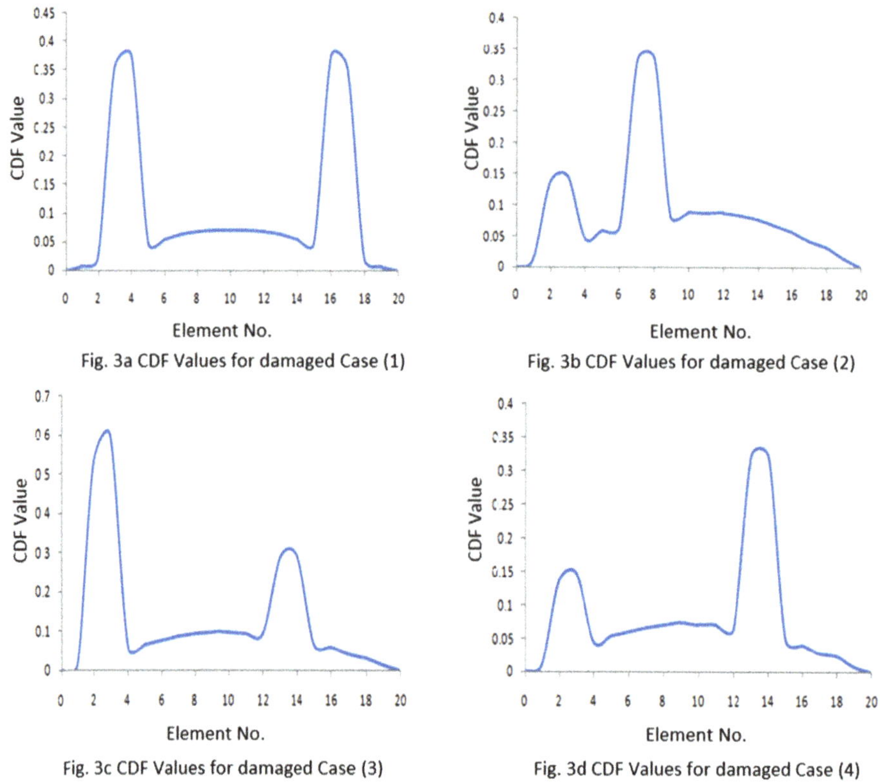

Figure 12.3. CDF values corresponding to various locations

Table 12.1. Comparison between damage proportion and CDF for various damage scenarios

Damaged cases	First Damaged Element No	% of damage	CDF Value	Second Damaged Element No	% of damage	CDF Value	Ratio of CDF Values	Ratio of % of damages	% of deviation
Case (1)	4	30	0.353	17	30	0.353	1.0	1.0	0.0
Case (2)	3	20	0.146	8	50	0.334	2.288	2.5	8.5
Case (3)	3	15	0.147	14	30	0.320	2.177	2.0	−8.9
Case (4)	3	30	0.559	14	15	0.291	0.521	0.5	−4.2

locations. If the extent of damage at one location is known, damage at other locations may be computed from CDF value plots. As can be seen, the deviation is zero in case (1) and for other cases the deviations obtained are attributed due to the noise introduced in the modal properties.

12.4 Experimental set-up and analysis

12.4.1 Description of beam specimens

1800 mm long RC beams of M40 grade of concrete and HYSD reinforcement (Fe 415) having sectional dimension of 100 × 150 mm have been considered for experimentation. Three types of beams with two samples of each beam were considered for the test. The beam types are varying due the different reinforcement types. Table 12.2 shows the reinforcement details of different types of beams considered. The design of beams was carried out following IS: 456-FX12000. The beam with ID B1 is a balanced reinforced beam having reinforcement percentage of 1.88%. The beams with ID U1 and U2 are under-reinforced beams. Beam U1 has a less tensile reinforcement and same compression reinforcement as compared to the balanced beam B1. Beam U2 has more compression reinforcement and less tensile reinforcement as compared to the balanced beam. The reinforcement percentages in beam U1 and U2 were of the order of 1.42% and 2.20%, respectively. Table 12.2 gives the description of the different beam elements.

Table 12.2. Description of tested RC beam specimens

Beam Samples	Reinforcement details (Fe 415-HYSD)	
Beam-B1		STIRRUPS 6mm Φ @ 90 mmc/c 2-6mm Φ bar 2-12mm Φ bar Concrete cover = 20mm
Beam-U1		STIRRUPS 6mm Φ @ 90 mmc/c 2-6mm Φ bar 2-10mm Φ bar Concrete cover = 20mm
Beam-U2		STIRRUPS 6mm Φ @ 90 mmc/c 2-8mm Φ bar 2-12mm Φ bar Concrete cover = 20mm

12.4.2 Experimental methodology and instrumentation

Similar types of experimental methodology and instrumentation are used for all the specimens of RC beam. The details of experimentation and the analysis for specimen U1 are discussed in the subsequent sections along with brief descriptions of other tested specimens.

The data acquisition system with four on-board tri-axial MEMs accelerometers is used for vibration data extraction and mounted as shown in Photo 12.1. The accelerometers used have the working range of ±2 g to ±10 g and resolution of 1.5 mg RMS (±2 g), 9 mg RMS (±10 g) and synchronized with a sampling rate of 1–512 Hz. A compatible base station (WSDA®-Base) is used to acquire data at a sampling rate of 256 Hz. The arrangement samples the extracted data from nodes placed at various locations of the beam in a synchronized manner.

A100 kN servo-hydraulic actuator was used to apply load on simple supported RC beam specimens for induction of cracks. The loading was a four-point type loading. As the beams are simply supported, there is a probability of vertical displacement at higher frequencies. This will cause weakness of supports resulting in changes in dynamic characteristics. Backing rollers have been used at the support points for the restriction of vertical displacement, as shown in Photo 12.2. This will reduce the monitoring error resulting due to the vibration of support.

The applied monotonic load is increased gradually on beam specimens and the vibration responses were recorded at each stage. The dynamic analysis of the extracted data was performed both for intact stage and progressively damaged state in 40 load steps. For damage detection of a particular RC beam specimen U1, the analysis of vibration signature was performed at different damage states viz. undamaged state, at first significant crack at loading—18 kN, at maximum loading—52 kN and at ultimate failure—42 kN. Photo 12.3 shows the crack pattern at ultimate load of beam U1.

Photo 12.1 Placement of 'wireless tri-axial accelerometers' on beam U1.

Photo 12.2 Experimental set-up of simply supported RC Beam subjected to 4-point loading.

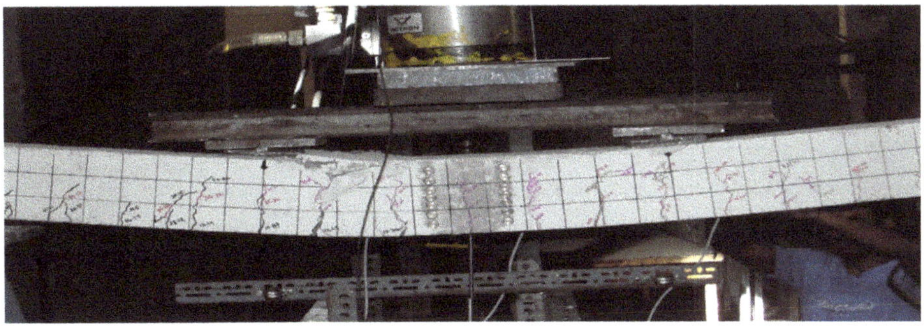

Photo 12.3 Ultimate failure of beam U1.

12.4.3 Health monitoring of RC beam

The analysis of the recorded vibration data was carried out for extraction of dynamic properties of undamaged and damage states of the experimented beam specimens. The frequencies and corresponding mode shapes computed for different states for beam U1 are given in table 12.3.

Figure 12.4 shows the change in frequency for each mode at different damage states for various load conditions. It is found that the damage increases with increase in load from zero to 50 kN which leads to frequency decrease. On attaining maximum load of 50 kN, there is substantial reduction in load with increase in displacement. There is an ultimate failure for RC beam specimen U1 at 42 kN load. So, a substantial increase in cracks was noticed leading to reduced frequency as a result of increase in load.

Figure 12.5 shows the mode shapes for RC beam specimen U1 at undamaged and damaged states, respectively. Some variations in mode shapes are observed for

Table 12.3. Experimentally obtained frequencies of RC Beam U1 at different damage states.

Damage states of beam	Load (kN)	1st Frequency (Hz)	2nd Frequency (Hz)	3rd Frequency (Hz)
Undamaged	0	3.22	7. 7.41	12.5
At first crack	18	3.14	7.31	12.37
At Max loading	50	3.02	7.22	12.15
At ultimate failure	42	2.91	7.06	11.96

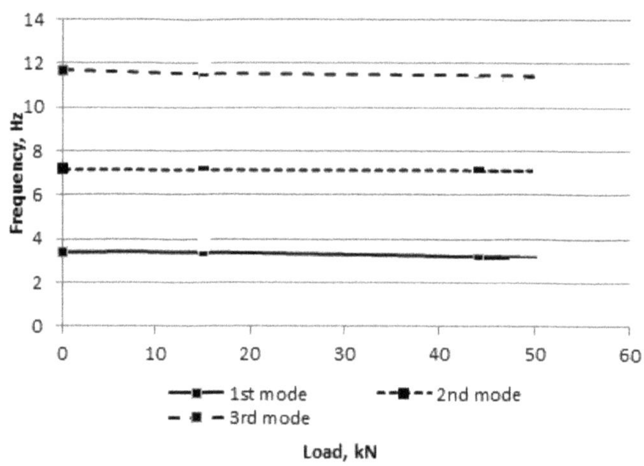

Figure 12.4. Variation of frequencies at different modes and loads

damaged state in comparison to undamaged state. For the case of the first mode shape (figure 12.5(a)), there is a significant variation in mode shape for 50 kN and 42 kN loaded situations. For the case of 18 kN, there is significant damage, with a slight variation in the graph. Identification of any change between the mode shape is difficult from lower mode towards higher mode side (figure 12.5(a)–(c)).

Modal curvatures are computed from the mode shape values using equation (12.1) considering the first three modes. Then CDFs using these modes are derived using equation (12.2) and are shown in figure 12.6. Figure 12.6 also shows the CDF values at various points of the beam at 18 kN, 50 kN and 42 kN respectively.

12.5 Results and discussion

Figure 12.6 shows how the CDF values changes for different locations with increase in load. For the case of 18 kN load, the CDF value is very small for the first element in comparison with that of the fourth element. It clearly shows that there is damage in the fourth element but it is difficult to identify the damage near the first element.

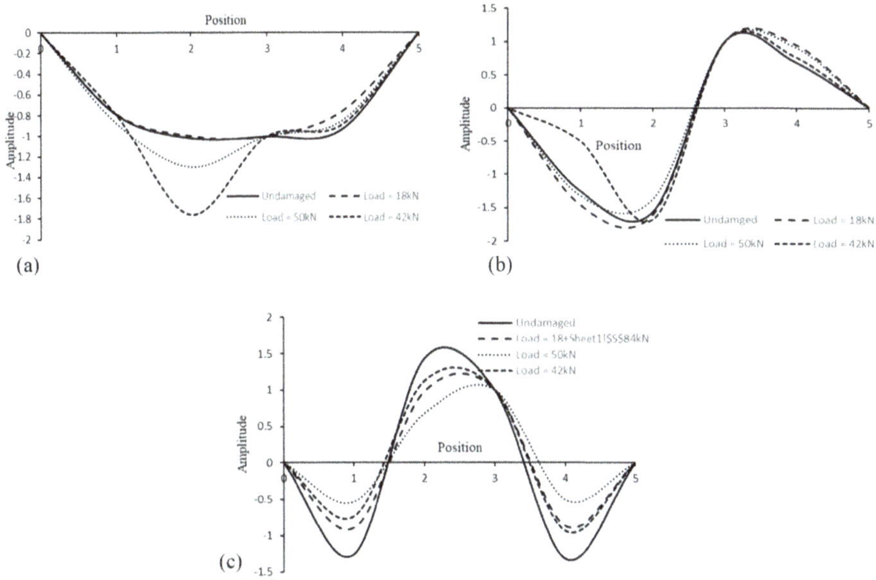

Figure 12.5. Mode shapes at intact and damaged states under various loadings. (a) First mode shape; (b) second mode shape; (c) third mode shape.

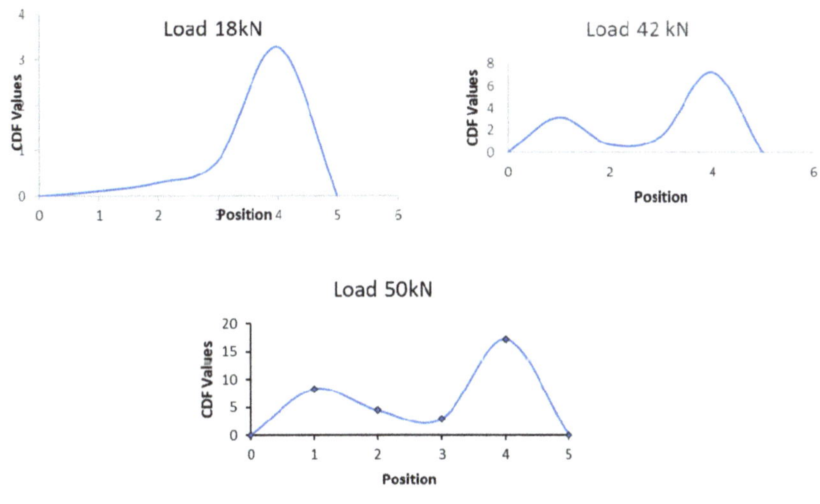

Figure 12.6. CDF for beam U1 at different loads.

However, CDF values near the first element keep on increasing with increase in load and become significant at maximum loading.

With the load of 18 kN, there is one peak of CDF value at the fourth element which indicates the crack near to the fourth element (figure 12.6). Two cracks are visible near first and fourth elements after application of maximum load of 50 kN and at failure. The matching of the real dense crack zone developed in the beam during the experimentation with the cracks identified from CDF plots, validates the CDF approach.

The observed variations in crack pattern during experiment are comparable to the variations observed in CDF values. The crack density near the fourth element is observed to be more than that near the first element after failure.

12.6 Conclusions

Studies by several researchers for damage detection in structural members using mode shape-based approaches have been described in brief. The present research mainly focuses on the damage identification procedure using mode shape-based approaches for damage assessment of a simply supported RC beam subjected to incrementally increasing load. Dynamic analysis by measuring vibrations using various mode shape approaches for damage identification of an RC beam at different load conditions in different damage states was carried out. It is observed from the results that the pure mode shape approach is not very successful in identification of damage. However, the modal curvature approach is very effective in damage location detection on RC beams. However, future research is needed to achieve the extent of damage exactly for life time prediction of a structure.

References

Bandyopadhyay D and Bhattacharyya S K 2007 *Statistical Health Monitoring of Structures Using Noisy Dynamic Responses* (IIT Kharagpur, India: ICTACEM) pp 27–9

Bandyopadhyay D and Bhattacharyya S K 2008 *An Inverse Dynamic Approach of Statistical Health Monitoring of Structure from Limited Noisy Data* (Adelaide: ICTAM) pp 25–9

Cao M and Qiao P 2009 Novel Laplacian scheme and multiresolution modal curvatures for structural damage identification *Mech. Syst. Sig. Process.* **23** 1223–42

Carden E and Fanning P 2004 Vibration based condition monitoring: a review *Struct. Health Monitor.* **3** 355–77

Chatterjee P K, Datta T K and Surana C S 1994 Vibration of continuous bridges under moving vehicles *J. Sound Vib.* **340** 619–32

Ciambella J, Vestroni F and Vidoli S 2011 Damage observability, localization and assessment based on eigen frequencies and eigen vectors curvatures *Smart Struct. Syst.* **9** 191–204

Ciambella J and Vestroni F 2015 The use of modal curvatures for damage localization in beam-type structures *J. Sound Vib.* **340** 126–37

Chaurasia A, Panigrahi S K and Patel S S 2018 Damage identification of reinforced concrete beam using modal curvature approach *J Sci. Ind. Res.* **77** 337–41

Darpe A K, Gupta K and Chawla A 2004 Coupled bending, longitudinal and torsional vibrations of a cracked rotor *J. Sound Vib.* **269** 33–60

Doebling S W, Farrar C R and Prime M B 1998 A summary review of vibration-based damage identification methods *Shock Vib. Digest* **30** 91–105

Dulieu-Barton J M, Brennan M J, Holford K M and Worden K 2003 Damage assessment of structures *Proceedings of DAMAS (5th Int. Conf. on Damage Assessment of Structures)* (Freienbach: Trans Tech Publications) pp 149–54

Foti D and Campian C 2007 Structural health monitoring through dynamic identification techniques: numerical simulation of a damage scenario *Buletinul, Institutului Politechnic Din Iasi, tomul LIII (LVII)* 3–4
Fan W and Qiao P 2011 Vibration-based damage identification methods: a review and comparative study *Struct. Health Monitor.* **10** 83–111
Farrar C and Jauregui D 1996 *Fault Detection Algorithms Applied to Experimental and Numerical Modal Data from The I-40 Bridge Technical Report LA-13074-MS* Los Alamos National Laboratory
Fox C H 1992 The location of defects in structures: a comparison of the use of natural frequency and mode shape data *Proc. of the 10th Int. Modal Analysis Conf.* pp 522–8
Gandomi A H, Sahab M G, Rahaei A and Safari G M 2008 Development in mode shape-based structural fault identification technique *World Appl. Sci. J.* **5** 29–38
Humar J, Bagchi A and Xu H 2006 Performance of vibration-based techniques for the identification of structural damage *Struct. Health Monitor.* **5** 215–41
Bureau of Indian Standards 2000 *IS 456 Plain and Reinforced Concrete Code* Bureau of Indian Standards, New Delhi
Ko J M, Sun Z G and Ni Y Q 2002 Multi-stage identification scheme for detecting damage in cable-stayed KapShuiMun Bridge *Eng. Struct.* **24** 857–68
Lee P K K, Ho E and Chung H W 1987 Static and dynamic tests of concrete bridges *ASCE J. Struct. Eng.* **113** 61–73
Pandey A K, Biswas M and Samman M M 1991 Damage detection from changes in curvature mode shapes *J. Sound Vib.* **145** 321–32
Panigrahi S K, Chakraverty S and Bhattacharyya S K 2015 Damage detection of multi-storeyed shear structure using sparse and noisy modal data *Smart Struct. Syst.* **15** 1215–32
Panigrahi S K, Chakraverty S and Mishra B K 2009 Vibration based damage detection in a uniform strength beam using genetic algorithm *Meccanica* **44** 697–710
Panigrahi S K, Chakraverty S and Mishra B K 2013 Damage identification of multistory shear structure from sparse modal information *J. Comp. Civ. Eng.* **27** 1–9
Ratcliffe C P 2000 A frequency and curvature based experimental method for locating damage in structures *J. Vib. Acoustics* **122** 324–29
Ren W-X and De Roeck G 2002 Structural damage identification using modal data. I: Simulation verification *J. Struct. Eng.* **128** 87–95
Rizos P F, Aspragathos N and Dimarogonas A D 1990 Identification of crack location and magnitude in a cantilever beam from the vibration modes *J. Sound Vib.* **138** 381–88
Saleh F, Supriyadi B, Suhendro B and Tran D 2004 Damage detection in non-prismatic reinforced concrete beams using curvature mode shapes *SIF2004 Structural Integrity and Fracture* (Melbourne: Australian Fracture Group Inc.) pp 331–37
Santos J V, Dos A, Mota S C M and Pina H L G 2000 Development of a numerical model for the damage identification on composite plate structure *Compos. Struct.* **48** 59–65
Srinivasa V, Sasmala S, Ramanjaneyulua K and Jeyasehar C A 2013 Influence of test conditions on modal characteristics of reinforced concrete structures under different damage scenarios *Arch. Civil Mech. Eng.* **13** 491–505
Stubbs N and Osegueda R 1990 Global damage detection in solids: experimental verification *Int. J. Anal. Exp. Modal Anal* **5** 81–97
Unger J F, Anne T and De R G 2005 Damage detection of a prestressed concrete beam using modal strains *J. Struct. Eng.* **131** 1456–63

Wahab M M A and Roeck G D 1999 Damage detection in bridges using modal curvatures: Application to a real damage scenario *J. Sound Vibration* **226** 217–35
Yu B, Shimao H, Wenjing N, Jing G and Xueqing S 2012 Plane grid structure damage location identification by model curvature *Proc. Eng.* **31** 534—40
Zhang J, Hue P, Hue Y C, Long D Y and Yu Z H 2011 Damage detection of plane member structures based on modal curvature difference method *Adv. Mater. Res.* **163** 2848–51

IOP Publishing

Modeling and Computation in Vibration Problems, Volume 1
Numerical and semi-analytical methods
S Chakraverty, F Tornabene and J N Reddy

Chapter 13

Modeling of honeycomb sandwich structure for spacecraft: analysis and testing

Ankur Kumar Gupta, Nazeer Ahmad, Akash Aditya and D Poomani

13.1 Introduction

Sandwich structures consist of a combination of different materials that are placed together so that the material properties of each one can be utilized for the structural advantage of the whole assembly. Sandwich panels generally consist of three significant components, two thin face sheets and a thick core. The adhesive films are placed between face sheets and core to bond them. A typical sandwich composite construction and an aluminum honeycomb sandwich panel are shown in figure 13.1. The face sheets have relatively high stiffness and strength to withstand high forces due to loading. The core material is light in weight and designed to have sufficient shear strength to withstand transverse shear stresses and also thick to provide high bending stiffness to resist buckling of the panel.

The bending stiffness and stiffness-to-weight ratio of the sandwich is greater than a single solid plate of the same total weight and the same material as that of the faces. As a result, sandwich construction results in lower lateral deformations, higher buckling resistance and higher natural frequencies than do other constructions. A major advantage of the sandwich construction compared to the solid plate design is that the core separates the stiff outer skins, placing them as far as possible from the neutral axis. The degree in which a structure prevents deflection in bending is known as the bending rigidity EI, where E is the Young's modulus of the material used and I is the second moment of area. The second moment of area I, which is the bending resistance of the cross-section, increases as more area is located away from the neutral axis. For example, the bending rigidity of a sandwich plate with 0.25 mm face sheet thickness and 25 mm honeycomb core is equivalent to a ~10 mm solid

doi:10.1088/978-0-7503-3483-9ch13

© IOP Publishing Ltd 2021

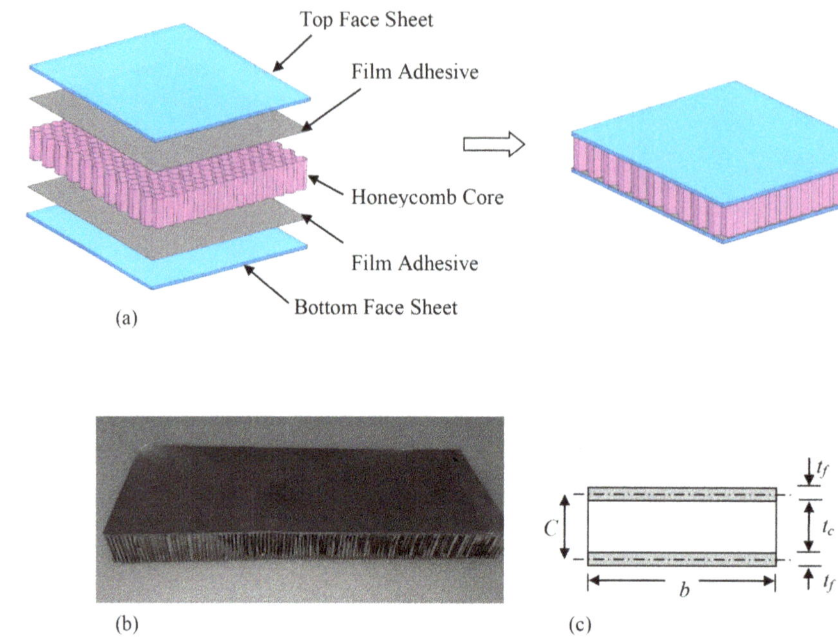

Figure 13.1. (a) A typical sandwich composite structure; (b) fabricated aluminum honeycomb sandwich panel; (c) cross-section of a sandwich beam.

plate of the same material and the mass ratio of solid plate to honeycomb sandwich plate is approximately 10.

Sandwich composite systems consist of advantages of miscellaneous materials having low density, high strength and high flexural resistance. They also possess the abilities of high-energy absorption and high load damping, sound and vibration insulation, shock mitigation and adaptive qualities. Nowadays, these materials are used as key technologies for innovative research and development. Sandwich composite materials use a homogeneous or non-homogeneous core of soft foams and other hard materials. Such composite constructions are widely used in aircraft, marine craft, spacecraft, and automobiles. Different newer materials are being tried for making cores, honeycombs, skin sheets; and the scope of applications is widening day-by-day. These are used to fabricate the structural components for large space stations and aeronautical applications [1, 2]. Honeycomb sandwich composites are commonly used in spacecraft structures due to their high specific stiffness and strength [3]. Sandwich composites with hexagonal aluminum core and aluminum or carbon fibre reinforced polymer (CFRP) face sheets are widely used in the primary and secondary structures of the spacecraft. Various antennas, solar arrays, payload support structures of the satellite are made of sandwich composites [4]. The honeycomb sandwich structure is widely used as a primary cylinder, payload mounting panels, shear webs, other support structures in spacecraft.

The equations for the optimization of sandwich construction for both strength and stiffness requirements, without significant computational expense can be derived analytically as explained here. The sandwich beam shown in figure 13.1(c) and its flexural rigidity with assumption $(E_c/E_f) \rightarrow 0$, $(t_f/C) \ll 1$ is given as:

$$D = E_f \frac{bt_f C^2}{2} \tag{13.1}$$

The combined weight of the two skins and the core is given by

$$W_t = \rho_c C + 2\rho_f t_f = \rho_c C + 2\rho_f \frac{2D}{E_f b C^2} \tag{13.2}$$

where, E_f and E_c are the modulus of elasticity; ρ_f and ρ_c are the density of the face and the core materials. For minimum value of weight (W_t) with respect to C:

$$\frac{dW_t}{dC} = \rho_c - \frac{8\rho_f D}{E_f b C^3} = 0 \quad \Rightarrow \rho_c C = 2(2\rho_f t) \tag{13.3}$$

This shows that the maximum flexural rigidity occurs when the mass of the core is equal to twice of the mass of face skin. Similarly for optimizing for the bending strength, the maximum bending stress is given by:

$$\sigma = \frac{My}{I} = \frac{M}{bt_f C} \tag{13.4}$$

The value of t_f from equation (13.4) can be put in equation (13.2) and for minimum value of weight (W_t) with respect to C:

$$\frac{dW_t}{dC} = \rho_c - \frac{2\rho_f M}{\sigma b C} = 0 \quad \Rightarrow \rho_c C = 2\rho_f t_f \tag{13.5}$$

This shows that the maximum bending strength occurs when the mass of the core is equal to the mass of face skin.

13.1.1 Types of sandwich core materials

Core materials are normally low strength materials, but their larger depth provides the sandwich construction with higher bending stiffness. Depending upon the nature of the support of a core to face sheets, the core materials can be divided into two categories: homogeneous and non-homogeneous core materials [5]. The various homogeneous and non-homogeneous cores are listed in table 13.1.

13.1.2 Honeycomb core

Honeycomb is most commonly used as the core of sandwich-structure composites. Honeycomb can be designed and manufactured with any number of cell shapes, sizes, and configurations. The honeycomb core is fabricated by three basic methods: expansion, corrugation and molding process. Figure 13.2 shows a typical hexagonal honeycomb that is widely used for many space and aerospace applications. In the honeycomb core, the ribbons (sheets) are attached to each other with the help of adhesive materials and made into repeated double walls. The direction of double-wall thickness is known as ribbon or 'L' direction and the other direction is

Table 13.1. Various types of cores in sandwich construction.

Types	Support	Cell orientation	Example
Homogeneous core	Continuous support	Open cells, closed cells or no cells	Foam, wood and balsa cores
Non-homogeneous core	Punctual support	Fully open	Textile/pin cores
	Regional support	Open to both direction	Cup shaped cores
	Unidirectional support	Open to one side	Corrugated cores
	Bi-directional support	Open in thickness direction	Honeycomb cores

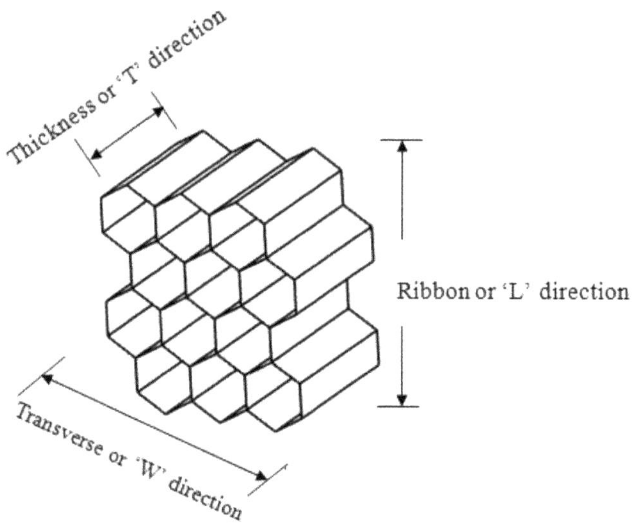

Figure 13.2. A typical hexagonal honeycomb core.

transverse or '*W*' direction. Due to their geometrical characteristics, the honeycomb cores are orthotropic in nature; hence, the stiffness and strength of the honeycomb panels significantly depend on the '*L*' and '*W*' directions [1].

13.2 Equivalent mechanical properties of honeycomb sandwich panels

The finite element method (FEM) is being widely used to analyze the composite sandwich. As discussed in the previous section, honeycomb sandwich composites consist of a large number of cells, so it is time-consuming and computationally demanding to create a detailed finite element (FE) model at the cell level in order to analyze their response. Cell-level simulations of sandwich composite structures would require modeling a large number of core cells which is computationally expensive. The cell-level modeling and simulation of sandwich structures require too much computational time and capability, which further increase for spacecraft level analysis. The honeycomb cell-level modeling in a sandwich panel is essential to find the failure modes involving core (e.g. transverse shear failure of the core) and estimation of transverse shear deformation in the case of a short beam. Except for some of the specific analyses, such as simulating failure propagation within the core, inserts bonding capacity, impact or crushing behavior of core, etc, analysis at spacecraft level may not require a detailed finite element modeling of the core. Thus computationally efficient modeling methods and constitutive laws are required to reduce time, whilst being accurate enough to realistically represent the overall structural behaviour.

The equivalent material modeling or homogenization techniques are the possible solutions to avoid individual unit cells level modeling. In past years, various homogenization methods have evolved to model the honeycomb core with a simplified continuous layer with effective (equivalent) material properties. These approaches eliminate the need for detailed finite element modeling of honeycomb-based sandwich structures. In these models, the core of the honeycomb sandwich panel is modeled as the equivalent single homogenous layer with the same thickness. Some of the methods are briefly discussed in the subsequent subsection. The equivalent model of a honeycomb sandwich panel is shown in figure 13.3.

Various researchers have developed analytical approaches for obtaining the equivalent orthotropic material properties of honeycomb cores [6, 7]. These analytical methods provide the equivalent material properties of honeycomb core which can avoid cell-level modeling of the honeycomb geometry. In general, the nine effective material constants for honeycomb structures, which need to be determined are: in-plane Young's moduli E_1, E_2, the out-of-plane Young's modulus E_3, in-plane shear modulus G_{12}, out-of-plane shear moduli G_{13}, G_{13}, and three Poisson ratios ν_{12}, ν_{13}, ν_{23}. The most popular models to determine the effective elastic properties of the hexagonal honeycomb core are proposed by Gibson *et al* [8], Gibson and Asby [9, 10], Grediac [11], Shi and Tong [12], Masters and Evans [13], Becker [14], Xu and Qiao [15].

Gibson *et al* [8] developed an equivalent model based on the Euler–Bernoulli beam theory for a two-dimensional honeycomb core. They investigated the

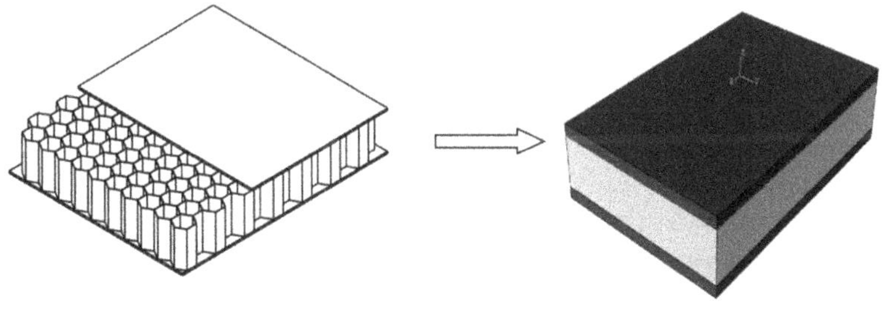

Figure 13.3. Equivalent model of a honeycomb sandwich panel.

equivalent mechanical properties of an equal-wall thickness hexagonal honeycomb. However, their equivalent method does not consider the shear deformation of the core layer, which affects the dynamic analysis accuracy once the ratio between the honeycomb wall thickness and the honeycomb side length increases. Further, Gibson and Ashby [9, 10] improved their model and presented all nine equivalent material constants for the honeycomb core. The in-plane properties were assumed to be isotropic and were calculated on the basis of representing the cell walls as beams. The in-plane Poisson's ratios were calculated from the formulas to be unity. The out-of-plane modulus was calculated based on the relative density of the material of the cell walls as a function of the core material. Zhang and Ashby [16] modified the formulae given by Gibson and Ashby [9] to include double thickness walls for the out-of-plane values, E_3, G_{13}, and G_{23}.

Grediac [11] calculated the transverse shear moduli of a honeycomb sandwich panel by using the finite element study of a representative unit cell. They provided the upper and lower bound of transverse shear moduli and investigated the effect of the core thickness on the transverse shear modulus. Shi and Tong [12] proposed a full model for the constitutive behavior of honeycomb structure. Their derivation was based upon a two-scale method for the homogenization of periodic media. They evaluated the equivalent transverse shear and in-plane moduli of honeycomb cellular structures. The equivalent two-dimensional constitutive equations are evaluated analytically in terms of their geometry and material properties. Masters and Evans [13] developed a model based on the deformation of the honeycomb cells by flexure, stretching and hinging. They considered the tensile, bending, and shear deformation of the Euler beam model to describe the displacement field of the honeycomb wall. Nast [6] presented the analytical formula for all the nine independent effective elastic moduli of the honeycomb core. They followed the Gibson and Ashby [9, 10] model for out-of-plane modulus calculation and in-plane moduli was calculated based on simple plate mechanics to the unit cell. They assumed that the straight cell walls remain un-deformed under in-plane loading. For the transverse shear moduli, they assumed the cell walls to behave as thick beams.

Becker [14] improved the model given by Gibson and Ashby [9, 10] by considering that the core deformations are dictated by the face sheets. Hence, the effective core stiffness becomes a nonlinear function of the total core thickness. Xu and Qiao [15] derived the eight elastic tensors of flexural, stretch, and transverse shear of honeycomb sandwich plates. They developed an effective approach to homogenize general honeycomb cells based on three aspects: the mathematical statement of sandwich homogenization theory, analytical solution of a multi-pass homogenization (MPH) technique, and a 3-D unit cell FEA homogenization technique. Further, Qiao and Xu [17] completed their study to derive formulas for equivalent torsional (warping flexural) and in-plane shear stiffness properties of sandwich plates with periodic core structures.

Mujika *et al* [18] investigated the out-of-plane shear modulus and the out-of-plane Young's modulus of the honeycomb core using the experimental, analytical and FEM. They concluded that FE characterization of the analytical model might be provided an overestimated result if the contact between double-cell walls is assumed to be perfectly bonded. Malek and Gibson [19] investigated all nine elastic constants more accurately by modifying the analysis of Gibson and Ashby [10] to account for the nodes at the intersection of the vertical and inclined members.

Mukhopadhyay and Adhikari [20] developed an analytical formulation predicting the equivalent elastic properties of irregular honeycombs with spatially random variations in cell angles. They proposed a novel analytical framework of a representative unit cell element (RUCE) approach to derive the closed-form expressions of equivalent in-plane elastic moduli of irregular honeycombs. Yuan *et al* [21] combined the analytical formulas given in the literature, and used an appropriate nine independent elastic parameters to study the behavior of honeycomb sandwich structure. They presented that the honeycomb core is equivalent to the interlayer with orthogonal anisotropic material properties. The introduction of interlayer makes the FE model represent the internal stress and local deformation of honeycomb sandwich structure, and provide more accurate simulation results.

13.2.1 Analytical formulations of the equivalent model of honeycomb core

Honeycomb core geometry and the coordinate system used to define the sandwich panel are shown in figure 13.4. It is assumed that the expansion method is used to manufacture the honeycomb cores leading to double thickness of the straight walls in the ribbon direction compared to the inclined walls. Figure 13.4 shows a unit cell of a conventional hexagonal honeycomb.

The dimensions 'l' and 'h' show the cell wall length, and they are equal for hexagonal honeycombs; 'd' is called cell size and 'θ' is core angle, and it is equal to 30° for hexagonal honeycombs. Moreover, 't' is the cell or foil wall thickness. The thickness of the core is perpendicular to the 'WL' plane and it is represented by 't_c'. The terms $\sin\theta$, $\cos\theta$, $\sin^2\theta$, $\cos^2\theta$, $\sin^3\theta$, and $\cos^3\theta$ in subsequent equations are denoted by s, c, s^2, c^2, s^3, and c^3, respectively. The material properties of equivalent core are represented as: in-plane Young's moduli E_1, E_2, out-of-plane

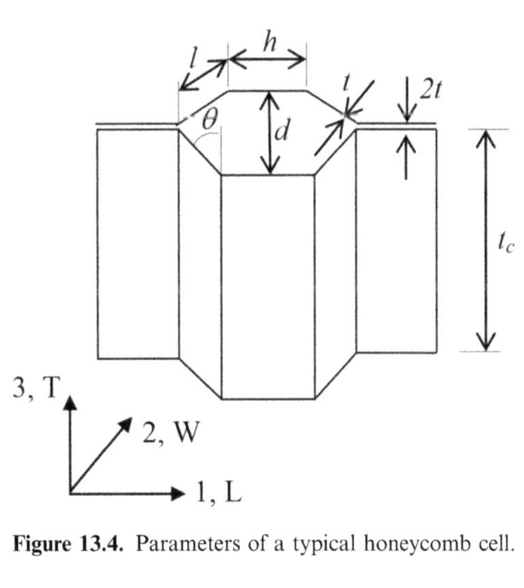

Figure 13.4. Parameters of a typical honeycomb cell.

Young's modulus E_3, in-plane shear modulus G_{12}, out-of-plane shear moduli G_{13}, G_{13}, and three Poisson ratios ν_{12}, ν_{13}, $\nu_{23,}$ and density $\rho_{ce.}$ The E_c, G_c, and ν_c are Young's moduli, shear modulus and Poisson ratios of the core material. The equations from some popular models for determining effective mechanical properties of honeycomb core are presented here.

(i) Gibson and Ashby [10]:

$$\begin{aligned}
&E_1 = \left(\frac{t}{l}\right)^3 \frac{(h/l+s)}{c^3} E_c; \quad E_2 = \left(\frac{t}{l}\right)^3 \frac{c}{(h/l+s)s^2} E_c \\
&\nu_{12} = \frac{(h/l+s)s}{c^2}; \quad G_{12} = \left(\frac{t}{l}\right)^3 \frac{(h/l+s)}{(h/l)^2(1+2h/l)c} E_c \\
&E_3 = \left(\frac{t}{l}\right) \frac{(h/l+2)}{2(h/l+s)c} E_c; \quad \nu_{13} = \nu_{23} = 0 \\
&G_{13} = \left(\frac{t}{l}\right) \frac{(h/l+s)}{(1+2h/l)c} G_c; \quad G_{23} = \left(\frac{t}{l}\right) \frac{c}{(h/l+s)} G_c; \\
&\rho_{ce} = \left(\frac{t}{l}\right) \frac{(h/l+2)}{2(h/l+s)c} \rho_c
\end{aligned} \tag{13.6}$$

(ii) Grediac [11];

$$\begin{aligned}
&R_1 = t/h; \quad R_2 = l/h; \quad R_3 = t_c/l \\
&G_{23} = R_1 \frac{c}{(1+R_2 s)} G_c \\
&G_{13_upper} = R_1 \frac{1+R_2 s^2}{(1+R_2 s)R_2 c} G_c; \quad G_{13_lower} = R_1 \frac{(1+R_2 s)}{R_2(1+R_2)c} G_c; \\
&G_{13} = G_{13_lower} + \frac{0.787}{R_3}(G_{13_upper} - G_{13_lower})
\end{aligned} \tag{13.7}$$

(iii) Shi and Tong [12]

$$\begin{aligned}
&E_1 = E_2 = \frac{\sqrt{3}}{5}\left(\frac{t}{l}\right) E_c; \quad G_{12} = \sqrt{3}\left(\frac{t}{l}\right)^3 E_c \\
&\nu_{12} = \nu_{21} = 1; \quad \nu_{13} = \nu_{23} = 0 \\
&G_{13} = \left(\frac{t}{l}\right)\frac{\sqrt{3}}{2} G_c; \quad G_{23} = \left(\frac{t}{l}\right)\frac{1}{\sqrt{3}} G_c
\end{aligned} \tag{13.8}$$

(iv) Masters and Evans [13]:

$$\begin{aligned}
&E_1 = \frac{1}{\frac{tcc}{(h/l+s)}\left[\frac{c^2}{K_f} + \frac{c^2}{K_h} + \frac{(2h/l+s^2)}{K_S}\right]}; \quad E_2 = \frac{1}{t_c(h/l+s)\left[\frac{s^2}{K_f c} + \frac{s^2}{K_h c} + \frac{c}{K_S}\right]} \\
&\nu_{12} = \frac{-s(h/l+s)\left[-\frac{1}{K_f} - \frac{1}{K_h} + \frac{1}{K_S}\right]}{\left[\frac{c^2}{K_f} + \frac{c^2}{K_h} + \frac{\left(2h/l+s^2\right)}{K_S}\right]} \quad G_{12} = \frac{1}{A + \frac{1}{K_h}\left[\frac{h^2+2l^2}{l(h+ls)}\right]t_c c + B\left[\frac{c}{(h+ls)} + \frac{s}{c}\right]}
\end{aligned} \tag{13.9}$$

where $A = \frac{t_c h^2(l+2h)c}{K_f l^2(h+ls)}$, $B = \frac{t_c(lc^2+(h+ls))s}{K_S}$,

$$K_f = \frac{E_c t_c t^3}{l^3},\ K_s = \frac{E_c t_c t}{l} \quad \text{and} \quad K_h = \frac{G_c t_c t}{l}$$

(v) Zhang and Ashby [16]:

$$\begin{gathered} G_{23} = \left(\frac{t}{l}\right)\frac{c}{(h/l+s)}G_c \\ G_{13\_upper} = \left(\frac{t}{l}\right)\frac{(h/l+s^2)}{(h/l+s)c}G_c;\ \ G_{13\_lower} = \left(\frac{t}{l}\right)\frac{(h/l+s)}{2c}G_c; \\ E_3 = \left(\frac{t}{l}\right)\frac{(1+h/l)}{(h/l+s)c}E_c \\ \rho_{ce} = \left(\frac{t}{l}\right)\frac{(1+h/l)}{(h/l+s)c}\rho_c \end{gathered} \tag{13.10}$$

(vi) Nast [6]

$$\begin{aligned} &E_1 = \left(\frac{t}{l}\right)^3 \frac{(s+1)}{12c^3\left[\frac{c}{3} - \frac{1+c}{8}\right]}\frac{E_c}{(1-\nu^2)}; \qquad E_2 = \left(\frac{t}{l}\right)^3 \frac{c}{(1+s)s^2}\frac{E_c}{(1-\nu^2)} \\ &\nu_{12} = \frac{s^2(1+s)^2}{12c^2\left[\frac{c}{3} - \frac{1+c}{8}\right]}; \qquad G_{12} = \left(\frac{t}{l}\right)^3 \frac{(s+1)}{c(6.25-6s)}\frac{E_c}{(1-\nu_c^2)} \\ &E_3 = \left(\frac{t}{l}\right)\frac{2E_c}{(1+s)c}; \qquad \nu_{13} = \left(\frac{t}{l}\right)^2 \frac{(s+1)^2}{24c\left[\frac{c}{3} - \frac{1+c}{8}\right]}\frac{\nu_c}{(1-\nu_c^2)} \\ &\nu_{23} = \left(\frac{t}{l}\right)^2 \frac{c^2}{2s^2}\frac{\nu_c}{(1-\nu_c^2)}; \qquad G_{13} = \left(\frac{t}{l}\right)\frac{2}{c(s+1)}G_c \\ &G_{23} = \left(\frac{t}{l}\right)\frac{10}{9c^3(s+1)}G_c \end{aligned} \tag{13.11}$$

(vii) Yuan *et al* [21]:

$$\begin{aligned} &E_1 = \left(\frac{t}{l}\right)^3 \frac{(h/l+s)}{c^3}E\left[1 - (h/l\sec^2\theta + \tan^2\theta)\frac{t^2}{l^2}\right] \\ &E_2 = \left(\frac{t}{l}\right)^3 \frac{c}{(h/l+s)s^2}E\left(1 - \cot^2\theta\frac{t^2}{l^2}\right) \\ &\nu_{12} = \frac{(h/l+s)s}{c^2}\left(1 - (h/l+1)\sec^2\theta\frac{t^2}{l^2}\right) \\ &G_{12} = \left(\frac{t}{l}\right)^3 \frac{(h/l+s)}{(h/l)^2(h/l/4+1)c}E \\ &E_3 = \left(\frac{t}{l}\right)\frac{(h/l+1)}{(h/l+s)c}E; \qquad \nu_{13} = \nu_{23} = \nu \\ &G_{13} = \left(\frac{t}{l}\right)\frac{(h/l+s)}{(h/l+1)c}G; \qquad G_{23} = \left(\frac{t}{l}\right)\frac{c}{(h/l+s)}G \\ &\rho_{ce} = \left(\frac{t}{l}\right)\frac{(h/l+1)}{(h/l+s)c}\rho \end{aligned} \tag{13.12}$$

Table 13.2. Material and geometrical parameters of honeycomb core given by Nast [6].

Item	Property	Value
Cell dimension	Cell thickness (t)	0.09 mm
	Cell size (d)	3.2 mm
	Cell side ($l = h$)	1.85 mm
	Cell angle (θ)	30°
	Core thickness (t_c)	19 mm
Elastic properties of core	Young's modulus (E_c)	1.1 GPa
	Poisson ratio (ν_c)	0.3
	Shear modulus (G_c)	0.42 GPa

Table 13.3. Comparison of equivalent properties of core.

Source	MPa						ν_{12}	ν_{13}	ν_{23}
	E_1	E_2	E_3	G_{12}	G_{13}	G_{23}			
Gibson and Ashby [10]	0.292	0.292	61.79	0.073	11.79	11.79	1.00	0.000	0.000
Grediac [11]					19.67–17.69	11.79			
Shi/Tong [12]	18.53	18.53	82.38	0.219	17.69	11.79	1.000	0.000	0.000
Masters and Evans [13]	0.288	0.288		0.073			0.991		
Zang and Ashby [16]			82.38		19.67–17.69	11.79			
Nast [6]	0.419	0.321	82.38	0.074	31.45	23.31	1.128	0.002	0.001
Yuan *et al* [21]	0.290	0.291	82.38	0.175	17.69	11.79	0.991	0.300	0.001

13.2.2 Equivalent properties prediction using analytical formulations

Further, all equivalent models of honeycomb core are compared for a given case. The material and geometrical parameters are taken from Nast [6] and presented in table 13.2. The equivalent properties of core are calculated using equations (13.6)–(13.12) given by various analytical models and summarized in table 13.3. It can be seen from the table that equivalent properties are in good agreement with those presented by Nast [6].

13.3 Finite element modeling of the honeycomb core sandwich laminates

The FEM is the most powerful tool which is widely used in structural analysis problems. Depending on the required output and accuracy, three basic FE modeling approaches are being used for honeycomb sandwich structural analysis. These FE modeling methods are detailed cell-level modeling, three-dimensional (3D) equivalent modeling and two-dimensional (2D) equivalent modeling approaches. A detailed description of these models is shown in figure 13.5. A summary of the FE modeling approaches is given in table 13.4:

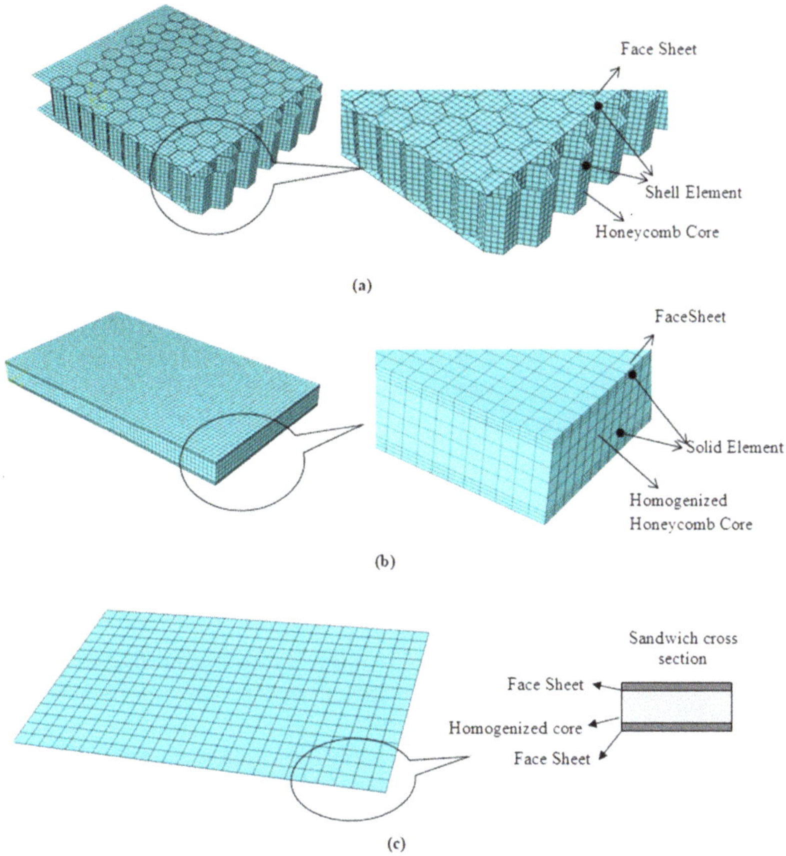

Figure 13.5. Finite element models used to model the honeycomb sandwich structure; (a) detailed cell-level modeling; (b) 3D equivalent modeling; (c) 2D equivalent modeling.

The detailed cell-level modeling approach (Model-1) involves modeling the core with the actual core geometry of the honeycomb sandwich. Model-1 is a computationally expensive method but useful in certain studies such as pull-out and in-plane capacity of an insert bonded in the sandwich panel, core crushing and impact behavior of sandwich composites, etc.

Model-1 does not require the equivalent core properties and gives accurate estimates of the modal frequencies of the sandwich laminate's modal frequencies. This model can be used as the reference model for comparison of the results obtained from the 2D model.

In Method-2, the cell-level FE modeling is replaced by an equivalent continuum modeling of the core. In this method, the core is modeled with 3D solid elements, and the equivalent orthotropic elastic properties obtained through analytical, numerical and/or experimental methods are used. The equivalent orthotropic material properties of the core can be obtained from the formulas explained in section 13.2.1. This method is less computationally expensive than Model-1, but the

Table 13.4. Description of FE modeling methods for honeycomb sandwich composites.

Model		Description
Model-1	Detailed cell-level modeling	• The actual geometry of the cells of honeycomb core modeled with 2D shell elements. • Facesheets are modeled with 3D solid element or 2D shell element. • Computationally expensive. • Can be used as the reference model for comparison.
Model-2	3D equivalent modeling	• Core is modeled as continuum with 3D solid element. • Elastic constants of the equivalent core are obtained by the various analytical, numerical and experimental approaches (discussed in section 13.2.1). • Facesheets are modeled with 3D solid element or 2D shell element. • Computationally less expensive compared to model-1. • Requires equivalent core properties.
Model-3	2D equivalent modeling	• Face sheets and core are modeled as different layers of the composite using shell elements. • Elastic constants of the equivalent core are estimated by the various analytical/numerical approaches (discussed in section 13.2.1). • Computationally less expensive compared to both models. • Requires equivalent core properties.

accuracy of the results depends on the predicted values of equivalent elastic constants. In Method-3, face sheets and core are modeled as different layers of the composite using layered shell elements. This method is significantly less computationally expensive than both models and requires the elastic constants of the equivalent core, which are estimated by the various methods. The accuracy of the results obtained using this method depends on the values of equivalent properties and efficiency of displacement theories [22, 23].

13.4 Modal study of honeycomb beam

In this section, a modal study of a honeycomb beam by the experimental test of the specimen and FE analysis is presented in detail. A comparison of the obtained results by these methods is shown here. This study aims to determine the Eigen frequencies and the mode shapes of this beam under the fixed at one end boundary condition. In this analysis, the adhesive is not taken into consideration.

A honeycomb sandwich beam of dimensions, length (L) = 1275 mm, width (W) = 100 mm and thickness (t_{tot}) = 25.5 mm was used for the study. The face sheets and core are made of aluminum alloy. The thickness of the honeycomb core and both face sheets (top and bottom) of the sandwich beam are 25 and 0.25 mm,

Table 13.5. Dimensions and mechanical properties of the honeycomb beam.

Item	Property	Value
Cell dimension	Cell thickness (t)	0.017 78 mm
	Cell size (d)	4.76 mm
	Cell angle (θ)	30°
Thickness of core and face sheet	Core thickness (t_c)	25 mm
	Face sheet thickness (t_f)	0.25 mm
Overall dimension	Length (L)	1275 mm
	Width (W)	100 mm
	Thickness (t_{tot})	25.5 mm
Elastic properties	Young's modulus (E_c and E_f)	72 GPa
	Poisson ratio (ν_c and ν_f)	0.33
	Shear modulus (G_c and G_f)	27 GPa

respectively. Dimensions and mechanical properties of the sandwich beam in detail are given in table 13.5.

13.4.1 Experimental investigation

The experimental setup used in this work is described first, and then responses obtained through experiments and FE computations are compared.

13.4.1.1 Experimental measurement setup

A sandwich cantilever beam with aluminium honeycomb core and face-sheets having dimensions (L) = 1275 mm, width (W) = 100 mm and thickness (t_{tot}) = 25.5 mm was used for the experimental study. A modal shaker from MB Dynamics with a force rating of 100 N was used to excite the beam. The stringer of the shaker was connected to the beam at a distance of 200 mm from the fixed as shown in figure 13.6. Choice of the location for mounting the shaker is to excite the multiple lower bending modes. The stinger is equipped with an impedance head (PCB-288D01) for input acceleration and force measurement. Five accelerometers (make: B & K, Model: 4517-002, sensitivity 10 mV/g) were mounted on the top-face center line of the beam. The accelerometers were kept at a distance of 200 mm from each other. The Leuven Measurement Systems (LMS) 32-channel DAS was used for data acquisition and LMS Test Lab for processing [24].

13.4.1.2 Experimental results

The modal shaker was excited with a broadband random Gaussian noise with a power spectral density (PSD) of 0.04 N^2/Hz and RMS14.14 N. The input signal has a bandwidth of 5–5000 Hz. A one-second slice of the random input force measured is given in figure 13.7. The acceleration of the beam was measured at 5 kHz sampling rate well above the requirements of the Nyquist criterion for capturing the structural frequencies within 1 kHz.

Figure 13.6. Experimental setup.

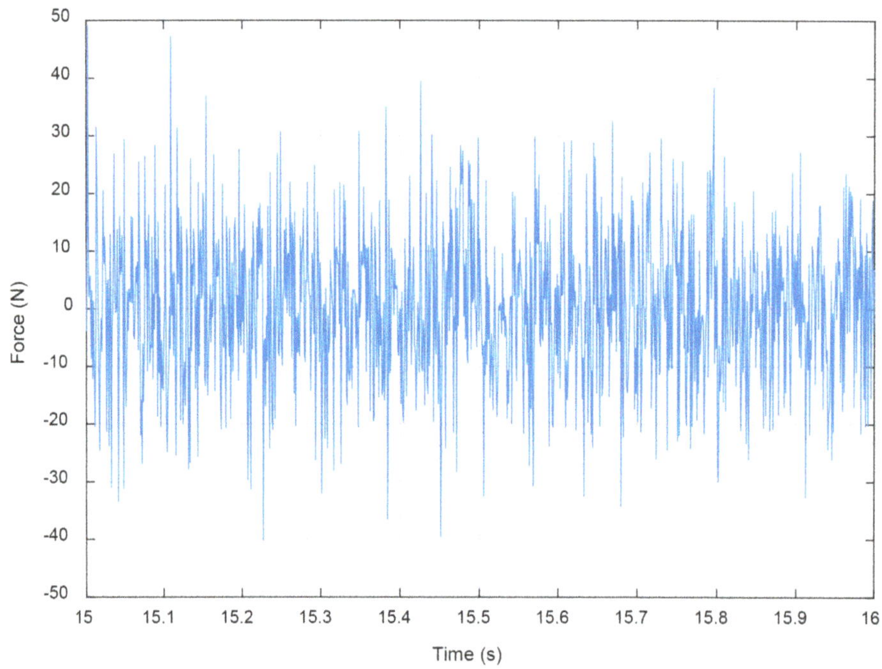

Figure 13.7. One-second slice of the excitation force.

The acceleration and force data for response and input was acquired for a duration of 40 s to get a frequency resolution of 0.25 Hz. The frequency response function called accelerance defined as the ratio of the output acceleration to input force was obtained from the acceleration measurements of five accelerometers

Table 13.6. Experiment model test results of sandwich beam.

Frequency (test) (Hz)	Mode shape description
17.1	1st bending mode
114.5	2nd bending mode
321.4	3rd bending mode
592.6	4th bending mode
915.1	5th bending mode

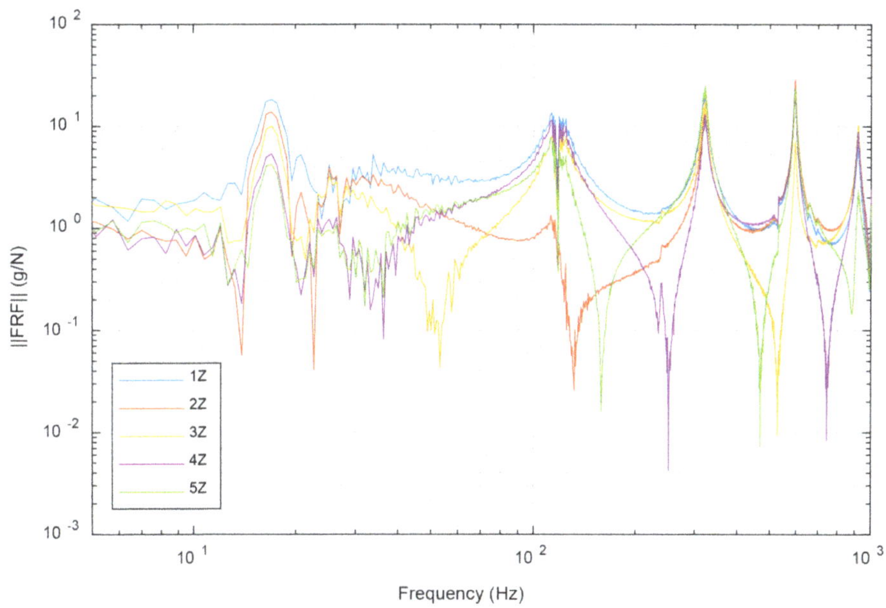

Figure 13.8. Transfer functions of the HC beam at locations 1Z–5Z.

mounted along the length of the beam. The transfer functions were obtained using following parameters: blocksize, 4s; window, Hanning; averages,16 and overlapping, 50% and shown in figure 13.8. The bending mode (BM) frequency from the experiment presented in table 13.6.

13.4.2 Modal analysis of honeycomb sandwich beam using finite element methods

In the present study, the modal analysis of a sandwich beam was carried out using detailed cell-level modeling and two-dimensional (2D) equivalent modeling approaches. Both types of FE analyses were carried using commercial ABAQUS

Table 13.7. Values of the orthotropic material constant for a sandwich beam.

Source	MPa						ν_{12}	ν_{13}	ν_{23}	ρ_{ce}
	E_1	E_2	E_3	G_{12}	G_{13}	G_{23}				Kg m^{-3}
Gibson and Ashby [10]	0.045	0.045	537.92	0.011	100.86	100.86	1.000	0.000	0.000	20.92
Grediac [11]					152.74	100.86				
Shi/Tong [12]	161.3	161.3	717.23	0.034	151.29	100.86	1.000	0.000	0.000	27.89
Masters and Evans [13]	0.045	0.045		0.011			1.000			
Zang and Ashby [16]			717.23		152.74	100.86				27.89
Nast [6]	0.066	0.051	717.23	0.012	268.96	199.23	1.128	0.002	0.003	27.89
Yuan *et al* [21]	0.045	0.045	717.23	0.027	151.29	100.86	1.000	0.330	0.002	27.89

Table 13.8. Comparison of cell-level model and 2D equivalent model with experimental modal results for sandwich beam.

	Frequency (Hz)						
			2D equivalent model				
Mode	Experimental	Cell-level model	Gibson and Ashby [10]	Zang and Ashby [16]	Shi/Tong [12]	Nast [6]	Yuan *et al* [21]
1st BM	17.1	18.61	19.61	21.64	22.762	18.04	17.08
2nd BM	114.5	114.27	123.68	133.07	139.6	112.16	105.76
3rd BM	321.4	310.23	334.7	361.93	378.42	310.18	290.62
4th BM	592.6	583.15	626.65	681.91	709.88	597.73	554.8
5th BM	915.1	918.11	982.4	1076	1114.9	966.58	888.75

software. The details of the FE modeling and results obtained from these methods are discussed subsequently.

13.4.2.1 Cell-level modeling

The cell-level detailed modeling of the honeycomb sandwich beam was carried out using an ABAQUS- Python script. The description of the FE modeling is given in section 13.3. The geometric and material parameter considered for modal analysis of the sandwich beam is presented in table 13.5. In this model, each unit cell of the honeycomb core was assumed to be a regular hexagon. The parallel wall of the honeycomb has a thickness twice that of the other that is incorporated through separate section definition for that wall. The adhesive layers are assumed to be very thin, so the effect of the adhesive layer was neglected. After the meshing operation, the nodes at the repeated cells were merged. The cell walls of single thickness are modeled with a single four-noded shell FE with reduced integration (S4R), having six degrees of freedom per node (three translations and three rotations), while those of double thicknesses are here modeled by superposing two elements; this simplifies, in particular, the number of elements properties. The surfaces of the cell and edges of the top and bottom face sheets are constrained fully (displacement of x, y and z are zero, and rotation of x, y and z are zero) in the cantilevered honeycomb sandwich beam.

The modal analysis results and mode shapes of the sandwich beam are shown in figure 13.9. The first five bending modes and their corresponding frequencies of the honeycomb sandwich beam are presented here. All bending frequencies, up to

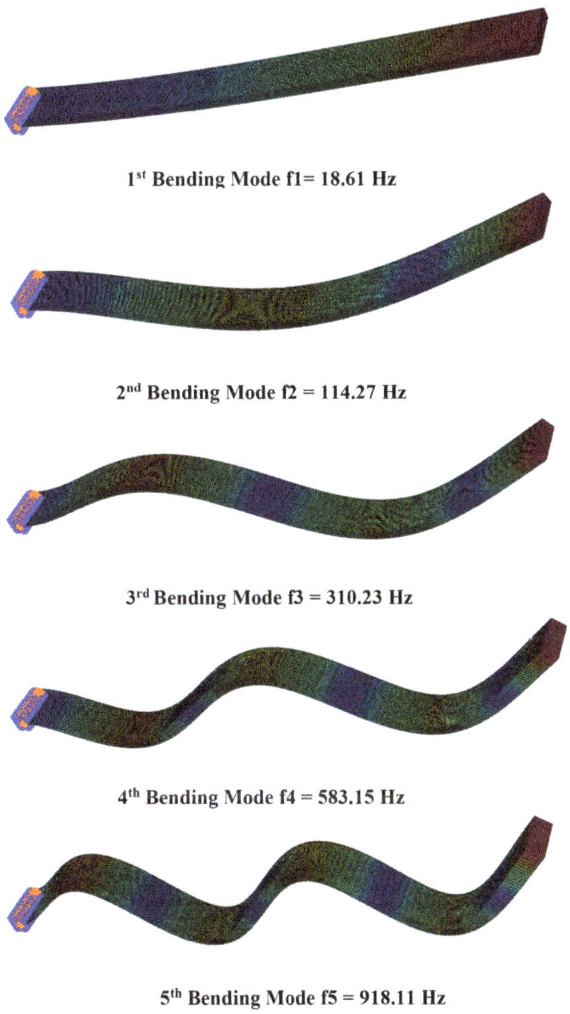

Figure 13.9. Various mode shapes of a honeycomb beam using cell-level modeling.

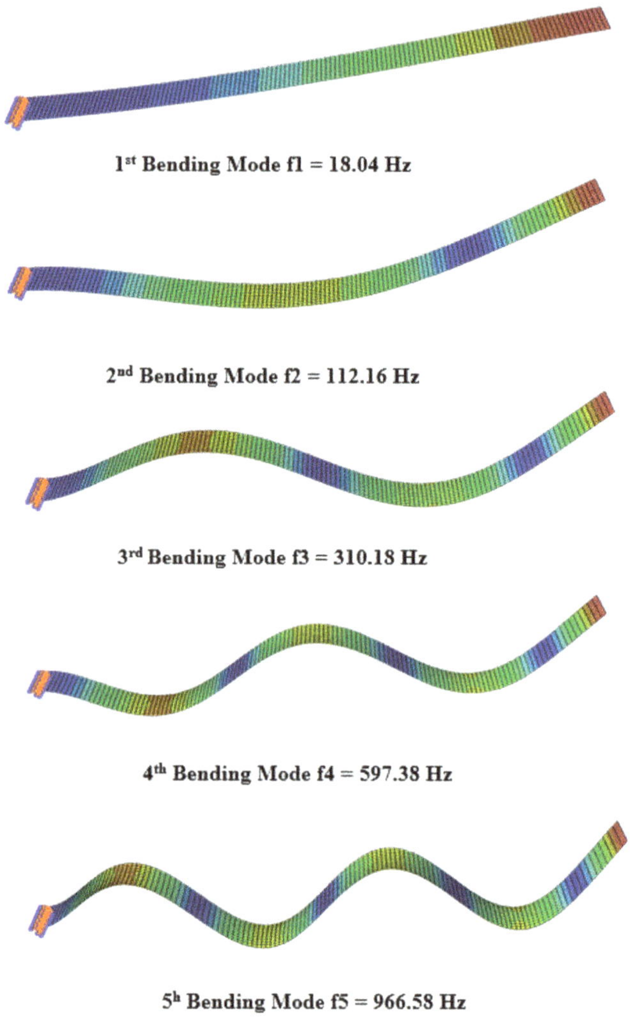

Figure 13.10. Various mode shapes of a honeycomb beam using 2D equivalent simulation based on the Nast [6] analytical model.

1000 Hz, of the honeycomb beam, are found to be in very good agreement with the frequencies obtained from experiments.

13.4.2.2 Two dimensional (2D) equivalent modeling

In the 2D equivalent FE modeling, the honeycomb core was modeled with shell elements just like the face sheets, and orthotropic elastic constants, are assigned to the core material. In the equivalent model, the honeycomb core, as well as the face sheets, were modeled with shell elements by defining the sectional properties.

Core and face sheets were modeled with the same thickness, length and width as the actual one. The sandwich composite was modeled as a laminate composed of two layers for the upper and the lower face sheets and single-layer honeycomb core. The elastic constants of the equivalent core layer were calculated from various analytical models given in section 13.2.1. The values of the orthotropic material constant for the sandwich beam calculated using equations (13.1)–(13.7) are summarized in table 13.7. Using the property summarized in table 13.7, the model analysis based on 2D modeling of the sandwich beam is carried out. The results are presented in table 13.8. In the table, the frequencies corresponding to the first five bending modes obtained from experiments, cell-level modeling and 2D equivalent modeling (for various analytical models) are compared. It can be seen from the table that the frequencies predicted using the Nast [6] model are closely matching with the cell-level model as well as experimental results. The mode shapes corresponding to the Nast [6] model are shown in figure 13.10.

References

[1] Zenkert D 1995 *An Introduction to Sandwich Construction* (Cradley Heath: EMAS Publishing)

[2] Castanié B, Bouvet C and Ginot M 2020 Review of composite sandwich structure in aeronautic applications *Compos. Part C: Open Access* **1** 100004

[3] Wijker J J 2008 *Spacecraft Structures* (Cham: Springer)

[4] Morozov E V, Lopatin A V and Taygin V B 2015 Design, fabrication and testing of composite sandwich integral structure of spacecraft antenna *Compos. Struct.* **134** 645–53

[5] Thomsen O T 2009 Sandwich materials for wind turbine blades—present and future *J. Sandw. Struct. Mater.* **11** 7–26

[6] Nast E 1997 On honeycomb-type core moduli *38th Structures, Structural Dynamics, and Materials Conference, Structures, Structural Dynamics, and Materials and Co-located Conferences* (Reston, VA: American Institute of Aeronautics and Astronautics)

[7] Schwingshackl C W, Aglietti G S and Cunningham P R 2006 Determination of honeycomb material properties: existing theories and an alternative dynamic approach *J. Aerosp. Eng.* **19** 177–83

[8] Gibson L J, Ashby M F, Schajer G S and Robertson C I 1982 The mechanics of two-dimensional cellular materials *Proc. Royal Soc. London A. Math. Phys. Sci.* **382** 25–42

[9] Gibson L J and Ashby M F 1988 *Cellular Solids: Structure and Properties* (Oxford: Pergamon)

[10] Gibson L J and Ashby M F 1997 *Cellular Solids: Structure and Properties* (Cambridge: Cambridge University Press) pp 175–231

[11] Grediac M 1993 A finite element study of the transverse shear in honeycomb cores *Int. J. Solids Struct.* **30** 1777–88

[12] Shi G and Tong P 1995 The derivation of equivalent constitutive equations of honeycomb structures by a two scale method *Comput. Mech.* **15** 395–407

[13] Masters I G and Evans K E 1996 Models for the elastic deformation of honeycombs *Compos. Struct.* **35** 403–22

[14] Becker W 1998 The in-plane stiffnesses of a honeycomb core including the thickness effect *Arch. Appl. Mech.* **68** 334–41

[15] Xu X F and Qiao P 2002 Homogenized elastic properties of honeycomb sandwich with skin effect *Int. J. Solids Struct.* **39** 2153–88

[16] Zhang J and Ashby M F 1992 The out-of-plane properties of honeycombs *Int. J. Mech. Sci.* **34** 475–89

[17] Qiao P and Xu X F 2005 Refined analysis of torsion and in-plane shear of honeycomb sandwich structures *J. Sandwich Struct. Mater.* **7** 289–305

[18] Mujika F, Pujana J and Olave M 2011 On the determination of out-of-plane elastic properties of honeycomb sandwich panels *Polym. Test.* **30** 222–28

[19] Malek S and Gibson L 2015 Effective elastic properties of periodic hexagonal honeycombs *Mech. Mater.* **91** 226–40

[20] Mukhopadhyay T and Adhikari S 2016 Free-vibration analysis of sandwich panels with randomly irregular honeycomb core *J. Eng. Mech.* **142** 06016008

[21] Yuan J, Zhang L and Huo Z 2020 An equivalent modeling method for honeycomb sandwich structure based on orthogonal anisotropic solid element *Int. J. Aeronaut. Space Sci.* **21** 957–69

[22] Irfan S and Siddiqui F 2019 A review of recent advancements in finite element formulation for sandwich plates *Chin. J. Aeronaut.* **32** 785–98

[23] Abrate S and Di M Sciuva 2017 Equivalent single layer theories for composite and sandwich structures: a review *Compos. Struct.* **179** 482–94

[24] Ahmad Nazeer, Ranganath R. and Ghosal Ashitava 2017 Modeling and experimental study of a honeycomb beam filled with damping particles *J. Sound Vibr.* **391** 20–34

IOP Publishing

Modeling and Computation in Vibration Problems, Volume 1
Numerical and semi-analytical methods
S Chakraverty, F Tornabene and J N Reddy

Chapter 14

Numerical analysis of Qutb Minar using non-linear plastic-damage macro model for constituent masonry

Ajay Chourasia and S K Panigrahi

This chapter deals with 3-D finite element (FE) modeling and analysis of the historical construction 'Qutb Minar', New Delhi, India. A plastic-damage macro model, originally proposed to model plasticity in concrete, has been adapted to simulate behavior of the masonry present in Qutb Minar. The natural frequencies obtained from an idealized FE-model show better correlation with experimental values from previous studies. The material model implemented is characterized by a bi-dissipative and isotropic degradation of material during cyclic loading. Thus, a seismic simulation of Qutb has been carried out for comparison between responses in two cases—elastic and non-elastic material models for constituent masonry. The instantaneous displacements were relatively lower than observed in the linear elastic analysis. Also, as stresses observed in masonry are very large, Qutb Minar needs to be strengthened against future seismic actions. The acceleration time-history is different in comparison with the seismic input at the base of the tower, for the filter effect of the whole structure. The need to take account of the dynamic parameters and material non-linearity for a realistic seismic response prevision has also been established.

14.1 Preamble

Ancient slender masonry structures may be traced to different structural typologies: lookout towers, bell towers, chimney stacks, minarets etc. At the time of historical seismic events, such constructions, even though intended for different functions and belonging to different epochs and architectural styles, have shown a structural response that was difficult to foresee in advance. Observation of post-earthquake damage has indeed led to the finding of many collapse modes when faced with the same typological features and, in some cases, an unexpected resistance if compared

 © IOP Publishing Ltd 2021

with their intrinsic vulnerability. Constructed in the 12th century AD, Qutb Minar is one of the tallest stone masonry minarets in the world, in New Delhi, India. It is 72.5 m high with 379 spiral steps and has a diameter of 14.3 m at the base and 2.8 m at the apex. Thus, the analysis was necessitated to assess safety of the tower in the incidence of an earthquake.

The geometric complexity, material complexity and sustained weathering deterioration make the historical structure unique in itself. Furthermore, from a computational point of view, modeling of masonry is also complex since mechanical behavior of historical masonry depends on the composite nature of the material, constituted by the fine dimensions, of discrete natural blocks connected by hydraulic lime mortar joints. Since discrete models of historical structures are computationally infeasible due to size and geometrical complexity, numerical analysis assuming a fictitious homogeneous and continuous equivalent material, the mechanical parameters of which can be determined on the basis of laboratory tests, needs to be conducted.

A plastic–damage material model, Barcelona Model, proposed by Lubliner *et al* (1988) and modified by Lee and Fenves (1998) has been adopted for simulating masonry present in Qutb Minar. Mechanical properties of the masonry were estimated using results from the *in situ* acquisition campgains by Peña *et al* (2010) for the dynamic identification of the tower. The subsequent dynamic analysis highlights the importance of amplification phenomena in seismic assessment of Qutb Minar.

14.2 Details of modeling

A three-dimensional (30-D) FE model of the structure was performed by using the general-purpose FE analysis program Abaqus [9]. It is based on a general 3-D state of stresses and is suited for modeling 3-D solid structure. The Qutb Minar directly rests on a 1.7 m deep square ashlar masonry platform with sides of approximately 16.5 m, which in turn overlies a 7.6 m deep lime mortar rubble masonry layer, also square, with sides of approximately 18.6 m. The bedrock is located around 50–65 m below the ground level. Thus the base of the minar is assumed to be fixed to the ground. This model does not consider the helical stairway and the Qutb Minar's outer and inner core is connected only by means of 27 beams of size 0.35 to 0.45 m × 0.35 to 0.50 m. In order to obtain reliable models, data that emerged from the investigation campaign were taken into account. In particular, the outcomes of the dynamic identification were considered, in the absence of direct definition of the mechanical characteristics of the different masonry layers constituting the tower. For this reason, the models were calibrated on the basis of the natural frequencies and mode shapes emerging from the experimental investigation by Lorenco *et al* (2006).

14.2.1 Geometric data

The geometry of the models is based on the survey provided by the Archaeological Survey of India. The Minar cross-section is circular/polilobed, with the base diameter equal to 14.07 m and tapering off to a diameter of 3.13 m at the top, over a height of 72.45 m. The tower is composed by an external shell corresponding to a three-leaf masonry wall and a cylindrical central core (figure 14.1).

The entire structure is sub-divided into 44 geometric parts. These parts comprise of discrete composite sections for outer veneer, rubble infill and inner veneer for the floors 1–3, discrete composite sections for outer veneer, rubble infill and inner veneer for the floors four–five, five balconies corresponding to each floor, 27 beams, the outer veneer for shaft and shaft infill. The outer veneers have been subject to geometrical simplification and are modeled as uniform frustums. The revolution of sections technique has been utilized to model the radially symmetrical parts, whereas each of the beams has been modeled individually (figure 14.2).

14.2.2 Structural idealization

To enable the geometric modeling of the Qutb Minar structure using the geometry based modeler available in Abaqus, the discrete components were merged together as a single part. The merged assembly has subsequently been meshed using quadratic tetrahedral elements with improved stress formulation (C3D10I in Abaqus library) through the Free Meshing technique, where the topology of a particular part of the mesh does not represent the global arrangement of the elements. A total of 62 455 elements have been utilized. The model comprises 22 656 distinct nodes. The mesh is optimized based on the number of seeds assigned to each edge and the aspect ratio

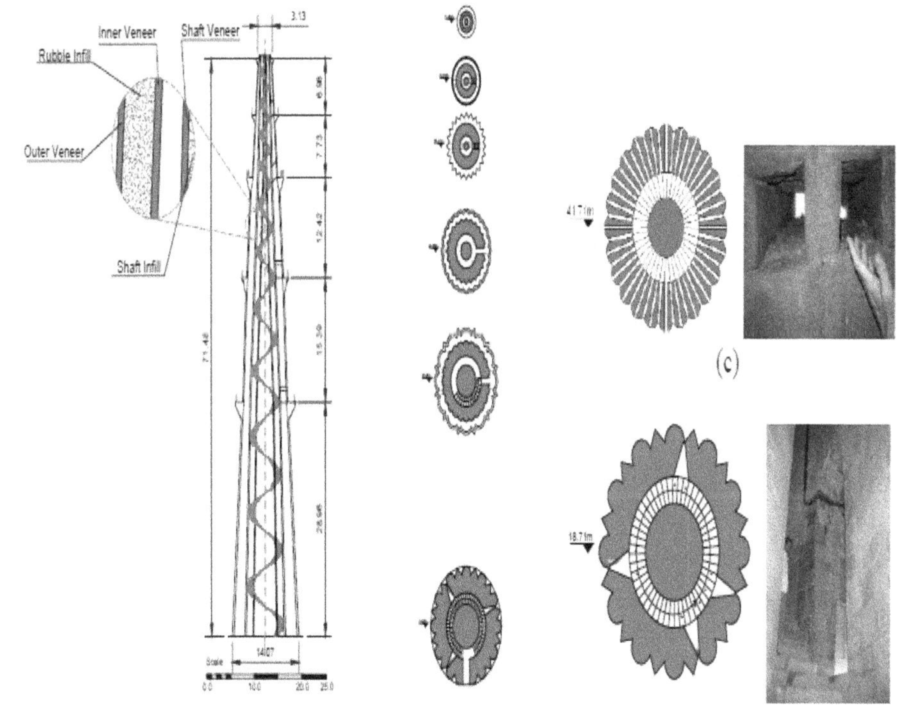

Figure 14.1. Geometrical features of Qutb minar.

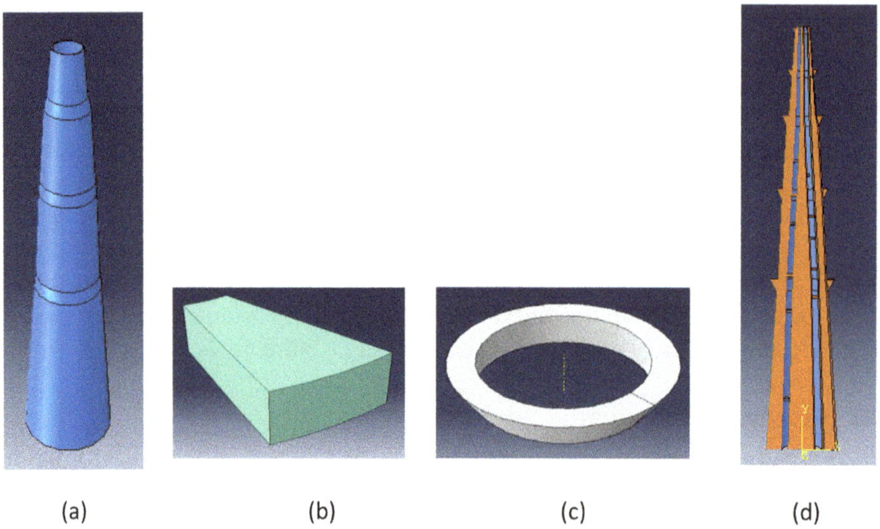

Figure 14.2. (a) Outer veneer for floors one–three, (b) balcony, (c) beam, (d) section of Qutb Minar based on idealized geometory.

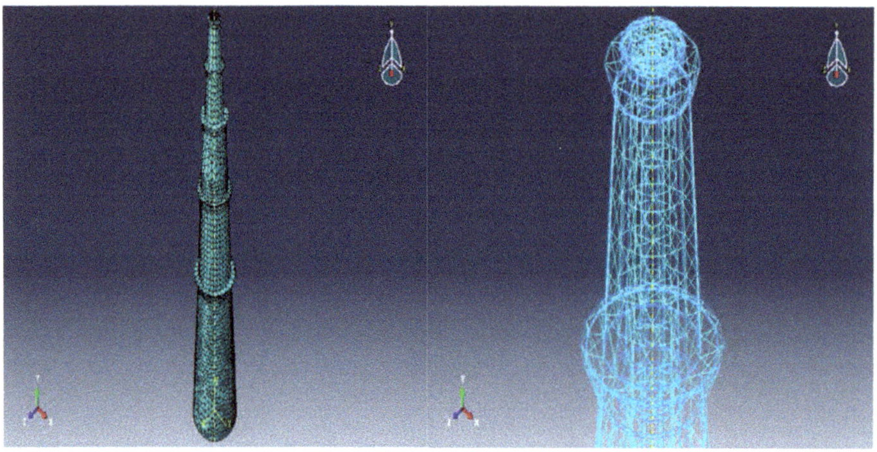

Figure 14.3. Finite element mesh developed in Abaqus/CAE.

specified. The same element is used throughout the masonry tower to eliminate the chances of nodal incompatibilities (figure 14.3).

14.2.3 Plastic damage macro model and material properties

The Barcelona Model has found application in predicting behavior of granular materials such as concrete. It is characterized by a yield surface which is an extension of the classical Drucker–Prager model with a non-circular deviatoric cross section, a non-associative flow rule and independent nonlinear isotropic hardening rules for tension and compression based on the hypothesis of strain hardening. The shape of the yield surface of the BM in the plane stress space that consists of four functions, equation '(i)' in the first quadrant, equation '(ii)' in the second quadrant equation '(iii)' in the first quadrant equation '(iv)' in the first quadrant.

$$f(\overline{\boldsymbol{\sigma}},\boldsymbol{\kappa}) = \sigma_{t0} \tag{14.1}$$

$$f(\overline{\boldsymbol{\sigma}},\boldsymbol{\kappa}) = \frac{1}{1-\alpha}(\overline{q} - 3\alpha\overline{p} + \beta\overline{\sigma}_1) - \overline{\sigma}_c(\kappa) \tag{14.2}$$

$$f(\overline{\boldsymbol{\sigma}},\boldsymbol{\kappa}) = \frac{1}{1-\alpha}(\overline{q} - 3\alpha\overline{p}) - \overline{\sigma}_c(\boldsymbol{\kappa}) \tag{14.3}$$

$f(\overline{\boldsymbol{\sigma}},\boldsymbol{\kappa}) = \frac{1}{1-\alpha}(\overline{q} - 3\alpha\overline{p} + \beta\overline{\sigma}_2) - \overline{\sigma}_c(\boldsymbol{\kappa})$
where,σ_{t0} ..... is the uniaxial tensile strength,
α, β are dimensionless parameters of the model,
p, q are stress tensor invariants.
Also,
σ_{c0} is the uniaxial yield strength,
σ_{cu} is the ultimate uniaxial strength,
σ_{b0} is the ultimate bi-axial strength.

The evolution of the yield surface is described by two hardening variables which are expressed by equivalent plastic strains $\tilde{\varepsilon}_c^p$ and $\tilde{\varepsilon}_t^p$, respectively for compression and tension, $\kappa = \{\tilde{\varepsilon}_c^p, \ \tilde{\varepsilon}_t^p\}^T$ (figure 14.4).

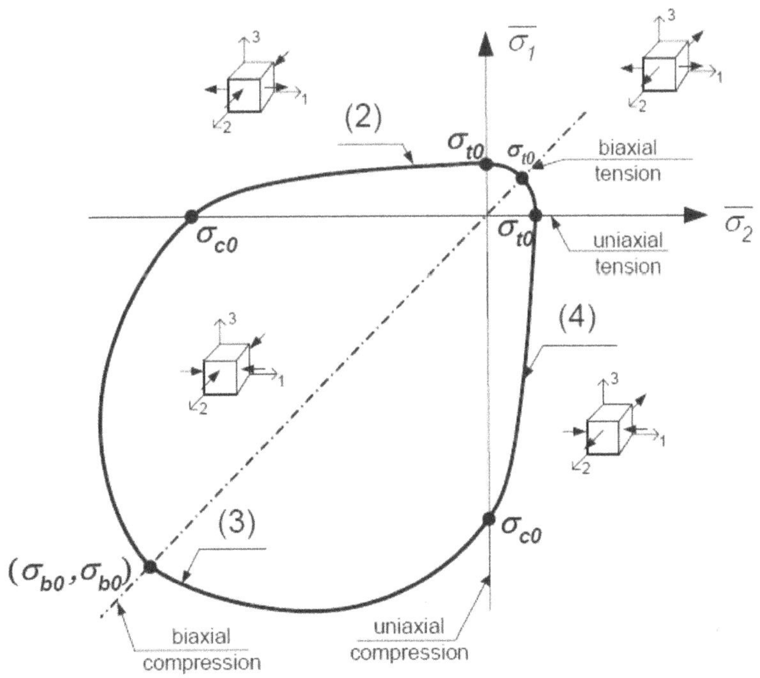

Figure 14.4. Yield surface of BM for the plane stress space.

However, the behavior of masonry can also be grouped in the behavior of granular materials, rendering it possible to adapt the BM model for masonry by making a few assumptions and approximations. Laboratory experiments show that the shape and dimensions of the biaxial envelope of the load capacity of masonry is strongly dependent on the angle between the stress principal directions and bed joints. So it is not possible to adapt concrete parameters of BM for masonry only by simple rescaling of its yield stresses.

The parameters of the model for masonry can hence be estimated based on propositions by Wawrzynek and Cińcio (2005) Based on the suggested simplifications we can assume $\sigma co = 0.67\ \sigma cu = 0.34 fmx$ for compression and $\sigma tu = \sigma t0 = 0.03\ fmx$, where fmx is the respective compressive or tensile strength of masonry in consideration. The values of ultimate compressive strengths and yield strengths in compression and tension have been presented in table 14.1.

Material degradation (and changes of elastic stiffness), which starts when the stress path reaches the yield surface, is described by two non-decreasing functions $dc = dc\ (\tilde{\varepsilon}_c^p)$ and $dt = dt\ (\tilde{\varepsilon}_t^p)$ which are presented in figure 14.5.

The hardening rule and degradation functions are determined for uniaxial material loading. Their specification can be approximated on the basis of the results of the cyclic compression of the masonry wall obtained by Naraine and Singha [7].

The function of material degradation dc is determined on the basis of changes of Young's modulus (depicted by the different slope of the dashed line in figure 14.6 during successive cycles of loading.

The hardening function for compression in the finite element software Abaqus is implemented by non-elastic strains $\varepsilon_c^{in} = \varepsilon_c - \varepsilon_c^e$.

Table 14.1. Comparison of properties for various geometrical parts.

Geometrical part	Young's modulus (MPa)	Specific mass kgm^{-3}	Poisson's coefficient	fmc (kPa)	fmt (kPa)	σco (kPa)	σcu (kPa)	$\sigma tu = \sigma t0$ (kPa)
Outer veneer 1–3	1.545	1800	0.2	2000	20	680	1014.924	0.6
Outer veneer 4–5	0.3	1800	0.2	600	20	204	304.4772	0.6
Masonry infill 1–3	6.171	2600	0.2	300	50	102	152.2386	1.5
Masonry infill 4–5	2	2300	0.2	5200	50	1768	2638.802	1.5
Inner veneer 1–3	0.785	1800	0.2	2000	20	680	1014.924	0.6
Inner veneer 4–5	0.3	1800	0.2	600	20	204	304.47	0.6
Shaft veneer	6.602	2600	0.2	2500	50	850	1268.655	1.5
Shaft infill	2	2600	0.2	3000	50	1020	1522.386	1.5
Beam	3.689	2000	0.2	5200	50	1768	2638.802	1.5

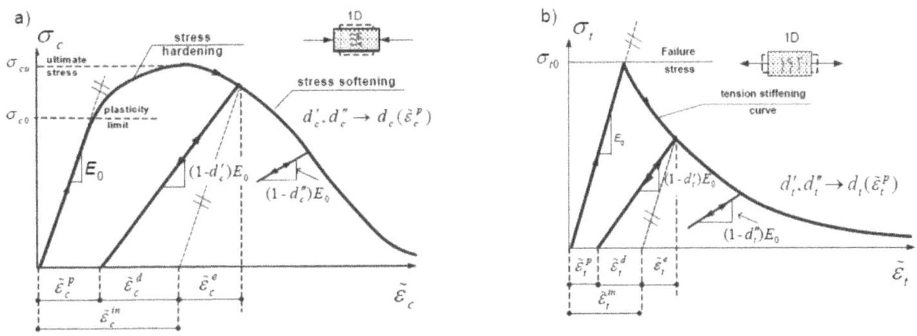

Figure 14.5. Description of the cyclic uniaxial test of BM for (a) compression, (b) tension.

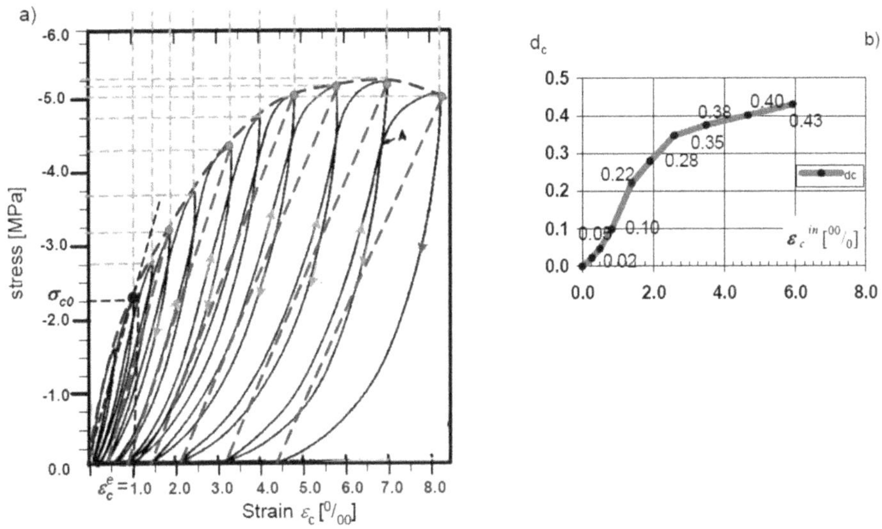

Figure 14.6. (a) Experimental results used for the calibration of BM parameters, (b) degradation function for compression (*dc*).

In the absence of degradation of stiffness under tension, constant stress cut-off was assumed. Therefore, there was no need to define the degradation function for tension stiffening of masonry.

14.3 Results and discussions

14.3.1 Free vibrational analysis

A free vibrational analysis of the numerical model was conducted through Lanczos subspace iteration. The results were compared with earlier experimental results. Actual experimentation using SYSCOM velocity sensors revealed there has not been much change in experimental data since the last experimental investigation (table 14.2).

Table 14.2. Comparisons between experimental results and numerical predictions.

Mode shape	Comment	Experimental (2011)	Uminho	Error [%]	Upadova	Error [%]	CBRI (NISA)	Error [%]	CBRI (Abaqus)	Error [%]
1	1st bending, z	0.781	0.71	−10.01	0.617	−23.02	0.680	−15.16	0.706 76	9.5
2	1st bending, x	0.806	0.71	−12.77					0.707 09	12.1
3	2nd bending, x	1.950	2.07	5.93	1.958	−1.21	1.933	−2.47	2.1948	12.5
4	2nd bending, z	2.030	2.09	3.98					2.1982	8.28
5	3rd bending, x	3.746	3.55	−5.10	3.280	0.48	3.660	−3.72	3.8033	1.53
6	3rd bending, z	3.860	3.59	−7.04					3.9318	1.86
7	1st torsion	4.109	4.8	8.06	4.286	−3.51	4.98	12.11	4.1087	0.007

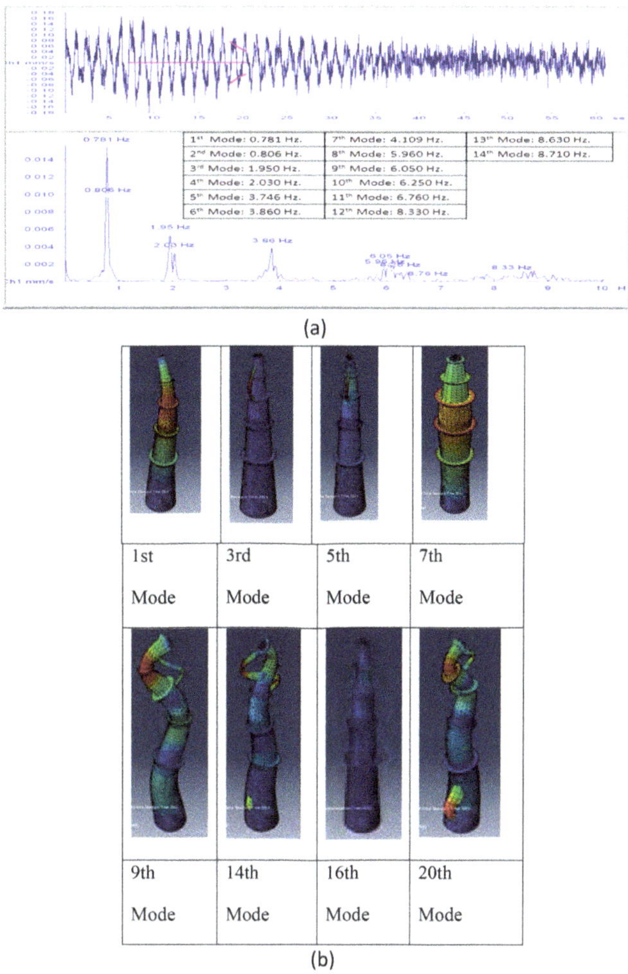

Figure 14.7. (a) Experimental results for natural frequency (b) Comparisons between experimental results and numerical predictions, (b) selected mode shapes of Qutb Minar.

The results show the importance of the stairway in the global behavior of the minaret. This model does not consider it and the connection between the external shell and the core is only done by 27 beams. Thus, the modal shapes (figure 14.7) show that the core has an independent vibration of the external shell. In fact, the core presents in all the modes an excessive torsional vibration around the fourth balcony, whereas the 3-D models by Lorenco *et al* (2006) do not present it because the stairway restricts this motion.

However, the correlation between numerical and experimental results suggests that although the helical staircase is an indispensable structural element in ascertaining the exact behavior of the structure, the overall stiffness of the system does not change too significantly. Finally, it is worth noting that at higher frequencies the top two storeys of the Minar seem to be vulnerable.

14.3.2 Simulation of seismic vibrations

Initially, a self-weight analysis was performed in order to evaluate the stress pattern. The total weight of the minaret is around 75 000 kN and the average compressive stress at the base is 0.54 MPa. The maximum compressive stress is 2.422 MPa in the external veneer of the central shaft. The stresses in the beams were observed to be tensile in nature, not exceeding 0.08 MPa, under the effect of gravity (figure 14.8).

This step was succeeded by exposing the base nodes of the structure to the acceleration time history recorded in the Koyna earthquake December 11, 1967 Maharashtra, India (figure 14.9).

The corresponding displacements for this time history were recorded at the topmost node within the shaft of the masonry tower. A similar time history analysis

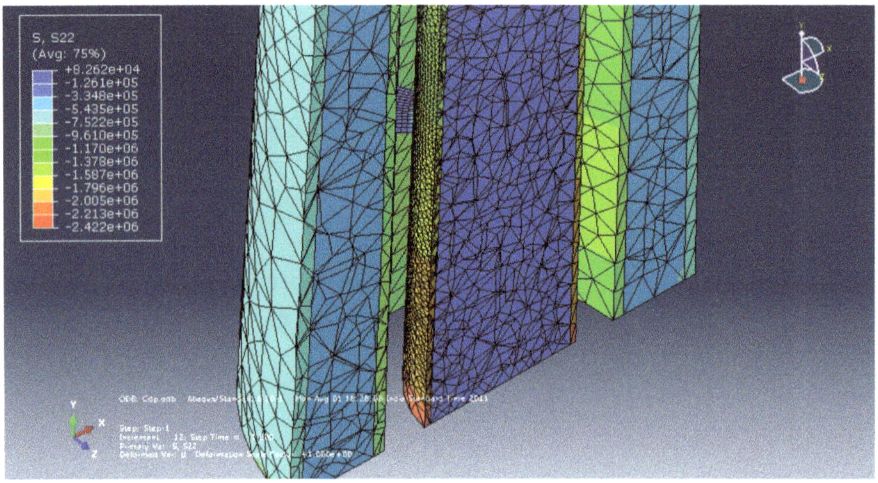

Figure 14.8. Stress distribution under gravity.

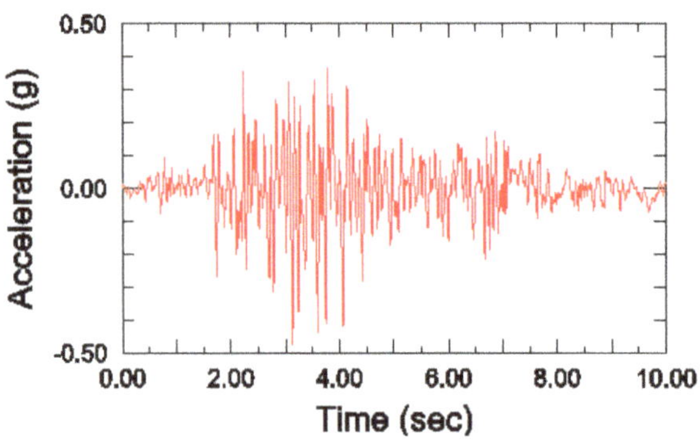

Figure 14.9. Time history applied at Qutb Minar.

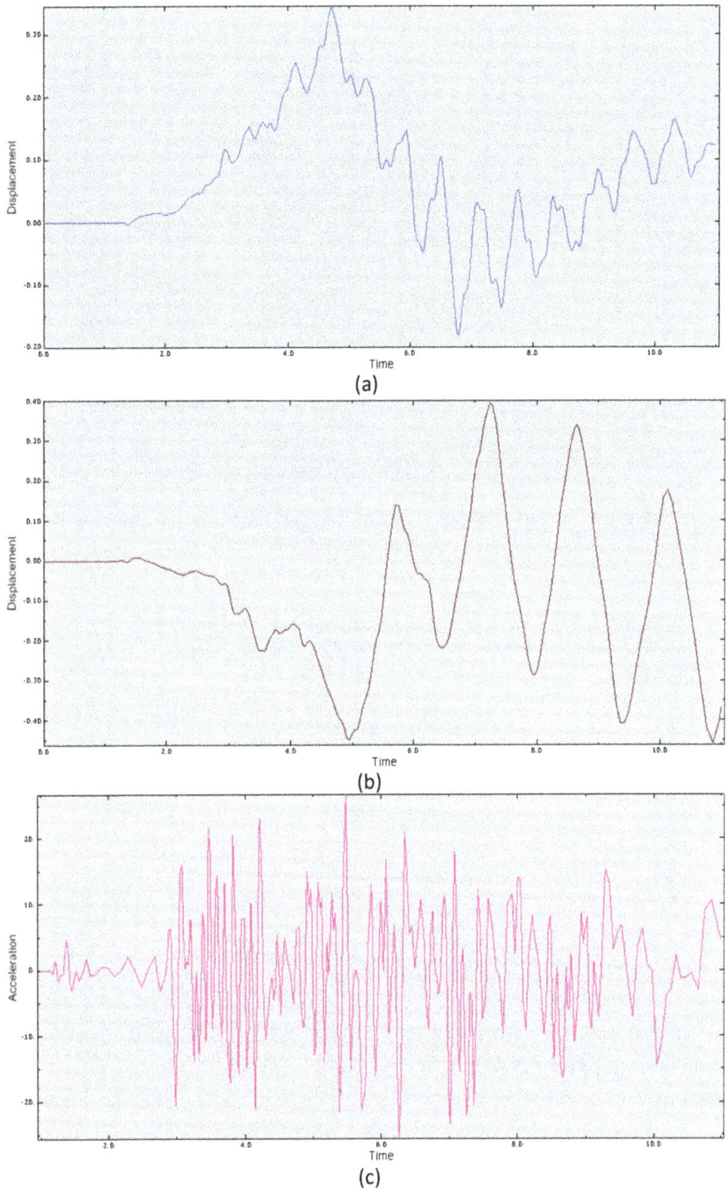

Figure 14.10. (a) Displacements observed in non-linear model, (b) Displacements observed in linear model, (c) acceleration profile at the top level for non-linear model.

is conducted assuming linear-elastic behavior of masonry and displacements recorded at the same node. The acceleration profile at this node was also recorded (figure 14.10).

The maximum observed instantaneous displacement of the crown of the Qutb Minar, for the earthquake measured at 7.5 on the Richter scale, was 0.32 m for the

Figure 14.11. Maximum observed tensile stress observed in y-direction.

Table 14.3. Maximum stresses at Qutb Minar.

Stresses	Compressive, MPa	Tensile, MPa
Sxx	−1.385	0.8846
Syy	−2.535	1.612
Szz	1.060	1.147
Sxy	−0.974	0.513
Syz	−0.606	0.929
Szx	−0.588	0.482

non-linear model and 0.45 m for the linear model. The maximum acceleration was experienced at the top of the masonry tower with a magnitude of 25.8 ms^{-2} or approximately 2.2 g (figure 14.10).

The maximum values of the stresses Sxx, Syy, Szz, Sxy, Syz, and Szx based on the critical times based on the maximum displacement magnitudes have been summarized in table 14.3 and the maximum observed tensile stresses are presented in figure 14.11.

14.4 Conclusions

The plastic damage macro model is useful in modelling masonry in historical structures like Qutb Minar. It was also observed that there is no fictitious stress concentration exceeding ultimate yield stresses for compression or tension during analysis. The mode shapes indicate that there are all kinds of vibrational modes present, i.e., longitudinal, translational, and rotational. In the lower modes, the inner and outer shaft moves in the same phase at which, at higher modes, both become out of phase. Also, drawing upon the non-linear dynamic analyses, probably due to lack of stiffeners connecting outer and inner shaft at the fourth and fifth storeys, the lower modes indicate that these storeys are vulnerable to seismic motions. The stiffening of inner shaft with outer shaft may obviate this problem. However, while studying higher modes, it is observed that the lower storey becomes vulnerable under seismic ground motions.

The non-linear earthquake dynamic analysis due to real earthquake time history with 0.49 g as PGA indicates that the results which are received with the application of masonry BM in comparison with linear elastic idealization of masonry, suggests the field of displacement is characterized by a lower instantaneous value of the displacement amplitude. The top two storeys may undergo severe damage owing to amplification of the order of 4 and hence experience acceleration levels of more than 2 g. The cross-section available at the fourth and fifth storeys would be insufficient to resist shear forces generated due to these high accelerations. The maximum stresses obtained during non-linear dynamic analysis are 2.535 MPa (compression) and 1.612 MPa (tension) and 0.929 MPa (shear). The maximum tensile stresses were concentrated in the beams connecting the inner shaft with the outer shell. These stresses are much higher than the permissible stresses for normal masonry as per IS 1893 : 2000 and hence, Qutb Minar needs strengthening against earthquake loads.

One of the main disadvantages of the BM is the assumption concerning the isotropy of masonry. Nowadays attempts are being made to modify the plane isotropic BM model for orthotropic material. Including the staircase within the model should further improve correlation of experimental data with the numerical results. Damage characteristics can be re-calibrated based on cyclic loading tests on chunks of masonry from contemporary structures of the Qutb Minar. The acceleration time-history is different in comparison with the seismic input at the base of the tower, for the filter effect of the whole structure. The need to take account of the dynamic parameters and material non-linearity for a realistic seismic response prevision has also been established.

References

[1] Giambanco G, Rizzo S and Spallino R 2001 Numerical analysis of masonry structures via interface models *Comput. Meth. Appl. Mech. Eng.* **190** 6493–511

[2] Lubliner J, Oliver S and Onate E 1989 A plastic-damage model for concrete *Int. J. Solids Struct.* **25** 299–329

[3] Lubliner J, Oliver S and Onate E 1990 Finite element nonlinear analysis of concrete structures using a plastic-damage model *Eng. Fract. Mech.* **35** 219–31

[4] Lourenço P B 1998 Experimental and numerical issues in the modelling of the mechanical behavior of masonry *Structural Analysis of Historical Constructions* ed P Roca *et al* (Barcelona: CIMNE) pp 57–91

[5] Lourenço P B 2000 A case study about using sophisticated masonry models in practice: the Lethes theatre (C. 1650) *Fifth Int. Conf. on Computational Structures Technology* ed B H V Topping and G de Roeck (Edinburgh: Civil-Comp Press) pp 433–42

[6] Naraine K and Sinha S 1989 Behavior of brick masonry under cyclic compressive loading *J. Constr. Eng. Manag. (ASCE)* **115** 1432–43

[7] Naraine K and Sinha S 1991 Cyclic behaviour of masonry in biaxial compression *J. Struct. Eng. (ASCE)* **117** 1336–55

[8] Naraine K and Sinha S 1992 Stress-strain curve for brick masonry in biaxial compression *J. Struct. Eng. (ASCE)* **118** 1451–61

[9] Abaqus/CAE documentation 2010 Section 18.5.1–18.5.3

[10] Lee J and Fenves G L 1998 A plastic-damage model for cyclic loading of concrete structures *J. Eng. Mech. ASCE* **124** 892–900

[11] Peña F, Lourenco P B, Mendes N and Oliveira D V 2010 Numerical models for the seismic assessment of an old masonry tower *Eng. Struct.* **32** 1466–78

www.ingramcontent.com/pod-product-compliance
Ingram Content Group UK Ltd.
Pitfield, Milton Keynes, MK11 3LW, UK
UKHW051908100926
8556IPUK00005B/814

9 780750 334815